AN ALGEBRAIC APPROACH TO NON-CLASSICAL LOGICS

STUDIES IN LOGIC

AND

THE FOUNDATIONS OF MATHEMATICS

VOLUME 78

NORTH-HOLLAND PUBLISHING COMPANY—AMSTERDAM • LONDON
AMERICAN ELSEVIER PUBLISHING COMPANY, INC.—NEW YORK

AN ALGEBRAIC APPROACH TO NON-CLASSICAL LOGICS

HELENA RASIOWA

University of Warsaw

N·H
P C

1974

NORTH-HOLLAND PUBLISHING COMPANY — AMSTERDAM • LONDON
AMERICAN ELSEVIER PUBLISHING COMPANY, INC. — NEW YORK

The English edition of this book has been published by PWN jointly with
NORTH-HOLLAND PUBLISHING COMPANY
Amsterdam

Library of Congress Catalogue Card Number 67-21972
ISBN 0 7204 2264 7

Sole distributors for the U.S.A. and Canada
American Elsevier Publishing Company, Inc.
52 Vanderbilt Avenue
New York, N.Y. 10017

PRINTED IN POLAND
(D.R.P.)

PREFACE

The search for relationships between logic and algebra goes back to the investigations of Boole and his followers. Those investigations yielded what we now call Boolean algebra. The close links between classical logic and the theory of Boolean algebras has been known for a long time. One of the turning points in the algebraic study of logic was the introduction by Lindenbaum and Tarski of the method of treating formulas, or equivalence classes of formulas as elements of an abstract algebra. Another milestone was the treatment of formulas as algebraic functions in certain algebras. This is a generalization of truth-tables in classical logic and can be found in the formulation by Łukasiewicz and Post of many-valued logics and in the metalogical study of those systems.

Studies by Stone and Tarski established the connections between intuitionistic propositional calculus and the algebra of open (closed) subsets of topological spaces. Similar connections between modal propositional calculus and the algebra of subsets of topological spaces were found by McKinsey, whose results in an algebraic form were later presented by McKinsey and Tarski.

The idea of extending the algebraic methods of Tarski and McKinsey to cover intuitionistic predicate calculus comes from Mostowski. At the same time it was taken up by Henkin, who applied it to the predicate calculus based on positive implicative logic, and by Rasiowa and Sikorski, who gave the first proof of Gödel's completeness theorem using algebra and topology. Another non-algebraic proof based on similar ideas was given by Henkin. Other proofs of Gödel's theorem using algebra and topology were later given by Rieger, Beth, Łoś, Reichbach and Scott. Rasiowa obtained results for intuitionistic and modal predicate logic, analogous to Gödel's completeness theorem for classical predicate logic. Further algebraic or topological proofs of various fundamental metamathematical theorems concerning classical predicate logic and analogues of those theorems for intuitionistic, modal, positive and minimal predicate logics as well as for constructive predicate logic with strong negation were given later by Rasiowa, Sikorski and other authors.

Mathematical investigations using the infinitistic methods of algebra, topology and set theory, undertaken by many authors, have produced a wealth of important and interesting results. Among these are Łoś's concept of an ultraproduct of models and its applications; the cylindric algebras of Henkin, Tarski and Thompson; Halmos's theory of polyadic algebras, Karp's theory of formalized languages with formulas of infinite length, which, using Hanf's work, Tarski has applied to the theory of prime filters in Boolean algebras; a simple proof by Scott of the independence of the continuum hypothesis using algebra and topology, and many others.

The attempt, in Rasiowa and Sikorski [3], to systematize the results concerning the algebras of various logics by formulating a general theory of an algebraic approach to logic, covers the class of extensional logics including positive logic, which is a considerable limitation. Moreover, in it additional conditions are imposed on the class of algebras determined by a given logic.

The Mathematics of Metamathematics presents, in a systematic and uniform manner, the algebraic approach to classical, intuitionistic, modal, and positive logic. The stress is laid on the first two, and the others are treated in a rather cursory way. The book concentrates on metalogical theorems concerning predicate calculi and elementary formalized theories and on methods directly connected with the authors' own work. Nevertheless, certain methods and theorems of a general nature, which can easily be adapted to special cases of other logics, including positive logic, can also be found in it (see especially Chapter VI).

The main aim of the present book is to formulate an algebraic approach to a carefully selected widest possible class of logics and to prove fundamental theorems for it, which previously have usually been proved for each of those logics separately. The class of logics presented in this book (Chapter VIII) covers cases which are outside the framework outlined in the previous work by Rasiowa and Sikorski ([3], [6]), such as logic weaker than positive implicative logic, positive and classical implicative logics, and constructive logic with strong negation. The basic concept is that of an implicative algebra, as introduced in Chapter II. The set of elements of such an algebra is ordered (partially ordered) by a relation determined by the operation corresponding to implication and a zero-argument operation in that algebra, i.e. by a constant element of it. This element is the greatest one with respect to the ordering. The definition of the class of logics described above and the theorems relating to it, as formulated in Chapter VIII, were presented in the Illinois Institute of Technology in November 1967 and at the Colloquium on Mathematics in the University of Illinois, Urbana, on December 21, 1967. Similar results have been obtained independently and presented in an elegant form by von Bummert [1].

The second aim of the book has been to give a number of examples of logics which belong to the class mentioned above. The intention has been to give examples of logics whose algebraic treatment has not so far been given in any book. They include positive implicative logic, classical implicative logic, minimal logic, positive logic with semi-negation, constructive logic with strong negation and some m-valued logics. Use has been made in this connection of results obtained by Diego, Abbott, Traczyk, Kirin and Rousseau. However, in order to give a more complete picture of the field, and also for the sake of continuity, positive intuitionistic, classical and modal logic, although described in *The Mathematics of Metamathematics*, have also been included here, but their treatment is rather cursory.

The crucial cases are these of propositional calculi and theories of zero order, which are treated in a fairly comprehensive manner, but the methods used can easily be generalized so as to cover first order predicate calculi and elementary formalized theories.

The Supplement contains a general introduction to the algebraic treatment of predicate calculi and elementary formalized theories based on logics which belong to the class of logics considered in this book. The completeness theorem for these predicate calculi and the theorem about the existence of models for consistent theories are proved.

The extension of algebraic methods so as to cover the systems specified above is possible because the logics under consideration determine classes of certain implicative algebras. As mentioned above, the set of elements of such an algebra is ordered, and so we can talk about greatest lower bounds and least upper bounds, which are interpretations, of the universal and the existential quantifiers, respectively, as is the case in classical, intuitionistic, modal and positive logic.

Certain repetitions from *The Mathematics of Metamathematics* have proved unavoidable, if the reader is not to be referred to that book too often. It seems reasonable to offer a complete exposition of the subject in one book.

The present book is intended for those mathematicians and students of mathematics who are interested in the algebraic aspects of logic. The philosophical aspects have been disregarded altogether.

The mathematics needed to comprehend the whole is expounded in Part One (Chapters I–VII), which includes a brief introduction to the elementary parts of topology and general algebra (Chapter I); the theory of implicative algebras, especially of positive implication algebras and implication algebras (Chapter II); and those branches of lattice theory which find application in the metamathematical analyses in Part Two, in particular quasi-Boolean algebras, quasi-pseudo-Boolean algebras ($\mathcal{N}$-lattices) and Post algebras which have not so far been discussed in any book.

Part Two forms the main body of the book. Chapter VIII deals with the algebraic study of the class of logics considered and gives the basic theorems. Chapters IX to XIV give examples of logics from that class, together with theorems relating to particular logics.

The exercises added to each chapter are intended to broaden the scope of the book, which, because of the limited space, omits many problems related to the main issues. Neither the book itself nor the bibliography pretend to be complete; the results presented here are meant just to make the reader familiar with a certain type of problems.

The terminology and symbols used here coincide in general with those used in *The Mathematics of Metamathematics* or in other papers by Rasiowa and Sikorski. For instance, "ordering relation" is used in the sense of "partial ordering relation".

Following the practice adopted in *The Mathematics of Metamathematics*, the same symbols stand for sentential connectives and the corresponding algebraic operations, and even set-theoretical operations, in order to bring out the close connections between them. This does not result in misunderstandings, since the context always shows in which sense a given symbol is used.

Theorems within one and the same chapter are referred to by numbers. Theorems from other chapters additionally have Roman numerals to indicate the chapter in question. Similarly, formulas within one and the same section are referred to by numbers, those from the same chapter but from other sections additionally have the number of the section, and those from other chapters are referred to by a Roman numeral indicating the chapter, the number of the section and that of the formula.

The author's special thanks are due to the late E. W. Beth, who inspired the present book. Acknowledgements are also due to A. Heyting and A. Mostowski for their valuable comments and suggestions.

HELENA RASIOWA

CONTENTS

PART ONE

IMPLICATIVE ALGEBRAS AND LATTICES

PART TWO

NON-CLASSICAL LOGICS

Chapter VIII. Implicative extensional propositional calculi

Chapter IX. Positive implicative logic and classical implicative logic

Chapter X. Positive logic

Chapter XI. Minimal logic, positive logic with semi-negation and intuitionistic logic

Chapter XII. Constructive logic with strong negation

Chapter XIII. Classical logic and modal logic

Chapter XIV. Many-valued logics

SUPPLEMENT

First order predicate calculi of non-classical logics

PART ONE

IMPLICATIVE ALGEBRAS AND LATTICES

CHAPTER I

PRELIMINARY SET-THEORETICAL, TOPOLOGICAL AND ALGEBRAIC NOTIONS

Introduction. This chapter lists the fundamental concepts of topology and abstract algebra needed to understand the ideas of this book and establishes the terminology and the notation, which almost completely coincides with that adopted in [MM] (1).

The definition of an abstract algebra in Section 4 assumes that all its operations are finite. However, the algebraic treatment of predicate calculi is more elegant when the concept of a generalized abstract algebra is used, which in addition to finite operations also admits some infinite generalized operations. That generalization as well as some related concepts and theorems are to be found in [MM] and, considering the scope of this book, are not repeated here (2).

1. Sets, mappings. We assume that the reader is familiar with the basic notions of set theory. We restrict ourselves to recalling the notation.

We write $a \in A$ if a is an element of A. Otherwise we write $a \notin A$. If a set A is a *subset* of a set B, we write $A \subset B$. The *empty set* is denoted by O. For any sets A, B the symbol $A \cup B$ ($A \cap B$) will denote the *union* (the *intersection*) of A and B. More generally, $\bigcup_{t \in T} A_t$ ($\bigcap_{t \in T} A_t$) will denote the union (the intersection) of sets A_t where $t \in T$. The difference of sets A, B will be denoted by $A - B$. We shall often consider only subsets of a fixed set X. The set X is then called a *space*. The difference $X - A$ (where $A \subset X$), i.e. the *complement* of A in X will be denoted by $-A$.

(1) The symbol [MM] will always be written instead of *The Mathematics of Metamathematics* (see Rasiowa and Sikorski [6]).

(2) For some ideas concerning generalized algebras see the exercises at the end of this chapter. For investigations of products of generalized algebras see Sikorski [4].

We shall write

$$f\colon X \to Y$$

to indicate that f is a mapping defined on X with values in Y. The set X is then called the *domain* of f. The value of a mapping f at x will be denoted by $f(x)$. Instead of $f(x)$ we shall also write fx or f_x. For any mapping $f\colon X \to Y$ the set $\{f(x)\}_{x \in X}$, i.e. the set of all elements $f(x)$ where $x \in X$, will be called the *range* of f. If the range of f is equal Y, we say that f *maps* X *onto* Y.

If f is a mapping of a set X into Y and $A \subset X$, $B \subset Y$, then $f(A)$ denotes the *image* of A (i.e. the set of all $f(x)$ where $x \in A$) and $f^{-1}(B)$ denotes the *inverse image* of B (i.e. the set of all elements $x \in X$ such that $f(x) \in B$). If f is a one-one mapping of X onto Y, then f^{-1} denotes the mapping inverse to f. If $f\colon X \to Y$ and $g\colon Y \to Z$, then $g \circ f\colon X \to Z$ denotes their composition. If $f\colon X \to Y$ and $X_0 \subset X$, then the restriction of f to X_0 will be denoted by $f|X_0$.

If to each integer n there is associated an element a_n, then (a_n) denotes the infinite sequence of these elements. An m-element sequence is denoted by $(a_1, \ldots, a_m)$. More generally, the mapping which to every $t \in T$ $(T \neq O)$ assigns an element a_t will be denoted by $(a_t)_{t \in T}$ or by (a_t).

The cartesian product of sets A_t where $t \in T$, i.e. the set of all mappings $(a_t)_{t \in T}$ such that $a_t \in A_t$ for every $t \in T$, will be denoted by $\prod_{t \in T} A_t$. If $T = \{1, \ldots, m\}$, we shall write $A_1 \times \ldots \times A_m$ instead of $\prod_{t \in T} A_t$. If $A_t = A$ for each $t \in T$, we shall write A^T instead of $\prod_{t \in T} A_t$ and A^m instead of $A \times \ldots \times A$ (m times).

The cardinal number or power of a set X will be denoted by card$X$. The cardinal number of enumerable sets (i.e. of one-one images of the set of all positive integers) will be denoted by $\aleph_0$.

2. Topological spaces (3). A set X is said to be a *topological space* if with every $A \subset X$ there is associated a set $\mathbf{I}A \subset X$, called the *interior of* A, in such a way that the following conditions are satisfied:

$$(\mathrm{i}_1) \qquad \mathbf{I}(A \cap B) = \mathbf{I}A \cap \mathbf{I}B,$$

$$(\mathrm{i}_2) \qquad \mathbf{I}A \subset A,$$

(3) For a detailed exposition of the theory of topological spaces see, for example, Kuratowski [2] or Kelley [1].

(i_3) $$\mathbf{I}\mathbf{I}A = \mathbf{I}A,$$

(i_4) $$\mathbf{I}X = X.$$

The operation **I** is then called the *interior operation.*

For every set $A \subset X$, the set $-\mathbf{I}-A$ is called the *closure of A* and will be denoted by $\mathbf{C}A$. Thus, by definition,

(1) $$\mathbf{C}A = -\mathbf{I}-A \quad \text{and} \quad \mathbf{I}A = -\mathbf{C}-A \quad \text{for every } A \subset X.$$

It follows from (1) and (i_1)–(i_4) that for every $A \subset X$:

(c_1) $$\mathbf{C}(A \cup B) = \mathbf{C}A \cup \mathbf{C}B,$$

(c_2) $$A \subset \mathbf{C}A,$$

(c_3) $$\mathbf{C}\mathbf{C}A = \mathbf{C}A,$$

(c_4) $$\mathbf{C}O = O\ [4].$$

Any operation **C** fulfilling conditions (c_1)–(c_4) is called a *closure operation.* Clearly any interior operation determines a closure operation by the first equation of (1) and every closure operation determines an interior operation by the second equation of (1).

By (i_1) and (c_1) for any subsets A, B of a topological space X

(2) $$\text{if } A \subset B, \text{ then } \mathbf{I}A \subset \mathbf{I}B \text{ and } \mathbf{C}A \subset \mathbf{C}B.$$

A subset A of a topological space X is said to be *open* (*closed*) if $A = \mathbf{I}A$ (if $A = \mathbf{C}A$). Clearly, by (1), for any $A \subset X$

(3) *A is open (closed) if and only if $-A$ is closed (open).*

It follows from (i_3) (from (c_3)) that the interior (the closure) of any subset A of a topological space X is an open set (is a closed set). By (i_2) and (i_4) (by (c_2) and (c_4)) the empty set O and the whole space X are open sets (are closed sets).

For any set $A \subset X$, the set $\mathbf{I}A$ (the set $\mathbf{C}A$) is the greatest open subset of A (is the least closed set containing A). This follows easily from (i_2), (2) (from (c_2), (2)).

The intersection (the union) of any finite number of open (of closed) sets is an open (a closed) set. This follows from (i_1) (from (c_1)) by an

[4] The axioms (c_1)–(c_4) are due to Kuratowski.

inductive argument. The union (the intersection) of any family $\{A_t\}_{t\in T}$, of open sets (of closed sets) is an open set (is a closed set). This is so because if A_t is open for every $t \in T$, then $A_t = \mathbf{I}A_t \subset \mathbf{I}\bigcup_{t\in T} A_t$ and consequently $\bigcup_{t\in T} A_t \subset \mathbf{I}\bigcup_{t\in T} A_t$. The converse inclusion follows from (i_2). The corresponding statement for closed sets can be obtained from the statement just proved by taking complements.

By a *base* of a topological space X we mean a class $\mathbf{B}$ of open subsets of X such that every open subset of X is the union of sets belonging to $\mathbf{B}$.

By a *subbase* of a topological space X we mean a class $\mathbf{B}_0$ of open subsets of X such that the class containing the empty set O, the whole space X and all finite intersections $B_1 \cap \ldots \cap B_m$ for $B_1, \ldots, B_m \in \mathbf{B}_0$ is a base of X.

We quote without proof the following simple statement.

2.1. *For every class $\mathbf{B}_0$ of subsets of a set X there exists exactly one interior operation $\mathbf{I}$ in X such that $\mathbf{B}_0$ is a subbase of the topological space X with the interior operation $\mathbf{I}$, namely if $\mathbf{B}$ is the class composed of the empty set O, of the whole space X and of all finite intersections $B_1 \cap \ldots \cap B_m$ for $B_1, \ldots, B_m \in \mathbf{B}$, then the operation $\mathbf{I}$ which with every $A \subset X$ associates the union of all sets $B \in \mathbf{B}$ such that $B \subset A$ is the required interior operation in X.*

A subset A of a topological space X is said to be *dense* if $\mathbf{C}A = X$. It follows from the definition of a dense set that a subset A of a topological space X is dense if and only if $-\mathbf{C}A = \mathbf{I}-A = O$, i.e. if and only if there does not exist a non-empty open subset $B \subset X$ such that $B \subset -A$. This is equivalent to the condition that the intersection of A with every non-empty open set B is a non-empty set.

A topological space X is said to be *compact* if for any family $\{A_t\}_{t\in T}$ of open subsets of X the condition $X = \bigcup_{t\in T} A_t$ implies that there exists a finite subset $T_0 \subset T$ such that $X = \bigcup_{t\in T_0} A_t$. It is easy to show by considering complements that a topological space X is compact if and only if for any family $\{B_t\}_{t\in T}$ of closed subsets of X the condition $\bigcap_{t\in T} B_t = O$ implies that there exists a finite subset $T_0 \subset T$ such that $\bigcap_{t\in T_0} B_t = O$.

If X is a compact space, then for any family $\{A_t\}_{t\in T}$ of open subsets of X and for an arbitrary closed set B the condition $B \subset \bigcup_{t\in T} A_t$

implies that there exists a finite subset $T_0 \subset T$ such that $B \subset \bigcup_{t\in T_0} A_t$. This is because if $B \subset \bigcup_{t\in T} A_t$ then $X = -B \cup \bigcup_{t\in T} A_t$. Thus, by compactness, there exists a finite subset $T_0 \subset T$ such that $X = -B \cup \cup \bigcup_{t\in T_0} A_t$, which implies that $B \subset \bigcup_{t\in T_0} A_t$.

A topological space X is said to be a T_0-*space* if for any two distinct points x, y in X there exists an open set containing exactly one of them.

A topological space X is said to be a *Hausdorff space* if for every pair of distinct points x, y there exist two open disjoint sets A, B such that $x \in A$ and $y \in B$.

A topological space X is said to be *totally disconnected* if for every two distinct points $x, y \in X$ there exist two open disjoint sets A, B such that $x \in A$, $y \in B$ and $A \cup B = X$.

Two topological spaces X, Y are said to be *homeomorphic* if there exists a one-one mapping f from X onto Y such that $f(\mathbf{I}A) = \mathbf{I}f(A)$ for each set $A \subset X$.

Homeomorphic spaces have the same topological properties.

3. Ordered sets and quasi-ordered sets. A binary relation $\leqslant$ defined on a set A is said to be an *ordering* [5] *on* A if it is *reflexive, transitive,* and *antisymmetric*, i.e. if for arbitrary elements x, y, z of A the following conditions are satisfied

(o₁) $$x \leqslant x,$$

(o₂) $$\text{if} \quad x \leqslant y \quad \text{and} \quad y \leqslant z, \quad \text{then} \quad x \leqslant z,$$

(o₃) $$\text{if} \quad x \leqslant y \quad \text{and} \quad y \leqslant x, \quad \text{then} \quad x = y.$$

An *ordered set* is a pair $(A, \leqslant)$ where A is a non-empty set and $\leqslant$ is an ordering on A.

Clearly, if $\leqslant$ is an ordering on A and $A_0 \subset A$, then $\leqslant|A_0$ (i.e. the relation $\leqslant$ restricted to A_0) is an ordering on A_0.

An element a of an ordered set $(A, \leqslant)$ is said to be *maximal* (*minimal*) if there is no element b in A such that $a \leqslant b$ ($b \leqslant a$) and $a \neq b$. An ordered set can have several maximal or minimal elements.

[5] The term "ordering" is used here in the sense of "partial ordering" adopted by many authors and "linear ordering" in the sense of "ordering" used by those authors.

An element a of an ordered set $(A, \leqslant)$ is said to be *greatest* (*least*) if for every $x \in A$, $x \leqslant a$ ($a \leqslant x$). It follows from this definition and (o_3) that an ordered set can have at most one greatest element and at most one least element. The greatest (least) element of an ordered set, if it exists, will be denoted by $\bigvee$ (by $\bigwedge$).

An element a of an ordered set $(A, \leqslant)$ is said to be *an upper* (*lower*) *bound* of a non-empty subset A_0 of A if $b \leqslant a$ ($a \leqslant b$) for every $b \in A_0$. If the set of all upper (lower) bounds of A_0 contains a least (greatest) element, then this element is said to be the *least upper bound of* A_0 (the *greatest lower bound of* A_0) and is denoted by l.u.b.A_0 (g.l.b.A_0). It follows from this definition that l.u.b.$A_0 = a$ (g.l.b.$A_0 = a$) if and only if the following conditions are satisfied

(1) $\qquad b \leqslant a \ (a \leqslant b) \quad$ for every $b \in A_0$,

(2) if $c \in A$ and $b \leqslant c$ ($c \leqslant b$) for every $b \in A_0$, then $a \leqslant c$ ($c \leqslant a$).

An ordering relation $\leqslant$ on a set A is called a *linear ordering* if the following condition is satisfied

(o_4) $\qquad x \leqslant y \quad$ or $\quad y \leqslant x, \quad$ for every $x, y \in A$.

A *chain* is a pair $(A, \leqslant)$, where A is a non-empty set and $\leqslant$ is a linear ordering on A. In every chain the notions of a greatest (least) element and a maximal (minimal) element coincide. Observe that if $\leqslant$ is a linear ordering on A and $A_0 \subset A$, then the relation $\leqslant$ restricted to A_0 is a linear ordering on A_0.

We quote without proof the following theorem.

3.1. (KURATOWSKI–ZORN LEMMA) [6]. *If $(A, \leqslant)$ is an ordered set and every chain $(B, \leqslant|B)$, where $B \subset A$, has an upper bound in A, then for every $a_0 \in A$ there exists a maximal element a in A such that $a_0 \leqslant a$.*

Given an equivalence relation $\approx$ on A, let $||x||$ denote the equivalence class determined by $x \in A$, i.e. the set of all $y \in A$ such that $x \approx y$. The set $\{||x||\}_{x \in A}$ of all equivalence classes $||x||$ ($x \in A$) of the equivalence relation $\approx$ will be denoted by $A/\approx$.

A binary relation $\leqslant$ defined on a set A is said to be a *quasi-ordering* on A, if it is *reflexive* and *transitive*, i.e. if it satisfies conditions (o_1)

[6] Kuratowski [1], Zorn [1].

and (o_2). A *quasi-ordered set* is a pair $(A, \leqslant)$, where A is a non-empty set and $\leqslant$ is a quasi-ordering on A.

We state without proof the following well-known theorem.

3.2. *Let $(A, \leqslant)$ be a quasi ordered set. Let $\approx$ be the relation on A defined as follows:*

(3) $$x \approx y \quad \textit{if and only if} \quad x \leqslant y \quad \textit{and} \quad y \leqslant x.$$

Then $\approx$ is an equivalence relation on A.

For any $x, y \in A$ let

(4) $$\|x\| \leqslant \|y\| \quad \textit{if and only if} \quad x \leqslant y.$$

The relation $\leqslant$ defined by (4) *on $A/\approx$ is an ordering on $A/\approx$.*

4. Abstract algebras. We recall in this section fundamental notions and theorems of abstract algebra.

By an *abstract algebra*, or briefly an *algebra*, we shall mean any pair

(1) $$(A, (o_\varphi)_{\varphi \in \Phi})$$

where A is a non-empty set and, for every $\varphi \in \Phi \neq O$, o_φ is, for some m, an m-argument operation on A $(m = 0, 1, 2, \ldots)$, i.e. a mapping $o_\varphi \colon A^m \to A$. In particular, if $m = 0$, by an *m-argument operation on A* we understand a constant element $o_\varphi \in A$. If $m = 1$, then instead of $o_\varphi(x)$ we shall usually write $o_\varphi x$ and if $m = 2$, then instead of $o_\varphi(x, y)$ we shall often write $x o_\varphi y$.

In the case where $\Phi = \{1, \ldots, m\}$, the algebra (1) will often be denoted by

(2) $$(A, o_1, \ldots, o_m).$$

If A contains only one element, then the algebra (1) is said to be *degenerate*.

A set $A_0 \subset A$ of elements of an abstract algebra (1) is said to be *closed under an m-argument operation* o_φ if for all $a_1, \ldots, a_m \in A_0$

(3) $$o_\varphi(a_1, \ldots, a_m) \in A_0.$$

In particular, A_0 is closed under a 0-argument operation o_φ if $o_\varphi \in A_0$.

Any non-empty set $A_0 \subset A$ of elements of an abstract algebra (1) closed under all operations o_φ, $\varphi \in \Phi$, and considered as an abstract algebra with respect to the operations $o_\varphi | A_0$, $\varphi \in \Phi$, is said to be a *subalgebra* of (1). It follows from this definition that every 0-argument operation in (1), i.e. every constant element in (1) belongs to every subalgebra of (1).

By the *subalgebra of* (1) *generated by a non-empty set* $A_0 \subset A$ we mean the least subalgebra containing A_0, i.e. the intersection of all subalgebras of (1) containing A_0. The set A_0 is said to *generate* this subalgebra or to be a *set of generators* of this subalgebra. It follows from this definition that a non-empty set $A_0 \subset A$ generates (1) or is a set of generators of (1) if the only subalgebra containing A_0 is the whole algebra (1).

Algebras $(A, (o_\varphi)_{\varphi\in\Phi})$ and $(B, (o'_\varphi)_{\varphi\in\Phi'})$ are said to be *similar* if $\Phi = \Phi'$, and for every $\varphi \in \Phi$ operations o_φ and o'_φ are of the same number of arguments. In particular, arbitrary subalgebras of an abstract algebra are similar. It will sometimes be convenient to denote corresponding operations in similar algebras by the same signs. This applies to the operations in an algebra and its subalgebras.

A mapping $h: A \to B$ of the set A of all elements of an abstract algebra $\mathfrak{A} = (A, (o_\varphi)_{\varphi\in\Phi})$ into the set B of all elements of a similar abstract algebra $\mathfrak{B} = (B, (o_\varphi)_{\varphi\in\Phi})$ is said to be a *homomorphism* of $\mathfrak{A}$ into $\mathfrak{B}$ provided it preserves all operations, i.e. if

(4) $h(o_\varphi(a_1, \ldots, a_m)) = o_\varphi(h(a_1), \ldots, h(a_m))$ for all $a_1, \ldots, a_m \in A$,

where $m \geqslant 0$ is the number of arguments of o_φ. In particular, if a homomorphism $h: A \to B$ of $\mathfrak{A}$ into $\mathfrak{B}$ maps A onto B, it is said to be an *epimorphism*.

If a homomorphism is one-one, it is called a *monomorphism*. If an epimorphism is one-one, it is said to be an *isomorphism*. If there exists an isomorphism of an algebra $\mathfrak{A}$ onto a similar algebra $\mathfrak{B}$, then $\mathfrak{A}$ and $\mathfrak{B}$ are called *isomorphic*.

We state without proofs the following simple results.

4.1. *If* $\mathfrak{A}$, $\mathfrak{B}$, $\mathfrak{C}$ *are similar algebras, h is a homomorphism of* $\mathfrak{A}$ *into* $\mathfrak{B}$ (*monomorphism of* $\mathfrak{A}$ *into* $\mathfrak{B}$, *epimorphism of* $\mathfrak{A}$ *onto* $\mathfrak{B}$, *isomorphism of* $\mathfrak{A}$ *onto* $\mathfrak{B}$) *and g is a homomorphism of* $\mathfrak{B}$ *into* $\mathfrak{C}$ (*monomorphism of* $\mathfrak{B}$ *into* $\mathfrak{C}$, *epimorphism of* $\mathfrak{B}$ *onto* $\mathfrak{C}$, *isomorphism of* $\mathfrak{B}$ *onto* $\mathfrak{C}$), *then the composition $g\circ h$ is a homomorphism of* $\mathfrak{A}$ *into* $\mathfrak{C}$ (*monomorphism of* $\mathfrak{A}$ *into* $\mathfrak{C}$, *epimorphism of* $\mathfrak{A}$ *onto* $\mathfrak{C}$, *isomorphism of* $\mathfrak{A}$ *onto* $\mathfrak{C}$).

4.2. *If h is a homomorphism of an algebra $\mathfrak{A}$ into a similar algebra $\mathfrak{B}$, then the image $h(A)$ of the set A of all elements of $\mathfrak{A}$ is closed under all operations in $\mathfrak{B}$*, i.e. *it is a subalgebra of $\mathfrak{B}$, with respect to all operations in $\mathfrak{B}$ restricted to $h(A)$. This subalgebra will be denoted in the sequel by $h(\mathfrak{A})$.*

4.3. *If a homomorphism h of an algebra $\mathfrak{A}$ into a similar algebra $\mathfrak{B}$ maps a set of generators of $\mathfrak{A}$ onto a set of generators of $\mathfrak{B}$, then h is an epimorphism of $\mathfrak{A}$ onto $\mathfrak{B}$.*

4.4. *If a mapping g of a set of generators of an algebra $\mathfrak{A}$ into a similar algebra $\mathfrak{B}$ can be extended to a homomorphism h of $\mathfrak{A}$ into $\mathfrak{B}$, then this extension h is unique.*

The following theorem easily follows from 4.1, 4.3 and 4.4.

4.5. *If a one-one mapping f from a set of generators of an algebra $\mathfrak{A}$ onto a set of generators of a similar algebra $\mathfrak{B}$ can be extended to a homomorphism h of $\mathfrak{A}$ into $\mathfrak{B}$ and f^{-1} can be extended to a homomorphism g of $\mathfrak{B}$ into $\mathfrak{A}$, then h is an isomorphism of $\mathfrak{A}$ onto $\mathfrak{B}$ and $h^{-1} = g$.*

By a *free algebra* in a class $\boldsymbol{K}$ of similar algebras we mean an algebra $\mathfrak{A} \in \boldsymbol{K}$ containing a set A_0 of generators such that every mapping f of A_0 into an arbitrary algebra $\mathfrak{B} \in \boldsymbol{K}$ can be extended to a homomorphism h of $\mathfrak{A}$ into $\mathfrak{B}$. A_0 is then said to be a *set of $\boldsymbol{K}$-free generators of* $\mathfrak{A}$, or, briefly, a set of free generators of $\mathfrak{A}$.

The following theorem can easily be deduced from 4.5.

4.6. *If $\mathfrak{A}$ and $\mathfrak{B}$ are free algebras in a class $\boldsymbol{K}$ of similar algebras and A_0, B_0 are sets of $\boldsymbol{K}$-free generators in these algebras such that* $\operatorname{card} A_0 = \operatorname{card} B_0$, *then $\mathfrak{A}$ and $\mathfrak{B}$ are isomorphic.*

The notion of epimorphism is closely related to the notion of a congruence relation in an abstract algebra.

By a *congruence* in an abstract algebra (1) we mean an equivalence relation $\approx$ defined on A satisfying the following condition for every operation o_φ

(5) if $a_1 \approx b_1, \ldots, a_m \approx b_m$, then $o_\varphi(a_1, \ldots, a_m) \approx o_\varphi(b_1, \ldots, b_m)$,

where m is the number of arguments of o_φ and all $a_1, \ldots, a_m, b_1, \ldots$ $\ldots, b_m \in A$.

For any congruence in an algebra (1) the equations

(6) $$o_\varphi(||a_1||, \ldots, ||a_m||) = ||o_\varphi(a_1, \ldots, a_m)||, \quad \varphi \in \Phi$$

define the corresponding operations on the set $A/\approx$ of all equivalence classes $||a||$, $a \in A$, of the relation $\approx$. Consequently, $(A/\approx, (o_\varphi)_{\varphi\in\Phi})$ is an algebra similar to algebra (1). The algebra defined above is said to be the *quotient algebra* of the given algebra (1) and $\approx$. The mapping

(7) $$h(a) = ||a||, \quad a \in A$$

is an epimorphism of the algebra (1) onto the quotient algebra. On the other hand, every epimorphism determines a congruence relation. In fact,

4.7. *If h is an epimorphism of an algebra $\mathfrak{A}$ onto a similar algebra $\mathfrak{B}$, then the relation $\approx$ defined as*

(8) $$a \approx b \quad \text{if and only if} \quad h(a) = h(b)$$

for arbitrary elements a, b of $\mathfrak{A}$ defines a congruence in $\mathfrak{A}$. The quotient algebra $\mathfrak{A}/\approx$ is isomorphic to $\mathfrak{B}$. The mapping

(9) $$g(||a||) = h(a), \quad \text{for every element } a \text{ of } \mathfrak{A},$$

is an isomorphism of $\mathfrak{A}/\approx$ onto $\mathfrak{B}$.

Let $\simeq$ be a congruence in a quotient algebra $\mathfrak{A}/\sim$ and let $\|a\|$ denote for every element a in $\mathfrak{A}$ the equivalence class of the relation $\simeq$, i.e. an element of the quotient algebra $(\mathfrak{A}/\sim)/\simeq$. Consider the following relation $\approx$ defined in the set of all elements of $\mathfrak{A}$:

(10) $$a \approx b \quad \text{if and only if} \quad |a| \simeq |b|,$$

where $|x|$ denotes an element of $\mathfrak{A}/\sim$ for every element x of the algebra $\mathfrak{A}$. The following theorem then holds:

4.8. *The relation $\approx$ defined by* (10) *is a congruence in the algebra $\mathfrak{A}$ and the equation*

(11) $$g(||a||) = \|a\|, \quad \text{for each element } a \text{ of } \mathfrak{A}$$

defines an isomorphism of $\mathfrak{A}/\approx$ onto $(\mathfrak{A}/\sim)/\simeq$.

A homomorphism h of an algebra $\mathfrak{A}$ into a similar algebra $\mathfrak{B}$ *preserves a congruence $\approx$ in $\mathfrak{A}$* provided

(12) $$a \approx b \quad \text{implies} \quad h(a) = h(b), \quad \text{for all } a, b \text{ in } \mathfrak{A}.$$

4.9. *If a homomorphism h of $\mathfrak{A}$ into a similar algebra $\mathfrak{B}$ preserves a congruence $\approx$ in $\mathfrak{A}$, then the equation*

$$g(||a||) = h(a) \quad \textit{for every element } a \textit{ in } \mathfrak{A} \tag{13}$$

defines a homomorphism of $\mathfrak{A}/\approx$ into $\mathfrak{B}$.

Given a set $\{\mathfrak{A}_t\}_{t\in T}$ of similar algebras

$$\mathfrak{A}_t = \big(A_t, (o_\varphi)_{\varphi\in\Phi}\big), \qquad t \in T,$$

we can form the product $\prod_{t\in T}\mathfrak{A}_t$ of algebras $\mathfrak{A}_t$, $t \in T$, i.e. the similar algebra $\big(A, (o_\varphi)_{\varphi\in\Phi}\big)$, where

$$A = \prod_{t\in T} A_t \tag{14}$$

and for every $\varphi \in \Phi$

$$o_\varphi\big((a_{1t})_{t\in T}, \ldots, (a_{mt})_{t\in T}\big) = \big(o_\varphi(a_{1t}, \ldots, a_{mt})\big)_{t\in T}, \tag{15}$$

where $(a_{1t})_{t\in T}, \ldots, (a_{mt})_{t\in T}$ are arbitrary elements of $\prod_{t\in T} A_t$ and m is the number of arguments of o_φ.

We make the following remark.

4.10. *If $\mathfrak{A} = \prod_{t\in T}\mathfrak{A}_t$ and, for every $t \in T$, h_t is a homomorphism of a similar algebra $\mathfrak{B}$ into $\mathfrak{A}_t$, then the equation*

$$h(a) = \big(h_t(a)\big)_{t\in T} \tag{16}$$

defines a homomorphism of $\mathfrak{B}$ into $\mathfrak{A}$.

A class $\mathbf{K}$ of similar algebras $\big(A, (o_\varphi)\big)_{\varphi\in\Phi}$ is said to be *equationally definable* if it is the class of all abstract algebras of this type satisfying a set of axioms each of which is an equation between two polynomials formed by means of the operations o_φ, $\varphi \in \Phi$.

4.11. *If $\mathbf{K}$ is an equationally definable class of similar algebras, then the following conditions are satisfied*:

(i) *if $\mathfrak{A} \in \mathbf{K}$ and $\mathfrak{B}$ is a subalgebra of $\mathfrak{A}$*, then $\mathfrak{B} \in \mathbf{K}$,

(ii) *if $\mathfrak{A} \in \mathbf{K}$ and h is an epimorphism of $\mathfrak{A}$ onto a similar algebra $\mathfrak{B}$, then $h(\mathfrak{A}) = \mathfrak{B} \in \mathbf{K}$,*

(iii) *if $\mathfrak{A}$ is the product of algebras $\mathfrak{A}_t$, $t \in T$, and for every $t \in T$ we have $\mathfrak{A}_t \in \mathbf{K}$, then $\mathfrak{A} \in \mathbf{K}$.*

Exercises

1. By a *generalized operation* on a set A we mean any mapping $\boldsymbol{O}\colon \mathbf{D} \to A$, where $\mathbf{D}$ is a class of non-empty subsets of A. The set $\mathbf{D}$ is then called the *domain* of $\boldsymbol{O}$. Give examples of generalized operations on the set R of real numbers whose domain is the class of all non-empty subsets of R.

2. A subset $A' \subset A$ will be said to be *closed under a generalized operation* $\boldsymbol{O}$: $:\mathbf{D} \to A$ on A if the conditions $S \subset A'$ and $S \in \mathbf{D}$ imply that $\boldsymbol{O}S \in A'$. Let A be the class of all subsets of a topological space X and let $\boldsymbol{O}\colon \mathbf{D} \to A$, where $\mathbf{D}$ is the class of all non-empty subsets of A, be the generalized operation on A defined as follows: for every $S \in \mathbf{D}$, $\boldsymbol{O}S = \bigcap_{Y \in S} Y$. Is the class $\mathbf{F}$ of all closed subsets of X (the class $\mathbf{G}$ of all open subsets of X) closed under the generalized operation $\boldsymbol{O}$?

3. By a *generalized algebra* we shall mean any system $(A, o_1, \ldots, o_s, \boldsymbol{O}_1, \ldots, \boldsymbol{O}_t)$, where $(A, o_1, \ldots, o_s)$ is an algebra and $\boldsymbol{O}_1, \ldots, \boldsymbol{O}_t$ are generalized operations on A. Generalize in the natural way the notions of a subalgebra and of a set of generators for generalized algebras.

4. A generalized algebra is said to be *complete* if all its generalized operations have as their common domain the class of all non-empty subsets of the set of its elements. Give examples of complete and of non-complete generalized algebras.

5. Prove or disprove that every generalized algebra can be extended to a complete generalized algebra.

6. Two generalized algebras $(A, o_1, \ldots, o_s, \boldsymbol{O}_1, \ldots, \boldsymbol{O}_t)$ and $(B, o_1', \ldots, o_{s'}', \boldsymbol{O}_1', \ldots, \boldsymbol{O}_{t'}')$ will be said to be *similar* provided that the algebras $(A, o_1, \ldots, o_s)$ and $(B, o_1', \ldots, o_{s'}')$ are similar and $t = t'$. Generalize the notions of homomorphism, epimorphism, endomorphism, monomorphism, isomorphism for generalized algebras.

Remark. It is necessary to restrict ourselves in the above generalizations to considering mappings $h\colon A \to B$ which satisfy the following condition: if a set S belongs to the domain $\mathbf{D}_i$ of a generalized operation $\boldsymbol{O}_i$, then hS belongs to the domain $\mathbf{D}_i'$ of the generalized operation $\boldsymbol{O}_i'$, $i = 1, \ldots, t$.

7. Prove or disprove Theorems 4.1, 4.2, 4.3, 4.4 for generalized algebras.

8. Show that Theorem 4.5 holds for generalized algebras.

9. Let $(A, o_1, \ldots, o_s, \boldsymbol{O}_1, \ldots, \boldsymbol{O}_t)$ be a generalized algebra with a set $A_0 \subset A$ of generators and let $\mathbf{K}$ be a class of complete generalized algebras similar to the one under consideration. We say that the generalized algebra in question is *free for the class* $\mathbf{K}$ and A_0 is the *set* of its *free generators* if for every generalized algebra in $\mathbf{K}$ each mapping h from A_0 into the set of elements of that generalized algebra can be extended to a homomorphism. Determine whether Theorem 4.6 holds for generalized algebras free for a class $\mathbf{K}$.

CHAPTER II

IMPLICATIVE ALGEBRAS

Introduction. A uniform algebraic treatment of various logics requires on the one hand the choice of an appropriate class of logics, and on the other the selection of suitable abstract algebras. The appropriate class of logics will be defined in Part Two. The propositional and predicate calculi of the logics we consider have algebraic interpretations in abstract algebras which, with respect to certain operations, are implicative algebras. Moreover, implicative algebras play, for the weakest logic in the class discussed here, a role analogous to that played by Boolean algebras for classical logic. This explains the importance of implicative algebras as the basic algebraic tool in the present book.

Implicative algebras are closely related to ordered sets with a greatest element V in the following sense. The only two-argument operation $\Rightarrow$ together with the only zero-argument operation V in any implicative algebra make it possible to define an ordering relation in such a way that V is the greatest element. Conversely, given any ordered set with a greatest element V, it is possible to define a two-argument operation $\Rightarrow$ in such a way as to obtain an implicative algebra. The concept of an implicative filter given here plays a part similar to that of a filter for lattices. Certain theorems concerning filters hold for implicative filters in implicative algebras. In particular, these two notions coincide in any relatively pseudo-complemented lattice, and thus in any pseudo-Boolean algebra, Boolean algebra, topological Boolean algebra, etc.

Note that the class of all implicative algebras is not equationally definable.

Section 1 is concerned with elementary properties of implicative algebras and implicative filters.

Sections 2, 3, 4 deal with positive implication algebras which characterize positive implicative logic (see Chapter IX). They are implicative algebras with certain additional properties. Algebras dual to positive implication algebras have been introduced by Henkin [2] under the name implicative models. The theory of positive implication algebras has been developed by Diego [1], [2], who calls them Hilbert algebras. Examples of such algebras can be constructed of open subsets of any topological space X. In that case the definition of the two-argument operation $\Rightarrow$ is the same as that of the relative pseudo-complement in the pseudo-Boolean algebra of all open subsets of X (see e.g. [MM]) and $V = X$. A representation theorem (Diego [1], [2]) states that each positive implication algebra is isomorphic to a subalgebra of the positive implication algebra of all open subsets of a topological space.

Sections 5, 6 are concerned with implication algebras which characterize classical implicative logic (see Chapter IX). They are positive implication algebras with an additional property. The theory of implication algebras has been developed by Abbott [1] (1). The algebras under consideration can be constructed of subsets of any space X. In that case the operation $\Rightarrow$ is defined in the same way as in the Boolean algebras of sets. On account of a representation theorem these examples cover all implication algebras up to an isomorphism.

Both classes of positive implication algebras and of implication algebras are equationally definable. The present author was told this by A. V. Kuznetsov when she visited Moscow in November 1964.

1. Definition and elementary properties. An abstract algebra $(A, V, \Rightarrow)$, where V is a 0-argument operation and $\Rightarrow$ is a two-argument operation, is said to be an *implicative algebra*, provided the following conditions are satisfied for all $a, b, c \in A$:

$$(\mathrm{i}_1) \qquad a \Rightarrow a = V,$$

$$(\mathrm{i}_2) \qquad \text{if} \quad a \Rightarrow b = V \quad \text{and} \quad b \Rightarrow c = V, \quad \text{then} \quad a \Rightarrow c = V$$

$$(\mathrm{i}_3) \qquad \text{if} \quad a \Rightarrow b = V \quad \text{and} \quad b \Rightarrow a = V, \quad \text{then} \quad a = b,$$

$$(\mathrm{i}_4) \qquad a \Rightarrow V = V.$$

(1) Professor J. C. Abbott presented his results at the Seminar on Foundations of Mathematics at the Institute of Mathematics of the Polish Academy of Sciences in November 1966.

The following is a direct consequence of the definitions of implicative algebras and ordering relations (I, 3).

1.1. *Let* $(A, V, \Rightarrow)$ *be an implicative algebra. Then*

$$a \leqslant b \quad \text{if and only if} \quad a \Rightarrow b = V \tag{1}$$

defines an ordering on A. *The element* V *is the greatest element in the ordered set* $(A, \leqslant)$.

Observe that in any ordered set $(A, \leqslant)$ with the greatest element V we can define a two-argument operation $\Rightarrow$ in such a way that $(A, V, \Rightarrow)$ is an implicative algebra, for instance the operation $\Rightarrow$ defined as follows:

$$a \Rightarrow b = \begin{cases} V & \text{if and only if} \quad a \leqslant b, \\ a_0 & \text{otherwise}, \end{cases}$$

where $a_0 \neq V$ is a fixed element in A, fulfils the conditions (i_1)–(i_4).

1.2. *In any implicative algebra* $(A, V, \Rightarrow)$ *the conditions* $a \Rightarrow b = V$ *and* $a = V$ *imply* $b = V$.

This follows from (i_3) and (i_4).

It is easy to see that every subalgebra of an implicative algebra is also an implicative algebra.

Consider the following two abstract algebras:

$$\mathfrak{A} = (A, V, \Rightarrow), \qquad \mathfrak{B} = (B, V', \Rightarrow')$$

where

$$A = \{V, a_1, a_2\}, \qquad B = \{V', b\}$$

and the operations $\Rightarrow$, $\Rightarrow'$ are defined as follows:

$$a_1 \Rightarrow V = a_2 \Rightarrow V = a_1 \Rightarrow a_1 = a_2 \Rightarrow a_2 = V \Rightarrow V = V,$$
$$V \Rightarrow a_1 = V \Rightarrow a_2 = a_1 \Rightarrow a_2 = a_2 \Rightarrow a_1 = a_1,$$
$$b \Rightarrow' b = b \Rightarrow' V' = V' \Rightarrow' b = V' \Rightarrow' V' = V'.$$

It is easy to verify that $\mathfrak{A}$ is an implicative algebra but $\mathfrak{B}$ is not. Let $h\colon A \to B$ be the mapping from A onto B defined as follows:

$$h(V) = h(a_1) = V', \quad h(a_2) = b.$$

The mapping h is an epimorphism of $\mathfrak{A}$ onto $\mathfrak{B}$. This shows that the epimorphic image of an implicative algebra need not be an implicative algebra. Consequently, the class of all implicative algebras is not equationally definable (I 4.11).

A subset ∇ of the set A of all elements of an implicative algebra $\mathfrak{A} = (A, V, \Rightarrow)$ will be said to be an *implicative filter*[2] provided the following conditions are satisfied:

(f$_1$) $V \in \nabla$,

(f$_2$) if $a \in \nabla$ and $a \Rightarrow b \in \nabla$, then $b \in \nabla$.

The simplest examples of implicative filters are $\{V\}$ and A. An implicative filter ∇ is said to be *proper* if $\nabla \neq A$.

Any non-empty class of implicative filters in an implicative algebra $\mathfrak{A}$ will be considered as an ordered set, the ordering relation being the set-theoretical inclusion. Thus by a *chain of implicative filters* we mean a non-empty class of implicative filters such that, for any ∇_1, ∇_2 in this class, either $\nabla_1 \subset \nabla_2$ or $\nabla_2 \subset \nabla_1$.

1.3. *The union of any chain of implicative filters in an implicative algebra $\mathfrak{A}$ is an implicative filter. The union of any chain of implicative filters which do not contain an element a_0 is an implicative filter which does not contain a_0.*

The proof is by an easy verification.

It is easy to see that the intersection of any non-empty class of implicative filters is an implicative filter. Consequently, for any set A_0 of elements of an implicative algebra $\mathfrak{A} = (A, V, \Rightarrow)$ there exists a least implicative filter containing A_0, namely the intersection of all implicative filters containing A_0. This filter is said to be the *implicative filter generated by* A_0. If A_0 is empty, then $\{V\}$ is the implicative filter generated by A_0.

A proper implicative filter ∇ is said to be *irreducible* provided that for any two proper implicative filters ∇_1, ∇_2 such that $\nabla = \nabla_1 \cap \nabla_2$ either $\nabla = \nabla_1$ or $\nabla = \nabla_2$. In other words, a proper implicative filter is irreducible if it is not the intersection of two proper implicative filters different from it.

1.4. *If ∇_0 is an implicative filter in an implicative algebra $\mathfrak{A}$ such that $a_0 \notin \nabla_0$ for some element a_0 in $\mathfrak{A}$, then there exists an irreducible implicative filter ∇^* such that $\nabla_0 \subset \nabla^*$ and $a_0 \notin \nabla^*$.*

[2] Besides this notion, which plays an important part for positive implication algebras and their extensions, it is useful to introduce a stronger notion of a special implicative filter (see Ex. 1, p. 36).

Let $\mathscr{A}$ be the ordered set of all implicative filters ∇ in $\mathfrak{A}$ containing ∇_0 and such that $a_0 \notin \nabla$. By 1.3 every chain of elements in $\mathscr{A}$ has an upper bound. By I 3.1 there exists in $\mathscr{A}$ a maximal element ∇^*. Clearly, $\nabla_0 \subset \nabla^*$ and $a_0 \notin \nabla^*$. Suppose that $\nabla^* = \nabla_1 \cap \nabla_2$ for some proper implicative filters ∇_1, ∇_2. Hence $\nabla_0 \subset \nabla_1$, $\nabla_0 \subset \nabla_2$ and either $a_0 \notin \nabla_1$ or $a_0 \notin \nabla_2$. Thus either $\nabla_1 \in \mathscr{A}$ or $\nabla_2 \in \mathscr{A}$. Since ∇^* is a maximal element in $\mathscr{A}$ and $\nabla^* \subset \nabla_1$, $\nabla^* \subset \nabla_2$, we infer that either $\nabla^* = \nabla_1$ or $\nabla^* = \nabla_2$.

Given an implicative filter ∇ of an implicative algebra $\mathfrak{A} = (A, V, \Rightarrow)$, we shall denote by $\approx_\nabla$ the binary relation on A defined as follows: for any $a, b \in A$

(2) $\qquad a \approx_\nabla b \quad$ if and only if $\quad a \Rightarrow b \in \nabla$ and $b \Rightarrow a \in \nabla$.

The *relation* $\approx_\nabla$ is then said to be *determined by* ∇.

Let us define the *kernel of an epimorphism* h of an implicative algebra $(A, V, \Rightarrow)$ onto an implicative algebra $(B, V', \Rightarrow)$ as the set of all elements of A that are mapped by h onto $V' \in B$.

1.5. *If $K(h)$ is the kernel of an epimorphism h of an implicative algebra $\mathfrak{A} = (A, V, \Rightarrow)$ onto an implicative algebra $\mathfrak{B} = (B, V', \Rightarrow)$, then $K(h)$ is an implicative filter*[(3)]. *Moreover, $h(a) = h(b)$ is equivalent to $a \Rightarrow b \in K(h)$ and $b \Rightarrow a \in K(h)$. Hence, by* I 4.7 *and* (2) *the relation $\approx_{K(h)}$ determined by $K(h)$ is a congruence in $\mathfrak{A}$. The quotient algebra $\mathfrak{A}/\approx_{K(h)}$ is isomorphic to $\mathfrak{B}$.*

The proof by an easy verification is left to the reader.

1.6. *Let $\mathfrak{A} = (A, V, \Rightarrow)$ be an implicative algebra and let ∇ be an implicative filter in $\mathfrak{A}$. If in $\mathfrak{A}$ the following conditions hold for any $a, b, c \in A$:*

(s_1) $\qquad (a \Rightarrow b) \Rightarrow ((b \Rightarrow c) \Rightarrow (a \Rightarrow c)) = V,$

(s_2) $\qquad (b \Rightarrow c) \Rightarrow ((a \Rightarrow b) \Rightarrow (a \Rightarrow c)) = V,$

then the relation $\approx_\nabla$ determined by ∇ is a congruence in $\mathfrak{A}$. If, moreover,

(3) $\qquad V \Rightarrow a = a, \quad$ *for all $a \in A$,*

then the quotient algebra $\mathfrak{A}/\approx_\nabla$, denoted later by $\mathfrak{A}/\nabla$, is an implicative algebra in which the equations corresponding to (s_1), (s_2) *and* (3) *hold.*

(3) It can be proved that kernels of epimorphisms are special implicative filters and conversely (see Ex. 1, Ex. 2, p. 36).

The mapping $h(a) = ||a|| \in A/\nabla$ *is an epimorphism of* $\mathfrak{A}$ *onto* $\mathfrak{A}/\nabla$ *and* ∇ *is the kernel of* h. *For each* $a \in A$, $a \in \nabla$ *if and only if* $a \approx_\nabla V$. *The quotient algebra* $\mathfrak{A}/\nabla$ *is degenerate if and only if* ∇ *is not proper.*

Observe that $a \approx_\nabla a$, since $a \Rightarrow a = V \in \nabla$. The symmetry of $\approx_\nabla$ follows directly from (2). The transitivity follows from (s_1) and properties (f_1), (f_2) of implicative filters. Let us suppose that $a \approx_\nabla b$ and $c \approx_\nabla d$. Thus $c \Rightarrow d \in \nabla$. Hence, by (s_2), (f_1) and (f_2)

$$(a \Rightarrow c) \Rightarrow (a \Rightarrow d) \in \nabla. \tag{4}$$

Since $a \approx_\nabla b$, we have $b \Rightarrow a \in \nabla$. Consequently, by (s_1), (f_1), (f_2),

$$(a \Rightarrow d) \Rightarrow (b \Rightarrow d) \in \nabla. \tag{5}$$

By (4), (5), (s_1), (f_1), (f_2) $(a \Rightarrow c) \Rightarrow (b \Rightarrow d) \in \nabla$. Similarly we prove that $(b \Rightarrow d) \Rightarrow (a \Rightarrow c) \in \nabla$. Thus $\approx_\nabla$ is a congruence in $\mathfrak{A}$.

Observe that if (3) is also satisfied, then

$$a \in \nabla \quad \textit{if and only if} \quad a \approx_\nabla V, \quad \textit{for all } a \in A. \tag{6}$$

If $a \approx_\nabla V$, then $V \Rightarrow a \in \nabla$ and $V \in \nabla$. Hence $a \in \nabla$. Conversely, if $a \in \nabla$, then by (3) $a = V \Rightarrow a \in \nabla$ and $a \Rightarrow V = V \in \nabla$. Thus $a \approx_\nabla V$. By (i_1) and I 4 (6), $||a|| \Rightarrow ||a|| = ||a \Rightarrow a|| = ||V||$. Suppose that $||a|| \Rightarrow ||b|| = ||V||$ and $||b|| \Rightarrow ||a|| = ||V||$. Hence, by (6) and I 4 (6), $a \Rightarrow b \in \nabla$ and $b \Rightarrow a \in \nabla$. Consequently $a \approx_\nabla b$, i.e. $||a|| = ||b||$. Suppose that $||a|| \Rightarrow ||b|| = ||V||$ and $||b|| \Rightarrow ||c|| = ||V||$. By I 4 (6), $a \Rightarrow b \approx_\nabla V$ and $b \Rightarrow c \approx_\nabla V$. By (6), $a \Rightarrow b \in \nabla$ and $b \Rightarrow c \in \nabla$. Hence, by (s_1), (f_1), (f_2), $a \Rightarrow c \in \nabla$. By (6), $a \Rightarrow c \approx_\nabla V$, i.e. $||a|| \Rightarrow ||c|| = ||a \Rightarrow c|| = ||V||$. The equation $||a|| \Rightarrow ||V|| = ||V||$ holds by (i_4) and properties of quotient algebras. Similarly we infer that the equations

$$(||a|| \Rightarrow ||b||) \Rightarrow ((||b|| \Rightarrow ||c||) \Rightarrow (||a|| \Rightarrow ||c||)) = ||V||,$$

$$(||b|| \Rightarrow ||c||) \Rightarrow ((||a|| \Rightarrow ||b||) \Rightarrow (||a|| \Rightarrow ||c||)) = ||V||,$$

$$||V|| \Rightarrow ||a|| = ||a||,$$

hold by (s_1), (s_2) and (3). By I 4 (7), the mapping $h(a) = ||a||$ for all $a \in A$ is an epimorphism of $\mathfrak{A}$ onto $\mathfrak{A}/\nabla$. We shall show that the kernel $K(h) = \nabla$. Indeed, $a \in K(h)$ if and only if $h(a) = ||V||$, i.e. if and only if $||a|| = ||V||$. This condition, by (6), is equivalent to $a \in \nabla$. It follows from (6) that $\mathfrak{A}/\nabla$ is degenerate, i.e. $\mathfrak{A}/\nabla$ contains only one element if and only if $\nabla = A$.

In an implicative algebra the set of all elements x, such that $a \leqslant x$ for a fixed element a, need not be an implicative filter. We shall show that

1.7. *If* $\mathfrak{A} = (A, V, \Rightarrow)$ *is an implicative algebra, then for every* $a \in A$ *the set* $\nabla(a) = \{x \in A\colon a \leqslant x\}$ *is an implicative filter if and only if the following condition is satisfied*:

(F_1) *if* $a \Rightarrow (b \Rightarrow c) = V$ *and* $a \Rightarrow b = V$, *then* $a \Rightarrow c = V$ *for all* a, b, c *in* A.

Let us suppose that (F_1) is satisfied. Observe that $a \leqslant V$ and hence $V \in \nabla(a)$. If $b, b \Rightarrow c \in \nabla(a)$, then $a \leqslant b$ and $a \leqslant b \Rightarrow c$. Consequently, $a \Rightarrow b = V$ and $a \Rightarrow (b \Rightarrow c) = V$. By ($F_1$), $a \Rightarrow c = V$, i.e. $a \leqslant c$. Then $c \in \nabla(a)$. Thus $\nabla(a)$ is an implicative filter. Conversely, if $\nabla(a)$ is an implicative filter for every $a \in A$, then for all b, c in A the following condition holds: if $b \in \nabla(a)$ and $b \Rightarrow c \in \nabla(a)$, then $c \in \nabla(a)$. In other words, if $a \Rightarrow (b \Rightarrow c) = V$ and $a \Rightarrow b = V$, then $a \Rightarrow c = V$.

If condition (F_1) is satisfied, then $\nabla(a)$ is the least implicative filter containing a. In fact, if ∇ is an implicative filter and $a \in \nabla$, then $\nabla(a) \subset \nabla$. If $b \in \nabla(a)$, then $a \Rightarrow b = V \in \nabla$. Since $a \in \nabla$, by (f_2) $b \in \nabla$. The implicative filter $\nabla(a)$ is said to be the *principal implicative filter generated by* a.

1.8. *If in an implicative algebra* $\mathfrak{A} = (A, V, \Rightarrow)$ *one of the following conditions is satisfied for all* $a, b, c \in A$:

(F_2) $$(a \Rightarrow (b \Rightarrow c)) \Rightarrow ((a \Rightarrow b) \Rightarrow (a \Rightarrow c)) = V,$$

(F_3) $$(a \Rightarrow b) \Rightarrow ((a \Rightarrow (b \Rightarrow c)) \Rightarrow (a \Rightarrow c)) = V,$$

then for every implicative filter ∇ *in* $\mathfrak{A}$ *and for every* $a \in A$, *the set* $\nabla^* = \{x \in A\colon a \Rightarrow x \in \nabla\}$ *is an implicative filter. If, moreover, for all* a, $b \in A$ *the following equation holds*:

(S) $$a \Rightarrow (b \Rightarrow a) = V$$

then ∇^* *is the least implicative filter containing* ∇ *and* a.

Of course, $V \in \nabla^*$ by (i_4) and (f_1). If $b, b \Rightarrow c \in \nabla^*$, then $a \Rightarrow b$, $a \Rightarrow (b \Rightarrow c) \in \nabla$. By ($F_2$) or ($F_3$) and ($f_1$), ($f_2$), $a \Rightarrow c \in \nabla$, i.e. $c \in \nabla^*$.

Thus ∇^* is an implicative filter. By (i_1) and (f_1), $a \in \nabla^*$. If $b \in \nabla$, then, by (S), $b \Rightarrow (a \Rightarrow b) \in \nabla$, and consequently $a \Rightarrow b \in \nabla$. Hence $b \in \nabla^*$. Thus $\nabla \subset \nabla^*$. Suppose that ∇_0 is an implicative filter such that $\nabla \subset \nabla_0$ and $a \in \nabla_0$. If $b \in \nabla^*$, then $a \Rightarrow b \in \nabla \subset \nabla_0$. Since $a \in \nabla_0$, by (f_2) $b \in \nabla_0$. Thus $\nabla^* \subset \nabla_0$.

An implicative filter in an implicative algebra $\mathfrak{A} = (A, V, \Rightarrow)$ is said to be *maximal* provided it is proper and is not a proper subset of any proper implicative filter in $\mathfrak{A}$. We shall show that

1.9. *If in an implicative algebra* $\mathfrak{A} = (A, V, \Rightarrow)$ *there is an element* Λ *such that* $\Lambda \Rightarrow a = V$ *for all* $a \in A$ (i.e. Λ is the least element in the ordered set $(A, \leqslant)$ where $\leqslant$ is defined by (1)), *then for every proper implicative filter* ∇ *there exists a maximal implicative filter* ∇^* *such that* $\nabla \subset \nabla^*$.

Since ∇ is proper, it does not contain Λ. Otherwise every element $a \in A$ would be in ∇. Consider the class of all proper implicative filters in $\mathfrak{A}$ containing ∇. No implicative filter in this class contains Λ. Thus the theorem follows from 1.3. and I 3.1.

2. Positive implication algebras. An abstract algebra $(A, V, \Rightarrow)$ with a 0-argument operation V and a two-argument operation $\Rightarrow$ will be called a *positive implication algebra* provided for all $a, b, c \in A$ the following conditions are satisfied [4]:

(p_1) $$a \Rightarrow (b \Rightarrow a) = V,$$

(p_2) $$(a \Rightarrow (b \Rightarrow c)) \Rightarrow ((a \Rightarrow b) \Rightarrow (a \Rightarrow c)) = V,$$

(p_3) $$\text{if} \quad a \Rightarrow b = V \quad \text{and} \quad b \Rightarrow a = V, \quad \text{then} \quad a = b,$$

(p_4) $$a \Rightarrow V = V.$$

Observe that in any positive implication algebra the following condition is also fulfiled

(1) $$\textit{if} \quad a \Rightarrow b = V \quad \textit{and} \quad a = V, \quad \textit{then} \quad b = V.$$

Indeed, if the assumptions hold, then $V \Rightarrow b = V$. By (p_4), $b \Rightarrow V = V$. Hence, using (p_3), we get $b = V$.

[4] This system of axioms is dual to that due to Henkin [2] for implicative models.

It follows directly from (p_4) that

$$\text{(2)} \qquad \textit{if} \quad a = V, \quad \textit{then} \quad b \Rightarrow a = V.$$

We shall show that

2.1. *Any positive implication algebra* $(A, V, \Rightarrow)$ *is an implicative algebra. Any implicative algebra* $(A, V, \Rightarrow)$ *in which* (p_1) *and* (p_2) *hold is a positive implication algebra.*

By (p_2)

$$\big(a \Rightarrow ((a \Rightarrow a) \Rightarrow a)\big) \Rightarrow \big((a \Rightarrow (a \Rightarrow a)) \Rightarrow (a \Rightarrow a)\big) = V.$$

Applying twice (p_1) and (1) we get

$$\text{(3)} \qquad a \Rightarrow a = V, \qquad \text{for all } a \in A.$$

Thus (i_1) (see Sec. 1) holds. To prove (i_2) let us suppose that $a \Rightarrow b = V$ and $b \Rightarrow c = V$. Applying (2) we obtain $a \Rightarrow (b \Rightarrow c) = V$ and $a \Rightarrow b = V$. Thus, by (p_2) and (1), we get $a \Rightarrow c = V$. Conditions (i_3), (i_4) are the same as (p_3) and (p_4), respectively. The second statement is obvious.

It follows directly from 2.1 and 1.1 that

2.2. *If* $(A, V, \Rightarrow)$ *is a positive implication algebra, then the relation* $\leqslant$ *defined on* A *as*

$$\text{(4)} \qquad a \leqslant b \quad \textit{if and only if} \quad a \Rightarrow b = V,$$

is an ordering on A. *The element* V *is the greatest element in the ordered set* $(A, \leqslant)$.

It is easy to verify that if $(A, \leqslant)$ is a chain with a greatest element V, then $(A, V, \Rightarrow)$, where

$$a \Rightarrow b = \begin{cases} V & \text{if} \quad a \leqslant b \\ b & \text{if} \quad b \leqslant a \text{ and } a \neq b \end{cases} \qquad \text{for all } a, b \in A$$

is a positive implication algebra. In particular, if $\mathfrak{A} = \{\Lambda, V\}$, $\Lambda \leqslant V$, $\Lambda \neq V$, then $(A, V, \Rightarrow)$ with the operation $\Rightarrow$ defined by the equations

$$\text{(5)} \qquad \Lambda \Rightarrow \Lambda = \Lambda \Rightarrow V = V \Rightarrow V = V, \qquad V \Rightarrow \Lambda = \Lambda$$

is an example of a positive implication algebra.

Consider a topological space X with an interior operation $\mathbf{I}$. Let $G(X)$ be the class of all open subsets of X. Then $\big(G(X), X, \Rightarrow\big)$, where

$$\text{(6)} \qquad Y \Rightarrow Z = \mathbf{I}\big((X - Y) \cup Z\big) \qquad \text{for all } Y, Z \in G(X),$$

is a positive implication algebra. In this algebra the ordering relation $\leqslant$ defined by (4) coincides with the set-theoretical inclusion. Clearly, any non-empty class $G_0(X) \subset G(X)$ closed under the operation $\Rightarrow$ defined by (6), satisfies the condition $X \in G_0(X)$ and is also a positive implication algebra. In other words, every subalgebra of a positive implication algebra $(G(X), X, \Rightarrow)$ is a positive implication algebra and will be called a *positive implication algebra of sets*. We shall see in Section 4 that each positive implication algebra is isomorphic to a positive implication algebra of sets.

2.3. *In any positive implication algebra* $(A, V, \Rightarrow)$ *the following conditions are satisfied*:

(7) $$a \leqslant b \Rightarrow a,$$

(8) $$a \leqslant b \Rightarrow c \quad \textit{implies} \quad b \leqslant a \Rightarrow c,$$

(9) $$a \leqslant (a \Rightarrow b) \Rightarrow b,$$

(10) $$V \Rightarrow a = a,$$

(11) $$\textit{if} \quad b \leqslant c, \quad \textit{then} \quad a \Rightarrow b \leqslant a \Rightarrow c,$$

(12) $$\textit{if} \quad a \leqslant b, \quad \textit{then} \quad b \Rightarrow c \leqslant a \Rightarrow c,$$

(13) $$a \Rightarrow (a \Rightarrow b) = a \Rightarrow b,$$

(14) $$a \Rightarrow (b \Rightarrow c) = b \Rightarrow (a \Rightarrow c),$$

(15) $$a \Rightarrow b \leqslant (b \Rightarrow c) \Rightarrow (a \Rightarrow c),$$

(16) $$b \Rightarrow c \leqslant (a \Rightarrow b) \Rightarrow (a \Rightarrow c).$$

(7) follows from (p_1) and 2.2. To prove (8), suppose that $a \leqslant b \Rightarrow c$, i.e. $a \Rightarrow (b \Rightarrow c) = V$. Consequently, by ($p_2$) and (1), we get $(a \Rightarrow b) \Rightarrow (a \Rightarrow c) = V$, i.e. $a \Rightarrow b \leqslant a \Rightarrow c$. On the other hand, $b \leqslant a \Rightarrow b$ by (7). Hence $b \leqslant a \Rightarrow c$. Applying (8) to $a \Rightarrow b \leqslant a \Rightarrow b$, we obtain (9). By (9) $V \leqslant (V \Rightarrow a) \Rightarrow a$. Since V is the greatest element, we have $(V \Rightarrow a) \Rightarrow a \leqslant V$. Thus $(V \Rightarrow a) \Rightarrow a = V$, i.e. $V \Rightarrow a \leqslant a$. On the other hand, $a \leqslant V \Rightarrow a$ by (7). Thus $V \Rightarrow a = a$ and (10) holds. To prove (11) suppose that $b \leqslant c$, i.e. $b \Rightarrow c = V$. Hence, by (2), $a \Rightarrow (b \Rightarrow c) = V$. Applying ($p_2$) and (1) we get $(a \Rightarrow b) \Rightarrow (a \Rightarrow c) = V$, i.e. $a \Rightarrow b \leqslant a \Rightarrow c$. To prove (12) suppose that $a \leqslant b$. Thus $a \Rightarrow b = V$. Applying (p_2), (4) and (10), we obtain $a \Rightarrow (b \Rightarrow c) \leqslant V \Rightarrow (a \Rightarrow c) = a \Rightarrow c$. Since

$b \Rightarrow c \leqslant a \Rightarrow (b \Rightarrow c)$ by (7), we have $b \Rightarrow c \leqslant a \Rightarrow c$. It also follows by ($p_2$), (4) (3) and (10) that $a \Rightarrow (a \Rightarrow b) \leqslant (a \Rightarrow a) \Rightarrow (a \Rightarrow b) = V \Rightarrow (a \Rightarrow b) = a \Rightarrow b$. By (7) $a \Rightarrow b \leqslant a \Rightarrow (a \Rightarrow b)$. Consequently (13) holds. It follows from (p_2), (4) and (8) that $a \Rightarrow b \leqslant \big(a \Rightarrow (b \Rightarrow c)\big) \Rightarrow (a \Rightarrow c)$. By (7) $b \leqslant a \Rightarrow b$. Hence $b \leqslant \big(a \Rightarrow (b \Rightarrow c)\big) \Rightarrow (a \Rightarrow c)$. Applying (8), we get $a \Rightarrow (b \Rightarrow c) \leqslant b \Rightarrow (a \Rightarrow c)$. The substitution of a for b and b for a gives $b \Rightarrow (a \Rightarrow c) \leqslant a \Rightarrow (b \Rightarrow c)$. Thus (14) holds. To prove (15) observe that, by (9), $b \leqslant (b \Rightarrow c) \Rightarrow c$. Applying (11), we obtain $(a \Rightarrow b) \Rightarrow b \leqslant (a \Rightarrow b) \Rightarrow \big((b \Rightarrow c) \Rightarrow c\big)$. Hence, by (9), $a \leqslant (a \Rightarrow b) \Rightarrow \big((b \Rightarrow c) \Rightarrow c\big)$. Applying (8) and (14), we get $a \Rightarrow b \leqslant a \Rightarrow \big((b \Rightarrow c) \Rightarrow c\big) = (b \Rightarrow c) \Rightarrow (a \Rightarrow c)$. (16) follows from (15) and (8).

The class of all positive implication algebras is equationally definable [5]. In order to show this we prove the following theorems.

2.4. *In every positive implication algebra* $(A, V, \Rightarrow)$ *the following equations hold*:

(17) $$(a \Rightarrow a) \Rightarrow b = b,$$

(18) $$a \Rightarrow (b \Rightarrow c) = (a \Rightarrow b) \Rightarrow (a \Rightarrow c),$$

(19) $$(a \Rightarrow b) \Rightarrow \big((b \Rightarrow a) \Rightarrow b\big) = (b \Rightarrow a) \Rightarrow \big((a \Rightarrow b) \Rightarrow a\big).$$

(17) follows from (10) and (3). To prove (18) observe, that by (7) $b \leqslant a \Rightarrow b$. Applying (12) and (14), we get

$$(a \Rightarrow b) \Rightarrow (a \Rightarrow c) \leqslant b \Rightarrow (a \Rightarrow c) = a \Rightarrow (b \Rightarrow c).$$

Hence (18) follows from (p_2) and (4). By (9) $b \leqslant (b \Rightarrow a) \Rightarrow a$. Applying (11), (14) and (13), we get

$$(a \Rightarrow b) \Rightarrow \big((b \Rightarrow a) \Rightarrow b\big) \leqslant (a \Rightarrow b) \Rightarrow \Big((b \Rightarrow a) \Rightarrow \big((b \Rightarrow a) \Rightarrow a\big)\Big)$$
$$= (a \Rightarrow b) \Rightarrow \big((b \Rightarrow a) \Rightarrow a\big) = (b \Rightarrow a) \Rightarrow \big((a \Rightarrow b) \Rightarrow a\big).$$

Hence, by symmetry,

$$(b \Rightarrow a) \Rightarrow \big((a \Rightarrow b) \Rightarrow a\big) \leqslant (a \Rightarrow b) \Rightarrow \big((b \Rightarrow a) \Rightarrow b\big)$$

follows. Thus (19) holds.

2.5. *Let* $(A, \Rightarrow)$ *be an algebra with a two-argument operation* $\Rightarrow$. *If for all* $a, b, c \in A$ *the equations* (17), (18) *and* (19) *are satisfied, then*

(20) *for all* $a, b \in A, \quad a \Rightarrow a = b \Rightarrow b,$

(21) $(A, V, \Rightarrow)$, *where* $V = a \Rightarrow a$, *is a positive implication algebra.*

[5] See Diego [1], [2] .

Substituting in (19) $a \Rightarrow a$ for a and $b \Rightarrow b$ for b and applying (17), we get (20). Let $V = a \Rightarrow a$ for all $a \in A$. From (18) it follows that $a \Rightarrow V = a \Rightarrow (a \Rightarrow a) = (a \Rightarrow a) \Rightarrow (a \Rightarrow a) = V$. Thus ($p_4$) holds, i.e.

$$a \Rightarrow V = V. \tag{22}$$

By (18) and (22) $a \Rightarrow (b \Rightarrow a) = (a \Rightarrow b) \Rightarrow V = V$, which proves ($p_1$). Equation ($p_2$) follows from (18) and the definition of V. Observe that by (17)

$$V \Rightarrow b = b \quad \text{for all } b \in A. \tag{23}$$

To prove (p_3) let us suppose that $a \Rightarrow b = V$ and $b \Rightarrow a = V$. Then, by (23) and (19), we get

$$\begin{aligned} a &= V \Rightarrow (V \Rightarrow a) = (b \Rightarrow a) \Rightarrow \big((a \Rightarrow b) \Rightarrow a\big) \\ &= (a \Rightarrow b) \Rightarrow \big((b \Rightarrow a) \Rightarrow b\big) = V \Rightarrow (V \Rightarrow b) = b. \end{aligned}$$

3. Implicative filters in positive implication algebras(6). Let $(A, V, \Rightarrow)$ be a positive implication algebra. By 2.1 it is an implicative algebra. By 2 $\big((p_1), (p_2), (15), (16)$ and $(10)\big)$ conditions 1 $\big((S), (F_1), (F_2), (s_1), (s_2)$ and $(3)\big)$ are satisfied in this algebra. Consequently, we get the following two *theorems on epimorphisms*.

3.1. *If $K(h)$ is the kernel of an epimorphism h of a positive implication algebra $\mathfrak{A} = (A, V, \Rightarrow)$ onto a similar algebra $\mathfrak{B} = (B, V', \Rightarrow)$, then $K(h)$ is an implicative filter. Moreover, the condition $h(a) = h(b)$ is equivalent to the condition $a \Rightarrow b, b \Rightarrow a \in K(h)$. Hence, by* I 4.7, *the relation $\approx_{K(h)}$ determined by $K(h)$* (see 2 (2)) *is a congruence in $\mathfrak{A}$. The quotient algebra $\mathfrak{A}/\approx_{K(h)}$ is isomorphic to $\mathfrak{B}$.*

This theorem follows from 1.5, 2.1, 2.4, 2.5 and I 4.11.

3.2. *Let $\mathfrak{A} = (A, V, \Rightarrow)$ be a positive implication algebra and let ∇ be an implicative filter in $\mathfrak{A}$. Then the relation $\approx_\nabla$ determined by ∇ is a congruence in $\mathfrak{A}$. The quotient algebra denoted by $\mathfrak{A}/\nabla$ is a positive implication algebra. Moreover, the following conditions are satisfied*:

(i) *the mapping h defined by the formula $h(a) = ||a||$, $a \in A$, is an epimorphism of $\mathfrak{A}$ onto $\mathfrak{A}/\nabla$ and ∇ is the kernel of h*;

(ii) *for every $a \in A$, $a \in \nabla$ if and only if $a \approx_\nabla V$*;

(iii) *the algebra $\mathfrak{A}/\nabla$ is degenerate if and only if $\nabla = A$.*

(6) The results in Sec. 3 are due to Diego [1], [2].

This theorem follows from 1.6, 2.1, 2.4, 2.5, I 4.11 and 2((15), (16)).

3.3. *If* $(A, V, \Rightarrow)$ *is a positive implication algebra, then for every element* $a \in A$ *the set* $\nabla(a) = \{x \in A\colon a \leqslant x\}$ *is the least implicative filter containing* a, i.e. *the principal filter generated by* a.

The above theorem follows from 1.7 and 2 ((p$_2$), (1)).

3.4. *If* $(A, V, \Rightarrow)$ *is a positive implication algebra, then for every implicative filter* ∇ *and for every element* $a \in A$ *the set* $\nabla^* = \{x \in A\colon a \Rightarrow x \in \nabla\}$ *is the least implicative filter containing* ∇ *and* a.

Theorem 3.4 follows from 1.8, 2.1 and 2 ((p$_1$), (p$_2$)).

The following theorem is a generalization of 3.4.

3.5. *Let* $(A, V, \Rightarrow)$ *be a positive implication algebra and let* $O \neq A_0 \subset A$. *Then the set* $\nabla(A_0)$ *composed of all* $x \in A$ *for which there exist* $a_1, \ldots, a_n \in A_0$ *such that* $a_1 \Rightarrow \big(a_2 \Rightarrow (\ldots (a_n \Rightarrow x)\ldots)\big) = V$ *is the implicative filter generated by* A_0.

Let $a \in A_0$. Since $a \Rightarrow V = V$, $V \in \nabla(A_0)$. Suppose that $a, a \Rightarrow b \in \nabla(A_0)$ for some $a, a \Rightarrow b$ in A. Then there exist $a_1, \ldots, a_n, b_1, \ldots, b_m \in A_0$ such that

$$a_1 \Rightarrow \big(a_2 \Rightarrow (\ldots(a_n \Rightarrow a)\ldots)\big) = V \tag{1}$$

and

$$b_1 \Rightarrow \big(b_2 \Rightarrow (\ldots(b_m \Rightarrow (a \Rightarrow b))\ldots)\big) = V. \tag{2}$$

By succesive applications of 2 ((7), (14)) to (1) and (2) we get

$$a_1 \Rightarrow \Big(a_2 \Rightarrow \Big(\ldots \Big(a_n \Rightarrow \big(b_1 \Rightarrow (\ldots(b_m \Rightarrow a)\ldots)\big)\Big)\ldots\Big)\Big) = V, \tag{3}$$

$$a_1 \Rightarrow \Big(a_2 \Rightarrow \Big(\ldots \Big(a_n \Rightarrow \big(b_1 \Rightarrow (\ldots(b_m \Rightarrow (a \Rightarrow b))\ldots)\big)\Big)\ldots\Big)\Big) = V. \tag{4}$$

Applying 2 (16) to 2 (p$_2$) and using (3), (4), we get

$$a_1 \Rightarrow \Big(a_2 \Rightarrow \Big(\ldots \Big(a_n \Rightarrow \big(b_1 \Rightarrow (\ldots(b_m \Rightarrow b)\ldots)\big)\Big)\ldots\Big)\Big) = V.$$

Since $a_1, \ldots, a_n, b_1, \ldots, b_m \in A_0$, we have $b \in \nabla(A_0)$. Thus $\nabla(A_0)$ is an implicative filter. The inclusion $A_0 \subset \nabla(A_0)$ is obvious. Let us suppose that ∇ is an implicative filter containing A_0. We shall show that $\nabla(A_0)$

$\subset \nabla$. If $a \in \nabla(A_0)$, then (1) holds for some $a_1, \ldots, a_n \in A_0$. Since $a_1, \ldots, a_n \in \nabla$ and ∇ is an implicative filter, by 1((f_1), (f_2)), $a \in \nabla$.

3.6. *If* ∇_0 *is an implicative filter in a positive implication algebra* $\mathfrak{A} = (A, V, \Rightarrow)$ *and* $a_0 \notin \nabla_0$, *then there exists an irreducible implicative filter* ∇^* *such that* $\nabla_0 \subset \nabla^*$ *and* $a \notin \nabla^*$.

3.6 follows directly from 1.4 and 2.1.

3.7. *If in a positive implication algebra* $\mathfrak{A} = (A, V, \Rightarrow)$ $a \leqslant b$ *does not hold for some* $a, b \in A$, *then there exists an irreducible implicative filter* ∇ *in* $\mathfrak{A}$ *such that* $a \in \nabla$ *and* $b \notin \nabla$.

Theorem 3.7 follows from 3.3 and 3.6.

3.8. *If in a positive implication algebra* $(A, V, \Rightarrow)$ *there is an element* Λ *such that* $\Lambda \Rightarrow a = V$ *for all* $a \in A$ (i.e. Λ is the least element in the ordered set $(A, \leqslant)$, where $\leqslant$ is defined by 2 (4)), *then for every proper implicative filter* ∇ *there exists a maximal implicative filter* ∇^* *containing* ∇.

Theorem 3.8 follows directly from 1.9 and 2.1.

4. Representation theorem for positive implication algebras. The aim of this section is to prove the following representation theorem [7].

4.1. *For every positive implication algebra* $\mathfrak{A} = (A, V, \Rightarrow)$ *there exists a monomorphism* h *of* $\mathfrak{A}$ *into a positive implication algebra* $(G(X), X, \Rightarrow)$ *of all open subsets of a topological* T_0*-space* X. *Hence, every positive implication algebra is isomorphic to a positive implication algebra of sets.*

Let X be the class of all irreducible implicative filters of $\mathfrak{A}$. Let h be the mapping of A into the class of all subsets of X defined as follows:

$$(1) \qquad h(a) = \{\nabla \in X \colon a \in \nabla\}, \qquad \text{for each } a \in A.$$

We shall consider X as a topological space, with the class $\{h(a)\}_{a \in A}$ of all sets $h(a)$ for $a \in A$ as a subbase (see I 2.1). Observe that X is then a T_0-space. Indeed, if $\nabla_1 \neq \nabla_2$, then there exists an element $a \in A$ such that a belongs exactly to one of the sets ∇_1, ∇_2, say $a \in \nabla_1$ and $a \notin \nabla_2$. Consequently, $\nabla_1 \in h(a)$ and $\nabla_2 \notin h(a)$, i.e. $h(a)$ is an open set containing exactly one of the points ∇_1, ∇_2 of the space X. Thus X is a T_0-space.

[7] Diego [1], [2].

Now observe that if $a \leqslant b$ with $a, b \in A$, then $h(a) \subset h(b)$. $a \leqslant b$ is equivalent to $a \Rightarrow b = V$. If $\nabla \in h(a)$, then, by (1), $a \in \nabla$. On the other hand, $a \Rightarrow b = V \in \nabla$. Thus, by 1 ($f_2$), $b \in \nabla$. Consequently, $\nabla \in h(b)$. Thus $h(a) \subset h(b)$. Conversely, if $a \leqslant b$ does not hold, then by 3.7 there exists an irreducible implicative filter ∇ such that $a \in \nabla$ and $b \notin \nabla$, i.e. $\nabla \in h(a)$ and $\nabla \notin h(b)$. Thus the condition $h(a) \subset h(b)$ does not hold. We have proved that

(2) $\quad a \leqslant b$ if and only if $h(a) \subset h(b)$ for all $a, b \in A$.

If $a \neq b$, then one of conditions $a \leqslant b$, $b \leqslant a$ does not hold. By (2) one of inclusions $h(a) \subset h(b)$, $h(b) \subset h(a)$ does not hold, i.e. $h(a) \neq h(b)$. Consequently, h is one-one.

Notice that

(3) $$h(V) = X.$$

This follows from the fact that $V \in \nabla$ for every ∇ in X; consequently, by (1) every ∇ in X belongs to $h(V)$.

It remains to prove that for all $a, b \in A$

(4) $$h(a \Rightarrow b) = \mathbf{I}\big((X - h(a)) \cup h(b)\big).$$

It follows from 2 (f_2) that $a \Rightarrow b \in \nabla$ implies that either $a \notin \nabla$ or $b \in \nabla$. Thus $\nabla \in h(a \Rightarrow b)$ implies that $\nabla \in (X - h(a)) \cup h(b)$. Hence

(5) $$h(a \Rightarrow b) \subset (X - h(a)) \cup h(b).$$

Thus $h(a \Rightarrow b)$ is an open set contained in $(X - h(a)) \cup h(b)$. In order to prove (4) it is sufficient to show that $h(a \Rightarrow b)$ is the greatest open set satisfying this condition. Since $\{h(a)\}_{a \in A}$ is a subbase of X containing X, it is sufficient to prove that

(6) if $\quad h(a_1) \cap \ldots \cap h(a_n) \subset (X - h(a)) \cup h(b)$, then
$$h(a_1) \cap \ldots \cap h(a_n) \subset h(a \Rightarrow b).$$

Suppose that $\nabla \in h(a_1) \cap \ldots \cap h(a_n)$, i.e. $a_1, \ldots, a_n \in \nabla$. This implies that either $a \notin \nabla$ or $b \in \nabla$. If $b \in \nabla$, then $a \Rightarrow b \in \nabla$ since $b \Rightarrow (a \Rightarrow b) = V \in \nabla$ (2 (p_1), 1 ((f_1), (f_2))). If $b \notin \nabla$, then $a \notin \nabla$. We shall show that also in this case $a \Rightarrow b \in \nabla$. For if $a \Rightarrow b \notin \nabla$, then by 3.4 $b \notin \nabla_1 = \{x \in A\colon a \Rightarrow x \in \nabla\}$, which is the least implicative filter containing ∇ and a. By 3.6 there exists an irreducible implicative filter $\nabla^* \in X$ such that $\nabla_1 \subset \nabla^*$ and $b \notin \nabla^*$. Consequently, $a_1, \ldots, a_n \in \nabla^*$, $a \in \nabla^*$ and $b \notin \nabla^*$. Thus $\nabla^* \in h(a_1) \cap \ldots \cap h(a_n)$ and $\nabla^* \in h(a) \cap (X - h(b))$

$= X-((X-h(a))\cup h(b))$, which contradicts our assumption that $h(a_1)\cap \ldots \cap h(a_n) \subset (X-h(a))\cup h(b)$. Consequently, $a \Rightarrow b \in \nabla$, i.e. $\nabla \in h(a \Rightarrow b)$. We have just proved that if $\nabla \in h(a_1)\cap \ldots \cap h(a_n)$, then $\nabla \in h(a \Rightarrow b)$. Hence (6) holds.

5. Implication algebras [8]. An abstract algebra $\mathfrak{A} = (A, V, \Rightarrow)$, where V is a zero-argument operation and $\Rightarrow$ is a two-argument operation, is said to be an *implication algebra* provided it is a positive implication algebra and in addition the following equation holds:

(P) $$(a \Rightarrow b) \Rightarrow a = a, \quad \text{for all } a, b \in A.$$

It follows from this definition that $(A, V, \Rightarrow)$ is an implication algebra if and only if conditions 2 ((p$_1$)–(p$_4$)) and condition (P) above are satisfied.

The class of all implicative algebras is equationally definable [9]. In fact, from 2.4 and 2.5 we deduce that

5.1. *In every implication algebra the following equations hold*:

(1) $$(a \Rightarrow a) \Rightarrow b = b,$$

(2) $$a \Rightarrow (b \Rightarrow c) = (a \Rightarrow b) \Rightarrow (a \Rightarrow c),$$

(3) $$(a \Rightarrow b) \Rightarrow ((b \Rightarrow a) \Rightarrow b) = (b \Rightarrow a) \Rightarrow ((a \Rightarrow b) \Rightarrow a),$$

(P) $$(a \Rightarrow b) \Rightarrow a = a.$$

On the other hand, if in an abstract algebra $(A, \Rightarrow)$ *equations* (1), (2), (3), (P) *hold, then*

(4) $$\text{for all } a, b \in A, \quad a \Rightarrow a = b \Rightarrow b,$$

(5) $(A, V, \Rightarrow)$, *where* $V = a \Rightarrow a$ *is an implication algebra.*

Obviously, Theorems 2 ((1), (2)), 2.2, 2.3 are true also for implication algebras.

As an example of an implication algebra we quote the two-element implication algebra $(\{\Lambda, V\}, V, \Rightarrow)$, where the operation $\Rightarrow$ is defined by means of equations 2 (5).

(8) The theory of implication algebras has been developed by Abbott [1] (cf. footnote (1), p. 16).

(9) Cf. the present author's remark concerning the equational definability of the class of all implication algebras in the Introduction to this chapter. See Abbott [1].

Consider a space $X \neq O$. Let $B(X)$ be the class of all subsets of X. Then $(B(X), X, \Rightarrow)$, where

$$Y \Rightarrow Z = (X - Y) \cup Z, \quad \text{for all } Y, Z \in B(X),$$

is an implication algebra. In this algebra the ordering relation $\leqslant$ coincides with the set-theoretical inclusion. Clearly, any non-empty class $B_0(X) \subset B(X)$ closed under the operation $\Rightarrow$ defined above satisfies the condition $X \in B_0(X)$ and is also an implication algebra. In other words, every subalgebra of $(B(X), X, \Rightarrow)$ is an implication algebra and will be called an *implication algebra of sets*. We shall show in Section 7 that the last examples are typical of implication algebras, i.e. that every implication algebra is isomorphic to an implication algebra of sets.

We shall prove that

5.2. *In any implication algebra* $(A, V, \Rightarrow)$ *the following equations are satisfied for all* $a, b, c \in A$:

$$(a \Rightarrow b) \Rightarrow b = (b \Rightarrow a) \Rightarrow a, \tag{6}$$

$$(a \Rightarrow c) \Rightarrow \big((b \Rightarrow c) \Rightarrow \big(((a \Rightarrow b) \Rightarrow b) \Rightarrow c\big)\big) = V, \tag{7}$$

$$\big(a \Rightarrow (a \Rightarrow b)\big) \Rightarrow (a \Rightarrow b) = V, \tag{8}$$

$$\big((a \Rightarrow b) \Rightarrow (b \Rightarrow a)\big) \Rightarrow (b \Rightarrow a) = V. \tag{9}$$

By 2.3 (9), $a \leqslant (a \Rightarrow b) \Rightarrow b$. Hence, by 2.3 (11), (14) and (P), we get

$$(b \Rightarrow a) \Rightarrow a \leqslant (b \Rightarrow a) \Rightarrow \big((a \Rightarrow b) \Rightarrow b\big) = (a \Rightarrow b) \Rightarrow \big((b \Rightarrow a) \Rightarrow b\big)$$
$$= (a \Rightarrow b) \Rightarrow b.$$

Thus $(b \Rightarrow a) \Rightarrow a \leqslant (a \Rightarrow b) \Rightarrow b$. Hence, by symmetry, $(a \Rightarrow b) \Rightarrow b \leqslant (b \Rightarrow a) \Rightarrow a$. Consequently, (6) holds.

By 2.3 (15)

$$(a \Rightarrow c) \leqslant (c \Rightarrow b) \Rightarrow (a \Rightarrow b),$$

and

$$(c \Rightarrow b) \Rightarrow (a \Rightarrow b) \leqslant \big((a \Rightarrow b) \Rightarrow b\big) \Rightarrow \big((c \Rightarrow b) \Rightarrow b\big).$$

Hence,

$$(a \Rightarrow c) \leqslant \big((a \Rightarrow b) \Rightarrow b\big) \Rightarrow \big((c \Rightarrow b) \Rightarrow b\big).$$

Applying (6) and 2.3 (14), we get

$$(a \Rightarrow c) \leqslant \big((a \Rightarrow b) \Rightarrow b\big) \Rightarrow \big((b \Rightarrow c) \Rightarrow c\big)$$
$$= (b \Rightarrow c) \Rightarrow \big(((a \Rightarrow b) \Rightarrow b) \Rightarrow c\big).$$

Hence, by 2.2 (4), we get (7).

Equation (8) follows directly from 2.3 (13) and from 5.1 (5). By 2.3 (7), $b \leqslant a \Rightarrow b$. Hence, applying 2.3 (12) and 2.3 (13), we get

$$(a \Rightarrow b) \Rightarrow (b \Rightarrow a) \leqslant b \Rightarrow (b \Rightarrow a) = b \Rightarrow a.$$

Hence (9) holds by 2.2.

We shall define a new two-argument operation in any implication algebra $(A, V, \Rightarrow)$ as follows:

$$(10) \qquad a \cup b = (a \Rightarrow b) \Rightarrow b, \quad \text{for all } a, b \in A.$$

5.3. *In any implication algebra* $(A, V, \Rightarrow)$ *and for all* $a, b \in A$

$$(11) \qquad a \cup b = \text{l.u.b.}\, \{a, b\},$$

where $\cup$ *is the operation defined by* (10) *and* l.u.b. $\{a, b\}$ *denotes the least upper bound of* $\{a, b\}$ (see I 3) *in the ordered set* $(A, \leqslant)$ (see 2.2).

By 2.3 (9), we get

$$(12) \qquad a \leqslant a \cup b.$$

By (6) and 2.3 (9), we get

$$(13) \qquad b \leqslant a \cup b.$$

We shall show that

$$(14) \qquad \text{if} \quad a \leqslant c \text{ and } b \leqslant c, \quad \text{then} \quad a \cup b \leqslant c.$$

If $a \leqslant c$ and $b \leqslant c$, then by (7), 2 (1), 2.1 we get

$$a \cup b = (a \Rightarrow b) \Rightarrow b \leqslant c.$$

From the definition (10) and (8) it follows that

$$(15) \qquad a \cup (a \Rightarrow b) = \big(a \Rightarrow (a \Rightarrow b)\big) \Rightarrow (a \Rightarrow b) = V.$$

From the definition (10) and (9) it follows that

$$(16) \qquad (a \Rightarrow b) \cup (b \Rightarrow a) = \big((a \Rightarrow b) \Rightarrow (b \Rightarrow a)\big) \Rightarrow (b \Rightarrow a) = V.$$

It follows directly from (11) that

$$(17) \qquad a \cup a = a,$$

$$(18) \qquad a \cup b = b \cup a,$$

$$(19) \qquad (a \cup b) \cup c = a \cup (b \cup c).$$

6. Implicative filters in implication algebras. Since every implication algebra is a positive implication algebra, Theorems 3.1, 3.3, 3.4, 3.5, 3.6, 3.7 in Section 3 on implicative filters in positive implication algebras hold also for implication algebras.

A proper implicative filter ∇ in an implication algebra $(A, V, \Rightarrow)$ will be said to be *prime* provided that for all $a, b \in A$

(1) $a \cup b = (a \Rightarrow b) \Rightarrow b \in \nabla$ *implies that either* $a \in \nabla$ *or* $b \in \nabla$.

The following theorem characterizes prime implicative filters in implication algebras.

6.1. *An implicative filter in an implication algebra is prime if and only if it is irreducible.*

Let us suppose that a proper implicative filter ∇ is not irreducible. Then there exist two proper implicative filters $\nabla_1 \neq \nabla$ and $\nabla_2 \neq \nabla$ such that $\nabla = \nabla_1 \cap \nabla_2$. It is easy to see that neither $\nabla_1 \subset \nabla_2$ nor $\nabla_2 \subset \nabla_1$. Consequently, there exist two elements, a, b, in the implication algebra under consideration such that $a \in \nabla_1$, $a \notin \nabla_2$ and $b \notin \nabla_1$, $b \in \nabla_2$. Let ∇_1^* be the implicative filter generated by ∇ and a, and ∇_2^* the implicative filter generated by ∇ and b. Obviously, $\nabla \subset \nabla_1^* \subset \nabla_1$ and $\nabla \subset \nabla_2^* \subset \nabla_2$. Hence $a \notin \nabla_2^*$ and $b \notin \nabla_1^*$. Consequently, by 3.4, $b \Rightarrow a \notin \nabla$ and $a \Rightarrow b \notin \nabla$. But, by 1 ($f_1$) and 5 (16), $(a \Rightarrow b) \cup (b \Rightarrow a) \in \nabla$. Thus ∇ is not prime.

Now, let us suppose that a proper implicative filter is not prime. Hence, there exist two elements a, b such that $a \neq b$ and $a \cup b = (a \Rightarrow b) \Rightarrow b \in \nabla$, but neither $a \in \nabla$ nor $b \in \nabla$. Let ∇_1 be the implicative filter generated by ∇ and a, and let ∇_2 be the implicative filter generated by ∇ and b. We shall show that $\nabla = \nabla_1 \cap \nabla_2$. The inclusion $\nabla \subset \nabla_1 \cap \nabla_2$ is obvious. Let us suppose that $x \in \nabla_1 \cap \nabla_2$. By 3.4, $a \Rightarrow x \in \nabla$ and $b \Rightarrow x \in \nabla$. Applying 5.2 (7), 1 (($f_1$), ($f_2$)) and 5 (10), we infer that $(a \cup b) \Rightarrow x \in \nabla$. Since $a \cup b \in \nabla$, we get $x \in \nabla$. This proves the inclusion $\nabla_1 \cap \nabla_2 \subset \nabla$. Therefore $\nabla = \nabla_1 \cap \nabla_2$, where $\nabla_1 \neq \nabla$ and $\nabla_2 \neq \nabla$. Of course, $a \notin \nabla_2$ and $b \notin \nabla_1$. Thus ∇_1 and ∇_2 are proper. Consequently, ∇ is not irreducible.

6.2. *An implicative filter in an implication algebra* $\mathfrak{A} = (A, V, \Rightarrow)$ *is prime if and only if it is maximal.*

Observe that any maximal implicative filter ∇ is irreducible and hence, by 2.1, is prime. Let us suppose that ∇ is prime. If $\nabla \subset \nabla_1$, $\nabla \neq \nabla_1$, where ∇_1 is an implicative filter, then there exists an element $a \in A$ such that $a \notin \nabla$ and $a \in \nabla_1$. Let ∇^* be the least implicative filter containing a and ∇. Clearly, $\nabla \subset \nabla^* \subset \nabla_1$. By 3.4, $\nabla^* = \{x \in A$:

$a \Rightarrow x \in \nabla\}$. By 5 (15) and 1 ($f_1$), for every $x \in A$, $a \cup (a \Rightarrow x) = V \in \nabla$. Since ∇ is prime and $a \notin \nabla$, by (1) $a \Rightarrow x \in \nabla$ for all $x \in A$. Consequently, $\nabla^* = A$. Thus $\nabla_1 = A$ and ∇ is maximal.

It follows from 3.2 and 5.1 that

6.3. *If* $\mathfrak{A} = (A, V, \Rightarrow)$ *is an implication algebra and* ∇ *is an implicative filter in* $\mathfrak{A}$, *then the relation* $\approx_\nabla$ *determined by* ∇ (*see* 1 (2)) *is a congruence in* $\mathfrak{A}$. *The quotient algebra* $\mathfrak{A}/\nabla$ *is an implication algebra and conditions* (i), (ii), (iii) *of* 3.2 *hold.*

We shall prove that

6.4. *The following conditions are equivalent for every implicative filter* ∇ *of an implication algebra* $\mathfrak{A} = (A, V, \Rightarrow)$:

(a) ∇ *is a maximal implicative filter,*

(b) ∇ *is a prime implicative filter,*

(c) ∇ *is an irreducible implicative filter,*

(d) $\mathfrak{A}/\nabla$ *is two-element implicative algebra.*

By 6.1 and 6.2 conditions (a), (b), (c) are equivalent. Let us suppose that ∇ is a prime implicative filter. It follows from 1 (6) that $a \in \nabla$ if and only if $a \approx_\nabla V$, i.e. $||a|| = ||V||$. We shall show that if $a \notin \nabla$ and $b \notin \nabla$, then $a \approx_\nabla b$. Since $a \notin \nabla$ and by 5 (15) $a \cup (a \Rightarrow b) = V \in \nabla$, we infer, applying (1), that $a \Rightarrow b \in \nabla$. Similarly, since $b \notin \nabla$ and $b \cup (b \Rightarrow a) = V \in \nabla$, we have by (1) $b \Rightarrow a \in \nabla$. Thus $a \approx_\nabla b$ and $||a|| = ||b||$. For any $a \notin \nabla$ let us denote the class $||a||$ by Λ', and for any $b \in \nabla$ the class $||b||$ by V'. Obviously $\Lambda' \neq V'$ and $A/\nabla = \{\Lambda', V'\}$. The easy verification of the fact that the equations 2 (5) hold is left to the reader. Conversely, let us suppose that $\mathfrak{A}/\nabla = (\{\Lambda', V'\}, V', \Rightarrow)$ is the two-element implication algebra. Thus ∇ is a proper implicative filter. Since the equivalence class $||V|| = V'$ contains exactly those elements which belong to ∇, we infer that, for every $a \notin \nabla$, $||a|| = \Lambda'$. Consequently, if $a \notin \nabla$ and $b \notin \nabla$, then $a \approx_\nabla b$. Thus $a \Rightarrow b \in \nabla$ and $b \Rightarrow a \in \nabla$. Consequently, $a \cup b = (a \Rightarrow b) \Rightarrow b \notin \nabla$, which proves by (1) that ∇ is a prime implicative filter.

7. Representation theorem for implication algebras [10]. In this section we shall prove the following representation theorem.

(10) This theorem is obtained by a slight modification of the proof of the representation Theorem 4.1.

7.1. *For every implication algebra* $\mathfrak{A} = (A, V, \Rightarrow)$ *there exists a monomorphism* h *of* $\mathfrak{A}$ *into an implication algebra* $(B(X), X, \Rightarrow)$ *of all subsets of a space* X. *Hence, every implication algebra is isomorphic to an implication algebra of sets.*

Let X be the class of all irreducible implicative filters of $\mathfrak{A}$. Let h be the mapping from A into the class $B(X)$ of all subsets of X defined as follows:

(1) $$h(a) = \{\nabla \in X\colon a \in \nabla\}, \quad \text{for all } a \in A.$$

Let us note that

(2) $$a \leqslant b \quad \text{if and only if} \quad h(a) \subset h(b), \quad \text{for all } a, b \in A.$$

The proof of (2) is the same as the proof of 4 (2).

If $a \neq b$, then one of the conditions $a \leqslant b$ and $b \leqslant a$ does not hold. By (2) one of the inclusions $h(a) \subset h(b)$ and $h(b) \subset h(a)$ does not hold, i.e. $h(a) \neq h(b)$. Consequently, h is one-one.

Clearly (see 4 (3)),

(3) $$h(V) = X.$$

Moreover,

(4) $$h(a \Rightarrow b) \subset (X - h(a)) \cup h(b).$$

The proof of (4) is the same as the proof of inclusion (5) in Section 4.

It remains to prove that

(5) $$(X - h(a)) \cup h(b) \subset h(a \Rightarrow b).$$

Suppose that $\nabla \in (X - h(a)) \cup h(b)$. If $\nabla \in h(b)$, then $b \in \nabla$. Since $b \Rightarrow (a \Rightarrow b) = V$ (see 2 (p_1)), by 1 ((f_1), (f_2)), we infer that $a \Rightarrow b \in \nabla$, i.e. $\nabla \in h(a \Rightarrow b)$. If $\nabla \in X - h(a)$, i.e. $a \notin \nabla$, then also $a \Rightarrow b \in \nabla$, i.e. $\nabla \in h(a \Rightarrow b)$. In fact, by 5 (15) and 1 (f_1), $a \cup (a \Rightarrow b) \in \nabla$. By 6.1, ∇ is prime. Hence, either $a \in \nabla$ or $a \Rightarrow b \in \nabla$. Since $a \notin \nabla$, we infer that $a \Rightarrow b \in \nabla$. Thus inclusion (5) is proved. It follows from (4) and (5) that

(6) $$h(a \Rightarrow b) = (X - h(a)) \cup h(b) = h(a) \Rightarrow h(b).$$

By (3) and (6), h is a homomorphism from $\mathfrak{A}$ into $(B(X), X, \Rightarrow)$. Since h is one-one, it is a monomorphism. Thus Theorem 7.1 is proved.

Exercises

1. An implicative filter ∇ in an implicative algebra $\mathfrak{A} = (A, V, \Rightarrow)$ will be said to be *special* provided the following conditions are satisfied: for any a, b, c, d in A

(f$_3$) if $a \in \nabla$, then $b \Rightarrow a \in \nabla$,

(f$_4$) if $a \Rightarrow b,\ b \Rightarrow c \in \nabla$, then $a \Rightarrow c \in \nabla$,

(f$_5$) if $b \Rightarrow a,\ c \Rightarrow d \in \nabla$, then $(a \Rightarrow c) \Rightarrow (b \Rightarrow d) \in \nabla$.

Prove that if h is an epimorphism of an implicative algebra $\mathfrak{A}$ onto an implicative algebra $\mathfrak{B}$, then the kernel $K(h)$ of h is a special implicative filter.

2. Prove that for each implicative algebra $\mathfrak{A} = (A, V, \Rightarrow)$ and each special implicative filter ∇ in $\mathfrak{A}$ the relation $\approx_\nabla$ on A defined by the equivalence

$$a \approx_\nabla b \quad \text{if and only if} \quad a \Rightarrow b \in \nabla \text{ and } b \Rightarrow a \in \nabla$$

is a congruence in $\mathfrak{A}$ such that for each $a \in A$ the following condition holds: $a \in \nabla$ if and only if $a \approx_\nabla V$. The quotient algebra $\mathfrak{A}/\nabla$ is an implicative algebra; the mapping h defined by the equation $h(a) = ||a||$, $a \in A$, is an epimorphism from $\mathfrak{A}$ onto $\mathfrak{A}/\nabla$ and ∇ is the kernel of h; the algebra $\mathfrak{A}/\nabla$ is degenerate if and only if $\nabla = A$.

3. Prove Theorems 1.3 and 1.4 for special implicative filters.

4. Prove that each implicative filter in every positive implication algebra is a special implicative filter.

5. An abstract algebra $\mathfrak{A} = (A, V, \Rightarrow, \sim_c)$ will be said to be a *contrapositionally complemented positive implication algebra* provided $(A, V, \Rightarrow)$ is a positive implication algebra and for each $a \in A$ the following condition is satisfied:

(c) $$a \Rightarrow \sim_c b = b \Rightarrow \sim_c a.$$

Prove that an abstract algebra $(A, V, \Rightarrow, \sim_c)$ is a contrapositionally complemented positive implication algebra if and only if $(A, V, \Rightarrow)$ is a positive implication algebra and, moreover,

(c*) $$\sim_c a = a \Rightarrow \sim_c V \quad \text{for each } a \in A.$$

6. Prove that in every contrapositionally complemented positive implication algebra the following conditions are satisfied: $a \leqslant \sim_c \sim_c a$, $a \Rightarrow b \leqslant \sim_c b \Rightarrow \sim_c v$. Moreover, if there exists a least element Λ, then $\sim_c V = \Lambda$.

7. Prove the analogue of 3.2 for contrapositionally complemented positive implication algebras.

8. Let X be a topological space and let $\mathfrak{G}_0(X)$ be a positive implication algebra of open subsets of X. Setting $\sim_c X = Y_0$ for some Y_0 in $\mathfrak{G}_0(X)$ and $\sim_c Z = Z \Rightarrow \sim_c X$ for each Z in $\mathfrak{G}_0(X)$, we define a contrapositional complementation in $\mathfrak{G}_0(X)$ and get a contrapositionally complemented positive implication algebra of open subsets of X. Prove the representation theorem analogous to 4.1 for contrapositionally complemented positive implication algebras.

9. By a *pseudo-complemented positive implication algebra* we shall understand any algebra $\mathfrak{A} = (A, V, \Rightarrow, \neg)$ such that $(A, V, \Rightarrow)$ is a positive implication algebra and $\neg$ is an one-argument operation on A satisfying the following conditions: $a \Rightarrow \neg b = b \Rightarrow \neg a$, $\neg(a \Rightarrow a) \Rightarrow b = V$. Prove the analogue of 3.2 for pseudo-complemented positive implication algebras.

10. Let $\mathfrak{G}(X)$ be the positive implication algebra of all open subsets of a topological space X. Setting $\neg Y = \mathbf{I}(X - Y)$ for each open subset Y of X, we get a pseudo-complemented positive implication algebra. Prove the analogue of Theorem 4.1 for pseudo-complemented positive implication algebras.

11. Examine the properties of *semi-complemented positive implication algebras* $(A, V, \Rightarrow, \dot{-})$, which are defined as follows: $(A, V, \Rightarrow)$ is a positive implication algebra and $\dot{-}$ is a one-argument operation on A satisfying the condition $\dot{-}(a \Rightarrow a) \Rightarrow b = V$.

12. Construct in a similar way the *contrapositionally complemented implication algebras* and *semi-complemented implication algebras* and examine properties of those algebras.

CHAPTER III

DISTRIBUTIVE LATTICES AND QUASI-BOOLEAN ALGEBRAS

Introduction. The first part of this chapter (Sections 1 and 2) lists fundamental notions and elementary theorems about lattices and distributive lattices including the representation theorem.

Distributive lattices with an additional one-argument operation $\sim$, such that the equations $\sim\sim a = a$ and $\sim(a \cup b) = \sim a \cap \sim b$ are satisfied by all elements, have been examined by Moisil [1] and termed De Morgan lattices. Quasi-Boolean algebras, to be examined in Section 3, are De Morgan lattices with a unit element and a zero element. Clearly, they are a slight generalization of Boolean algebras. Each set lattice of all subsets of an arbitrary space $X \neq O$ is a quasi-Boolean algebra if we define $\sim Y$ to be $X - g(Y)$ for every $Y \subset X$, $g\colon X \to X$ being a fixed involution of X. Similarly, each subalgebra of such an algebra is also a quasi-Boolean algebra, called a quasi-field of sets. The Representation Theorem 3.2, due to Białynicki-Birula and Rasiowa [1], states that each quasi-Boolean algebra is isomorphic to a quasi-field of sets.

The four-element quasi-Boolean algebra $\mathfrak{A}_0$ (see p. 47) plays a role analogous to that of the two-element Boolean algebra for the class of all Boolean algebras. Indeed, each quasi-Boolean algebra is isomorphic to a subalgebra of a product $\prod_{t \in T} \mathfrak{A}_t$, where $\mathfrak{A}_t = \mathfrak{A}_0$ for all $t \in T$ (see 3.3). This theorem, due to Białynicki-Birula [1], implies that $\mathfrak{A}_0$ is functionally free in the class of all quasi-Boolean algebras, i.e. that any two quasi-Boolean polynomials are identically equal in each quasi-Boolean algebras if and only if they are identically equal in $\mathfrak{A}_0$. Certain further analogies between quasi-Boolean algebras and Boolean algebras are indicated in the Exercises.

The theory of quasi-Boolean algebras presented here will be applied in Chapter V, which deals with quasi-pseudo-Boolean algebras. They characterize algebraically constructive logic with strong negation, which is examined in Chapter XII.

1. Lattices [1]. An abstract algebra $(A, \cup, \cap)$ with two binary operations is said to be a *lattice* provided the following equations are satisfied for all $a, b, c \in A$:

$$(l_1) \qquad a \cup b = b \cup a, \quad a \cap b = b \cap a,$$

$$(l_2) \qquad a \cup (b \cup c) = (a \cup b) \cup c, \quad a \cap (b \cap c) = (a \cap b) \cap c,$$

$$(l_3) \qquad (a \cap b) \cup b = b, \quad a \cap (a \cup b) = a.$$

For all $a, b \in A$, $a \cup b$ is called the *join* of a and b, and $a \cap b$ the *meet* of a and b.

We quote without proof the following theorem:

1.1. *If $(A, \cup, \cap)$ is a lattice, then for all $a, b \in A$*

$$(1) \qquad a \cup b = b \quad \textit{if and only if} \quad a \cap b = a.$$

The relation $\leqslant$ on A, defined as follows

$$(2) \qquad a \leqslant b \quad \textit{if and only if one of the equations (1) holds,}$$

is an ordering on A, called the lattice ordering on A. Moreover,

$$(3) \qquad a \cup b = \text{l.u.b.}\,\{a, b\}, \quad a \cap b = \text{g.l.b.}\,\{a, b\},$$

where l.u.b. $\{a, b\}$ *and* g.l.b. $\{a, b\}$ *are, respectively, the least upper bound of $\{a, b\}$ and the greatest lower bound of $\{a, b\}$ in the ordered set $(A, \leqslant)$.*

The following properties of join and meet follow directly from Theorem 1.1:

$$(4) \qquad a \cup a = a, \quad a \cap a = a,$$

$$(5) \qquad a \leqslant a \cup b, \quad a \cap b \leqslant a,$$

$$(6) \qquad b \leqslant a \cup b, \quad a \cap b \leqslant b,$$

(7) $a \leqslant c$ *and* $b \leqslant c$ *imply* $a \cup b \leqslant c$, $c \leqslant a$ *and* $c \leqslant b$ *imply* $c \leqslant a \cap b$,

(8) $a \leqslant c$ *and* $b \leqslant d$ *imply* $a \cup b \leqslant c \cup d$, $a \leqslant c$ *and* $b \leqslant d$ *imply* $a \cap b \leqslant c \cap d$.

We also state without proof the following theorem:

[1] For a detailed exposition of lattice theory see Birkhoff [2].

1.2. *If in an ordered set* $(A, \leqslant)$ *for all* $a, b \in A$, *there exist a* l.u.b. $\{a, b\}$ *and a* g.l.b. $\{a, b\}$, *then the abstract algebra* $(A, \cup, \cap)$, *where* $\cup$ *and* $\cap$ *are binary operations defined by* (3), *is a lattice, i.e. the equations* (l_1), (l_2), (l_3) *hold for all* $a, b, c \in A$. *The relation* $\leqslant$ *is the lattice ordering in this lattice.*

A non-empty class $L(X)$ of subsets of a space X such that the union and the intersection of two sets belonging to $L(X)$ also belong to $L(X)$ is an example of a lattice, called a *lattice of sets* or *set lattice.* The ordering determined in a set lattice by Theorem 1.1 coincides with the set-theoretical inclusion. Every chain is also a lattice.

Observe that every homomorphism h of a lattice $(A, \cup, \cap)$ into a lattice $(B, \cup, \cap)$ preserves the lattice ordering, i.e.

(9) $$a \leqslant b \quad \textit{implies} \quad h(a) \leqslant h(b), \quad \textit{for} \quad a, b \in A.$$

This follows from the fact that the lattice ordering is defined by one of the equations (1), which are preserved under every homomorphism. Note also that:

1.3. *A one-one mapping* h *of a lattice* $(A, \cup, \cap)$ *onto a lattice* $(B, \cup, \cap)$ *is an isomorphism if and only if*

(10) $$a \leqslant b \quad \textit{if and only if} \quad h(a) \leqslant h(b), \quad \textit{for all } a, b \in A.$$

This follows from the fact that the join and the meet are uniquely determined by the lattice ordering (see 1.2).

Let $\mathfrak{A} = (A, \cup, \cap)$ be a lattice. If in the ordered set $(A, \leqslant)$, where $\leqslant$ is the lattice ordering on A, there exists a greatest (a least) element, it is called the *unit element* (the *zero element*) and is denoted by $\bigvee$ ($\bigwedge$). By definition

(11) $$x \leqslant \bigvee, \quad \bigwedge \leqslant x,$$

(12) $$x \cup \bigvee = \bigvee, \quad x \cap \bigvee = x,$$

(13) $$x \cup \bigwedge = x, \quad x \cap \bigwedge = \bigwedge,$$

for every $x \in A$. In any lattice the notion of a maximal element and of the greatest element coincide. The same applies for the notion of a minimal element and that of the least element in any lattice.

If we add to a lattice $\mathfrak{A} = (A, \cup, \cap)$ a new element $\bigvee$ ($\bigwedge$) and extend the operations $\cup$, $\cap$ on $\bigvee$ ($\bigwedge$) by means of (12) (of (13)), where $x \in A \cup$

$\cup \{V\} = B$ (where $x \in A \cup \{\Lambda\} = B$), then we obtain the lattice $\mathfrak{B} = (B, \cup, \cap)$ and V (Λ) is the unit element (the zero element) of $\mathfrak{B}$. This remains true if we simultaneously add V and Λ.

A non-empty subset ∇ of the set A of all elements of a lattice $(A, \cup, \cap)$ is said to be a *filter* if for all elements $a, b \in A$

(f) $\qquad a \cap b \in \nabla \quad$ if and only if $\quad a \in \nabla$ and $b \in \nabla$.

The condition (f) is equivalent to the following two conditions:

(f_1) $\qquad$ if $a \in \nabla$ and $b \in \nabla$, $\quad$ then $\quad a \cap b \in \nabla$,

(f_2) $\qquad$ if $a \in \nabla$ and $a \leqslant b$, $\quad$ then $\quad b \in \nabla$.

The condition (f_2) can be replaced by the following condition:

(f_2') $\qquad$ if $a \in \nabla$, $\quad$ then for all $b \in A$, $\quad a \cup b \in \nabla$.

A non-empty subset Δ of the set A of all elements of a lattice $(A, \cup, \cap)$ is said to be an *ideal* if for all elements $a, b \in A$

(i) $\qquad a \cup b \in \Delta \quad$ if and only if $\quad a \in \Delta$ and $b \in \Delta$.

The condition (i) is equivalent to the following two conditions:

(i_1) $\qquad$ if $a \in \Delta$ and $b \in \Delta$, $\quad$ then $\quad a \cup b \in \Delta$,

(i_2) $\qquad$ if $a \in \Delta$ and $b \leqslant a$, $\quad$ then $\quad b \in \Delta$.

The condition (i_2) can be replaced by the following condition:

(i_2') $\qquad$ if $a \in \Delta$, $\quad$ then $\quad$ for all $b \in A$, $\quad a \cap b \in \Delta$.

The whole lattice A is a filter (an ideal). If a lattice has a unit element V (a zero element Λ), then $\{V\}$ ($\{\Lambda\}$) is a filter (an ideal), called the *unit filter* (the *zero ideal*). For every fixed element $a \in A$ the set $\nabla(a) = \{x \in A\colon a \leqslant x\}$ (the set $\Delta(a) = \{x \in A\colon x \leqslant a\}$) is the least filter (ideal) containing a and is called the *principal filter* (*principal ideal*) *generated by a*.

We mention without proof the following simple results.

1.4. *If ∇ is a filter (Δ is an ideal) in a lattice $\mathfrak{B}$ and h is a homomorphism of a lattice $\mathfrak{A}$ into $\mathfrak{B}$, then $h^{-1}(\nabla)$ (the set $h^{-1}(\Delta)$) is a filter in $\mathfrak{A}$ (is an ideal in $\mathfrak{A}$) provided it is not empty. In particular, if $\mathfrak{B}$ contains the unit element V (the zero element Λ), then $h^{-1}(V)$ (the set $h^{-1}(\Lambda)$) is a filter (an ideal) in $\mathfrak{A}$ provided it is not empty.*

1.5. *If a lattice* $\mathfrak{A}$ *contains the unit element* $\bigvee$ *(the zero element* $\bigwedge$*), then* $\bigvee \in \nabla$ $(\bigwedge \in \Delta)$ *for every filter* ∇ *(ideal* Δ*) in* $\mathfrak{A}$.

1.6. *The filter (ideal) generated by a non-empty set* A_0 *of elements of a lattice* $\mathfrak{A}$ (i.e. *the least filter (ideal) in* $\mathfrak{A}$ *containing* A_0*) is the set of all elements* a *in* $\mathfrak{A}$ *such that there exist* $a_1, \ldots, a_n$ *in* A_0 *for which* $a_1 \cap \ldots \cap a_n \leqslant a$ *(for which* $a \leqslant a_1 \cup \ldots \cup a_n$*).*

1.7. *The filter (ideal) generated by a fixed element* a_0 *and a filter* ∇ *(an ideal* Δ*) in a lattice* $\mathfrak{A}$ *is the set of all elements* a *in* $\mathfrak{A}$ *such that* $a_0 \cap c \leqslant a$ *(such that* $a \leqslant a_0 \cup c$*) for some element* $c \in \nabla$ *(for some element* $c \in \Delta$*).*

A filter ∇ (an ideal Δ) in a lattice $\mathfrak{A}$ is said to be *proper* provided there exists an element a in $\mathfrak{A}$ such that $a \notin \nabla$ (such that $a \notin \Delta$). In particular, if $\mathfrak{A}$ has the zero element $\bigwedge$ (the unit element $\bigvee$), then a filter ∇ (an ideal Δ) is proper if and only if $\bigwedge \notin \nabla$ $(\bigvee \notin \Delta)$.

A filter ∇ (an ideal Δ) in a lattice $\mathfrak{A}$ is said to be *maximal* provided it is proper and it is not a proper subset of any proper filter (ideal) in $\mathfrak{A}$.

A filter ∇ (an ideal Δ) in a lattice $\mathfrak{A}$ is said to be *irreducible* provided it is proper and for any two proper filters ∇_1, ∇_2 (two proper ideals Δ_1, Δ_2) the condition $\nabla = \nabla_1 \cap \nabla_2$ (the condition $\Delta = \Delta_1 \cap \Delta_2$) implies either $\nabla = \nabla_1$ or $\nabla = \nabla_2$ (implies either $\Delta = \Delta_1$ or $\Delta = \Delta_2$). In other words, a filter (an ideal) is irreducible provided it is proper and it is not the intersection of two proper filters (of two proper ideals) different from it.

A filter ∇ (an ideal Δ) is said to be *prime* provided it is proper and $a \cup b \in \nabla$ $(a \cap b \in \Delta)$ implies that either $a \in \nabla$ or $b \in \nabla$ (implies that either $a \in \Delta$ or $b \in \Delta$).

We note without proof the following simple results.

1.8. *Every maximal filter (maximal ideal) in a lattice is irreducible. Every prime filter (prime ideal) in a lattice is irreducible.*

1.9. *The union of any chain of filters (ideals) in a lattice* $\mathfrak{A}$ *is a filter (ideal) in* $\mathfrak{A}$. *The union of any chain of proper filters (ideals) in a lattice* $\mathfrak{A}$ *which contains the zero element (the unit element) is a proper filter (ideal).*

1.10. *If a lattice* $\mathfrak{A}$ *has a zero element (a unit element), then every proper filter (ideal) in* $\mathfrak{A}$ *is contained in a maximal filter (ideal).*

The proof follows from 1.9 by the application of I 3.1.

1.11. *If a lattice* $\mathfrak{A}$ *contains the zero element* $\wedge$ *(the unit element* $\vee$*), then for every element* $a \neq \wedge$ $(a \neq \vee)$ *there exists a maximal filter* ∇ *(maximal ideal* Δ*) such that* $a \in \nabla$ $(a \in \Delta)$.

This follows from 1.10.

1.12. *If* ∇_0 (Δ_0) *is a filter (an ideal) in a lattice* $\mathfrak{A}$ *and* $a \notin \nabla_0$ $(a \notin \Delta_0)$, *then there exists an irreducible filter* ∇ *(an irreducible ideal* Δ*) in* $\mathfrak{A}$, *such that* $\nabla_0 \subset \nabla$ $(\Delta_0 \subset \Delta)$ *and* $a \notin \nabla$ *(and* $a \notin \Delta$*).*

The proof is similar to that of II 1.4 and depends on 1.9 and I 3.1.

1.13. *If* Δ *and* ∇ *are two disjoint subsets of the set* A *of all elements of a lattice* $(A, \cup, \cap)$ *and the union of* Δ *and* ∇ *is equal to* A, *then the set* Δ *is a prime ideal if and only if the set* ∇ *is a prime filter.*

The proof is by an easy verification.

2. Distributive lattices. A lattice $(A, \cup, \cap)$ is said to be *distributive* if for all $a, b, c \in A$ the following equations hold

(d) $a \cap (b \cup c) = (a \cap b) \cup (a \cap c), \quad a \cup (b \cap c) = (a \cup b) \cap (a \cup c).$

If one of the distributive laws (d) is satisfied for all $a, b, c \in A$, then the remaining law is also satisfied (see e.g. [MM], p. 48).

Note that

2.1. *In a distributive lattice every irreducible filter (irreducible ideal) is prime.*

Indeed, if ∇ is a proper filter which is not prime, then there are a_1, a_2 such that $a_1 \cup a_2 \in \nabla$, $a_1 \notin \nabla$, $a_2 \notin \nabla$. Let ∇_i be the filter generated by ∇ and a_i, $i = 1, 2$. It is easy to show making use of 1.7, 1 (5), 1 (7) and (d) that $\nabla = \nabla_1 \cap \nabla_2$, i.e. ∇ is not irreducible. The proof for ideals is dual.

2.2. *For any elements* a, b *of a distributive lattice, if the relation* $a \leqslant b$ *does not hold, then there exists a prime filter* ∇ *(a prime ideal* Δ*) such that* $a \in \Delta$ *and* $b \notin \Delta$ *(such that* $a \notin \nabla$ *and* $b \in \nabla$*).*

Let ∇_0 be the principal filter generated by a. Then $a \in \nabla$ and $b \notin \nabla_0$. Thus 2.2 follows from 1.12 and 2.1. The proof for ideals is dual.

Given a distributive lattice $\mathfrak{A} = (A, \cup, \cap)$, let $X(\mathfrak{A})$ be the set of all prime filters in $\mathfrak{A}$. For each $a \in A$, let

(1) $$h(a) = \{\nabla \in X(\mathfrak{A}) \colon a \in \nabla\}.$$

Consider $X(\mathfrak{A})$ as a topological space, the class $P(\mathfrak{A}) = \{h(a)\}_{a \in A}$ being taken as a subbase (see I 2.1). This topological space is called the *Stone space of* $\mathfrak{A}$.

We quote without proof the following well-known representation theorem for distributive lattices.

2.3. *Each distributive lattice* $\mathfrak{A} = (A, \cup, \cap)$ *is isomorphic to a set lattice*[2]. *More exactly,* $P(\mathfrak{A})$ *is a set lattice and the mapping* h, *defined by* (1), *is an isomorphism of* $\mathfrak{A}$ *onto the set lattice* $\mathfrak{P}(\mathfrak{A}) = (P(\mathfrak{A}), \cup, \cap)$. *The Stone space* $X(\mathfrak{A})$ *of* $\mathfrak{A}$ *is a* T_0*-space. If* $\mathfrak{A}$ *has the unit element* V, *then* $X(\mathfrak{A})$ *is compact*[3].

For a proof of 2.3 see e.g. [MM], pp. 50–51.

3. Quasi-Boolean algebras. An abstract algebra $(A, V, \cup, \cap, \sim)$ is said to be a *quasi-Boolean algebra*[4] if $(A, \cup, \cap)$ is a distributive lattice with a unit element V and $\sim$ is a unary operation on A satisfying the following conditions:

(q_1) $$\sim\sim a = a, \quad \text{for all } a \in A,$$

(q_2) $$\sim(a \cup b) = \sim a \cap \sim b, \quad \text{for all } a, b \in A.$$

Observe that

3.1. *In any quasi-Boolean algebra* $(A, V, \cup, \cap, \sim)$

(1) *there is a zero element* Λ,

(2) $$\sim\Lambda = V \quad \textit{and} \quad \sim V = \Lambda,$$

(3) $$\sim(a \cap b) = \sim a \cup \sim b.$$

(2) The theorem which states that every distributive lattice is isomorphic to a set lattice is due to Birkhoff [1].

(3) Stone [4].

(4) This notion was introduced in Białynicki-Birula and Rasiowa [1]. The algebras $(A, \cup, \cap, \sim)$, where $(A, \cap, \cup)$ is a distributive lattice and $\sim$ is a unary operation satisfying the conditions (q_1), (q_2), have been investigated by Moisil [1] and termed *De Morgan lattices*. This type of algebras has also been examined by Kalman [1] under the term *distributive i-lattices*. A generalization of De Morgan lattices has been discussed by Henkin [3].

The proof is by an easy verification.

Let X be a non-empty set and let g be an involution of X, i.e. a mapping from X into X such that

$$g(g(x)) = x, \quad \text{for all } x \in X. \tag{4}$$

Clearly, every involution g of X is a one-one mapping from X onto X.

Let us put for each $Y \subset X$

$$\sim Y = X - g(Y). \tag{5}$$

Let $Q(X)$ be a non-empty class of subsets of X, containing X and closed under set-theoretical union and intersection as well as under the operation $\sim$ defined above, which is called *quasi-complementation*. It is easy to verify that $(Q(X), X, \cup, \cap, \sim)$ is an example of a quasi-Boolean algebra, called a *quasi-field of subsets of X*. We shall see that quasi-fields of sets are typical examples of quasi-Boolean algebras, in the sense that every quasi-Boolean algebra is isomorphic to a quasi-field of sets.

Given a quasi-Boolean algebra $(A, V, \cup, \cap, \sim)$, let us put for any $S \subset A$

$$\tilde{S} = \{\sim a \in A \colon a \in S\}. \tag{6}$$

Observe that, for any $S \subset A$,

$$\widetilde{A-S} = A - \tilde{S}, \quad \tilde{\tilde{S}} = S. \tag{7}$$

Let us note the following simple fact:

(8) *If ∇ is a prime filter in the lattice $(A, \cup, \cap)$, then $\tilde{\nabla}$ is a prime ideal in this lattice.*

The proof of (8) is by an easy verification.

Now we shall prove the following representation theorem for quasi-Boolean algebras:

3.2. *Every quasi-Boolean algebra is isomorphic to a quasi-field of certain open subsets of a topological, compact T_0-space* (5).

Let $(A, V, \cup, \cap, \sim)$ be a quasi-Boolean algebra and let X be the set of all prime filters in the lattice $(A, \cup, \cap)$.

(5) Białynicki-Birula and Rasiowa [1], [2].

We put for every $a \in A$:

$$h(a) = \{\nabla \in X\colon a \in \nabla\}.$$

It follows from 2.3 that X can be considered a topological space, the class $\{h(a)\}_{a\in A}$ being taken as a subbase. Then X is a compact T_0-space. The mapping h is one-one and satisfies the conditions:

$$\begin{aligned} h(a \cup b) &= h(a) \cup h(b), \\ h(a \cap b) &= h(a) \cap h(b), \\ h(V) &= X. \end{aligned} \tag{9}$$

Let $g\colon X \to X$ be the mapping defined by

$$g(\nabla) = A - \tilde{\nabla} \quad \text{for each } \nabla \in X. \tag{10}$$

By (8), for every $\nabla \in X$, the set $\tilde{\nabla}$ defined by (6) is a prime ideal. Thus, by 1.13, for every $\nabla \in X$, the set $A - \tilde{\nabla} = g(\nabla)$ is a prime filter, i.e. g is a mapping from X into X.

We shall show that

$$g(g(\nabla)) = \nabla, \quad \text{for all } \nabla \in X. \tag{11}$$

Indeed,

$$g(g(\nabla)) = A - \widetilde{g(\nabla)} = A - (\widetilde{A - \tilde{\nabla}}) = A - (A - \tilde{\tilde{\nabla}}) = A - (A - \nabla) = \nabla.$$

The mapping g defined by (9) determines the operation of quasi-complementation in the way given by (5). We shall prove that for every $a \in A$

$$\sim h(a) = h(\sim a). \tag{12}$$

By (9) the condition $\nabla \in g(h(a))$ is equivalent to the existence of a prime filter ∇_0 belonging to $h(a)$ and such that $\nabla = g(\nabla_0)$, i.e. $\nabla = A - \tilde{\nabla}_0$. Observe that $\nabla_0 \in h(a)$ if and only if $a \in \nabla_0$. But $a \in \nabla_0$ if and only if $\sim a \in \tilde{\nabla}_0$. The last condition is equivalent to $\sim a \notin A - \tilde{\nabla}_0$. Hence $\nabla \in g(h(a))$ if and only if there exists a prime filter ∇_0 such that $\sim a \notin A - \tilde{\nabla}_0 = \nabla$. Obviously, for the prime filter $\nabla_0 = A - \tilde{\nabla}$ we have $A - \tilde{\nabla}_0 = \nabla$. Thus we infer that

$$\nabla \in g(h(a)) \quad \textit{if and only if} \quad \sim a \notin \nabla. \tag{13}$$

Hence $\nabla \in \sim h(a) = X - g(h(a))$ if and only if $\sim a \in \nabla$, i.e. if and only if $\nabla \in h(\sim a)$. This equivalence proves (12). It follows from the equations (9) and (12) that the class $Q(X) = \{h(a)\}_{a\in A}$ is a quasi-field of open subsets of X and that the mapping h is a homomorphism from $\mathfrak{A}$ onto $Q(X)$. Since h is one-one, it is an isomorphism.

We shall denote by $\mathfrak{A}_0$ the following four-element quasi-Boolean algebra: $\mathfrak{A}_0 = (A_0, V, \cup, \cap, \sim)$, where $A_0 = \{V, a, b, \Lambda\}$ and the operations $\cup, \cap, \sim$ are defined by means of the equations:

(14) $\Lambda \cup x = x \cup \Lambda = x, \quad V \cup x = x \cup V = V$ for every $x \in A_0$,

(15) $a \cup a = a, \quad b \cup b = b, \quad a \cup b = V$,

(16) $\Lambda \cap x = x \cap \Lambda = \Lambda, \quad V \cap x = x \cap V = x$ for every $x \in A_0$,

(17) $a \cap a = a, \quad b \cap b = b, \quad a \cap b = \Lambda$,

(18) $\sim V = \Lambda, \quad \sim a = a, \quad \sim b = b, \quad \sim \Lambda = V$.

We denote by $\mathfrak{B}_0$ and $\mathfrak{C}_0$ the following subalgebras of $\mathfrak{A}_0$:

$$\mathfrak{B}_0 = (B_0, V, \cup, \cap, \sim), \quad \text{where} \quad B_0 = \{V, \Lambda\}$$

and

$$\mathfrak{C}_0 = (C_0, V, \cup, \cap, \sim), \quad \text{where} \quad C_0 = \{V, a, \Lambda\}.$$

Observe that $\mathfrak{B}_0$ and $\mathfrak{C}_0$ are the only non-degenerate subalgebras of $\mathfrak{A}_0$ if we identify the two isomorphic subalgebras: $\mathfrak{C}_0$ and $(\{V, b, \Lambda\}, V, \cup, \cap, \sim)$.

The algebras $\mathfrak{A}_0, \mathfrak{B}_0, \mathfrak{C}_0$ are important examples of finite quasi-Boolean algebras.

We shall show that

3.3. *Every quasi-Boolean algebra is isomorphic to a subalgebra of a product* $\prod_{t \in T} \mathfrak{A}_t$, *where* T *is a set of indices and* $\mathfrak{A}_t = \mathfrak{A}_0$, *for every* $t \in T$ [6].

On account of 3.2 it is sufficient to prove that every quasi-field of sets is isomorphic to a subalgebra of a product $\prod_{t \in T} \mathfrak{A}_t$, where T is a set of indices and $\mathfrak{A}_t = \mathfrak{A}_0$ for each $t \in T$. Let $\mathfrak{Q}(X) = (Q(X), X, \cup, \cap, \sim)$ be a quasi-field of subsets of a space X and let $g: X \to X$ be an involution of X determining the operation $\sim$ of quasi-complementation, i.e.

(19) $\sim Y = X - g(Y)$, for every $Y \in Q(X)$.

[6] Białynicki-Birula [1].

Let us put

(20) $$T = \{(x, y) \in X \times X \colon y = g(x)\} = \{(x, g(x))\}_{x \in X}.$$

For every $t \in T$, i.e. for every $(x, g(x))$ where $x \in X$, we define a mapping $h_t \colon Q(X) \to A_0$ as follows

(21) $$h_t(Y) = \begin{cases} \wedge & \text{if } t = (x, g(x)) \text{ and } \{x, g(x)\} \cap Y = O, \\ a & \text{if } t = (x, g(x)) \text{ and } \{x, g(x)\} \cap Y = \{x\} \neq \{g(x)\}, \\ b & \text{if } t = (x, g(x)) \text{ and } \{x, g(x)\} \cap Y = \{g(x)\} \neq \{x\}, \\ \vee & \text{if } t = (x, g(x)) \text{ and } \{x, g(x)\} \cap Y = \{x, g(x)\}. \end{cases}$$

Each mapping h_t, $t \in T$, is a homomorphism from $Q(X)$ into $\mathfrak{A}_0$. The proof, by an easy verification, is left to the reader. Hence, by I 4.10, the mapping $h \colon Q(X) \to \prod_{t \in T} A_t$, where $A_t = A_0$ for every $t \in T$, defined by the equation

$$h(Y) = (h_t(Y))_{t \in T} \quad \text{for every } Y \in Q(X),$$

is a homomorphism from $\mathfrak{Q}(X)$ into $\prod_{t \in T} \mathfrak{A}_t$.

The mapping h is one-one. Indeed, if $Y_1 \neq Y_2$, where $Y_1, Y_2 \in Q(X)$, then there exists an $x \in X$ belonging to exactly one of the sets Y_1, Y_2. Suppose that $x \in Y_1$ and $x \notin Y_2$. Then $h_{(x,g(x))}(Y_1)$ is equal to either $\vee$ or a and $h_{(x,g(x))}(Y_2)$ is equal to either $\wedge$ or b. Hence $h(Y_1) \neq h(Y_2)$. Thus h is a monomorphism from $\mathfrak{Q}(X)$ into $\prod_{t \in T} \mathfrak{A}_t$.

Exercises

1. By a *Peirce algebra* we mean any algebra $\mathfrak{A} = (A, \vee, \Rightarrow, \cup, \cap)$, where $(A, \vee, \Rightarrow)$ is an implication algebra (see II 5) and the following conditions are satisfied: $a \cup b = (a \Rightarrow b) \Rightarrow b$, $(a \cap b) \Rightarrow a = \vee$, $(a \cap b) \Rightarrow b = \vee$, $(a \Rightarrow b) \Rightarrow ((a \Rightarrow c) \Rightarrow (a \Rightarrow (b \cap c))) = \vee$ for all $a, b, c \in A$. Prove that every Peirce algebra is a distributive lattice with the unit element $\vee$.

2. Prove that the notion of an implicative filter in Peirce algebras coincides with the notion of a filter.

3. Prove the analogue of II 6.3 for Peirce algebras.

4. Let $B(X)$ be the class of all subsets of X. Then $(B(X), X, \Rightarrow, \cup, \cap)$, where $Y \Rightarrow Z = (X - Y) \cup Z$ for all $Y, Z \in B(X)$ and $\cup$, $\cap$ are the set-theoretical union and intersection, is a Peirce algebra. Prove the analogue of the Theorems II 7.1 and 2.3 for Peirce algebras.

5. An algebra $(A, \cup, \cap, \sim)$ is said to be a *De Morgan lattice* if $(A, \cup, \cap)$ is a distributive lattice and the equations 3 ((q_1), (q_2)) are satisfied for all a, b in A. Let us define in De Morgan lattices a new operation $\Rightarrow$ as follows: $a \Rightarrow b = \sim a \cup b$ for all elements a, b. Prove that the following equations are identically satisfied:

(1) $a \Rightarrow \sim b = b \Rightarrow \sim a$,

(2) $(a \Rightarrow \sim b) \Rightarrow b = b$,

(3) $(a \Rightarrow b) \Rightarrow c = \sim((\sim a \Rightarrow c) \Rightarrow \sim(b \Rightarrow c))$ (7).

6. Let $(A, \Rightarrow, \sim)$ be an abstract algebra satisfying the conditions (1), (2), (3) of Ex. 5. Put $a \cup b = \sim a \Rightarrow b$, $a \cap b = \sim(a \Rightarrow \sim b)$. Prove that $(A, \cup, \cap)$ is a De Morgan lattice (8).

7. Define in each De Morgan lattice $(A, \cup, \cap, \sim)$ a new two-argument operation $|$ as follows:

$$a|b = \sim a \cap \sim b.$$

We shall write ab instead of $a|b$ and a^2 instead of $a|a$. Prove that the following equations hold identically:

(1) $a^2(ab) = a$,

(2) $(a(bc))^2 = (b^2a)(c^2a)$ (9).

8. Let $(A, |)$ be an algebra satisfying the conditions (1), (2) of Ex. 7 for all a, b in A. Define operations $\sim, \cup, \cap$ as follows:

$$\sim a = a^2, \quad a \cup b = (ab)^2, \quad a \cap b = a^2b^2.$$

Prove that $(A, \cup, \cap, \sim)$ is a De Morgan lattice (10).

9. Let $D = \{0, 1\}$ and $B = D \times D$. For each $(x, y) \in B$ let $\varphi(x, y) = (1-y, 1-x)$. Let I be a set of cardinal $\mathfrak{m}$ and let $E = B^I$. The elements of E are mappings $e = (e_i)_{i \in I}$ such that $e_i \in B$, for $i \in I$. Put $\varphi(e) = (\varphi(e_i))_{i \in I}$. The mapping $\varphi: E \to E$ is an involution of E. Let $(2^E, E, \cup, \cap, \sim)$, where 2^E is the class of all subsets of E and $\sim X = E - \varphi(X)$ for all $X \subset E$, be the quasi-field of all subsets of E. Put $G_i = \{e \in E: e_i = (1, y)\}$, $y \in D$, $i \in I$. Let A be the quasi-field generated in $(2^E, E, \cup, \cap, \sim)$ by $\{G_i\}_{i \in I}$. Show that A is a free algebra in the class of all quasi-Boolean algebras and $\{G_i\}_{i \in I}$ is the set of free generators for $\mathfrak{A}$, card $\{G_i\}_{i \in I} = \mathfrak{m}$ (11).

10. A quasi-Boolean algebra $\mathfrak{A}$ is said to be *functionally free* for a non-empty class K of quasi-Boolean algebras provided that the following condition is satisfied: any two quasi-Boolean polynomials are identically equal in each quasi-Boolean algebra in K if and only if they are identically equal in $\mathfrak{A}$. Prove that the four-element

(7) Gastaminza, M. L., and Gastaminza, S. [1].
(8) Gastaminza, M. L., and Gastaminza, S. [1].
(9) Monteiro, L., and Picco [1].
(10) Monteiro, L., and Picco [1].
(11) Chateaubriand and Monteiro [1].

quasi-Boolean algebra $\mathfrak{A}_0$ (see p. 47) is functionally free for the class K of all quasi-Boolean algebras[12].

11. Prove that the three-element quasi-Boolean algebra $\mathfrak{C}_0$ (see p. 47) is functionally free for the class K_0 of all quasi-Boolean algebras satisfying the condition $a \cap \sim a \leqslant b \cup \sim b$[13].

12. Prove that the following equations can be chosen as the system of axioms for distributive lattices:

$$a \cap (a \cup b) = a, \quad a \cap (b \cup c) = (c \cap a) \cup (b \cap a)^{(14)}.$$

(12) Białynicki-Birula [1].

(13) Białynicki-Birula and Rasiowa [2].

(14) Sholander [1].

CHAPTER IV

RELATIVELY PSEUDO-COMPLEMENTED LATTICES, CONTRAPOSITIONALLY COMPLEMENTED LATTICES, SEMI-COMPLEMENTED LATTICES AND PSEUDO-BOOLEAN ALGEBRAS

Introduction. The present chapter is concerned with four types of abstract algebras, each of these types being related others.

Relatively pseudo-complemented lattices characterize, from the algebraic point of view, positive logic, to be discussed in Chapter X. These abstract algebras, as introduced by Birkhoff [2], are lattices satisfying the following condition: for any two elements a, b there exists a l.u.b. $\{x: a \cap x \leqslant b\}$ called the pseudo-complement of a relative to b and denoted by $a \Rightarrow b$. Each relatively pseudo-complemented lattice has a unit element V (but does not need to have a zero element) and is a positive implication algebra with respect to the operations $V, \Rightarrow$ (cf. II 2). Set lattices of all subsets of a topological space which are both open and dense, set lattices of all open subsets of a topological space and their subalgebras are typical examples of relatively pseudo-complemented lattices. The operation $\Rightarrow$ is then defined as in positive implication algebras (see II 2 (6)). Sections 1, 2, 3 summarize the fundamental properties of relatively pseudo-complemented lattices, including the representation theorem (Rasiowa and Sikorski [3]). A much more comprehensive exposition is to be found in [MM].

The term contrapositionally complemented lattices is here used for algebras which characterize minimal logic, to be analysed in Chapter X. These algebras have been examined by Rasiowa and Sikorski [3] who called them $\mathcal{S}_\mu$-algebras. They are relatively pseudo-complemented lattices with an additional unary operation $\sim_c$ satisfying the contraposition law: $a \Rightarrow \sim_c b = b \Rightarrow \sim_c a$. In any relatively pseudo-complemented set lattice we can define various unary operations $\sim_c$ satisfying this law. The basic properties of contrapositionally complemented lattices and the representation theorem (Rasiowa and Sikorski [3]) are given in Section 4.

Section 5 deals with semi-complemented lattices. This term is here used for algebras which characterize positive logic with semi-negation, to be discussed in Chapter X. That logic and the above-mentioned algebras were introduced by Rasiowa and Sikorski [3], who called the latter $\mathcal{S}_\nu$-algebras. They are relatively pseudo-complemented lattices with a zero element $\wedge$ and with an additional unary operation $\dot{-}$ characterized by the equation $\dot{-} \vee = \wedge$. Clearly, in any relatively pseudo-complemented lattice with $\wedge$ we can define various unary operations satisfying this condition.

Pseudo-Boolean algebras characterize algebraically intuitionistic logic (see Chapter X). These algebras, called also Heyting algebras or pseudo-complemented lattices, are dual to Brouwerian algebras, which have been investigated in detail by McKinsey and Tarski [2]. Pseudo-Boolean algebras are relatively pseudo-complemented lattices with a zero element $\wedge$ and a unary operation $\urcorner$ defined by the equation $\urcorner a = a \Rightarrow \wedge$. Each pseudo-Boolean algebra is simultaneously a contrapositionally complemented lattice and a semi-complemented lattice; conversely, each contrapositionally complemented lattice which is simultaneously semi-complemented (with respect to the same operation) is a pseudo-Boolean algebra. This remark explains the relationships between the four types of algebras to be discussed in the present chapter.

Section 6 lists the basic properties of pseudo-Boolean algebras with the representation theorem. A much more comprehensive exposition of pseudo-Boolean algebras is to be found in [MM].

1. Relatively pseudo-complemented lattices. An abstract algebra $(A, \Rightarrow, \cup, \cap)$ is said to be a *relatively pseudo-complemented lattice* [1] if $(A, \cup, \cap)$ is a lattice and, moreover, for all $a, b, x \in A$ the following condition holds:

(r) $$a \cap x \leqslant b \quad \text{if and only if} \quad x \leqslant a \Rightarrow b.$$

(1) See Birkhoff [2].

The element $a \Rightarrow b$ is called the *pseudo-complement of a relative to b*. Let us note that

1.1. *Every relatively pseudo-complemented lattice includes a unit element* V.

Indeed, for all elements a, b in such a lattice we have $b \leqslant a \Rightarrow a$ since $a \cap b \leqslant a$ (see III 1 (5)). Thus, for every element a, the element $a \Rightarrow a$ is the unit element, i.e.

(1) $$a \Rightarrow a = V.$$

Therefore, in relatively pseudo-complemented lattices we shall distinguish the unit element and usually write $(A, V, \Rightarrow, \cup, \cap)$ instead of $(A, \Rightarrow, \cup, \cap)$.

We quote without proofs the following two facts.

1.2. *Every relatively pseudo-complemented lattice is distributive.*

1.3. *In every relatively pseudo-complemented lattice*

(2) $a \Rightarrow b = V$ *if and only if* $a \leqslant b$,

(3) $a = b$ *if and only if* $a \Rightarrow b = V$ *and* $b \Rightarrow a = V$,

(4) $a \leqslant b \Rightarrow a$,

(5) $a \cap (a \Rightarrow b) \leqslant b$,

(6) $a \Rightarrow (b \Rightarrow c) \leqslant (a \Rightarrow b) \Rightarrow (a \Rightarrow c)$,

(7) $a \Rightarrow V = V$,

(8) $(a \Rightarrow c) \cap (b \Rightarrow c) \leqslant (a \cup b) \Rightarrow c$,

(9) $(a \Rightarrow b) \cap (a \Rightarrow c) \leqslant a \Rightarrow (b \cap c)$.

For proofs of 1.2 and 1.3 see e.g. [MM] (I 12.1 and I 12.2).

The following theorem states a connection between relatively pseudo-complemented lattices and positive implication algebras.

1.4. *An abstract algebra* $(A, V, \Rightarrow, \cup, \cap)$ *is a relatively pseudo-complemented lattice if and only if* $(A, V, \Rightarrow)$ *is a positive implication algebra and, moreover, the following conditions are satisfied:*

(p_5) $a \Rightarrow (a \cup b) = V$,

(p_6) $b \Rightarrow (a \cup b) = V$,

(p_7) $(a \Rightarrow c) \Rightarrow \big((b \Rightarrow c) \Rightarrow ((a \cup b) \Rightarrow c)\big) = V$,

(p_8) $(a \cap b) \Rightarrow a = V$,

(p_9) $(a \cap b) \Rightarrow b = V$,

(p_{10}) $(a \Rightarrow b) \Rightarrow \big((a \Rightarrow c) \Rightarrow (a \Rightarrow (b \cap c))\big) = V$,

for all elements a, b, c *in this algebra.*

Suppose that the algebra in question is a relatively pseudo-complemented lattice. We shall prove that the axioms (p_1)–(p_4) in II 2 for positive implication algebras and the axioms (p_5)-(p_{10}) are fulfilled. Indeed, (p_1) follows from (4) and (2); (p_2) results from (6) and (2); (p_3) is contained in (2); (p_4) is formulated in (7); (p_5) follows from (2) and $a \leqslant a \cup b$ (III 1 (5)); (p_6) follows from (2) and $b \leqslant a \cup b$ (III 1 (6)); (p_7) results from (8), III 1 (1_1), (r) and (2); (p_8) follows from (2) and $a \cap b \leqslant a$ (III 1 (5)); (p_9) follows from (2) and $a \cap b \leqslant b$ (III 1 (6)); (p_{10}) results from (9), III 1 (1_1), (r) and (2).

Conversely, let us suppose that the algebra in question is a positive implication algebra satisfying the conditions (p_5)–(p_{10}). Then, by II 2.2, the relation $\leqslant$, defined by means of the formula

(10) $a \leqslant b$ *if and only if* $a \Rightarrow b = V$,

is an ordering of A and V is the greatest element in $(A, \leqslant)$. Hence, by (p_5) and (p_6), we get $a \leqslant a \cup b$ and $b \leqslant a \cup b$. Suppose that $a \leqslant c$ and $b \leqslant c$. Then $a \Rightarrow c = V$ and $b \Rightarrow c = V$. Hence, by (p_7) and II 2 (1), we get $(a \cup b) \Rightarrow c = V$, i.e. $a \cup b \leqslant c$. Thus $a \cup b = \text{l.u.b.}\,\{a, b\}$ in $(A, \leqslant)$. Similarly, by (p_8), (p_9) and (10), $a \cap b \leqslant a$ and $a \cap b \leqslant b$. Suppose that $c \leqslant a$ and $c \leqslant b$. Then $c \Rightarrow a = V$ and $c \Rightarrow b = V$. Consequently, by (p_{10}) and II 2 (1), we get $c \Rightarrow (a \cap b) = V$, i.e. $c \leqslant a \cap b$. Thus $a \cap b = \text{g.l.b.}\,\{a, b\}$ in $(A, \leqslant)$. By III 1.2, $(A, \cup, \cap)$ is a lattice with the unit element V. In order to prove that the axiom (r) is satisfied we shall show that for all $a, b \in A$

(11) $a \Rightarrow (b \Rightarrow (a \cap b)) = V$.

By (p_{10}) and (10) we have $b \Rightarrow a \leqslant (b \Rightarrow b) \Rightarrow \big(b \Rightarrow (a \cap b)\big)$. Applying II 2.3 (10) and II 2.1 (3), we get $b \Rightarrow a \leqslant b \Rightarrow (a \cap b)$. Hence, by ($p_1$) and II 2.3 (11), we obtain $V = a \Rightarrow (b \Rightarrow a) \leqslant a \Rightarrow \big(b \Rightarrow (a \cap b)\big)$. Thus (11) holds.

Let us suppose that $a \cap b \leqslant c$. Then, applying II 2.3 (11) twice, we get $a \Rightarrow (b \Rightarrow (a \cap b)) \leqslant a \Rightarrow (b \Rightarrow c)$. Consequently, by (11), $a \Rightarrow (b \Rightarrow c) = V$ and, by II 2.3 (14), $b \Rightarrow (a \Rightarrow c) = V$, i.e. $b \leqslant a \Rightarrow c$. Conversely, suppose that $b \leqslant a \Rightarrow c$. Since $a \cap b \leqslant b$, we get $a \cap b \leqslant a \Rightarrow c$. By II 2.3 (8) we get $a \leqslant (a \cap b) \Rightarrow c$. Consequently, since $a \cap b \leqslant a$, we have $a \cap b \leqslant (a \cap b) \Rightarrow c$, i.e. $(a \cap b) \Rightarrow ((a \cap b) \Rightarrow c) = V$. By II 2.3 (13), this implies that $(a \cap b) \Rightarrow c = V$, i.e. $a \cap b \leqslant c$. The equivalence (r) is proved.

It follows from 1.4, II 2.4 and II 2.5 that *the class of all relatively pseudo-complemented lattices is equationally definable.*

Consider a topological space X with an interior operation $\mathbf{I}$. Let $G(X)$ and $G^*(X)$ be the class of all open subsets of X and the class of all dense open subsets of X, respectively. Then $(G(X), X, \Rightarrow, \cup, \cap)$ and $(G^*(X), X, \Rightarrow, \cup, \cap)$, where

(12) $\quad Y \Rightarrow Z = \mathbf{I}((X - Y) \cup Z)$ for all subsets Y, Z of X

and $\cup$, $\cap$ are set-theoretical union and intersection, respectively, are relatively pseudo-complemented lattices. Clearly, all subalgebras of these algebras are also relatively pseudo-complemented lattices. They will be said to be *relatively pseudo-complemented set lattices.* We shall see in Section 3 that these are typical examples of relatively pseudo-complemented lattices.

We quote without proof the following result.

1.5. *In any relatively pseudo-complemented lattice*

(13) $\quad V \Rightarrow b = b,$

(14) $\quad a \cap (a \Rightarrow b) = a \cap b,$

(15) $\quad (a \Rightarrow b) \cap b = b,$

(16) $\quad (a \Rightarrow b) \cap (a \Rightarrow c) = a \Rightarrow (b \cap c),$

(17) $\quad (a \Rightarrow c) \cap (b \Rightarrow c) = (a \cup b) \Rightarrow c,$

(18) $\quad a \Rightarrow (b \Rightarrow c) = (a \cap b) \Rightarrow c,$

(19) $\quad c \Rightarrow a \leqslant (c \Rightarrow (a \Rightarrow b)) \Rightarrow (c \Rightarrow b),$

(20) $\quad (a \Rightarrow b) \cap (b \Rightarrow c) \leqslant a \Rightarrow c,$

(21) $\quad a \leqslant b \Rightarrow (a \cap b),$

(22) $\quad c \cap ((c \cap a) \Rightarrow (c \cap b)) = c \cap (a \Rightarrow b).$

For a proof of 1.5 see e.g. [MM] (I 12.2).

A relatively pseudo-complemented lattice always contains a unit element but it need not contain a zero element. However, every relatively pseudo-complemented lattice can be extended to a relatively pseudo-complemented lattice with the zero element. More precisely:

1.6. *Let* $\mathfrak{A} = (A, \vee, \Rightarrow, \cup, \cap)$ *be a relatively pseudo-complemented lattice without a zero element. Let* A^* *be the set composed of all elements in* A *and of a new element* $\wedge$. Then, extending the operations $\cup, \cap, \Rightarrow$ *in* $\mathfrak{A}$ *onto* A^* *by the following rules*:

(24) $a \cup \wedge = \wedge \cup a = a, \quad a \cap \wedge = \wedge \cap a = \wedge,$ *for every* $a \in A^*$,

(25) $a \Rightarrow \wedge = \wedge$ *for every* $a \in A$, $\wedge \Rightarrow a = \vee$ *for every* $a \in A^*$,

we get a relatively pseudo-complemented lattice $\mathfrak{A}^* = (A^*, \vee, \Rightarrow, \cup, \cap)$ *with the zero element* $\wedge$ *such that* $\mathfrak{A}$ *is a subalgebra of* $\mathfrak{A}^*$.

The proof is by an easy verification.

2. Filters in relatively pseudo-complemented lattices. Any relatively pseudo-complemented lattice is a distributive lattice with a unit element but it is also a positive implication algebra. Therefore, it is possible to consider not only the notion of a filter but also the notion of an implicative filter. We shall prove, however, that these two notions coincide in relatively pseudo-complemented lattices.

2.1. *A non-empty subset* ∇ *of the set of all elements of a relatively pseudo-complemented lattice is a filter if and only if it is an implicative filter.*

Suppose that ∇ is a filter. Then, by III 1.5, $\vee \in \nabla$. If $a \in \nabla$ and $a \Rightarrow b \in \nabla$, then $a \cap (a \Rightarrow b) \in \nabla$. Hence, by 1.3 (5) and III 1 (f_2), $b \in \nabla$. Thus ∇ is an implicative filter. Now let us suppose that ∇ is an implicative filter. Then, by 1 (11), $a \in \nabla$ and $b \in \nabla$ imply that $a \cap b \in \nabla$. If $a \in \nabla$ and $a \leqslant b$, then $a \Rightarrow b = \vee \in \nabla$. Thus $b \in \nabla$. Consequently, ∇ is a filter.

2.2. *For every filter* ∇ *in any relatively pseudo-complemented lattice the following conditions are satisfied*:

(1) *if* ∇ *is maximal, then* ∇ *is prime,*

(2) ∇ *is irreducible if and only if* ∇ *is prime.*

(1) follows from 1.2, III 1.8 and III 2.1; (2) results from 1.2, III 2.1 and III 1.8.

The following theorem follows from 1.4, 2.1 and II 3.1.

2.3. *If $K(h)$ is the kernel of an epimorphism h of a relatively pseudo-complemented lattice $\mathfrak{A} = (A, \vee, \Rightarrow, \cup, \cap)$ onto a similar algebra $\mathfrak{B} = (B, \vee', \Rightarrow, \cup, \cap)$*, i.e. $K(h) = h^{-1}(\vee')$, *then $K(h)$ is a filter. Moreover, $h(a) = h(b)$ is equivalent to $a \Rightarrow b$, $b \Rightarrow a \in K(h)$. Consequently the relation $\approx_{K(h)}$ determined by $K(h)$* (see II 1) *is a congruence in $\mathfrak{A}$. The quotient algebra $\mathfrak{A}/\approx$ is isomorphic to $\mathfrak{B}$.*

2.4. *Let $\mathfrak{A} = (A, \vee, \Rightarrow, \cup, \cap)$ be a relatively pseudo-complemented lattice and let ∇ be a filter in $\mathfrak{A}$. Then the relation $\approx_\nabla$ determined by ∇ is a congruence in $\mathfrak{A}$. The quotient algebra $\mathfrak{A}/\nabla$ is a relatively pseudo-complemented lattice and the conditions* (i), (ii), (iii) *of* II 3.2 *are satisfied.*

The easy proof is left to the reader; it can also be found in [MM] (I 13.2).

3. Representation theorem for relatively pseudo-complemented lattices. The aim of this section is to prove the following representation theorem for relatively pseudo-complemented lattices.

3.1. *For every relatively pseudo-complemented lattice $\mathfrak{A} = (A, \vee, \Rightarrow, \cup, \cap)$ there exists a monomorphism h of $\mathfrak{A}$ into the relatively pseudo-complemented set lattice of all open subsets of a compact topological T_0-space. Moreover, if $\mathfrak{A}$ does not contain a zero element, then there exists a monomorphism h of $\mathfrak{A}$ into the relatively pseudo-complemented set lattice of all open and dense subsets of a compact topological T_0-space* (2).

Let X be the class of all prime filters in $\mathfrak{A}$. By 2.2, X is the class of all irreducible filters. Let h be the mapping from A into the class of all subsets of X defined as follows:

(1) $$h(a) = \{\nabla \in X\colon a \in \nabla\}, \quad \text{for all } a \in A.$$

We shall consider X as a topological space, the class $\{h(a)\}_{a\in A}$ being taken as a subbase. Since $\mathfrak{A}$ is a distributive lattice with a unit element

(2) Rasiowa and Sikorski [3].

(see 1.1, 1.2), it follows from III 2.3 that X is a compact T_0-space. Moreover, h is one-one and

$$h(a \cup b) = h(a) \cup h(b), \quad h(a \cap b) = h(a) \cap h(b), \quad \text{and} \quad h(V) = X.$$

On the other hand, since $(A, V, \Rightarrow)$ is a positive implication algebra, it follows from the proof of II 4.1 that

$$h(a \Rightarrow b) = \mathbf{I}\big((X - h(a)) \cup h(b)\big) = h(a) \Rightarrow h(b).$$

Thus h is a monomorphism from $\mathfrak{A}$ into the relatively pseudo-complemented set lattice of all open subsets of X.

If $\mathfrak{A}$ does not contain a zero element, then for every $a \in A$, the set $h(a)$ is both open and dense. In fact, the intersection of $h(a)$ with an arbitrary non-empty open subset G of X is non-empty. For let us suppose $G = \bigcup_{t \in T} h(a_t)$. Then $h(a) \cap G = \bigcup_{t \in T} h(a \cap a_t) \neq O$, since for every $b \in A$, $h(b) \neq O$. This proves the second part of the theorem.

4. Contrapositionally complemented lattices [(3)]. By a *contrapositionally complemented lattice* we shall mean an abstract algebra $(A, V, \Rightarrow, \cup, \cap, \sim_c)$ such that $(A, V, \Rightarrow, \cup, \cap)$ is a relatively pseudo-complemented lattice and $\sim_c$ is a one-argument operation characterized by the following *contraposition law*:

(c) $$a \Rightarrow \sim_c b = b \Rightarrow \sim_c a, \quad \text{for all } a, b \in A.$$

The operation $\sim_c$ will be called *contrapositional complementation* and for every $a \in A$, the element $\sim_c a$ will be called the *contrapositional complement of a*.

It follows from the above definition and from the fact that the class of all relatively pseudo-complemented lattices is equationally definable that the class of all contrapositionally complemented lattices is also equationally definable.

The following theorem gives a method of constructing contrapositionally complemented lattices.

4.1. *An abstract algebra $(A, V, \Rightarrow, \cup, \cap, \sim_c)$ is a contrapositionally*

(3) Contrapositionally complemented lattices have been studied by Rasiowa and Sikorski [3] under the term S_μ-algebras.

complemented lattice if and only if $(A, V, \Rightarrow, \cup, \cap)$ *is a relatively pseudo-complemented lattice and the following equation holds*:

(c*) $$\sim_c a = a \Rightarrow \sim_c V, \quad \textit{for every } a \in A\text{ [4]}.$$

If (c) holds, then by 1.5 (13)

$$\sim_c a = V \Rightarrow \sim_c a = a \Rightarrow \sim_c V, \quad \text{for every } a \in A.$$

If (c*) is satisfied, then by 1.4 and II 2.3 (14) we get

$$a \Rightarrow \sim_c b = a \Rightarrow (b \Rightarrow \sim_c V) = b \Rightarrow (a \Rightarrow \sim_c V) = b \Rightarrow \sim_c a.$$

It follows from the above theorem that the operation $\sim_c$ is completely determined by the operations V, $\Rightarrow$, $\cup$, $\cap$ and by the element $\sim_c V$. Clearly, the $c_0 = \sim_c V$ may be an arbitrary fixed element. In particular, it is possible that $V = \sim_c V$.

The following theorem follows from 4.1 and 1.3 (7).

4.2. *If a contrapositionally complemented lattice has a zero element* Λ, *then*

(1) $$\sim_c \Lambda = V.$$

By 4.1, 4.2 and 1.6 we get the following theorem.

4.3. *Let* $\mathfrak{A} = (A, V, \Rightarrow, \cup, \cap, \sim_c)$ *be a contrapositionally complemented lattice without a zero element, and let* A^* *be the set consisting of all elements in* A *and of a new element* Λ. *Then extending the operations* $\cup$, $\cap$, $\Rightarrow$ *in* $\mathfrak{A}$ *onto* A^* *by rules* 1 (24), (25) *and setting* $\sim_c \Lambda = V$, *we get a contrapositionally complemented lattice* $\mathfrak{A}^* = (A^*, V, \Rightarrow, \cup, \cap, \sim_c)$ *such that* $\mathfrak{A}$ *is a subalgebra of* $\mathfrak{A}^*$ *and* Λ *is the zero element in* $\mathfrak{A}^*$ [5].

Let X be a topological space and let $\mathfrak{G}_0(X)$ be a relatively pseudo-complemented lattice of open subsets of X. Then, setting

(2) $$\sim_c X = Y_0 \quad \text{for some } Y_0 \text{ belonging to } \mathfrak{G}_0(X),$$

(3) $$\sim_c Z = Z \Rightarrow \sim_c X,$$

we define a contrapositional complementation in $\mathfrak{G}_0(X)$ and get a contrapositionally complemented lattice of open subsets of X. The following representation theorem follows from 4.1 and 3.1.

4.4. *For every contrapositionally complemented lattice* $\mathfrak{A} = (A, V, \Rightarrow, \cup, \cap, \sim_c)$ *there exists a monomorphism* h *from* $\mathfrak{A}$ *into a contra-*

[4] Rasiowa and Sikorski [3].

[5] Rasiowa and Sikorski [3].

positionally complemented set lattice of all open subsets of a compact topological T_0-space. Moreover, if $\mathfrak{A}$ does not contain a zero element, then there exists a monomorphism h of $\mathfrak{A}$ into a contrapositionally complemented set lattice of all open and dense subsets of a compact T_0-space (6).

By 3.1 there exists a monomorphism h from $(A, V, \Rightarrow, \cup, \cap)$ into the relatively pseudo-complemented lattice $\mathfrak{G}(X) = (G(X), X, \Rightarrow, \cup, \cap)$ of all open subsets of a compact T_0-space X. Suppose that $\sim_c V = c_0$. Then we put

$$\sim_c Y = Y \Rightarrow h(c_0), \qquad \text{for every } Y \text{ in } G(X).$$

Then $(G(X), X, \Rightarrow, \cup, \cap, \sim_c)$ is a contrapositionally complemented set lattice and for every $a \in A$,

$$\sim_c h(a) = h(a) \Rightarrow h(c_0) = h(a \Rightarrow c_0) = h(a \Rightarrow \sim_c V) = h(\sim_c a).$$

Thus h is a monomorphism from $\mathfrak{A}$ into this contrapositionally complemented set lattice. The proof of the second part of the theorem is analogous.

4.5. *In every contrapositionally complemented lattice the following equations hold:*

(4) $$a \leqslant \sim_c \sim_c a,$$

(5) $$a \Rightarrow b \leqslant \sim_c b \Rightarrow \sim_c a.$$

By 1.4, II 2.3 (9) and (c*) we get

$$a \leqslant (a \Rightarrow \sim_c V) \Rightarrow \sim_c V = \sim_c \sim_c a.$$

By (4), 1.4, II 2.3 (11) and (c) we have

$$a \Rightarrow b \leqslant a \Rightarrow \sim_c \sim_c b = \sim_c b \Rightarrow \sim_c a.$$

We shall use the above facts to prove the following theorem:

4.6. *Let $\mathfrak{A} = (A, V, \Rightarrow, \cup, \cap, \sim_c)$ be a contrapositionally complemented lattice and let ∇ be a filter in $\mathfrak{A}$. Then the relation $\approx_\nabla$ determined by ∇ is a congruence in $\mathfrak{A}$. The quotient algebra $\mathfrak{A}/\nabla$ is a contrapositionally complemented lattice and the conditions* (i), (ii), (iii) *of* II 3.2 *hold.*

(6) Rasiowa and Sikorski [3].

Because of Theorem 2.4 and the fact the class of all contrapositionally complemented lattices is equationally definable it is sufficient to prove that $\approx_\nabla$ preserves the operation $\sim_c$, i.e. that the condition $a \approx_\nabla b$ implies that $\sim_c a \approx_\nabla \sim_c b$. Suppose that $a \approx_\nabla b$. Then $a \Rightarrow b \in \nabla$ and $b \Rightarrow a \in \nabla$. By (5), $\sim_c b \Rightarrow \sim_c a \in \nabla$ and $\sim_c a \Rightarrow \sim_c b \in \nabla$, which proves that $\sim_c a \approx_\nabla \sim_c b$.

5. Semi-complemented lattices [7]. By a semi-complemented lattice we mean an abstract algebra $(A, V, \Rightarrow, \cup, \cap, \dot{-})$ such that $(A, V, \Rightarrow, \cup, \cap)$ is a relatively pseudo-complemented lattice and $\dot{-}$ is a one-argument operation on A satisfying the following condition:

(s) $$\dot{-}(a \Rightarrow a) \Rightarrow b = V \quad \text{for all } a, b \text{ in } A.$$

The operation $\dot{-}$ is then called *semi-complementation*, and for every $a \in A$ the element $\dot{-}a$ will be called the *semi-complement* of a. It follows from (s) and 1 (1), 1.3 (2) that $\dot{-}(a \Rightarrow a) = \dot{-}V$ is the zero element in every semi-complemented lattice. Hence every semi-complemented lattice has a zero element.

If $\mathfrak{G}_0(X) = (G_0(X), X, \Rightarrow, \cup, \cap)$ is a relatively pseudo-complemented lattice of open subsets of a topological space X and the empty set belongs to $G_0(X)$, then we can introduce in $G_0(X)$ a semi-complementation by defining the operation $\dot{-}$ on $G_0(X) - \{X\}$ in an arbitrary way and setting $\dot{-}X = O$. Every semi-complemented lattice of this type will be said to be a *semi-complemented set lattice.*

5.1. *For every semi-complemented lattice* $\mathfrak{A} = (A, V, \Rightarrow, \cup, \cap, \dot{-})$ *there exists a monomorphism* h *from* $\mathfrak{A}$ *into the semi-complemented set lattice* $\mathfrak{G}(X)$ *of all open subsets of a compact* T_0*-space* X [8].

By 3.1 there exists a monomorphism h from $(A, V, \Rightarrow, \cup, \cap)$ into the relatively pseudo-complemented lattice $(G(X), X, \Rightarrow, \cup, \cap)$ of all open subsets of a compact T_0-space X. We define on $G(X)$ the operation $\dot{-}$ as follows:

$$(1) \quad \begin{array}{lllll} \text{if} & Y = h(a) & \text{for some } a \in A, & \text{then} & \dot{-}Y = h(\dot{-}a), \\ \text{if} & Y \neq h(a) & \text{for every } a \in A, & \text{then} & \dot{-}Y = O. \end{array}$$

[7] Semi-complemented lattices have been studied by Rasiowa and Sikorski [3] under the term S_ν-algebras.

[8] Rasiowa and Sikorski [3].

Since h is one-one, this is a good definition. Observe that

$$\dot{-}X = \dot{-}h(V) = h(\dot{-}V) = h(\Lambda) = O.$$

Thus the operation $\dot{-}$ defined above on $G(X)$ is a semi-complementation operation and consequently $(G(X), X, \Rightarrow, \cup, \cap, \dot{-})$ is a semi-complemented set lattice. Moreover, for every $a \in A$

$$h(\dot{-}a) = \dot{-}h(a),$$

which proves that h is the required monomorphism.

6. Pseudo-Boolean algebras [9]. An abstract algebra $\mathfrak{A} = (A, V, \Rightarrow, \cup, \cap, \urcorner)$ is said to be a *pseudo-Boolean algebra* provided $(A, V, \Rightarrow, \cup, \cap)$ is a relatively pseudo-complemented lattice and, moreover, the following equations hold:

(c) $$a \Rightarrow \urcorner b = b \Rightarrow \urcorner a,$$

(s) $$\urcorner(a \Rightarrow a) \Rightarrow b = V.$$

The operation $\urcorner$ is then called a *pseudo-complementation* and for every $a \in A$ the element $\urcorner a$ is said to be the *pseudo-complement* of a. Note that the class of all pseudo-Boolean algebras is equationally definable since the class of all relatively pseudo-complemented lattices is equationally definable.

It follows from the above definition, 4.1, 4.5 and 1.3 (2) that

6.1. *An abstract algebra $(A, V, \Rightarrow, \cup, \cap, \urcorner)$ is a pseudo-Boolean algebra if and only if it is a contrapositionally complemented lattice and a semi-complemented lattice. Consequently, every pseudo-Boolean algebra has a zero element Λ and the following equations hold*:

(1) $\urcorner V = \urcorner(a \Rightarrow a) = \Lambda,$

(2) $\urcorner a = a \Rightarrow \Lambda,$

(3) $a \leqslant \urcorner\urcorner a,$

(4) $a \Rightarrow b \leqslant \urcorner b \Rightarrow \urcorner a.$

6.2. *In every pseudo-Boolean algebra*

(5) *if* $a \leqslant b$, *then* $\urcorner b \leqslant \urcorner a$,

[9] For a more comprehensive exposition of pseudo-Boolean algebras or their duals see [MM] and McKinsey and Tarski [2], respectively.

(6) $\neg\Lambda = V$, *if* $\neg a = V$ *then* $a = \Lambda$,

(7) $a \cap \neg a = \Lambda$,

(8) $\neg(a \cap \neg a) = V$,

(9) $\neg\neg\neg a = \neg a$,

(10) $\neg(a \cup b) = \neg a \cap \neg b$,

(11) $\neg a \cup \neg b \leqslant \neg(a \cap b)$,

(12) $\neg a \cup b \leqslant a \Rightarrow b$,

(13) $\neg\neg(a \cup \neg a) = V$.

(5) follows from (4) and 1.3 (2); (6) follows from (2), 1.1 (1) and 1 (r); by (2) and 1.3 (5) we get (7); (8) results from (7) and (6); by (3) $\neg a \leqslant \neg\neg\neg a$ and, on the other hand, by (3) and (5) $\neg\neg\neg a \leqslant \neg a$, which proves (9); (10) follows from (2) and 1.5 (17); from the distributive law and (7) we get $(\neg a \cup \neg b) \cap (a \cap b) \leqslant \Lambda$ and hence by 1 (r) $\neg a \cup \neg b \leqslant (a \cap b) \Rightarrow \Lambda = \neg(a \cap b)$, which proves (11); by the distributive law and (7) $(\neg a \cup b) \cap a = (\neg a \cap a) \cup (b \cap a) \leqslant b$ and hence by 1 (r) we get (12); (13) follows by (10), (7) and (6).

The following theorem follows directly from 6.1 and 4.6:

6.3. *Let* $\mathfrak{A} = (A, V, \Rightarrow, \cup, \cap, \neg)$ *be a pseudo-Boolean algebra and let* ∇ *be a filter in* $\mathfrak{A}$. *Then the relation* $\approx_\nabla$ *determined by* ∇ *is a congruence in* $\mathfrak{A}$. *The quotient algebra* $\mathfrak{A}/\nabla$ *is a pseudo-Boolean algebra, and the conditions* (i), (ii), (iii) *of* II 3.2 *hold.*

6.4. *Let* ∇ *be a filter in a pseudo-Boolean algebra and let* a_0 *be an element of this algebra. Then the filter generated by* ∇ *and* a_0 *is proper if and only if* $\neg a_0 \notin \nabla$.

In fact, the condition that the filter ∇^* generated by ∇ and a_0 is not proper is equivalent to the condition that $\Lambda \in \nabla^*$ and consequently, by III 1.7, to the condition that there exists a $c \in \nabla$ such that $a_0 \cap c \leqslant \Lambda$. This holds by 1 (r) if and only if $c \leqslant a_0 \Rightarrow \Lambda = \neg a_0$. This condition is equivalent to the condition that $\neg a_0 \in \nabla$.

6.5. *The following conditions are equivalent for every filter* ∇ *in a pseudo-Boolean algebra* $\mathfrak{A}$

(14) ∇ *is maximal*,

(15) *for every a in this algebra exactly one of the elements a,* $\neg a$ *belongs to* ∇,

(16) *the quotient algebra* $\mathfrak{A}/\nabla$ *has exactly two elements.*

(14) is equivalent to (15). If ∇ is maximal, then it is proper and consequently, on account of (7), for every element a at most one of the elements a, $\neg a$ can be in ∇. Suppose that $\neg a \notin \nabla$; then by 6.4 the filter ∇^* generated by ∇ and a is proper and contains the maximal filter ∇. Hence $\nabla^* = \nabla$, which proves that $a \in \nabla$. Thus (14) implies (15). On the other hand, if (15) holds, then obviously ∇ is proper and because of (7) it is not a proper subset of any proper filter, i.e. ∇ is maximal.

(15) is equivalent to (16). It follows from 6.3 that for every $b \in A$, $b \in \nabla$ if and only if $||b|| = ||V||$. Hence (15) holds if and only if, for every $a \in A$, exactly one of the conditions: $||a|| = ||V||$, $||\neg a|| = \neg ||a|| = ||V||$ is satisfied. By (6) this holds if and only if exactly one of the conditions $||a|| = V_{\mathfrak{A}/\nabla}$, $||a|| = \Lambda_{\mathfrak{A}/\nabla}$ is fulfilled, i.e. if and only if the quotient algebra $\mathfrak{A}/\nabla$ has exactly two elements: $V_{\mathfrak{A}/\nabla}$, $\Lambda_{\mathfrak{A}/\nabla}$.

Note that

6.6. *If* $(A, V, \Rightarrow, \cup, \cap)$ *is a relatively pseudo-complemented lattice with a zero element* Λ, *then putting*

(17) $$\neg a = a \Rightarrow \Lambda, \quad \textit{for every } a \in A,$$

we define a one-argument operation on A such that the axioms (c) *and* (s) *are satisfied. Consequently,* $(A, V, \Rightarrow, \cup, \cap, \neg)$ *is a pseudo-Boolean algebra.*

By (17), 1.4 and II 2.3 (14)

$$a \Rightarrow \neg b = a \Rightarrow (b \Rightarrow \Lambda) = b \Rightarrow (a \Rightarrow \Lambda) = b \Rightarrow \neg a.$$

By (17), 1.1 (1), 1.5 (13), 1.3 (2)

$$\neg(a \Rightarrow a) \Rightarrow b = ((a \Rightarrow a) \Rightarrow \Lambda) \Rightarrow b = (V \Rightarrow \Lambda) \Rightarrow b = \Lambda \Rightarrow b = V.$$

Let $\mathfrak{G}_0(X) = (G_0(X), X, \Rightarrow, \cup, \cap)$ be a relatively pseudo-complemented lattice of open subsets of a topological space X and let $O \in G_0(X)$. Then, by 6.6 and 1 (12), putting

(18) $$\neg Y = Y \Rightarrow O = \mathbf{I}(X - Y), \quad \text{for every } Y \in G_0(X),$$

we define the pseudo-complementation in $G_0(X)$, i.e. the algebra $(G_0(X), X, \Rightarrow, \cup, \cap, \neg)$ is a pseudo-Boolean algebra of sets called a *pseudo-field* of open subsets of X.

The following representation theorem follows from 3.1 and (2).

6.7. *For every pseudo-Boolean algebra* $\mathfrak{A} = (A, V, \Rightarrow, \cup, \cap, \neg)$ *there exists a monomorphism h from* $\mathfrak{A}$ *into the pseudo-field of all open subsets of a compact* T_0*-space* (10).

By 3.1 there exists a monomorphism h from $(A, V, \Rightarrow, \cup, \cap)$ into the relatively pseudo-complemented set lattice $(G(X), X, \Rightarrow, \cup, \cap)$ of all open subsets of a compact T_0-space X. Defining by the formula (18) the pseudo-complementation on $G(X)$, we get the pseudo-field $(G(X), X, \Rightarrow, \cup, \cap, \neg)$ of all open subsets of X. By (2) and (18) we have

$$h(\neg a) = h(a \Rightarrow \wedge) = h(a) \Rightarrow h(\wedge) = h(a) \Rightarrow O = \neg h(a).$$

Thus h is the required monomorphism.

Exercises

1. Prove that the following equations can be taken as a system of axioms for relatively pseudo-complemented lattices: $a \cap (a \cup b) = a$, $a \cap (b \cup c) = (c \cap a) \cup (b \cap a)$ $(a \Rightarrow a) \cap b = b$, $a \cap (a \Rightarrow b) = a \cap b$, $(a \Rightarrow b) \cap b = b$, $(a \Rightarrow b) \cap (a \Rightarrow c) = a \Rightarrow (b \cap c)$ Apply Ex. 12 in Chap. III.

2. Prove that every finite distributive lattice is a pseudo-Boolean algebra; the operations $\Rightarrow$ and $\neg$ are defined as follows: $a \Rightarrow b = \text{l.u.b.}\{c\colon a \cap c \leqslant b\}$, $\neg a = a \Rightarrow \wedge$.

3. Following Tarski [8] an element a of a pseudo-Boolean algebra $\mathfrak{A}$ is said to be dense provided $\neg a = \wedge$. Prove that the following conditions are equivalent for each element a in a pseudo-Boolean algebra: (i) a is dense, (ii) $\neg\neg a = V$, (iii) $a \cap c \neq \wedge$ for every element $c \neq \wedge$, (iv) $a = b \cup \neg b$ for an element b of the pseudo-Boolean algebra (11).

4(12)**.** Prove that a pseudo-Boolean algebra is a Boolean algebra if and only if the unit element V is the only dense element.

5. Prove that the set of all dense elements in a pseudo-Boolean algebra is a filter.

(10) Stone [4], McKinsey and Tarski [2].
(11) Cf. [MM].
(12) For Ex.4–Ex.6 see [MM].

6. Prove that if a filter ∇ in a pseudo-Boolean algebra contains all dense elements. then the quotient algebra $\mathfrak{A}/\nabla$ is a Boolean algebra.

7. An element a of of a pseudo-Boolean algebra $\mathfrak{A}$ is said to be *regular* provided $a = \neg\neg a$[13]. Show that the set of all regular elements in a pseudo-Boolean algebra is a Boolean algebra with the join $\cup^*$ defined by $a \cup^* b = \neg\neg(a \cup b)$.

8. Prove that the Boolean algebra of all regular elements in a pseudo-Boolean algebra $\mathfrak{A}$ is isomorphic to the quotient algebra $\mathfrak{A}/\nabla$, where ∇ is the filter of all dense elements in $\mathfrak{A}$.

9. A pseudo-Boolean algebra is said to be *linear* provided the following condition is satisfied: $(a \Rightarrow b) \cup (b \Rightarrow a) = V$ for all elements a, b of that algebra. Prove that for each linear pseudo-Boolean algebra $\mathfrak{A}$, the quotient algebra $\mathfrak{A}/\nabla$, where ∇ is a prime filter, is a chain.

[13] For the study of regular elements see McKinsey and Tarski [2].

CHAPTER V

QUASI-PSEUDO-BOOLEAN ALGEBRAS

Introduction. Quasi-pseudo-Boolean algebras correspond to constructive logic with strong negation which is discussed in Chapter XII. These algebras were introduced and studied under the name $\mathcal{N}$-lattices by Rasiowa alone ([6], [7]) and jointly with Białynicki-Birula [1]. In later papers of other authors these algebras have also been called Nelson algebras. Since quasi-pseudo-Boolean algebras cover both quasi-Boolean algebras and pseudo-Boolean algebras, the new term has been chosen to bring out the close relationship between them.

The aim of this chapter is to give an exposition of the elementary properties of these algebras, the theory of filters and the representation theorem, which makes it possible to construct quasi-pseudo-Boolean algebras of sets from pseudo-Boolean algebras of open subsets of topological spaces. McKinsey and Tarski [2] have proved that the pseudo-Boolean algebra of all open subsets of n-dimensional Euclidean space or the Cantor discontinuum is functionally free for the class of all pseudo-Boolean algebras, i.e., that any two pseudo-Boolean polynomials are identically equal in each pseudo-Boolean algebra if and only if they are identically equal in the above-mentioned pseudo-Boolean algebra. The representation theorem for quasi-pseudo-Boolean algebras makes it possible to generalize McKinsey and Tarski's result (Białynicki-Birula and Rasiowa [2]), which is given in Chapter XII.

The non-equational definition of a quasi-pseudo-Boolean algebra, as adopted in Section 1, is a slight modification of that given by the present author ([6], [7]). However, the class of quasi-pseudo-Boolean algebras is equationally definable (Brignole and Monteiro [1]) and this is proved

[1] Białynicki-Birula and Rasiowa [2].

in Section 2. Simple examples of these algebras and a general method of constructing quasi-pseudo-Boolean algebras of sets from a pseudo-Boolean algebra of open subsets of a topological space are given in Section 3. Filters in these algebras are examined in Section 4. Among all filters in any quasi-pseudo-Boolean algebra there are distinguished two kinds. Filters of the first kind are kernels of epimorphisms and conversely. Prime filters of both kinds cover all prime filters and play an important role in the proof of the Representation Theorem 5.6, given in Section 5.

1. Definition and elementary properties. By a *quasi-pseudo-Boolean algebra* (2) we shall mean an abstract algebra $\mathfrak{A} = (A, V, \Rightarrow, \cup, \cap, \rightarrow, \sim, \urcorner)$ such that

(qpB$_1$) $(A, V, \cup, \cap, \sim)$ is a quasi-Boolean algebra,

and for all $a, b, c, x \in A$ the following conditions are satisfied:

(qpB$_2$) the relation $\prec$ defined by the formula

(1) $$a \prec b \quad \text{if and only if} \quad a \rightarrow b = V,$$

is a quasi-ordering on A,

(qpB$_3$) $a \cap x \prec b$ if and only if $x \prec a \rightarrow b$,

(qpB$_4$) $a \Rightarrow b = (a \rightarrow b) \cap (\sim b \rightarrow \sim a)$,

(qpB$_5$) $a \Rightarrow b = V$ if and only if $a \cap b = a$,

(qpB$_6$) $a \prec c$ and $b \prec c$ imply $a \cup b \prec c$,

(qpB$_7$) $a \prec b$ and $a \prec c$ imply $a \prec b \cap c$,

(qpB$_8$) $(a \cap \sim b) \prec \sim (a \rightarrow b)$,

(qpB$_9$) $\sim (a \rightarrow b) \prec (a \cap \sim b)$,

(qpB$_{10}$) $a \prec \sim \urcorner a$,

(qpB$_{11}$) $\sim \urcorner a \prec a$,

(qpB$_{12}$) $a \cap \sim a \prec b$,

(qpB$_{13}$) $\urcorner a = a \rightarrow \sim V$.

(2) Rasiowa [6], [7].

The operations $\Rightarrow$, $\rightarrow$, $\sim$, $\neg$ are then called the *implicative operation, weak relative pseudo-complementation, quasi-complementation* and *weak pseudo-complementation*, respectively.

Observe that

1.1. *In any quasi-pseudo-Boolean algebra for any elements* a, b *the following conditions are equivalent*:

(2) $a \leqslant b$,

(3) $a \Rightarrow b = V$,

(4) $a \prec b$ *and* $\sim b \prec \sim a$.

This follows from (qpB_5), (qpB_4) and (1).

1.2. *In every quasi-pseudo-Boolean algebra* $(A, V, \Rightarrow, \cup, \cap, \rightarrow, \sim, \neg)$ *the following conditions are satisfied for all* $a, b, c \in A$:

(5) $a \Rightarrow a = V$,

(6) *if* $a \Rightarrow b = V$ *and* $b \Rightarrow c = V$, *then* $a \Rightarrow c = V$,

(7) *if* $a \Rightarrow b = V$ *and* $b \Rightarrow a = V$, *then* $a = b$,

(8) $a \Rightarrow V = V$,

i.e. $(A, V, \Rightarrow)$ *is an implicative algebra.*

Theorem 1.2 follows easily from 1.1.

1.3. *In any quasi-pseudo-Boolean algebra*:

(9) $a \rightarrow a = V$,

(10) *if* $a \rightarrow b = V$ *and* $b \rightarrow c = V$, *then* $a \rightarrow c = V$,

(11) *if* $b \prec c$, *then* $a \cup b \prec a \cup c$ *and* $a \cap b \prec a \cap c$,

(12) *if* $a \prec b$ *and* $c \prec d$, *then* $a \cup c \prec b \cup d$ *and* $a \cap c \prec b \cap d$,

(13) $a \cap (a \rightarrow b) \prec b$,

(14) $a \rightarrow (b \rightarrow c) = (a \cap b) \rightarrow c$,

(15) $a \rightarrow (b \rightarrow c) = b \rightarrow (a \rightarrow c)$,

(16) $a \prec (a \rightarrow b) \rightarrow b$,

(17) $a \cap \sim a \leqslant b \cup \sim b$,

(18) $a \leqslant b \Rightarrow (a \cap b)$,

(19) $a \to \urcorner b = b \to \urcorner a,$

(20) $a \to \vee = \vee,$

(21) $\wedge \to a = \vee,$

(22) $\vee \to b = b,$

(23) *if* $a = \vee$ *and* $a \to b = \vee$, *then* $b = \vee$,

(24) $a \to (b \to a) = \vee,$

(25) $\big(a \to (b \to c)\big) \to \big((a \to b) \to (a \to c)\big) = \vee,$

(26) $a \to (a \cup b) = \vee,$

(27) $b \to (a \cup b) = \vee,$

(28) $(a \to c) \to \big((b \to c) \to \big((a \cup b) \to c\big)\big) = \vee,$

(29) $(a \cap b) \to a = \vee,$

(30) $(a \cap b) \to b = \vee,$

(31) $(a \to b) \to \Big((a \to c) \to \big(a \to (b \cap c)\big)\Big) = \vee,$

(32) $(a \to \urcorner b) \to (b \to \urcorner a) = \vee,$

(33) $\urcorner(a \to a) \to b = \vee,$

(34) $\sim a \to (a \to b) = \vee,$

(35) $\big(\sim (a \to b) \to (a \cap \sim b)\big) \cap \big((a \cap \sim b) \to \sim (a \to b)\big) = \vee,$

(36) $(\sim \urcorner a \to a) \cap (a \to \sim \urcorner a) = \vee,$

(37) $\sim \sim a = a,$

(38) $\sim (a \cup b) = \sim a \cap \sim b,$

(39) $\sim (a \cap b) = \sim a \cup \sim b,$

(40) *the condition* $a \prec b$ *implies* $b \to c \prec a \to c$ *and* $c \to a \prec c \to b$,

(41) $(a \to b) \to \big((c \to d) \to (a \cap c \to b \cap d)\big) = \vee,$

(42) $(a \to b) \to \big((c \to d) \to (a \cup c \to b \cup d)\big) = \vee,$

(43) $(b \to a) \to \big((c \to d) \to \big((a \to c) \to (b \to d)\big)\big) = \vee.$

(9) and (10) follow from (qpB_2); since by 1.1 $a \prec a \cup c$, $c \prec a \cup c$, the condition $b \prec c$ implies by (qpB_6) that $a \cup b \prec a \cup c$; similarly,

by 1.1, $a \cap b \prec a$ and $a \cap b \prec b$, and hence the condition $b \prec c$ implies by (qpB_7) that $a \cap b \prec a \cap c$, which proves (11); (12) follows from (11) and the transitivity of the relation $\prec$; since $a \rightarrow b \prec a \rightarrow b$, by ($qpB_3$) we get (13);

the proof of (14): by (13) $a \cap (a \rightarrow (b \rightarrow c)) \prec b \rightarrow c$, hence by (11) and (13) $a \cap b \cap (a \rightarrow (b \rightarrow c)) \prec b \cap (b \rightarrow c) \prec c$, thus by ($qpB_3$)

$$a \rightarrow (b \rightarrow c) \prec (a \cap b) \rightarrow c; \tag{44}$$

on the other hand, by (qpB_9), (qpB_8) and (11) we easily get

$$\sim ((a \cap b) \rightarrow c) \prec \sim (a \rightarrow (b \rightarrow c)); \tag{45}$$

thus by 1.1

$$a \rightarrow (b \rightarrow c) \leqslant (a \cap b) \rightarrow c; \tag{46}$$

on the other hand, by (13) $a \cap b \cap ((a \cap b) \rightarrow c) \prec c$, which by ($qpB_3$) is equivalent, successively, to $a \cap ((a \cap b) \rightarrow c) \prec b \rightarrow c$, and $(a \cap b) \rightarrow c \prec a \rightarrow (b \rightarrow c)$, thus

$$(a \cap b) \rightarrow c \prec a \rightarrow (b \rightarrow c); \tag{47}$$

by (qpB_8), (qpB_9) and (11) we easily get

$$\sim (a \rightarrow (b \rightarrow c)) \prec \sim ((a \cap b) \rightarrow c), \tag{48}$$

which together with (46) gives

$$(a \cap b) \rightarrow c \leqslant a \rightarrow (b \rightarrow c); \tag{49}$$

now (14) follows from (46) and (49);

(15) follows directly from (14); (16) follows from (13) by (qpB_3); on account of (qpB_{12}) $a \cap \sim a \prec b \cup \sim b$ and $\sim (b \cup \sim b) = \sim b \cap \cap b \prec \sim (a \cap \sim a)$, which proves (17);

the proof of (18): since, on account of 1.1, $b \cap a \prec a \cap b$, by (qpB_3),

$$a \prec b \rightarrow (a \cap b); \tag{50}$$

by 1.1 $\sim b \cap a \prec \sim b$ and by (qpB_{12}) $\sim a \cap a \prec \sim b$; hence, applying (11), we get $(\sim a \cap a) \cup (\sim b \cap a) \prec \sim b$; by distributivity this implies

$$(\sim a \cup \sim b) \cap a \prec \sim b, \tag{51}$$

which is equivalent by (qpB_3) to $a \prec (\sim a \cup \sim b) \rightarrow \sim b$; hence, applying ($qpB_1$) and III 3.1 (3), we get

$$a \prec \sim (a \cap b) \rightarrow \sim b; \tag{52}$$

applying (qpB_7) and (qpB_4) to (50) and (52), we obtain

(53) $$a \prec b \Rightarrow (a \cap b);$$

substituting in (51) a for b and b for a, we get $(\sim b \cup \sim a) \cap b \prec \sim a$; hence by ($qpB_1$) and III 3.1 (3)

(54) $$\sim (a \cap b) \cap b \prec \sim a \quad \text{and} \quad b \cap \sim (a \cap b) \prec \sim a;$$

from the first of these results if follows by (qpB_9), (qpB_1) and III 3 (q_1) that $\sim \big(\sim (a \cap b) \rightarrow \sim b\big) \prec \sim a$; from the second result of (54) it follows by (qpB_9) that $\sim \big(b \rightarrow (a \cap b)\big) \prec \sim a$; applying ($qpB_6$) to the last two results, we get

$$\sim \big(b \rightarrow (a \cap b)\big) \cup \sim \big(\sim (a \cap b) \rightarrow \sim b\big) \prec \sim a;$$

hence by (qpB_1), III 3.1 (3) and (qpB_4) we get

(55) $$\sim \big(b \Rightarrow (a \cap b)\big) \prec \sim a;$$

now (18) follows from (53) and (55);

the proof of (19): by (15), $a \rightarrow (b \rightarrow \sim V) \prec b \rightarrow (a \rightarrow \sim V)$; hence, by ($qpB_{13}$),

(56) $$a \rightarrow \neg b \prec b \rightarrow \neg a;$$

on the other hand, by (qpB_9), (qpB_{11}), (11), (qpB_{10}), (qpB_8), we get

$$\sim (b \rightarrow \neg a) \prec b \cap \sim \neg a \prec b \cap a \prec \sim \neg b \cap a$$
$$\prec a \cap \sim \neg b \prec \sim (a \rightarrow \neg b);$$

thus

(57) $$\sim (b \rightarrow \neg a) \prec \sim (a \rightarrow \neg b);$$

it follows from (56) and (57) that $a \rightarrow \neg b \leqslant b \rightarrow \neg a$ and consequently, by symmetry, $b \rightarrow \neg a \leqslant a \rightarrow \neg b$, which proves (19);

(20) follows from (5) by 1.1 and (1); (21) follows from $\Lambda \leqslant a$ by 1.1 and (1); by (qpB_1), (qpB_2), (qpB_3) and (qpB_9), $b \prec V \rightarrow b$ and $\sim (V \rightarrow b) \prec V \cap \sim b = \sim b$, thus $b \leqslant V \rightarrow b$; on the other hand, by (13), $V \rightarrow b = V \cap (V \rightarrow b) \prec b$ and by (qpB_8) $\sim b = V \cap \sim b \prec \sim (V \rightarrow b)$; thus $V \rightarrow b \leqslant b$, which with $b \leqslant V \rightarrow b$ proves (22); (23) results from (22); it follows from $b \cap a \leqslant a$, by 1.1, that $b \cap a \prec a$; this is equivalent by (qpB_3) to $a \prec b \rightarrow a$, which implies (24);

the proof of (25): by (13), (11), (15), we get

$$a \cap (a \rightarrow b) \cap \big(a \rightarrow (b \rightarrow c)\big) \prec b \cap \big(a \rightarrow (b \rightarrow c)\big)$$
$$\prec b \cap \big(b \rightarrow (a \rightarrow c)\big) \prec a \rightarrow c;$$

hence, by (qpB_3) and $a \cap a = a$, we get

$$a \cap (a \to b) \cap (a \to (b \to c)) \prec c;$$

applying (qpB_3) twice, we get $a \to (b \to c) \prec (a \to b) \to (a \to c)$, which implies (25);

(26) and (27) follow from $a \leqslant a \cup b$ and $b \leqslant a \cup b$ by 1.1 and (1);

the proof of (28): by (13) $a \cap (a \to c) \prec c$ and $b \cap (b \to c) \prec c$; consequently, $a \cap (b \to c) \cap (a \to c) \prec c$ and $b \cap (b \to c) \cap (a \to c) \prec c$; now applying ($qpB_6$), we get

$$(a \cap (b \to c) \cap (a \to c)) \cup (b \cap (b \to c) \cap (a \to c)) \prec c;$$

hence by distributivity we obtain

$$(a \cup b) \cap (b \to c) \cap (a \to c) \prec c;$$

applying (qpB_3) twice, and (1), we get (28);

(29) and (30) follow from $a \cap b \leqslant a$ and $a \cap b \leqslant b$ by 1.1 and (1);

the proof of (31): by (13) $a \cap (a \to c) \prec c$ and $a \cap (a \to b) \prec b$; hence, by (12), $a \cap (a \to c) \cap (a \to b) \prec b \cap c$; applying ($qpB_3$) twice, we get $a \to b \prec (a \to c) \to (a \to (b \cap c))$, which by (1) implies (31);

(32) follows from (19) and (9); by (9), (qpB_{13}), (qpB_1), III 3.1 (2), (22) and (21) we have $\neg(a \to a) \to b = \neg V \to b = (V \to \sim V) \to b = (V \to \Lambda) \to b = \Lambda \to b = V$, i.e. (33) holds; it follows from (qpB_{12}) by (qpB_3) that $\sim a \prec a \to b$, which by (1) gives (34); (35) results from (qpB_9) and (qpB_8) by (1); (36) follows from (qpB_{11}) and (qpB_{10}) by (1); (37), (38) and (39) follow from (qpB_1) and III 3 ((q_1), (q_2), (3));

the proof of (40): suppose that $a \prec b$, i.e. $a \to b = V$; then by (22) $(a \to b) \to (a \to c) = V \to (a \to c) = a \to c$ and hence by (24), (25) and (1) we get $b \to c \prec a \to (b \to c) \prec a \to c$, which proves the first part of (40); on the other hand, if $a \prec b$, i.e. $a \to b = V$, then by (20) $c \to (a \to b) = c \to V = V$; hence by (25), (23) we get $(c \to a) \to (c \to b) = V$, i.e. $c \to a \prec c \to b$, which proves the second part of (40);

the proofs of (41), (42), (43), based on (qpB_3), (13), (12), (qpB_1), are left to the reader.

It follows from 1.3 that if $(A, V, \Rightarrow, \cup, \cap, \to, \sim, \neg)$ is a quasi-pseudo-Boolean algebra, then

1° axioms II 2 ((p_1), (p_2), (p_4)) for positive implication algebras with respect to the operations $V, \to$ are fulfilled ((24), (25), (20)) but because of 1.1 and (qpB_2) we should expect that the axiom (p_3) is in general not satisfied,

2° axioms IV 1 ((p_5)–(p_{10})), which together with (p_1)–(p_4) characterize relatively pseudo-complemented lattices, are satisfied ((26), (27), (28), (29), (30), (31)),

3° axioms IV 6 ((c), (s)), which together with the above-mentioned axioms (p_1)–(p_{10}) characterize pseudo-Boolean algebras, are satisfied ((19), (33)).

However, the relation $\prec$ defined by (1) in quasi-pseudo-Boolean algebras is a quasi-ordering only, whereas the analogously defined relation in pseudo-Boolean algebras is an ordering.

Since $V \to b = b$ holds in quasi-pseudo-Boolean algebras (see (22)), and this equation together with II 2 (19) enables us to prove that the conditions $a \to b = V$ and $b \to a = V$ imply $a = b$, we should expect that the equation II 2 (19) does not hold in general in quasi-pseudo-Boolean algebras. However, the next theorem shows that an analogous equation, in which the operations $\Rightarrow$, $\to$ both appear, holds in quasi-pseudo-Boolean algebras.

1.4. *In every quasi-pseudo-Boolean algebra*

$$(a \Rightarrow b) \to \big((b \Rightarrow a) \to b\big) = (b \Rightarrow a) \to \big((a \Rightarrow b) \to a\big).$$

Note that by (14)

$$a \to (a \to b) = (a \cap a) \to b = a \to b. \tag{58}$$

By (16), (24), (1), (14), (qpB_4)

$$b \prec (b \to a) \to a \prec (\sim a \to \sim b) \to \big((b \to a) \to a\big)$$
$$= \big((\sim a \to \sim b) \cap (b \to a)\big) \to a = (b \Rightarrow a) \to a.$$

Thus

$$b \prec (b \Rightarrow a) \to a. \tag{59}$$

Applying (40) to (59) twice and later (58), (15), we get

$$(a \Rightarrow b) \to \big((b \Rightarrow a) \to b\big) \prec (a \Rightarrow b) \to \big((b \Rightarrow a) \to ((b \Rightarrow a) \to a)\big)$$
$$= (a \Rightarrow b) \to \big((b \Rightarrow a) \to a\big) = (b \Rightarrow a) \to \big((a \Rightarrow b) \to a\big),$$

which proves that

$$(a \Rightarrow b) \to \big((b \Rightarrow a) \to b\big) \prec (b \Rightarrow a) \to \big((a \Rightarrow b) \to a\big). \tag{60}$$

Observe that by (13)

$$\sim a \cap (\sim a \to \sim b) \prec \sim b.$$

Hence by (12)

$$(a \Rightarrow b) \cap (b \Rightarrow a) \cap (\sim a \to \sim b) \cap \sim a \prec (a \Rightarrow b) \cap (b \Rightarrow a) \cap \sim b.$$

On the other hand, by (qpB_4)

$$(a \Rightarrow b) \cap (b \Rightarrow a) \cap (\sim a \to \sim b) \cap \sim a = (a \Rightarrow b) \cap (b \Rightarrow a) \cap \sim a.$$

Consequently

$$(61) \qquad (a \Rightarrow b) \cap (b \Rightarrow a) \cap \sim a \prec (a \Rightarrow b) \cap (b \Rightarrow a) \cap \sim b.$$

By (qpB_9), (11)

$$(62) \qquad \sim \big((b \Rightarrow a) \to ((a \Rightarrow b) \to a)\big) \prec (b \Rightarrow a) \cap (a \Rightarrow b) \cap \sim a,$$

and by (qpB_8), (11)

$$(63) \qquad (a \Rightarrow b) \cap (b \Rightarrow a) \cap \sim b \prec \sim \big((a \Rightarrow b) \to ((b \Rightarrow a) \to b)\big).$$

It follows from (62), (61) and (63) by the transitivity of $\prec$ that

$$(64) \qquad \sim \big((b \Rightarrow a) \to ((a \Rightarrow b) \to a)\big) \prec \sim \big((a \Rightarrow b) \to ((b \Rightarrow a) \to b)\big).$$

By (60), (64) and 1.1 we have

$$(65) \qquad (a \Rightarrow b) \to \big((b \Rightarrow a) \to b\big) \leqslant (b \Rightarrow a) \to \big((a \Rightarrow b) \to a\big),$$

which on account of symmetry proves 1.4.

2. Equational definability of quasi-pseudo-Boolean algebras. The aim of this section is to prove that the class of all quasi-pseudo-Boolean algebras is equationally definable (3).

2.1. *An abstract algebra* $(A, V, \Rightarrow, \cup, \cap, \to, \sim, \neg)$ *is a quasi-pseudo-Boolean algebra if and only if for all* $a, b, c \in A$ *the following equations hold:*

(qpB_1^*) $\quad a \to (b \to a) = V,$

(qpB_2^*) $\quad \big(a \to (b \to c)\big) \to \big((a \to b) \to (a \to c)\big) = V,$

(3) The first equational system of axioms for quasi-pseudo-Boolean algebras has been given by Brignole and Monteiro [1]. The system (qpB_1^*)-(qpB_{19}^*) contains many more axioms, but it is convenient for the applications to logic.

(qpB$_3^*$) $V \to b = b$,

(qpB$_4^*$) $a \Rightarrow b = (a \to b) \cap (\sim b \to \sim a)$,

(qpB$_5^*$) $(a \Rightarrow b) \to \big((b \Rightarrow a) \to b\big) = (b \Rightarrow a) \to \big((a \Rightarrow b) \to a\big)$,

(qpB$_6^*$) $a \to (a \cup b) = V$,

(qpB$_7^*$) $b \to (a \cup b) = V$,

(qpB$_8^*$) $(a \to c) \to \big((b \to c) \to ((a \cup b) \to c)\big) = V$,

(qpB$_9^*$) $(a \cap b) \to a = V$,

(qpB$_{10}^*$) $(a \cap b) \to b = V$,

(qpB$_{11}^*$) $(a \to b) \to \big((a \to c) \to (a \to (b \cap c))\big) = V$,

(qpB$_{12}^*$) $(a \to \neg b) \to (b \to \neg a) = V$,

(qpB$_{13}^*$) $\neg(a \to a) \to b = V$,

(qpB$_{14}^*$) $\sim a \to (a \to b) = V$,

(qpB$_{15}^*$) $\big(\sim (a \to b) \to (a \cap \sim b)\big) \cap \big((a \cap \sim b) \to \sim (a \to b)\big) = V$,

(qpB$_{16}^*$) $\sim (a \cap b) = \sim a \cup \sim b$,

(qpB$_{17}^*$) $\sim (a \cup b) = \sim a \cap \sim b$,

(qpB$_{18}^*$) $\sim \sim a = a$,

(qpB$_{19}^*$) $(\sim \neg a \to a) \cap (a \to \sim \neg a) = V$.

It follows from (qpB$_1$), (qpB$_4$), 1.3 (24), (25), (22), (26), (27), (28), i29), (30), (31), (32), (33), (34), (35), (36) and 1.4 that in every quasi-pseudo-Boolean algebra the equations (qpB$_1^*$)–(qpB$_{19}^*$) hold. It remains to prove that every algebra in which the above equations hold s a quasi-pseudo-Boolean algebra. Suppose that in an algebra $(A, V, \Rightarrow, \cup, \cap, \to, \sim, \neg)$ the equations (qpB$_1^*$)–(qpB$_{19}^*$) hold.

It follows from (qpB$_3^*$) that

(1) *if* $a \to b = V$, *and* $a = V$, *then* $b = V$.

By (qpB$_3^*$) and (qpB$_1^*$) $a \to V = V \to (a \to V) = V$; thus

(2) $$a \to V = V.$$

By (qpB$_2^*$) $\big(a \to ((a \to a) \to a)\big) \to \big((a \to (a \to a)) \to (a \to a)\big) = V$. Applying (1) twice, we get by (qpB$_1^*$)

(3) $$a \to a = V.$$

Suppose that $a \to b = V$ and $b \to c = V$. Then by (qpB_3^*), (2) and (qpB_2^*) $a \to c = V \to (V \to (a \to c)) = (a \to V) \to (V \to (a \to c)) = (a \to (b \to c)) \to ((a \to b) \to (a \to c)) = V$. Thus

(4) $a \to b = V$ *and* $b \to c = V$ *imply* $a \to c = V$.

It follows from (3) and (4) that the relation $\prec$ defined by the formula

$$a \prec b \quad \text{if and only if} \quad a \to b = V$$

is the quasi-ordering relation, i.e. (qpB_2) is fulfilled. Note that

(5) if $V \prec b$, then $b = V$,

because if $V \prec b$, then $V \to b = V$ and by (qpB_3^*) $V \to b = b$.

Observe that by (qpB_1^*) and (qpB_2^*)

(6) $$a \prec b \to a,$$

(7) $$a \to (b \to c) \prec (a \to b) \to (a \to c).$$

We shall prove that

(8) *if* $a \prec b$, *then* $b \to c \prec a \to c$ *and* $c \to a \prec c \to b$.

Suppose that $a \prec b$, i.e. $a \to b = V$. Then by (6), (7) and (qpB_3^*)

$$b \to c \prec a \to (b \to c) \prec (a \to b) \to (a \to c) \prec V \to (a \to c) = a \to c.$$

On the other hand, by (qpB_2^*), (2) and (qpB_3^*)

$$V = (c \to (a \to b)) \to ((c \to a) \to (c \to b)) = (c \to V) \to ((c \to a) \to (c \to b)) = V \to ((c \to a) \to (c \to b)) = (c \to a) \to (c \to b),$$

hence $c \to a \prec c \to b$.

We shall prove that

(9) $$a \to (b \to (a \cap b)) = V.$$

By (qpB_{11}^*), (3) and (qpB_3^*)

$$b \to a \prec (b \to b) \to (b \to (a \cap b)) = V \to (b \to (a \cap b)) = b \to (a \cap b).$$

Hence, by (8) and (qpB_1^*),

$$V = a \to (b \to a) \prec a \to (b \to (a \cap b)),$$

which on account of (5) implies (9).

It is easy to show that

(10) *if* $a \prec b \to c$, *then* $b \prec a \to c$.

Indeed, if $a \prec b \to c$, then $a \to (b \to c) = V$. Hence, by (7) and (5), $(a \to b) \to (a \to c) = V$, i.e. $a \to b \prec a \to c$ and consequently by (6) $b \prec a \to b \prec a \to c$, which proves (10).

By (7), (3) and (qpB$_3^*$)

$$a \to (a \to c) \prec (a \to a) \to (a \to c) = V \to (a \to c) = a \to c$$

i.e.

(11) $$a \to (a \to c) \prec a \to c\,.$$

Now we shall prove that

(12) $$a \cap x \prec b \quad \textit{if and only if} \quad x \prec a \to b\,,$$

i.e. that the axiom (qpB$_3$) holds. Suppose that $a \cap x \prec b$. Applying (8) twice, we get $a \to (x \to (a \cap x)) \prec a \to (x \to b)$. Hence, by (9) and (5), we get $a \to (x \to b) = V$, i.e. $a \prec x \to b$, which by (10) implies $x \prec a \to b$. Suppose that $x \prec a \to b$. Since by (qpB$_{10}^*$) $a \cap x \prec x$, we get $a \cap x \prec a \to b$. Hence, by (10), $a \prec (a \cap x) \to b$ and by (qpB$_9^*$) $a \cap x \prec a$, thus $a \cap x \prec (a \cap x) \to b$, i.e. $(a \cap x) \to ((a \cap x) \to b) = V$. Consequently, by (11) and (5), we get $(a \cap x) \to b = V$, i.e. $a \cap x \prec b$, which proves (12).

By (qpB$_3^*$), (3) and (qpB$_{11}^*$)

$$V \cap V = V \to (V \to (V \to (V \cap V)))$$
$$= (V \to V) \to ((V \to V) \to (V \to (V \cap V))) = V\,.$$

Thus

(13) $$V \cap V = V\,.$$

Hence, by (qpB$_4^*$) and (3), follows

(14) $$a \Rightarrow a = V\,.$$

We shall prove that

(15) $$a = b \quad \textit{if and only if} \quad a \Rightarrow b = V \textit{ and } b \Rightarrow a = V\,.$$

If $a \Rightarrow b = V$ and $b \Rightarrow a = V$, then by (qpB$_3^*$) and (qpB$_5^*$)

$$a = V \to (V \to a) = (b \Rightarrow a) \to ((a \Rightarrow b) \to a)$$
$$= (a \Rightarrow b) \to ((b \Rightarrow a) \to b) = V \to (V \to b) = b\,.$$

Conversely, if $a = b$, then by (14) $a \Rightarrow b = b \Rightarrow a = a \Rightarrow a = V$.

Note that

(16) $$\textit{if} \quad a \cap b = V, \quad \textit{then} \quad a = V \textit{ and } b = V\,.$$

In fact, by (qpB_9^*) and (qpB_{10}^*)

(17) $$a \cap b \prec a,$$

(18) $$a \cap b \prec b.$$

Hence, by (5), we get (16).

Now we shall prove that

(19) *if* $a \Rightarrow b = V$ *and* $b \Rightarrow c = V$, *then* $a \Rightarrow c = V$.

Suppose that $a \Rightarrow b = V$ and $b \Rightarrow c = V$. Now by (qpB_4^*) and (16) $a \to b = V$, $\sim b \to \sim a = V$, $b \to c = V$ and $\sim c \to \sim b = V$. This implies by (4) that $a \to c = V$ and $\sim c \to \sim a = V$. Hence, by (13) and (qpB_4^*), we get $a \Rightarrow c = V$.

It follows from (14), (15) and (19) that the relation $\leqslant$ defined by the formula

$$a \leqslant b \quad \textit{if and only if} \quad a \Rightarrow b = V,$$

is an ordering in any algebra satisfying the equations (qpB_1^*)-(qpB_{19}^*). Moreover,

(20) $a \leqslant b$ *if and only if* $a \prec b$ *and* $\sim b \prec \sim a$.

This follows easily by (qpB_4^*), (13) and (16).

We shall show that

(21) $$a \cup b = \text{l.u.b.}\, \{a, b\},$$

(22) $$a \cap b = \text{g.l.b.}\, \{a, b\}.$$

By (qpB_6^*) and (qpB_7^*)

(23) $$a \prec a \cup b \quad \text{and} \quad b \prec a \cup b.$$

On the other hand, by (qpB_{17}^*), (17) and (18)

(24) $\sim (a \cup b) = \sim a \cap \sim b \prec \sim a$ *and*

$$\sim (a \cup b) = \sim a \cap \sim b \prec \sim b.$$

It follows from (23), (24) and (20) that

(25) $$a \leqslant a \cup b \quad \text{and} \quad b \leqslant a \cup b.$$

Suppose that $a \leqslant c$ and $b \leqslant c$. Then $a \to c = V$, $\sim c \to \sim a = V$, $b \to c = V$ and $\sim c \to \sim b = V$. By (qpB_8^*) and (1) we get $(a \cup b) \to c = V$, i.e. $a \cup b \prec c$ and by (qpB_{11}^*), (qpB_{17}^*) and (1) $V = \sim c \to (\sim a \cap \sim b) = \sim c \to \sim (a \cup b)$, i.e. $\sim c \prec \sim (a \cup b)$, which proves that

$a \cup b \leqslant c$ and that (21) holds. The analogous easy proof of (22) is left to the reader.

It follows from (21) and (22) that $\leqslant$ is the lattice ordering and that every algebra fulfilling (qpB_1^*)-(qpB_{19}^*) is a lattice. Thus (qpB_5) holds by III 1.1 and III 1.2.

We shall show that V is the unit element and that $\sim V$ is the zero element. By (2), $a \prec V$ for every element a in the algebra. By (qpB_{14}^*) and (qpB_3^*) $\sim V \prec V \to \sim a = \sim a$. Thus $\sim V \prec \sim a$. Hence

(26) $a \leqslant V$ for every element a in the algebra under consideration.

On the other hand, by (qpB_{14}^*), $\sim V \prec V \to a = a$ and by (26), (qpB_{18}^*), we have $\sim a \prec V = \sim \sim V$. Hence

(27) $\sim V \leqslant a$ for every element a in the algebra under consideration.

Thus V is the unit element and $\sim V$ is the zero element.

The axioms (qpB_6) and (qpB_7) follow from (qpB_8^*) and (qpB_{11}^*) by (1). The axioms (qpB_8) and (qpB_9) follow from (qpB_{15}^*) by (16). The axioms (qpB_{10}) and (qpB_{11}) follow from (qpB_{19}^*) by (16). The axiom (qpB_{12}) is equivalent by (12) to $\sim a \prec a \to b$, which holds on account of (qpB_{14}^*).

Observe that by (27) $\sim V \leqslant \neg V$. By (qpB_{13}^*) and (3)

(28) $$\neg V \prec \sim V.$$

On account of (qpB_{18}^*), since (qpB_{10}) holds, we get $\sim \sim V = V \prec \sim \neg V$. Hence and from (28) we get $\neg V \leqslant \sim V$. Consequently,

(29) $$\neg V = \sim V.$$

By (qpB_{12}^*) $a \to \neg b \prec b \to \neg a$. Since ($qpB_8$), ($qpB_9$), ($qpB_{10}$) and ($qpB_{11}$) hold, we get $\sim (b \to \neg a) \prec b \cap \sim \neg a \prec b \cap a = a \cap b \prec a \cap \cap \sim \neg b \prec \sim (a \to \neg b)$. Thus $a \to \neg b \leqslant b \to \neg a$. Hence by symmetry $b \to \neg a \leqslant a \to \neg b$ and consequently, $a \to \neg b = b \to \neg a$. From this equation it follows by (qpB_3^*) and (29) that

$$\neg a = V \to \neg a = a \to \neg V = a \to \sim V,$$

which proves that the axiom (qpB_{13}) holds.

It remains to show that (qpB_1) is also satisfied. Since $(A, V, \cup, \cap)$ is a lattice with unit element V, it is sufficient to show, because of (qpB_{17}^*)

and (qpB^*_{18}), that the lattice is distributive, i.e. that $a \cap (b \cup c) = (a \cap b) \cup (a \cap c)$ (see III 2). In every lattice

$$(30) \qquad (a \cap b) \cup (a \cap c) \leqslant a \cap (b \cup c).$$

Observe that $a \cap b \prec (a \cap b) \cup (a \cap c)$ and $a \cap c \prec (a \cap b) \cup (a \cap c)$. Hence, by (12), $b \prec a \to (a \cap b) \cup (a \cap c)$ and $c \prec a \to (a \cap b) \cup (a \cap c)$. Consequently, since ($qpB_6$) holds, $b \cup c \prec a \to (a \cap b) \cup (a \cap c)$. Hence, by (12),

$$(31) \qquad a \cap (b \cup c) \prec (a \cap b) \cup (a \cap c).$$

Hence, by III 1 (l_3),

$$(a \cup b) \cap (a \cup c) \prec \big((a \cup b) \cap a\big) \cup \big((a \cup b) \cap c\big) \prec a \cup \big((a \cap c) \cup (b \cap c)\big)$$
$$= \big(a \cup (a \cap c)\big) \cup (b \cap c) = a \cup (b \cap c).$$

Thus

$$(32) \qquad (a \cup b) \cap (a \cup c) \prec a \cup (b \cap c).$$

Consequently,

$$(\sim a \cup \sim b) \cap (\sim a \cup \sim c) \prec \sim a \cup (\sim b \cap \sim c).$$

Hence, by (qpB^*_{16}), (qpB^*_{17}) and (qpB^*_{18}), we get

$$\sim \big((a \cap b) \cup (a \cap c)\big) \prec \sim \big(a \cap (b \cup c)\big),$$

which together with (31) implies that

$$a \cap (b \cup c) \leqslant (a \cap b) \cup (a \cap c).$$

Consequently, by (30), $a \cap (b \cup c) = (a \cap b) \cup (a \cap c)$, which proves the distributivity of the lattice and thus Theorem 2.1.

3. Examples of quasi-pseudo-Boolean algebras. In this section we present some examples of quasi-pseudo-Boolean algebras.

Let $B_0 = \{\Lambda, V\}$ and let $\Rightarrow, \cup, \cap, \to, \sim, \neg$ be operations on B_0 defined by means of the equations

$$\Lambda \Rightarrow \Lambda = \Lambda \Rightarrow V = V \Rightarrow V = V, \qquad V \Rightarrow \Lambda = \Lambda,$$

$$\Lambda \cup \Lambda = \Lambda, \qquad \Lambda \cup V = V \cup \Lambda = V \cup V = V,$$

$$\Lambda \cap \Lambda = \Lambda \cap V = V \cap \Lambda = \Lambda, \qquad V \cap V = V,$$

$$a \to b = a \Rightarrow b \quad \text{for all } a, b \in B_0,$$

$$\sim \Lambda = V, \qquad \sim V = \Lambda,$$

$$\neg a = \sim a, \quad \text{for every} \quad a \in B_0.$$

Then $\mathfrak{B}_0 = (B_0, \vee, \Rightarrow, \cup, \cap, \Rightarrow, \sim, \neg)$ is a two-element quasi-pseudo-Boolean algebra. It is easy to show that every two-element quasi-pseudo-Boolean algebra is isomorphic to $\mathfrak{B}_0$.

Let $C_0 = \{\wedge, a, \vee\}$ and let $\Rightarrow, \cup, \cap, \rightarrow, \sim, \neg$ be the operations on C_0 defined by means of the following tables:

$\Rightarrow$	$\wedge$	a	$\vee$
$\wedge$	$\vee$	$\vee$	$\vee$
a	a	$\vee$	$\vee$
$\vee$	$\wedge$	a	$\vee$

$\cup$	$\wedge$	a	$\vee$
$\wedge$	$\wedge$	a	$\vee$
a	a	a	$\vee$
$\vee$	$\vee$	$\vee$	$\vee$

$\cap$	$\wedge$	a	$\vee$
$\wedge$	$\wedge$	$\wedge$	$\wedge$
a	$\wedge$	a	a
$\vee$	$\wedge$	a	$\vee$

$\rightarrow$	$\wedge$	a	$\vee$
$\wedge$	$\vee$	$\vee$	$\vee$
a	$\vee$	$\vee$	$\vee$
$\vee$	$\wedge$	a	$\vee$

x	$\sim x$
$\wedge$	$\vee$
a	a
$\vee$	$\wedge$

x	$\neg x$
$\wedge$	$\vee$
a	$\vee$
$\vee$	$\wedge$

Then the algebra $\mathfrak{C}_0 = (C_0, \vee, \Rightarrow, \cup, \cap, \rightarrow, \sim, \neg)$ is a three-element quasi-pseudo-Boolean algebra. It is easy to prove that every three-element quasi-pseudo-Boolean algebra is isomorphic to $\mathfrak{C}_0$.

Now we give a general method for constructing *quasi-pseudo-Boolean algebras of sets* (4).

Let X_1 be a topological space with an interior operation $\mathbf{I}_1$ and let $G(X_1)$ be a pseudo-field of some open subsets of X_1 (see IV 6) constituting a base of X_1. Then $(G(X_1), X_1, \Rightarrow_1, \cup, \neg_1)$ is a pseudo-Boolean algebra of some open subsets of X_1, where $\cup, \cap$ are the set-theoretical operations of union and intersection and $\Rightarrow_1, \neg_1$ are defined by means of the equations

(1) $$Y \Rightarrow_1 Z = \mathbf{I}_1((X_1 - Y) \cup Z),$$

(2) $$\neg_1 Y = Y \Rightarrow_1 O = \mathbf{I}_1(X_1 - Y),$$

for any $Y, Z \in G(X_1)$.

(4) Rasiowa [6], [7], Białynicki-Birula and Rasiowa [2].

Let X_2 be an arbitrary set such that $\operatorname{card} X_1 = \operatorname{card} X_2$ and let f be a one-one mapping from X_1 onto X_2 such that

(3) $$f(x) = x \quad \text{for every } x \in X_1 \cap X_2.$$

We assume in addition that

(4) $$X_1 \cap X_2 \subset \bigcap_{Y \in G(X_1)} \big(Y \cup f(\mathbf{I}_1(X_1 - Y))\big),$$

i.e. that

(5) $$X_1 \cap X_2 \subset \bigcap_{Y \in G(X_1)} (Y \cup f(\neg_1 Y)).$$

Clearly, if X_1 and X_2 are two disjoint sets, then condition (4) is satisfied for every one-one mapping f from X_1 onto X_2.

Let

(6) $$X = X_1 \cup f(X_1) = X_1 \cup X_2.$$

Then the mapping f determines the following mapping $g\colon X \to X$

(7) $$g(x) = \begin{cases} f(x) & \text{if } x \in X_1, \\ f^{-1}(x) & \text{if } x \in X_2. \end{cases}$$

It is easy to see that g is an involution on X, i.e. that

(8) $$g(g(x)) = x \quad \text{for every } x \in X.$$

Clearly, $g(X_1) = X_2$ and $g(X_2) = X_1$.

Let $B(X)$ be the class of all subsets Y of X satisfying the following three conditions:

(b_1) $Y \cap X_1 \in G(X_1)$,

(b_2) $Y \cap X_2 = X_2 - g(Z)$ for some $Z \in G(X_1)$,

(b_3) $Y \cap X_1 \subset g(Y) \cap X_1$.

Then the following theorem holds.

3.1. *$B(X)$ is a set lattice of subsets of X containing the empty set O and the whole space X.*

The proof by an easy verification is left to the reader.

It follows from 3.1 that X can be considered as a topological space with the base $B(X)$.

For any $Y \in B(X)$ let

$$(9) \qquad \sim Y = X - g(Y).$$

We shall prove that

3.2. *The class $B(X)$ of open subsets of X is a quasi-field of subsets of X.*

Because of 3.1, in order to prove 3.2 it is sufficient to show that for any $Y \in B(X)$ the set $\sim Y \in B(X)$ (see III 3), i.e. to show that if Y satisfies conditions (b_1), (b_2) and (b_3), then $\sim Y$ also satisfies these conditions. Since g is a one-one mapping from X onto X, we get by (9) and (b_2)

$$\sim Y \cap X_1 = (X - g(Y)) \cap X_1 = X_1 - g(Y \cap X_2) = X_1 - g(X_2 - g(Z))$$

for some $Z \in G(X_1)$. Hence $\sim Y \cap X_1 = Z \in G(X_1)$, i.e. $\sim Y$ satisfies (b_1).

$$\sim Y \cap X_2 = (X - g(Y)) \cap X_2 = X_2 - g(Y \cap X_1), \quad \text{where} \quad Y \cap X_1 \in G(X_1);$$

thus $\sim Y$ satisfies (b_2).

Because of (b_3), we have

$$X_1 - g(Y) \cap X_1 \subset X_1 - Y \cap X_1.$$

Since, on the other hand, we have

$$\sim Y \cap X_1 = (X - g(Y)) \cap X_1 = X_1 - g(Y) \cap X_1$$

and

$$g(\sim Y) \cap X_1 = g(X - g(Y)) \cap X_1 = (X - Y) \cap X_1 = X_1 - Y \cap X_1,$$

the set $\sim Y$ satisfies the condition (b_3). Thus 3.2 is proved.

Let us put, for any $Y, Z \in B(X)$,

$$(10) \quad Y \to Z = (Y \cap X_1 \Rightarrow_1 Z \cap X_1) \cup \big((X_2 - g(Y \cap X_1)) \cup Z \cap X_2\big).$$

We shall prove that

3.3. *For any sets $Y, Z \in B(X)$, we have $Y \to Z \in B(X)$.*

Suppose $Y, Z \in B(X)$, i.e. Y and Z satisfy conditions (b_1), (b_2), (b_3). From (3), (4), (7) and (b_1) we get

$$X_1 \cap X_2 = g(X_1 \cap X_2) \subset g(Y \cap X_1) \cup \mathbf{I}_1(X_1 - Y \cap X_1).$$

Hence we obtain

$$(11) \qquad X_1 \cap X_2 - g(Y \cap X_1) \subset \mathbf{I}_1(X_1 - Y \cap X_1).$$

On the other hand, since $Z \cap X_1 \in G(X_1)$, the set $Z \cap X_1$ is an open subset of X_1 and therefore $Z \cap X_1 = \mathbf{I}_1(Z \cap X_1)$. Hence, using the fact that in any topological space the union of the interiors of two sets is contained in the interior of the union of those sets, we obtain

$$\mathbf{I}_1(X_1 - Y \cap X_1) \cup Z \cap X_1 = \mathbf{I}_1(X_1 - Y \cap X_1) \cup \mathbf{I}_1(Z \cap X_1)$$
$$\subset \mathbf{I}_1\big((X_1 - Y \cap X_1) \cup Z \cap X_1\big) = Y \cap X_1 \Rightarrow_1 Z \cap X_1 .$$

Thus

$$(12) \qquad \mathbf{I}_1(X_1 - Y \cap X_1) \cup Z \cap X_1 \cap X_2 \subset Y \cap X_1 \Rightarrow_1 Z \cap X_1 .$$

It follows from (11) and (12) that

$$(13) \qquad (X_1 \cap X_2 - g(Y \cap X_1)) \cup Z \cup X_1 \cap X_2 \subset Y \cap X_1 \Rightarrow_1 Z \cap X_1 .$$

Now observe that by (10)

$$(Y \to Z) \cap X_1$$
$$= (Y \cap X_1 \Rightarrow_1 Z \cap X_1) \cap X_1 \cup \Big(\big((X_2 - g(Y \cap X_1)) \cap X_1\big) \cup Z \cap X_1 \cap X_2\Big)$$
$$= (Y \cap X_1 \Rightarrow_1 Z \cap X_1) \cup \big((X_1 \cap X_2 - g(Y \cap X_1)) \cup Z \cap X_1 \cap X_2\big) .$$

Since by (13) the second part of the last union is contained in the first one, we infer that

$$(14) \qquad (Y \to Z) \cap X_1 = (Y \cap X_1 \Rightarrow_1 Z \cap X_1) \in G(X_1) .$$

Thus the set $Y \to Z$ satisfies the condition (b_1).

Since by (b_3) $Y \cap X_1 \subset g(Y) \cap X_1$, we infer that

$$g(Y \cap X_1) \subset g\big(g(Y) \cap X_1\big) = Y \cap X_2 .$$

Hence

$$(X_1 - Y \cap X_1) \cap X_2 = (X_1 - Y) \cap X_2 = X_1 \cap X_2 - Y \cap X_2$$
$$\subset X_1 \cap X_2 - g(Y \cap X_1) \subset X_2 - g(Y \cap X_1),$$

i.e.

$$(15) \qquad (X_1 - Y \cap X_1) \cap X_2 \subset X_2 - g(Y \cap X_1) .$$

On the other hand,

$$\mathbf{I}_1\big((X_1 - Y \cap X_1) \cup Z \cap X_1\big) \cap X_2 \subset \big((X_1 - Y \cap X_1) \cup Z \cap X_1\big) \cap X_2$$
$$= \big((X_1 - Y \cap X_1) \cap X_2\big) \cup Z \cap X_1 \cap X_2 .$$

Hence, by (15),

$$(16) \quad \mathbf{I}_1\big((X_1 - Y \cap X_1) \cup Z \cap X_1\big) \cap X_2 \subset (X_2 - g(Y \cap X_1)) \cup Z \cap X_2 .$$

It follows from (10) and (1) that

$$(17)\qquad (Y\to Z)\cap X_2 = \big(\mathbf{I}_1((X_1-Y\cap X_1)\cup Z\cap X_1)\cap X_2\big)\cup \\ \cup\big((X_2-g(Y\cap X_1))\cup Z\cap X_2\big).$$

Since by (16) the first part of the union on the right side of (17) is contained in the second one, we infer that

$$(18)\qquad (Y\to Z)\cap X_2 = (X_2-g(Y\cap X_1))\cup Z\cap X_2.$$

By (b_2), $Z\cap X_2 = X_2-g(Z^*)$, where $Z^*\in G(X_1)$. Hence, by (18),

$$(Y\to Z)\cap X_2 = (X_2-g(Y\cap X_1))\cup(X_2-g(Z^*)) \\ = X_2-g((Y\cap X_1)\cap Z^*).$$

Since $Z^*\in G(X_1)$ and by (b_1), $Y\cap X_1\in G(X_1)$, we infer that $(Y\cap X_1)\cap \cap Z^*\in G(X_1)$. Thus $(Y\to Z)\cap X_2 = X_2-g\big((Y\cap X_1)\cap Z^*\big)$, where $(Y\cap \cap X_1)\cap Z^*$ belongs to $G(X_1)$. This proves that the set $Y\to Z$ satisfies the condition (b_2).

Observe that by (14) and (1)

$$(Y\to Z)\cap X_1 = \mathbf{I}_1\big((X_1-Y\cap X_1)\cup Z\cap X_1\big)\subset(X_1-Y\cap X_1)\cup Z\cap X_1$$

i.e.

$$(19)\qquad (Y\to Z)\cap X_1\subset(X_1-Y\cap X_1)\cup Z\cap X_1.$$

On the other hand, by (18)

$$g(Y\to Z)\cap X_1 = g\big((Y\to Z)\cap X_2\big) = g\big((X_2-g(Y\cap X_1))\cup Z\cap X_2\big) \\ = (X_1-Y\cap X_1)\cup\big(g(Z)\cap X_1\big),$$

i.e.

$$(20)\qquad g(Y\to Z)\cap X_1 = (X_1-Y\cap X_1)\cup\big(g(Z)\cap X_1\big).$$

Since by (b_3), $Z\cap X_1\subset g(Z)\cap X_1$, we infer by (19) and (20) that

$$(Y\to Z)\cap X_1\subset g(Y\to Z)\cap X_1,$$

which proves that the set $Y\to Z$ satisfies the condition (b_3). Consequently, the set $Y\to Z$ belongs to $B(X)$ if Y and Z belong to $B(X)$.

We shall prove that

3.4. *The algebra* $\mathfrak{B}(X) = \big(B(X), X, \Rightarrow, \cup, \cap, \to, \sim, \neg\big)$ *where* $\cup$, $\cap$ *are set-theoretical operations of union and intersection, the operations*

$\sim$ and $\rightarrow$ are defined by means of equations (9) and (10) respectively and the operations $\Rightarrow, \neg$ are defined by the following equations

(21) $\quad Y \Rightarrow Z = (Y \rightarrow Z) \cap (\sim Z \rightarrow \sim Y)$ *for any* Y, Z *in* $B(X)$,

(22) $\quad \neg Y = Y \rightarrow O$ *for every* Y *in* $B(X)$,

is a quasi-pseudo-Boolean algebra of some open subsets of the topological space X. It will be called a quasi-pseudo-Boolean algebra of sets connected with the pseudo-field of sets $G(X_1)$.

It follows from 3.2 that $(B(X), X, \cup, \cap, \sim)$ is a quasi-Boolean algebra. Thus the axiom 1 (qpB_1) is satisfied. We shall show that

(23) $$X_1 \subset X - \bigcup_{Y \in B(X)} Y \cap \sim Y.$$

Suppose that $x \in X_1$ and $x \in Y \in B(X)$. Then $x \in Y \cap X_1$. Hence by (b_3), $x \in g(Y) \cap X_1$. Consequently $x \notin X - g(Y) = \sim Y$. This proves that for any x in X_1 and for every $Y \in B(X)$, $x \notin Y \cap \sim Y$, i.e. $x \notin \bigcup_{Y \in B(X)} Y \cap \cap \sim Y$. Thus (23) holds.

Let us put for any Y, Z in $B(X)$

(24) $$Y \prec Z \quad \text{if and only if} \quad Y \rightarrow Z = X.$$

We shall prove that

(25) $$Y \prec Z \quad \text{if and only if} \quad Y \cap X_1 \subset Z \cap X_1.$$

If $Y \prec Z$, then by (24), $Y \rightarrow Z = X$. Hence by (14) and (1)

$$X_1 = X \cap X_1 = (Y \rightarrow Z) \cap X_1 = Y \cap X_1 \Rightarrow_1 Z \cap X_1$$
$$= \mathbf{I}_1((X_1 - Y \cap X_1) \cup Z \cap X_1).$$

Thus $(X_1 - Y \cap X_1) \cup Z \cap X_1 = X_1$, i.e. $Y \cap X_1 \subset Z \cap X_1$. Suppose that $Y \prec Z$ does not hold. Then by (24) $Y \rightarrow Z \neq X$. Since by (6)

$$Y \rightarrow Z = (Y \rightarrow Z) \cap (X_1 \cup X_2) = (Y \rightarrow Z) \cap X_1 \cup (Y \rightarrow Z) \cap X_2,$$

we infer that either $(Y \rightarrow Z) \cap X_1 \neq X_1$ or $(Y \rightarrow Z) \cap X_2 \neq X_2$. The first inequality implies by (14) that $Y \cap X_1 \Rightarrow_1 Z \cap X_1 \neq X_1$, i.e. that $Y \cap X_1 \not\subset Z \cap X_1$. The second inequality implies by (18) that $(X_2 - -g(Y \cap X_1)) \cup X_2 \cap Z \neq X_2$. Hence $g(Y \cap X_1) \not\subset Z \cap X_2$. Thus $Y \cap X_1 \not\subset g(Z \cap X_2) = g(Z) \cap X_1$. Since by ($b_3$) $Z \cap X_1 \subset g(Z) \cap X_1$, we get $Y \cap X_1 \not\subset Z \cap X_1$. Thus in both cases under consideration $Y \cap X_1 \not\subset Z \cap \cap X_1$. This proves that if $Y \prec Z$ does not hold, then $Y \cap X_1 \not\subset Z \cap X_1$.

It follows from (25) that the relation $\prec$ is a quasi-ordering on $B(X)$, i.e. that axiom 1 (qpB_2) is satisfied.

In order to prove that axiom 1 (qpB_3) is satisfied let us observe that by (25) the condition $Z_1 \cap Y \prec Z_2$ for any Y, Z_1, Z_2 in $B(X)$ is equivalent to the condition $(Z_1 \cap Y) \cap X_1 \subset Z_2 \cap X_1$, i.e. to the condition

$$(26) \qquad (Z_1 \cap X_1) \cap (Y \cap X_1) \subset Z_2 \cap X_1.$$

By (b_1) the sets $Z_1 \cap X_1$, $Y \cap X_1$, $Z_2 \cap X_1$ belong to the pseudo-field $G(X_1)$. Since the relation of inclusion is the lattice ordering in $G(X_1)$, we infer by IV 1 (r) and by the definition of pseudo-Boolean algebras that condition (26) is equivalent to the following one $(Y \cap X_1) \subset Z_1 \cap \cap X_1 \Rightarrow_1 Z_2 \cap X_1$. By (14) the last condition is satisfied if and only if $Y \cap X_1 \subset (Z_1 \rightarrow Z_2) \cap X_1$, i.e. by (25) if and only if $Y \prec Z_1 \rightarrow Z_2$. We have proved that $Z_1 \cap Y \prec Z_2$ if and only if $Y \prec Z_1 \rightarrow Z_2$. Thus the axiom ($qpB_3$) holds.

Axiom 1 (qpB_4) is satisfied by definition (21).

Now we shall show that

$$(27) \qquad \sim Z \prec \sim Y \quad \text{if and only if} \quad Y \cap X_2 \subset Z \cap X_2.$$

If follows from (25) that $\sim Z \prec \sim Y$ if and only if $\sim Z \cap X_1 \subset \sim Y \cap \cap X_1$, i.e. by (9) if and only if $(X - g(Z)) \cap X_1 \subset (X - g(Y)) \cap X_1$. The last inclusion holds if and only if $g(Y) \cap X_1 \subset g(Z) \cap X_1$, which is equivalent to $g(g(Y) \cap X_1) \subset g(g(Z) \cap X_1)$ and hence to $Y \cap X_2 \subset Z \cap \cap X_2$.

If follows from (25), (27) and (6) that for any Y, Z in $B(X)$

$$(28) \qquad Y \subset Z \quad \text{if and only if} \quad Y \prec Z \text{ and } \sim Z \prec \sim Y.$$

In order to prove that axiom 1 (qpB_5) holds observe that by (21) $Y \Rightarrow Z = X$ if and only if $Y \rightarrow Z = X$ and $\sim Z \rightarrow \sim Y = X$. Hence by (24), $Y \Rightarrow Z = X$ if and only if $Y \prec Z$ and $\sim Z \prec \sim Y$. Applying (28) we deduce that $Y \Rightarrow Z = X$ if and only if $Y \subset Z$, i.e. if and only if $Y \cap Z = Y$. Thus (qpB_5) holds.

It follows from (25) that axioms 1 ((qpB_6) and (qpB_7)) are satisfied.

Now we shall prove that for any Y, Z in $B(X)$

$$(29) \qquad (\sim Y \cup Z) \cap X_2 = (Y \rightarrow Z) \cap X_2.$$

Indeed, by (9), (18) and $X_2 = g(X_1)$ we get

$$(\sim Y \cup Z) \cap X_2 = \big((X - g(Y)) \cup Z\big) \cap X_2 = (X_2 - g(Y \cap X_1)) \cup Z \cap X_2$$
$$= (Y \to Z) \cap X_2 .$$

Hence, by (27) and III 3 $\big((q_1), (q_2)\big)$, we obtain

$$Y \cap \sim Z = \sim(\sim Y \cup Z) \prec \sim(Y \to Z)$$

and

$$\sim(Y \to Z) \prec \sim(\sim Y \cup Z) = Y \cap \sim Z,$$

which proves that axioms 1 $\big((qpB_8), (qpB_9)\big)$ are satisfied.

Now we shall prove that for every Y in $B(X)$

$$(30) \qquad \sim Y \cap X_2 = \neg Y \cap X_2 .$$

In fact, by (9), (22) and (18) we get

$$\sim Y \cap X_2 = \big(X - g(Y)\big) \cap X_2 = X_2 - g(Y \cap X_1) = (Y \to O) \cap X_2$$
$$= \neg Y \cap X_2 .$$

It follows from (30) and III 3 (q_1) that

$$Y = \sim \sim Y \prec \sim \neg Y \quad \text{and} \quad \sim \neg Y \prec \sim \sim Y = Y,$$

i.e. that axioms 1 $\big((qpB_{10}), (qpB_{11})\big)$ are satisfied.

It follows from (23) that for every Y in $B(X)$

$$X_1 \subset X - (Y \cap \sim Y).$$

Hence $(Y \cap \sim Y) \cap X_1 = O$. Thus for every Z in $B(X)$

$$(Y \cap \sim Y) \cap X_1 \subset Z \cap X_1 ,$$

which proves by (25) that

$$Y \cap \sim Y \prec Z,$$

i.e. that axiom 1 (qpB_{12}) holds. Axiom 1 (qpB_{13}) is satisfied by definition (22) and the equation $O = \sim X$.

Thus $B(X)$ is a quasi-pseudo-Boolean algebra. Since, by the definition of the topology in X, all sets in $B(X)$ are open, we have constructed a quasi-pseudo-Boolean algebra of some open subsets of the topological space X.

It follows from the equational definability of the class of quasi-pseudo-Boolean algebras that every subalgebra of $B(X)$ is a quasi-pseudo-Boolean algebra and we shall also say that it is *connected* with $G(X_1)$. We shall see in Section 5 that quasi-pseudo-Boolean algebras of sets connected with pseudo-fields of sets are typical examples of quasi-pseudo-Boolean algebras.

Theorem 3.4 gives a general method of constructing quasi-pseudo-Boolean algebras. We shall demonstrate this method on the following example. Let X_1 be the set of all odd positive integers. We shall treat X_1 as a topological space with the interior operation $\mathbf{I}_1$ such that for every $Y_1 \subset X_1$, $\mathbf{I}_1 Y_1 = Y_1$. Let $G(X_1)$ be the class of all subsets of X_1, i.e. of all open subsets of X_1. Let us take as X_2 the set of all even positive integers. Let $f(2k-1) = 2k$ for every $k = 1, 2, \ldots$. Then

$$g(2k-1) = 2k \quad \text{and} \quad g(2k) = 2k-1 \qquad \text{for every } k = 1, 2, \ldots .$$

It is easy to prove that the class $B(X)$ is then the class of all subsets Y of $X = X_1 \cup X_2$ satisfying the condition:

$$\text{if} \quad 2k-1 \in Y, \quad \text{then} \quad 2k \in Y.$$

According to Theorem 3.4 the algebra $(B(X), X, \Rightarrow, \cup, \cap, \rightarrow, \sim, \urcorner)$ where $\cup$, $\cap$ are set-theoretical operations of union and intersection and where the operations $\sim$, $\rightarrow$, $\Rightarrow$, $\urcorner$ are defined by means of equations (9), (10), (21), (22) is a quasi-pseudo-Boolean algebra of sets.

4. Filters in quasi-pseudo-Boolean algebras [5]. Every quasi-pseudo-Boolean algebra $(A, V, \Rightarrow, \cup, \cap, \rightarrow, \sim, \urcorner)$ is a distributive lattice under the operations $\cup$, $\cap$ with unit element V and zero element $\sim V$. Consequently, it is possible to consider in any quasi-pseudo-Boolean algebra the notion of a filter. Among all the filters in any quasi-pseudo-Boolean algebra we shall distinguish two kinds of filters which play an important role.

A non-empty subset ∇ of the set A of all elements of a quasi-pseudo-Boolean algebra will be said to be a *special filter of the first kind* (s.f.f.k.) provided the following conditions are satisfied:

(s.f.$_1$) *if* $a \in \nabla$ *and* $b \in \nabla$, *then* $a \cap b \in \nabla$,

(s.f.$_2$) *if* $a \in \nabla$ *and* $a \prec b$, *then* $b \in \nabla$.

Note that

[5] For the results in Sec. 4 see Rasiowa [7], Białynicki-Birula and Rasiowa [2].

4.1. *If* ∇ *is a special filter of the first kind, then* ∇ *is a filter.*

This follows immediately from the definition of s.f.f.k. and 1.1. In fact, the condition III 1 (f_1) is satisfied because it is identical to (s.f.$_1$) and the condition III 1 (f_2) is also satisfied, since by 1.1 $a \leqslant b$ implies $a \prec b$.

4.2. *A subset* ∇ *of the set* A *of all elements of a quasi-pseudo-Boolean algebra is a special filter of the first kind if and only if the following conditions hold*:

(s.f.$_1^*$) $$V \in \nabla,$$

(s.f.$_2^*$) $$\text{if} \quad a \in \nabla \text{ and } a \to b \in \nabla, \quad \text{then} \quad b \in \nabla.$$

Since $a \prec V$, for every element a in a quasi-pseudo-Boolean algebra, V belongs to every s.f.f.k. If ∇ is a special filter of the first kind and $a \in \nabla$, $a \to b \in \nabla$, then by (s.f.$_1$) $a \cap (a \to b) \in \nabla$. Hence, by 1.3 (13) and (s.f.$_2$), $b \in \nabla$. On the other hand, if (s.f.$_1^*$) and (s.f.$_2^*$) are satisfied then ∇ is non empty. However, by 1 (46) $a \to (b \to (a \cap b)) = V \in \nabla$. Hence if $a \in \nabla$ and $b \in \nabla$, then $a \cap b \in \nabla$. If $a \prec b$ then $a \to b = V \in \nabla$. Hence the condition $a \in \nabla$ and $a \prec b$ implies by (s.f.$_1^*$) and (s.f.$_2^*$) that $b \in \nabla$.

4.3. *Every special filter* ∇ *of the first kind is an implicative filter,* i.e.

(1) $$V \in \nabla,$$

(2) $$\text{if} \quad a \in \nabla \text{ and } a \Rightarrow b \in \nabla, \quad \text{then} \quad b \in \nabla.$$

(1) follows from 4.2. If $a \Rightarrow b \in \nabla$, then by 1 ($qpB_4$) $(a \to b) \cap (b \to a) \in \nabla$ and by 1.3 (29), 4.2 we infer that $a \to b \in \nabla$. Consequently, if $a \in \nabla$ and $a \Rightarrow b \in \nabla$, then by 4.2 $b \in \nabla$, which proves (2).

It is possible to prove that every implicative filter is a filter in every quasi-pseudo-Boolean algebra and that the converse statement does not hold.

4.4 [6]. *If* $K(h)$ *is the kernel of an epimorphism* h *of a quasi-pseudo-Boolean algebra* $\mathfrak{A}$ *onto a similar algebra* $\mathfrak{B}$ (i.e. $K(h) = h^{-1}(V')$, where V' is the unit element in $\mathfrak{B}$), *then* $K(h)$ *is a special filter of the first kind. Moreover,* $h(a) = h(b)$ *is equivalent to* $a \Rightarrow b \in K(h)$

[6] For 4.4 and 4.5 cf. Monteiro [2].

and $b \Rightarrow a \in K(h)$. *Hence by* I 4.7 *the relation* $a \approx b$ *if and only if* $a \Rightarrow b \in K(h)$ *and* $b \Rightarrow a \in K(h)$ *is a congruence in* $\mathfrak{A}$. *The quotient algebra* $\mathfrak{A}/\approx$ *is isomorphic to* $\mathfrak{B}$.

Suppose that $K(h) = h^{-1}(V')$. Since h is an epimorphism, $h(V) = V'$, which proves that $V \in K(h)$. Moreover, if $a \in K(h)$ and $a \to b \in K(h)$, then $h(a) = V'$ and $h(a) \to h(b) = h(a \to b) = V'$. Hence by 2.1, I 4.11, 1.3 (23), $h(b) = V'$, i.e. $b \in K(h)$. Thus by 4.2 $K(h)$ is a special filter of the first kind. The condition $h(a) = h(b)$ is satisfied if and only if $h(a) \leqslant h(b)$ and $h(b) \leqslant h(a)$, which by 1.1 is equivalent to $h(a \Rightarrow b) = h(a) \Rightarrow h(b) = V'$ and $h(b \Rightarrow a) = h(b) \Rightarrow h(a) = V'$, i.e. to the condition $a \Rightarrow b \in K(h)$ and $b \Rightarrow a \in K(h)$.

4.5. *Let* $\mathfrak{A} = (A, V, \Rightarrow, \cup, \cap, \to, \sim, \neg)$ *be a quasi-pseudo-Boolean algebra and let* ∇ *be a special filter of the first kind. Then the relation defined on* A *by*

(3) $\quad a \approx_\nabla b$ *if and only if* $a \Rightarrow b \in \nabla$ *and* $b \Rightarrow a \in \nabla$,

is a congruence in $\mathfrak{A}$. *The quotient algebra denoted by* $\mathfrak{A}/\nabla$ *is a quasi-pseudo-Boolean algebra. The mapping* $h(a) = ||a|| \in A/\nabla$ *is an epimorphism of* $\mathfrak{A}$ *onto* $\mathfrak{A}/\nabla$ *and* ∇ *is the kernel of* h. *Moreover,* $a \in \nabla$ *if and only if* $a \approx_\nabla V$. *The algebra* $\mathfrak{A}/\nabla$ *is degenerate if and only if* $\nabla = A$.

By 1.2 (5) the relation $\approx_\nabla$ is reflexive. It follows directly from (3) that it is symmetrical. On account of 1.3 ((24), (25)) and (s.f.$_1^*$), (s.f.$_2^*$)

(4) $\quad$ *if* $a \to b \in \nabla$ *and* $b \to c \in \nabla$, *then* $a \to c \in \nabla$.

Hence, by 1 (qpB$_4$), 1.3 ((29), (30)) and (s.f.$_1$),

(5) $\quad$ *if* $a \Rightarrow b \in \nabla$ *and* $b \Rightarrow c \in \nabla$, *then* $a \Rightarrow c \in \nabla$.

Consequently the relation $\approx_\nabla$ is transitive. Thus it is an equivalence relation.

By (3), (qpB$_5$), (qpB$_1$), III 3(q$_1$)

(6) $\quad$ *if* $a \approx_\nabla b$, *then* $\sim a \approx_\nabla \sim b$.

We shall prove that

(7) $\quad$ *if* $a \approx_\nabla b$ *and* $c \approx_\nabla d$, *then* $a \cup c \approx_\nabla b \cup d$.

If $a \approx_\nabla b$ and $c \approx_\nabla d$, then $a \to b \in \nabla$ and $c \to d \in \nabla$. Consequently, by 1.3 (42) and 4.2, $a \cup c \to b \cup d \in \nabla$. On the other hand, it follows from our assumptions that $(\sim b \to \sim a) \cap (\sim d \to \sim c) \in \nabla$. Hence, by 1.3

(41), $(\sim b \cap \sim d) \to (\sim a \cap \sim c) \in \nabla$ and consequently, by 1.3 (38), $\sim(b \cup d) \to \sim(a \cup c) \in \nabla$. Thus $(a \cup c) \Rightarrow (b \cup d) \in \nabla$. By symmetry $(b \cup d) \Rightarrow (a \cup c) \in \nabla$, which proves (7).

Now we shall prove that

(8) *if* $a \approx_\nabla b$ *and* $c \approx_\nabla d$, *then* $a \cap c \approx_\nabla b \cap d$.

It follows from our assumptions that $(a \to b), (c \to d) \in \nabla$. Hence, by 1.3 (41) $(a \cap c) \to (b \cap d) \in \nabla$. On the other hand, since also $(\sim b \to \sim a), (\sim d \to \sim c) \in \nabla$, it follows from 1.3 (42) and 4.2 that $(\sim b \cup \cup \sim d) \to (\sim a \cup \sim c) \in \nabla$. Consequently, by 1.3 (39), $\sim(b \cap d) \to \sim(a \cap \cap c) \in \nabla$. Thus $(a \cap c) \Rightarrow (b \cap d) \in \nabla$. By symmetry $(b \cap d) \Rightarrow (a \cap c) \in \nabla$. Thus (8) holds. Now we shall prove that

(9) *if* $a \approx_\nabla b$ *and* $c \approx_\nabla d$, *then* $a \to c \approx_\nabla b \to d$.

It follows from the assumptions that $(b \to a), (c \to d) \in \nabla$. Consequently, by 1.3 (43), $(a \to c) \to (b \to d) \in \nabla$. On the other hand, since also $(b \to a), (\sim d \to \sim c) \in \nabla$, we infer by 1.3 (41) that $(b \cap \sim d) \to (a \cap \sim c) \in \nabla$. Hence, by 1 ((qp$B_8$), (qp$B_9$)) and (4), $\sim(b \to d) \to \sim(a \to c) \in \nabla$ and consequently $(a \to c) \Rightarrow (b \to d) \in \nabla$. By the symmetry of the assumptions $(b \to d) \Rightarrow (a \to c) \in \nabla$, which together with the previous condition proves that (9) holds.

It follows from (6), (9), (8), 1 ((qpB_4), (qpB_{13})) that

(10) *if* $a \approx_\nabla b$ *and* $c \approx_\nabla d$, *then* $a \Rightarrow c \approx_\nabla b \Rightarrow d$,

(11) *if* $a \approx_\nabla b$, *then* $\neg a \approx_\nabla \neg b$.

Thus $\approx_\nabla$ is a congruence relation. Since the class of all quasi-pseudo-Boolean algebras is equationally definable, the quotient algebra $\mathfrak{A}/\nabla$ is a quasi-pseudo-Boolean algebra. Clearly, the mapping $h(a) = ||a||$ is an epimorphism of $\mathfrak{A}$ onto $\mathfrak{A}/\nabla$. We shall prove that $a \in \nabla$ if and only if $a \approx_\nabla V$. In fact, if $a \in \nabla$, then by 1.3 ((24), (34), (15)) $V \Rightarrow a \in \nabla$ and by 1.2 (8) $a \Rightarrow V \in \nabla$. Thus $a \approx_\nabla V$. Conversely, if $a \approx_\nabla V$, then $V \Rightarrow a \in \nabla$ and consequently, by 4.3, $a \in \nabla$. This proves that ∇ is the kernel of h. The last statement is obvious.

A non-empty subset ∇ of the set of all elements of a quasi-pseudo-Boolean algebra is said to be a *special filter of the second kind* (s.f.s.k.) provided the following conditions are satisfied:

(s.f.s.$_1$) *if* $a \in \nabla$ *and* $b \in \nabla$, *then* $a \cap b \in \nabla$,

(s.f.s.$_2$) *if* $a \in \nabla$ *and* $\sim b \prec \sim a$, *then* $b \in \nabla$.

4.6. *Every* s.f.s.k. *is a filter.*

This follows from the definition and 1.1.

The whole quasi-pseudo-Boolean algebra is a s.f.f.k. and a s.f.s.k. The set $\{V\}$ is also a s.f.f.k. and a s.f.s.k.

4.7. *For every fixed element* a *in a quasi-pseudo-Boolean algebra* $\mathfrak{A} = (A, V, \Rightarrow, \cup, \cap, \rightarrow, \sim, \neg)$ *the set* $\nabla_1(a) = \{x \in A\colon\ a \prec x\}$ $(\nabla_2(a) = \{x \in A\colon\ \sim x \prec \sim a\})$ *is the least* s.f.f.k. (s.f.s.k.) *containing* a. *It is called* the *principal* s.f.f.k. (s.f.s.k.) *generated by* a.

The easy proof based on 1 (qpB_2), 1.3 (31), 1.3 (23), 1.3 (28), 1.3 (39) is left to the reader.

4.8. *The* s.f.f.k. (s.f.s.k.) *generated by a non-empty set* A_0 *of elements of a quasi-pseudo-Boolean algebra* $\mathfrak{A}$, i.e. *the least* s.f.f.k. (s.f.s.k.) *containing* A_0, *is the set of all elements* a *in* $\mathfrak{A}$ *such that* $a_1 \cap \ldots \cap a_n \prec a$ $(\sim a \prec \sim(a_1 \cap \ldots \cap a_n))$ *for some* $a_1, \ldots, a_n$ *in* A_0.

The easy proof by verification based on 1 (qpB_2), 1.3 (12) and 1.3 (39) is left to the reader.

4.9. *The* s.f.f.k. (s.f.s.k.) *generated by a fixed element* a_0 *and a* s.f.f.k. ∇ (s.f.s.k. ∇) *in a quasi-pseudo-Boolean algebra* $\mathfrak{A}$ *is the set of all elements* a *in* $\mathfrak{A}$ *such that* $a_0 \cap c \prec a$ (*such that* $\sim a \prec \sim(a_0 \cap c)$) *for an element* $c \in \nabla$.

The easy proof by verification based on 1 (qpB_2), 1.3 (12), 1.3 (39) is left to the reader.

A s.f.f.k. (s.f.s.k.) is said to be *proper* provided there exists an element a in the quasi-pseudo-Boolean algebra under consideration which does not belong to it.

4.10. *A* s.f.f.k. (s.f.s.k.) *is proper if and only if* $\Lambda = \sim V$ *does not belong to it.*

This follows from 1.3 (21).

A s.f.f.k. (s.f.s.k.) is said to be *maximal* provided it is proper and it is not a proper subset of any proper s.f.f.k. (s.f.s.k.).

A s.f.f.k. (s.f.s.k.) is said to be *irreducible* provided it is proper and it is not the intersection of two proper special filters of the first kind (or two special filters of the second kind) different from it.

A s.f.f.k. ∇ (s.f.s.k. ∇) is said to be *prime* provided it is proper and the condition $a \cup b \in \nabla$ implies that either $a \in \nabla$ or $b \in \nabla$.

It is easy to see that

4.11. *Every maximal* s.f.f.k. (*maximal* s.f.s.k.) *is irreducible.*

4.12. *For every* s.f.f.k. ∇ (s.f.s.k. ∇) *the following conditions are equivalent*:

(12) ∇ *is irreducible,*

(13) ∇ *is prime.*

Suppose that a s.f.f.k. (s.f.s.k.) ∇ is not irreducible. Then $\nabla = \nabla_1 \cap \nabla_2$, where ∇_1, ∇_2 are two special filters of the first kind (of the second kind), $\nabla_1 \neq \nabla$ and $\nabla_2 \neq \nabla$. Thus $\nabla_1 \not\subset \nabla_2$ and $\nabla_2 \not\subset \nabla_1$. Consequently, there exists an $a \in \nabla_1$ such that $a \notin \nabla_2$ and there exists a $b \in \nabla_2$ such that $b \notin \nabla_1$. Hence $a \notin \nabla$ and $b \notin \nabla$. Since $a \prec a \cup b$ $(\sim(a \cup b) = \sim a \cap \sim b \prec \sim a)$ and $b \prec a \cup b$ $(\sim(a \cup b) = \sim a \cap \sim b \prec \sim b)$, we infer from (s.f.$_2$) (from (s.f.s.$_2$)) that $a \cup b \in \nabla_1$ and $a \cup b \in \nabla_2$. Consequently, $a \cup b \in \nabla$. Thus ∇ is not prime.

Suppose that a s.f.f.k. ∇ is not prime. Then there exist a, b such that

(14) $$a \cup b \in \nabla, \quad a \notin \nabla,\ b \notin \nabla.$$

Let ∇_1 be the s.f.f.k. generated by a and ∇, and let ∇_2 be the s.f.f.k. generated by b and ∇_2. We shall show that $\nabla = \nabla_1 \cap \nabla_2$. Clearly, $\nabla \subset \nabla_1 \cap \nabla_2$. Suppose that $x \in \nabla_1$ and $x \in \nabla_2$. Then by 4.9 $c_1 \cap a \prec x$ and $c_2 \cap b \prec x$ for some $c_1, c_2 \in \nabla$. Let $c = c_1 \cap c_2$. Then $c \cap a \prec x$ and $c \cap b \prec x$, where $c \in \nabla$. Hence by 1 ((qpB$_6$) and (qpB$_1$)) $c \cap (a \cup b) = (c \cap a) \cup (c \cap b) \prec x$. Thus $x \in \nabla$. This proves that the equation above holds and consequently that ∇ is not irreducible. Suppose that a s.f.s.k. ∇ is not prime. Then there exist a, b such that (14) holds. Let ∇_1 be the s.f.s.k. generated by a and ∇ and let ∇_2 be the s.f.s.k. generated by b and ∇. We shall show that $\nabla = \nabla_1 \cap \nabla_2$. Clearly, $\nabla \subset \nabla_1 \cap \nabla_2$. Suppose that $x \in \nabla_1$ and $x \in \nabla_2$. Then by 4.9 $\sim x \prec \sim(c_1 \cap a) = \sim c_1 \cup \sim a$ and $\sim x \prec \sim(c_2 \cap b) = \sim c_2 \cup \sim b$ for some $c_1, c_2 \in \nabla$. Let $c = c_1 \cap c_2 \in \nabla$. Then $\sim x \prec \sim c_1 \cup \sim c_2 \cup \sim a = \sim c \cup \sim a = \sim(c \cap a)$ and $\sim x \prec \sim c_1 \cup \sim c_2 \cup \sim b = \sim c \cup \sim b = \sim(c \cap b)$. Hence, by 1 ((qpB$_7$), (qpB$_1$)),

$$\sim x \prec \sim(c \cap a) \cap \sim(c \cap b) = \sim((c \cap a) \cup (c \cap b)) = \sim(c \cap (a \cup b)).$$

Consequently, $x \in \nabla$. This proves that $\nabla = \nabla_1 \cap \nabla_2$ holds and therefore that ∇ is not irreducible.

We state without proof that

4.13. *The union of any chain of proper special filters of the first kind (of the second kind) is a proper* s.f.f.k. (*is a proper* s.f.s.k.).

4.14. *Every proper* s.f.f.k. (*proper* s.f.s.k.) *is contained in a maximal* s.f.f.k. (*in a maximal* s.f.s.k.).

The proof follows from 4.13 by the application of I 3.1.

It follows from 4.14, 4.7 and 4.10 that

4.15. *For every element* $a \neq \wedge$ *in a quasi-pseudo-Boolean algebra there exists a maximal* s.f.f.k. ∇ (*a maximal* s.f.s.k. ∇) *such that* $a \in \nabla$.

4.16. *If* ∇ *is a* s.f.f.k. (s.f.s.k.) *in a quasi-pseudo-Boolean algebra* $\mathfrak{A}$ *and* $a \notin \nabla$, *then there exists an irreducible* s.f.f.k.∇^* (an irreducible s.f.s.k. ∇^*) *such that* $\nabla \subset \nabla^*$ *and* $a \notin \nabla^*$.

The proof, analogous to the proof of I 1.12 based on 4.13, is left to the reader.

For each prime filter ∇ in a quasi-pseudo-Boolean algebra $\mathfrak{A} = (A, V, \Rightarrow, \cup, \cap, \rightarrow, \sim, \neg)$ let

(15) $g(\nabla) = A - \tilde{\nabla}$, where $\tilde{\nabla} = \{\sim a \in A \colon a \in \nabla\}$ (see III 3 (6)).

Then, by III 3 (8) and by III 1.13, $g(\nabla)$ is also a prime filter in $\mathfrak{A}$. It follows from (15) and 1.3 (37) that for every $a \in A$

(16) $a \in g(\nabla)$ *if and only if* $\sim a \notin \nabla$,

and consequently

(17) $\sim a \in g(\nabla)$ *if and only if* $a \notin \nabla$.

Moreover, let us note (III 3 (11)) that

(18) $g(g(\nabla)) = \nabla$.

4.17. *If* ∇ *is a prime* s.f.f.k., *then* $g(\nabla)$ *is a prime* s.f.s.k., *and if* ∇ *is a prime* s.f.s.k., *then* $g(\nabla)$ *is a prime* s.f.f.k.

If ∇ is a prime s.f.f.k., then $g(\nabla)$ is a prime filter. Suppose that $a \in g(\nabla)$ and $\sim b \prec \sim a$. Then by (16) $\sim a \notin \nabla$. Hence $\sim b \notin \nabla$ and consequently, by (16), $b \in g(\nabla)$. Thus $g(\nabla)$ is a prime s.f.s.k. Now suppose that ∇ is a prime s.f.s.k. Then $g(\nabla)$ is a prime filter. Suppose that $a \in g(\nabla)$

and $a \prec b$. Then by (16) $\sim a \notin \nabla$ and $\sim \sim a \prec \sim \sim b$. Thus $\sim b \notin \nabla$, i.e. $b \in g(\nabla)$. This proves that $g(\nabla)$ is a prime s.f.f.k.

4.18. *For every prime filter* ∇ *in a quasi-pseudo-Boolean algebra* $\mathfrak{A}$ *either* $\nabla \subset g(\nabla)$ *or* $g(\nabla) \subset \nabla$.

Suppose that $\nabla \not\subset g(\nabla)$. Then there exists an element a in $\mathfrak{A}$ such that $a \in \nabla$ and $a \notin g(\nabla)$. Hence, by (16), $a \in \nabla$ and $\sim a \in \nabla$. Consequently, $a \cap \sim a \in \nabla$. By 1.3 (17) $a \cap \sim a \leqslant b \cup \sim b$ for every b in $\mathfrak{A}$. Thus $b \cup \sim b \in \nabla$ for every b in $\mathfrak{A}$. Since ∇ is a prime filter, it follows from the last condition that for every b in $\mathfrak{A}$ either $b \in \nabla$ or $\sim b \in \nabla$. Suppose that $b \in g(\nabla)$. Then by (16) $\sim b \notin \nabla$. Consequently $b \in \nabla$. This proves that $g(\nabla) \subset \nabla$.

4.19. *For every prime filter* ∇ *in a quasi-pseudo-Boolean algebra* $\mathfrak{A}$ *the following conditions are equivalent*:

(i) ∇ *is a prime* s.f.f.k.,

(ii) $\nabla \subset g(\nabla)$.

Suppose that ∇ is a prime s.f.f.k. and $\nabla \not\subset g(\nabla)$. Then there exists an a in $\mathfrak{A}$ such that $a \in \nabla$ and $a \notin g(\nabla)$. Hence $a \in \nabla$ and by (16) $\sim a \in \nabla$. Consequently, $a \cap \sim a \in \nabla$ and, by 1 (qpB_{12}), $b \in \nabla$ for every b in $\mathfrak{A}$, which contradicts the assumption that ∇ is a prime filter and therefore proper. Thus (i) implies (ii). Suppose that $\nabla \subset g(\nabla)$, $a \in \nabla$, $a \prec b$ and $b \notin \nabla$. Then $a \in g(\nabla)$ and by (17) $\sim b \in g(\nabla)$. Thus $a \cap \sim b \in g(\nabla)$. Since $a \prec b$, we get by 1.3 (12) and 1 (qpB_{12})

$$a \cap \sim b \prec b \cap \sim b \prec \sim a$$

and by 1.3 (27), 1.3 (37), 1.3 (39)

$$\sim \sim a = a \prec \sim a \cup b = \sim (a \cap \sim b).$$

Hence $a \cap \sim b \leqslant \sim a$. Consequently, $\sim a \in g(\nabla)$, i.e., by (17), $a \notin \nabla$, which contradicts our assumption that $a \in \nabla$. Thus (ii) implies (i).

4.20. *For every prime filter* ∇ *in a quasi-pseudo-Boolean algebra* $\mathfrak{A}$ *the following conditions are equivalent*:

(i) ∇ *is a prime* s.f.s.k.,

(ii) $g(\nabla) \subset \nabla$.

Suppose that ∇ is a prime s.f.s.k. Then by 4.17 $g(\nabla)$ is a prime s.f.f.k. and hence, by 4.19 and (18), we get $g(\nabla) \subset g(g(\nabla)) = \nabla$. Now suppose that $g(\nabla) \subset \nabla$. Then for every a in $\mathfrak{A}$ either $a \in \nabla$ or $\sim a \in \nabla$. Indeed, if $a = \sim \sim a \notin \nabla$, then by (16) $\sim a \in g(\nabla) \subset \nabla$, i.e. $\sim a \in \nabla$. Suppose that $a \in \nabla$, $\sim b \prec \sim a$ and $b \notin \nabla$. Then $\sim b \in \nabla$. Hence $a \cap \sim b \in \nabla$. Since $\sim b \prec \sim a$, we get by 1.3 (12), 1 (qpB_{12})

$$a \cap \sim b \prec a \cap \sim a \prec b$$

and by 1.3 (27), 1.3 (37), 1.3 (39)

$$\sim b \prec \sim a \cup b = \sim (a \cap \sim b).$$

Thus $a \cap \sim b \leqslant b$. Hence $b \in \nabla$, which contradicts our hypothesis that $b \notin \nabla$. Consequently, (ii) implies (i).

The following theorem follows from 4.18, 4.19 and 4.20:

4.21. *Every prime filter* ∇ *in a quasi-pseudo-Boolean algebra is either a prime* s.f.f.k. *or a prime* s.f.s.k.

4.22 [7]. *For every* s.f.f.k. ∇ *in a quasi-pseudo-Boolean algebra* $\mathfrak{A}$ *the following conditions are equivalent*:

(i) ∇ *is maximal,*

(ii) *for every element a in* $\mathfrak{A}$ *exactly one of the elements* a, $\neg a$ *belongs to* ∇,

(iii) *for every element a in* $\mathfrak{A}$ *exactly one of the following conditions is satisfied*:

(a) $a \in \nabla$, $\neg a \notin \nabla$, $\sim a \notin \nabla$, $\neg \sim a \in \nabla$,

(b) $a \notin \nabla$, $\neg a \in \nabla$, $\sim a \in \nabla$, $\neg \sim a \notin \nabla$,

(c) $a \notin \nabla$, $\neg a \in \nabla$, $\sim a \notin \nabla$, $\neg \sim a \in \nabla$,

(iv) *the quotient algebra* $\mathfrak{A}/\nabla$ *is non-degenerate and has at most three elements*, i.e. $\mathfrak{A}/\nabla$ *is isomorphic either to* $\mathfrak{C}_0$ *or to* $\mathfrak{B}_0$ (see Sec. 3).

(i) *implies* (ii). If ∇ is maximal, then ∇ is proper. Hence, for any a in $\mathfrak{A}$ at most one of the elements a, $\neg a$ is in ∇. Indeed, by 1 ((qpB_{13}), (qpB_1)), III 3 (2), 1.3 (13)

$$(19) \qquad a \cap \neg a = a \cap (a \rightarrow \Lambda) \prec b \quad \text{for each } b \text{ in } \mathfrak{A}.$$

Suppose that $\neg a \notin \nabla$. Then the s.f.f.k. ∇^* generated by ∇ and a is proper. Otherwise, $\Lambda \in \nabla^*$. Hence, by 4.9, there exists a $c \in \nabla$ such

[7] Cf. Monteiro [2].

that $a \cap c \prec \Lambda$. By 1 (qpB_3), $c \prec a \to \Lambda = \neg a$. Thus $\neg a \in \nabla$, which contradicts our hypothesis. Since ∇^* is proper, $\nabla \subset \nabla^*$ and ∇ is maximal, we get $\nabla = \nabla^*$. Hence $a \in \nabla$.

(ii) *implies* (iii). Observe that

(20) $$\sim a \prec \neg a \quad \text{for every } a \text{ in } A.$$

Indeed, by 1.3 (34), $\sim a \to \neg a = \sim a \to (a \to \Lambda) = V$, which by 1 (1) is equivalent to (20). Consequently, if $\neg a \notin \nabla$, then $\sim a \notin \nabla$. Hence (ii) implies (iii).

(iii) *implies* (i). It is easy to see that if (iii) holds, then ∇ is proper and by (19) it is not a proper subset of any proper s.f.f.k.

(iii) *implies* (iv). Suppose that (iii) holds and a is an arbitrary element in $\mathfrak{A}$. If (a) is satisfied, then by 4.5, $||a|| = V_{\mathfrak{A}/\nabla}$. If (b) holds, then $\sim ||a|| = V_{\mathfrak{A}/\nabla}$ and consequently $||a|| = \Lambda_{\mathfrak{A}/\nabla}$. Suppose that the condition (c) is satisfied for two elements a, b in $\mathfrak{A}$. Then

(21) $$a \to \Lambda \in \nabla, \quad \sim a \to \Lambda \in \nabla, \quad b \to \Lambda \in \nabla, \quad \sim b \to \Lambda \in \nabla.$$

On the other hand, by 1.3 (20),

(22) $$\Lambda \to a \in \nabla, \quad \Lambda \to \sim a \in \nabla, \quad \Lambda \to b \in \nabla, \quad \Lambda \to \sim b \in \nabla.$$

It follows from (21), (22) and (4), that

$$a \to b \in \nabla, \quad \sim b \to \sim a \in \nabla, \quad b \to a \in \nabla, \quad \sim a \to \sim b \in \nabla.$$

Thus $a \Rightarrow b \in \nabla$ and $b \Rightarrow a \in \nabla$, i.e. $a \approx_\nabla b$. Hence $||a|| = ||b||$.

(iv) *implies* (iii). Suppose that $\mathfrak{A}/\nabla$ is non-degenerate and has at most three elements. If $||a|| = V_{\mathfrak{A}/\nabla}$, then $||\neg a|| = \Lambda_{\mathfrak{A}/\nabla}$, $||\sim a|| = \Lambda_{\mathfrak{A}/\nabla}$, $||\neg \sim a|| = V_{\mathfrak{A}/\nabla}$. Hence, by 4.5, (a) holds. If $||a|| = \Lambda_{\mathfrak{A}/\nabla}$, then $||\neg a|| = V_{\mathfrak{A}/\nabla}$, $||\sim a|| = V_{\mathfrak{A}/\nabla}$, $||\neg \sim a|| = \Lambda_{\mathfrak{A}/\nabla}$. Hence, by 4.5, (b) holds. Suppose that $||a|| \neq V_{\mathfrak{A}/\nabla}$ and $||a|| \neq \Lambda_{\mathfrak{A}/\nabla}$. Then $||\sim a|| \neq V_{\mathfrak{A}/\nabla}$ and $||\sim a|| \neq \Lambda_{\mathfrak{A}/\nabla}$. Since $\mathfrak{A}/\nabla$ has at most three elements, $||a|| = ||\sim a||$. Hence $a \approx_\nabla \sim a$, i.e. $a \Rightarrow \sim a \in \nabla$ and $\sim a \Rightarrow a \in \nabla$. Thus $a \to \sim a \in \nabla$ and $\sim a \to a \in \nabla$. On the other hand, by 1.3 (9), $a \to a \in \nabla$ and $\sim a \to \sim a \in \nabla$. Hence, by 1.3 (31), $a \to (a \cap \sim a) \in \nabla$ and $\sim a \to (a \cap \sim a) \in \nabla$. Applying 1 ($qpB_{12}$) and (4), we infer that $a \to \Lambda \in \nabla$ and $\sim a \to \Lambda \in \nabla$. Thus $\neg a \in \nabla$, $\neg \sim a \in \nabla$, $a \notin \nabla$, $\sim a \notin \nabla$, i.e. (c) is satisfied.

5. Representation theorem for quasi-pseudo-Boolean algebras [8]. The aim of this section is to prove that every quasi-pseudo-Boolean algebra is isomorphic to a quasi-pseudo-Boolean algebra of sets (of subsets of a compact topological T_0-space) connected with a pseudo-field.

Given an arbitrary quasi-pseudo-Boolean algebra $\mathfrak{A} = (A, V, \Rightarrow, \cup, \cap, \rightarrow, \sim, \neg)$, let X_1 be the set of all prime special filters of the first kind in $\mathfrak{A}$. For every $a \in A$ let

(1) $$h_1(a) = \{\nabla \in X_1 \colon a \in \nabla\}.$$

Then the following theorem holds.

5.1. *For any elements a, b in a quasi-pseudo-Boolean algebra* $\mathfrak{A}$:

(2) $h_1(a) \subset h_1(b)$ *if and only if* $a \prec b$,

(3) $h_1(V) = X_1$,

(4) $h_1(\Lambda) = O$,

(5) $h_1(a \cup b) = h_1(a) \cup h_1(b)$,

(6) $h_1(a \cap b) = h_1(a) \cap h_1(b)$.

Suppose that $a \prec b$, i.e. that $a \rightarrow b = V$ (see 1 (1)). Then, by 4.2, for every $\nabla \in X_1$, we have $a \rightarrow b \in \nabla$ and consequently if $a \in \nabla$, then $b \in \nabla$. Thus $h_1(a) \subset h_1(b)$. Conversely, let us suppose that $a \prec b$ does not hold. Then, by 4.7, b does not belong to the principal special filter of the first kind (s.f.f.k.) $\nabla_1(a)$ generated by a. Hence, by 4.16 and 4.12, there exists a prime s.f.f.k. ∇ such that $\nabla_1(a) \subset \nabla$ and $b \notin \nabla$, i.e. such that $a \in \nabla$ and $b \notin \nabla$. Consequently, by (1), $\nabla \in h_1(a)$ and $\nabla \notin h_1(b)$. Thus $h_1(a) \not\subset h_1(b)$. This proves (2).

Condition (3) is satisfied since by 4.2 the unit element V belongs to every ∇ in X_1, and so, by (1), every ∇ in X_1 belongs to $h_1(V)$.

In order to prove (4) let us observe that by 1 (qpB_1), 1.1 and III 3.1 $\Lambda \prec a$ for every a in $\mathfrak{A}$. Since every ∇ in X_1 is a proper s.f.f.k., we infer by 4.2 that $\Lambda \notin \nabla$, i.e. $h_1(\Lambda) = O$.

Condition (5) is satisfied since for every ∇ in X_1 and for any a, b in $\mathfrak{A}$

(7) $$a \cup b \in \nabla \quad \text{if and only if} \quad \text{either } a \in \nabla \text{ or } b \in \nabla.$$

[8] The first representation theorem for quasi-pseudo-Boolean algebras was formulated and proved by Rasiowa [6], [7].

Indeed, by 1 (1) and 1.3 (26), (27) we have $a \prec a \cup b$ and $b \prec a \cup b$. Hence by 4 (s.f.$_2$) if either $a \in \nabla$ or $b \in \nabla$, then $a \cup b \in \nabla$. The converse statement follows from the definition of a prime s.f.f.k.

Condition (6) follows from the following statement: for every ∇ in X_1 and for any a, b in $\mathfrak{A}$

$$(8) \qquad a \cap b \in \nabla \quad \text{if and only if} \quad a \in \nabla \text{ and } b \in \nabla.$$

To prove (8) observe that by 1 (1) and 1.3 (29), (30) we have $a \cap b \prec a$ and $a \cap b \prec b$. Hence, by 4 (s.f.$_2$), the condition $a \cap b \in \nabla$ implies that $a \in \nabla$ and $b \in \nabla$. The converse statement follows from 4 (s.f.$_1$).

Let us consider X_1 as a topological space with the base $\{h_1(a)\}_{a \in A}$. For any a, b in $\mathfrak{A}$ put

$$(9) \qquad h_1(a) \Rightarrow_1 h_1(b) = \mathbf{I}_1\big((X_1 - h_1(a)) \cup h_1(b)\big),$$

where $\mathbf{I}_1$ is the interior operation in the topological space X_1, and let

$$(10) \qquad \neg_1 h_1(a) = \mathbf{I}_1(X_1 - h_1(a)).$$

By the above assumptions the following theorem holds.

5.2. *For any a, b in the quasi-pseudo-Boolean algebra $\mathfrak{A}$*

$$(11) \qquad h_1(a \to b) = h_1(a) \Rightarrow_1 h_1(b),$$

$$(12) \qquad h_1(\neg a) = \neg_1 h_1(a).$$

To prove (11) suppose that $\nabla \in h_1(a \to b)$, i.e. that $a \to b \in \nabla$. Then, by 4 (s.f.$_2^*$), either $a \notin \nabla$ or $b \in \nabla$. Hence either $\nabla \in X_1 - h_1(a)$ or $\nabla \in h_1(b)$, i.e. $\nabla \in (X_1 - h_1(a)) \cup h_1(b)$. Thus

$$(13) \qquad h_1(a \to b) \subset (X_1 - h_1(a)) \cup h_1(b).$$

Suppose that

$$(14) \qquad h_1(c) \subset (X_1 - h_1(a)) \cup h_1(b), \quad \text{for some } c \text{ in } \mathfrak{A}.$$

Then $h_1(a) \cap h_1(c) \subset h_1(b)$. Hence, by (6), $h_1(a \cap c) \subset h_1(b)$. Consequently by (2) we get $a \cap c \prec b$. By 1 (qpB$_3$) this condition is equivalent to $c \prec a \to b$. Hence, by (2),

$$(15) \qquad h_1(c) \subset h_1(a \to b).$$

Thus (14) implies (15), which proves that $h_1(a \to b)$ is the largest open set in X_1 contained in $(X_1 - h_1(a)) \cup h_1(b)$, i.e. that $h_1(a \to b)$

$= \mathbf{I}_1\big((X_1 - h_1(a)) \cup h_1(b)\big)$. Hence, by (9), we get (11). By 1 (qpB_{13}) and (11), (4), (9), (10)

$$h_1(\neg a) = h_1(a \to \sim V) = h_1(a \to \Lambda) = h_1(a) \Rightarrow_1 h_1(\Lambda)$$
$$= h_1(a) \Rightarrow_1 O = \mathbf{I}_1(X_1 - h_1(a)) = \neg_1 h_1(a).$$

Thus (12) holds.

Let us denote by $G(X_1)$ the family $\{h_1(a)\}_{a \in A}$ of open subsets of the topological space X_1. Then by 5.1 and 5.2 we get the following theorem.

5.3. $\mathfrak{G}(X_1) = (G(X_1), X_1, \Rightarrow_1, \cup, \cap, \neg_1)$ *is a pseudo-field of some open subsets in* X_1 *constituting a base of* X_1.

We shall prove later that the quasi-pseudo-Boolean algebra $\mathfrak{A}$ under consideration is isomorphic to a quasi-pseudo-Boolean algebra of sets connected with the pseudo-field $\mathfrak{G}(X_1)$ constructed above.

Let us take as X_2 the set of all prime special filters of the second kind (s.f.s.k.) in $\mathfrak{A}$ and let X be the set of all prime filters in $\mathfrak{A}$. It follows from 4.21 that

(16) $$X = X_1 \cup X_2.$$

For any ∇ in X let

(17) $$g(\nabla) = A - \tilde{\nabla}, \quad \text{where} \quad \tilde{\nabla} = \{\sim a \in A : a \in \nabla\}.$$

Then by 1 (qpB_1) and III 3 (11) we get

(18) $$g(g(\nabla)) = \nabla, \quad \text{i.e. } g \text{ is an involution on } X.$$

Hence, by 4.17,

(19) $$g(X_1) = X_2 \quad \text{and} \quad g(X_2) = X_1.$$

It follows from (18), 4.19 and 4.20 that

(20) the mapping $f = g|X_1$ satisfies conditions 3 (3), 3 (6), 3 (7).

We shall show that

(21) $$X_1 \cap X_2 \subset \bigcap_{h_1(a) \in G(X_1)} \Big(h_1(a) \cup f\big(\mathbf{I}_1(X_1 - h_1(a))\big)\Big),$$

i.e. that the condition 3 (4) holds too.

Since by (10)

$$h_1(a) \cup f\big(\mathbf{I}_1(X_1 - h_1(a))\big) = h_1(a) \cup f(h_1(\neg a)), \tag{22}$$

in order to prove (21) it is sufficient to show that for every a in $\mathfrak{A}$

$$X_1 \cap X_2 \subset h_1(a) \cup f(h_1(\neg a)). \tag{23}$$

Let us note that the conditions $\nabla \in X_1 \cap X_2$ and $a \notin \nabla$ imply that $\neg a \in \nabla$. Indeed, if $a \notin \nabla$, then $\sim a \notin \tilde{\nabla}$ and hence $\sim a \in A - \tilde{\nabla} = f(\nabla) \in X_2$. Hence, on account of $\sim \neg a \prec a = \sim \sim a$ (see 1 ((qp$\mathbf{B}_{11}$), (qp$\mathbf{B}_1$))), we get $\neg a \in f(\nabla)$. The condition $\nabla \in X_1 \cap X_2$ implies $f(\nabla) = \nabla$. Thus $\neg a \in \nabla$. Consequently, if $\nabla \in X_1 \cap X_2$ and $\nabla \notin h_1(a)$, then $\nabla = f(\nabla)$ and $\nabla \in h_1(\neg a)$, i.e. $\nabla \in f(h_1(\neg a))$. This proves (23) and consequently the inclusion (21).

Let $B(X)$ be the class of all subsets Y of X satisfying the following conditions:

(b$_1$) $Y \cap X_1 \in G(X_1)$,

(b$_2$) $Y \cap X_2 = X_2 - g(Z)$, for some $Z \in G(X_1)$,

(b$_3$) $Y \cap X_1 \subset g(Y) \cap X_1$,

and let us put

$$\sim Y = X - g(Y), \tag{24}$$

$$Y \to Z = (Y \cap X_1 \Rightarrow_1 Z \cap X_1) \cap \big((X_2 - g(Y \cap X_1)) \cup Z \cap X_2\big), \tag{25}$$

$$Y \Rightarrow Z = (Y \to Z) \cap (\sim Z \to \sim Y), \tag{26}$$

$$\neg Y = Y \to O. \tag{27}$$

It follows from 5.3, (16), (18), (19), (20), (21) and 3.4 that under the above hypotheses and with the notation described above the following theorem holds.

5.4. $\mathfrak{B}(X) = (B(X), X, \Rightarrow, \cup, \cap, \to, \sim, \neg)$ *is a quasi-pseudo-Boolean algebra of sets connected with the pseudo-field* $\mathfrak{G}(X_1)$.

Let

$$B_0(X) = \{h(a)\}_{a \in A}, \tag{28}$$

where

$$h(a) = \{\nabla \in X \colon a \in \nabla\}. \tag{29}$$

We shall consider X as a topological space with the subbase $B_0(X)$. Then the following theorem holds.

5.5. *The topological space X of all prime filters in a quasi-pseudo-Boolean algebra $\mathfrak{A} = (A, V, \Rightarrow, \cup, \cap, \rightarrow, \sim, \neg)$ with the subbase $B_0(X)$ is a compact T_0-space. The class $B_0(X)$ is a subalgebra of $\mathfrak{B}(X)$,* i.e. $\mathfrak{B}_0(X) = (B_0(X), X, \Rightarrow, \cup, \cap, \rightarrow, \sim, \neg)$ *is a quasi-pseudo-Boolean algebra of some open subsets of X, connected with the pseudo-field $\mathfrak{G}(X_1)$. The mapping h is an isomorphism from $\mathfrak{A}$ onto $\mathfrak{B}_0(X)$* [9].

Since $(A, V, \cup, \cap, \sim)$ is a quasi-Boolean algebra, it follows from the proof of III 3.2 that X is a compact T_0-space and h is an isomorphism from $(A, V, \cup, \cap, \sim)$ onto $(B_0(X), X, \cup, \cap, \sim)$, i.e. h is one-one and

(30) $$h(V) = X,$$

(31) $$h(a \cup b) = h(a) \cup h(b),$$

(32) $$h(a \cap b) = h(a) \cap h(b),$$

(33) $$h(\sim a) = X - g(h(a)) = \sim h(a).$$

We shall prove that $B_0(X) \subset B(X)$.

By (29) and (1) for every $a \in A$ we have $h(a) \cap X_1 = h_1(a) \in G(X_1)$, i.e. $h(a)$ satisfies the condition (b_1). Moreover, for any $a \in A$

(34) $$h(a) \cap X_2 = X_2 - g(h_1(\sim a)), \quad h_1(\sim a) \in X_1.$$

To prove (34), let us observe that $\nabla \in g(h_1(\sim a))$ if and only if $\nabla \in X_2$, $\nabla = g(\nabla_1)$, where $\nabla_1 \in X_1$ and $\sim a \in \nabla_1$. This holds if and only if $\nabla = A - \tilde{\nabla}_1$ and $\sim a \in \nabla_1$. The last condition holds if and only if $\nabla \in X_2$ and $a \notin \nabla$. Consequently, for every ∇ in X_2, $\nabla \notin g(h_1(\sim a))$ if and only if $a \in \nabla$, i.e. if and only if $\nabla \in h(a)$. This proves (34). Thus for every $a \in A$, the set $h(a)$ satisfies (b_2). In order to prove that

(35) $$h(a) \cap X_1 \subset g(h(a)) \cap X_1,$$

suppose that $\nabla \in h(a) \cap X_1$, i.e. $a \in \nabla \in X_1$. Then it follows by 1 (qpB_{12}) that $\sim a \notin \nabla$. Hence $a \notin \tilde{\nabla}$. Thus $a \in A - \tilde{\nabla} = g(\nabla)$, i.e. $g(\nabla) \in h(a)$. Hence $\nabla = g(g(\nabla)) \in g(h(a))$. Inclusion (35) shows that for every $a \in A$, the set $h(a)$ satisfies the condition (b_3). Thus $B_0(X) \subset B(X)$.

[9] Białynicki-Birula and Rasiowa [2].

Now, observe that by (29), (1) and 5.2 we have

$$(36) \qquad h(a \to b) \cap X_1 = h_1(a \to b) = h_1(a) \Rightarrow_1 h_1(b) = h(a) \cap X_1 \Rightarrow_1 h(b) \cap X_1.$$

We shall prove that

$$(37) \qquad h(a \to b) \cap X_2 = \big(X_2 - g(h(a) \cap X_1)\big) \cup h(b) \cap X_2.$$

Indeed, $\nabla \in h(a \to b) \cap X_2$ if and only if $\nabla \in X_2$ and $a \to b \in \nabla$. By 1 ((qpB$_8$), (qpB$_9$)) and 1.3 (37) we have

$$\sim\sim(a \cap \sim b) = a \cap \sim b \prec \sim(a \to b),$$

$$\sim(a \to b) \prec a \cap \sim b = \sim\sim(a \cap \sim b).$$

Hence, by 4 (s.f.s.$_2$) and 1 (qpB$_1$), we get

$$(38) \quad a \to b \in \nabla \quad \text{if and only if} \quad \sim a \cup b = \sim(a \cap \sim b) \in \nabla.$$

Since ∇ is a prime s.f.s.k., the condition $\sim a \cup b \in \nabla$ implies that either $\sim a \in \nabla$ or $b \in \nabla$. On the other hand, since by 4.6 every s.f.s.k is a filter and $\sim a \leqslant \sim a \cup b$, $b \leqslant \sim a \cup b$, we infer (see III 1 (f$_2$)) that if either $\sim a \in \nabla$ or $b \in \nabla$, then $\sim a \cup b \in \nabla$. Thus $\sim a \cup b \in \nabla$ if and only if either $\sim a \in \nabla$ or $b \in \nabla$. Hence, by (38) for every $\nabla \in X_2$, $a \to b \in \nabla$ if and only if either $\sim a \in \nabla$ or $b \in \nabla$. Consequently, by (33), (19)

$$\begin{aligned} h(a \to b) \cap X_2 &= h(\sim a) \cap X_2 \cup h(b) \cap X_2 \\ &= \big(X_2 - g(h(a)) \cap X_2\big) \cup h(b) \cap X_2 \\ &= \big(X_2 - g(h(a) \cap X_1)\big) \cup h(b) \cap X_2. \end{aligned}$$

Thus (37) holds.

By (36), (37) and (16) we get

$$h(a \to b) = \big(h(a) \cap X_1 \Rightarrow_1 h(b) \cap X_1\big) \cup \cup\Big(\big(X_2 - g(h(a) \cap X_1)\big) \cup h(b) \cap X_2\Big).$$

Hence, applying (25), we get

$$(39) \qquad h(a \to b) = h(a) \to h(b).$$

By 1 (qpB$_4$), (32), (39), (33), (26)

$$(40) \qquad h(a \Rightarrow b) = h(a) \Rightarrow h(b).$$

By 1 (qpB_{13}), (39), (33), (30), (24), (27)

(41) $$h(\neg a) = \neg h(a).$$

Since $B_0(X) \subset B(X)$, the equations (30), (31), (32), (33), (39), (40), (41) show that $B_0(X)$ is a subalgebra of $\mathfrak{B}(X)$ and that h is a homomorphism from $\mathfrak{A}$ onto $\mathfrak{B}_0(X)$. Since h is one-one, it is an isomorphism.

The following theorem follows from 5.5.

5.6. *Every quasi-pseudo-Boolean algebra is isomorphic to a quasi-pseudo-Boolean algebra of sets connected with a pseudo-field of sets, more exactly with a quasi-pseudo-Boolean algebra of open subsets of a compact topological T_0-space.*

We shall complete this section by proving the following theorem, which gives the necessary and sufficient condition for a quasi-Boolean algebra to be extendable to a quasi-pseudo-Boolean algebra.

5.7. *A quasi-Boolean algebra $\mathfrak{A} = (A, V, \cup, \cap, \sim)$ can be extended to a quasi-pseudo-Boolean algebra if and only if*

(42) $$a \cap \sim a \leqslant b \cup \sim b \quad \textit{for all } a, b \in A$$ [10].

The necessity of (42) follows from 1.3 (17). To prove the sufficiency let us suppose that in $\mathfrak{A}$ the condition (42) is satisfied. For every prime filter ∇ in $\mathfrak{A}$ let

(43) $$g(\nabla) = A - \tilde{\nabla}, \quad \text{where} \quad \tilde{\nabla} = \{\sim a \in A \colon a \in \nabla\}.$$

First of all we shall prove that

(44) *for every prime filter ∇ in $\mathfrak{A}$, either $\nabla \subset g(\nabla)$ or $g(\nabla) \subset \nabla$.*

Indeed, let us suppose that $\nabla \subset g(\nabla)$ does not hold. Then there exists an $a \in A$, such that $a \in \nabla$ and $a \notin g(\nabla)$. It follows from $a \notin g(\nabla)$ that $\sim a \in \nabla$. Thus $a \cap \sim a \in \nabla$. Hence, by (42) for every $b \in A$, $b \cup \sim b \in \nabla$. Since ∇ is a prime filter, this implies that for every $b \in A$, either $b \in \nabla$ or $\sim b \in \nabla$. We shall show that $g(\nabla) \subset \nabla$. In fact, suppose $b \in g(\nabla)$. This implies that $\sim b \notin \nabla$. Hence $b \in \nabla$.

Let X_1 be the set of all prime filters ∇ in $\mathfrak{A}$ such that $\nabla \subset g(\nabla)$. We shall consider X_1 as a topological space with the discrete topology, i.e. we set $\mathbf{I}_1 Y = Y$ for every $Y \subset X_1$. Let $G(X_1)$ be the class of all open

[10] Białynicki-Birula and Rasiowa [2].

subsets in X_1, i.e. the class of all subsets of X_1. Let $\mathfrak{G}(X_1) = (G(X_1), X_1, \Rightarrow_1, \cup, \cap, \neg_1)$, where $\cup$ and $\cap$ are set-theoretical union and intersection and $\Rightarrow_1, \neg_1$ are defined by (1), (2) in Section 3. Then for every $Y, Z \in G(X_1)$ we have

$$(45) \qquad Y \Rightarrow_1 Z = (X_1 - Y) \cup Z,$$

$$(46) \qquad \neg_1 Y = X_1 - Y.$$

Clearly, $\mathfrak{G}(X_1)$ is a pseudo-field of sets. Let X_2 be the set of all prime filters ∇ in $\mathfrak{A}$ such that $g(\nabla) \subset \nabla$. Moreover, let

$$f(\nabla) = g(\nabla) \quad \text{for every } \nabla \in X_1.$$

It is easy to verify that f is a one-one mapping from X_1 onto X_2 satisfying conditions (3), (4), (6), (7) in Section 3 and by (44) the set $X = X_1 \cup X_2$ is the set of all prime filters in $\mathfrak{A}$. For every $a \in A$ let

$$(47) \qquad h(a) = \{\nabla \in X \colon a \in \nabla\}.$$

Moreover, let $B(X)$ be the class of all subsets Y of X satisfying condition

$$(48) \qquad Y \cap X_1 \subset g(Y) \cap X_1.$$

It is easy to verify that $B(X)$ coincides with the class of all subsets of X fulfilling conditions (b_1), (b_2), (b_3) in Section 3, and that $h(a) \in B(X)$ for every $a \in A$. Thus by 3.4, the algebra $(B(X), X, \Rightarrow, \cup, \cap, \rightarrow, \sim, \neg)$, where $\sim, \rightarrow, \Rightarrow, \neg$ are defined by (9), (10), (21), (22) in Section 3 and $\cup, \cap$ are set-theoretical union and intersection, is a quasi-pseudo-Boolean algebra of sets connected with $G(X_1)$. On the other hand, it follows from the proof of III 3.2 that $\mathfrak{Q}(X) = (\{h(a)\}_{a \in A}, X, \cup, \cap, \sim)$ is a quasi-field of sets and h is an isomorphism from $\mathfrak{A}$ onto $\mathfrak{Q}(X)$. Thus Theorem 5.7 is proved.

Exercises

1. Show that the following equations can be chosen as a system of axioms for quasi-pseudo-Boolean algebras: N1) $a \cup V = V$, N2) $a \cap (a \cup b) = a$, N3) $a \cap (b \cup c) = (c \cap a) \cup (b \cap a)$, N4) $\sim \sim a = a$, N5) $\sim (a \cap b) = \sim a \cup \sim b$, N6) $a \cap \sim a = (a \cap \sim a) \cap (b \cup \sim b)$, N7) $a \rightarrow a = V$, N8) $(a \rightarrow b) \cap b = b$, N9) $a \cap (a \rightarrow b) = a \cap (\sim a \cup b)$, N10) $a \rightarrow (b \cap c) = (a \rightarrow b) \cap (a \rightarrow c)$, N11) $a \rightarrow (b \rightarrow c) = (a \cap b) \rightarrow c$, N12) $\neg a = a \rightarrow \sim V$, N13) $a \Rightarrow b = (a \rightarrow b) \cap (\sim b \rightarrow \sim a)$[11]. Apply Ex. 12 in Chap. III.

[11] Brignole and Monteiro [1].

2. We shall say that a lattice $\mathfrak{A} = (A, \cup, \cap)$ contains a *pseudo-complement* of $a \in A$ *relative* to $b \in A$ if there exists an element $c \in A$ such that the following conditions are satisfied: 1) $a \cap c \leqslant b$, 2) if $d \in A$ and $a \cap d \leqslant b$, then $d \leqslant c$. If such an element exists it is uniquely determined and will be denoted by $a \Rightarrow^* b$. Prove that each quasi-pseudo-Boolean algebra $\mathfrak{A} = (A, V, \Rightarrow, \cup, \cap, \sim, \rightarrow, \neg)$ contains for all $a, b \in A$ the pseudo-complement $a \Rightarrow^* (\sim a \cup b)$ of a relative to $\sim a \cup b$ which is equal to $a \rightarrow b$ (12).

3. Let $\mathfrak{A} = (A, V, \cup, \cap)$ be a finite distributive lattice. An element $a \in A$ is said to be *prime* provided that $a \neq \Lambda$ and the condition $a = b \cup c$ implies that either $a = b$ or $a = c$. Let P be the set of all prime elements in $\mathfrak{A}$. The set P is ordered by the lattice ordering relation $\leqslant$. Prove that $\mathfrak{A}$ can be extended to a quasi-pseudo-Boolean algebra $(A, V, \Rightarrow, \cup, \cap, \sim, \rightarrow, \neg)$ if and only if there exists a one-one mapping I from P onto P satisfying the following conditions: 1) $p \leqslant q$ if and only if $Iq \leqslant Ip$; 2) $IIp = p$; 3) $p \leqslant Ip$ or $Ip \leqslant p$; 4) if $Ip' \leqslant p'$, $Ip' \leqslant p''$, $Ip'' \leqslant p'$ and $Ip'' \leqslant p''$, then there exists a $q \in P$ such that $q \leqslant p'$, $q \leqslant p''$, $Ip' \leqslant q$ and $Ip'' \leqslant q$ (13). In this case the operations $\sim, \rightarrow, \neg, \Rightarrow$ are defined as follows: $\sim a = \bigcup_{\substack{b \in P \\ \text{non } Ib \leqslant a}} b$, $a \rightarrow b = a \Rightarrow^* (\sim a \cup b)$ (see Ex. 2) $\neg a = a \rightarrow \Lambda$, $a \Rightarrow b = (a \rightarrow b) \cap (\sim b \rightarrow \sim a)$.

4. Applying Ex. 3, construct a finite quasi-pseudo-Boolean algebra from the distributive lattice determined by the following diagram (14).

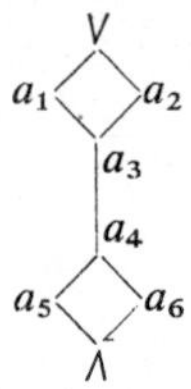

5. By the *radical* *Rad* $\mathfrak{A}$ of a quasi-pseudo-Boolean algebra $\mathfrak{A} = (A, V, \Rightarrow, \cup, \cap, \sim, \rightarrow, \neg)$ we mean the intersection of all maximal s.f.f.k. in $\mathfrak{A}$. Prove, that *Rad* $\mathfrak{A}$ is identical with each of the following sets: $\{a \in A\colon \neg\neg a = V\}$, the s.f.f.k. generated by the set $\{a \cup \neg a\}_{a \in A}$, the s.f.f.k. generated by the set $\{((a \rightarrow b) \rightarrow a) \rightarrow a\}_{a \in A}$ (15).

6. Show that a prime s.f.f.k. is maximal if and only if it contains *Rad* $\mathfrak{A}$ (16).

7. A quasi-pseudo-Boolean algebra $\mathfrak{A}$ is said to be *semi-simple* provided Rad $\mathfrak{A} = \{V\}$. Prove that $\mathfrak{A}$ is semi-simple if and only if one of the following conditions

(12) Monteiro [3].
(13) Monteiro [3].
(14) Monteiro [3].
(15) Monteiro [2].
(16) Monteiro [2].

is satisfied: 1) $a \cup \neg a = V$, for each a in $\mathfrak{A}$; 2) $(a \to b) \to a = a$ for all a, b in $\mathfrak{A}$; 3) $a \to b = \neg a \cup b$ for all a, b in $\mathfrak{A}$; 4) each prime s.f.f.k. is maximal (17).

8. Prove that each semi-simple quasi-pseudo-Boolean algebra is isomorphic to a subalgebra of a product of the algebras $\mathfrak{C}_0$ (see Sec. 3), i.e. of the three-element quasi-pseudo-Boolean algebras (18).

9. Prove that each semi-simple quasi-pseudo-Boolean algebra is isomorphic to a quasi-pseudo-Boolean algebra of sets connected with a pseudo-Boolean algebra of simultaneously open and closed subsets of a compact totally disconnected Hausdorff space.

(17) Monteiro [2].
(18) Monteiro [2].

CHAPTER VI

BOOLEAN ALGEBRAS AND TOPOLOGICAL BOOLEAN ALGEBRAS

Introduction. The theory of Boolean algebras has been presented in detail in various books and is rather well known. Therefore, we assume that the reader is familiar with this theory. However, for the sake of continuity and because of the applications to topological Boolean algebras and Post algebras, explained in this chapter and Chapter VII, respectively, we describe here some basic properties of Boolean algebras. These include Stone's representation theorem.

Topological Boolean algebras were introduced and studied by McKinsey and Tarski [1] under the term closure algebras. Following Nöbeling, we use here the term topological Boolean algebras, which is adopted by many authors. These algebras characterize modal logic, to be analysed in Chapter XIII.

Topological Boolean algebras are, roughly speaking, Boolean algebras with, in addition, a one-argument operation **I**, called the interior operation and characterized by axioms analogous to those of the interior operations in topological spaces. The class of all subsets of any topological space is a topological Boolean algebra with respect to the set-theoretical operations (of union, intersection, complementation) and the interior operation. These algebras and their subalgebras cover all topological Boolean algebras up to isomorphism.

This chapter sums up the basic properties of these algebras, the theory of filters and the representation theorem included. Moreover, theorems which establish the relationship between topological Boolean algebras, on the one hand, and relatively pseudo-complemented lattices, contrapositionally complemented lattices, semi-complemented lattices and pseudo-Boolean algebras, on the other hand, are also included. These theorems have their metalogical analogues, to be given in Part Two. This chapter contains also lemmas on imbeddings for all the above-mentioned types of algebras. They will be applied in Part Two to the decision problem for the propositional calculi of logics which are algebraically characterized by algebras of these types.

1. Definition and elementary properties of Boolean algebras (1). We recall that an abstract algebra $(A, V, \Rightarrow, \cup, \cap, -)$ is said to be a *Boolean algebra* provided the following two conditions are satisfied:

(b_1) $(A, V, \Rightarrow, \cup, \cap, -)$ is a pseudo-Boolean algebra,

(b_2) $a \cup -a = V$ for each $a \in A$.

The operations $\Rightarrow$, $-$ are then called the *implication operation* and *complementation*, respectively.

The two-element Boolean algebra $(\{\Lambda, V\}, V, \Rightarrow, \cup, \cap, -)$ will be denoted by $\mathfrak{B}_0$. The Boolean algebras $\mathfrak{B}(X) = \big(B(X), X, \Rightarrow, \cup, \cap, -\big)$ of all subsets of any space $X \neq O$ and their subalgebras are called *fields of subsets of X*. The operations $\cup$, $\cap$, $-$ in these algebras are the set-theoretical operations of union, intersection and complementation, respectively. The operation $\Rightarrow$ is defined as in the implication algebras of sets (cf. II 5).

Note that

1.1. *In every Boolean algebra*

(1) $-\Lambda = V$,

(2) $--a = a$,

(3) $-(a \cap b) = -a \cup -b$,

(4) $a \Rightarrow b = -b \Rightarrow -a$,

(5) $a \Rightarrow b = -a \cup b$,

(6) $a \leqslant b$ *if and only if* $-b \leqslant -a$.

We state without proof that:

1.2. *Every Boolean algebra* $(A, V, \Rightarrow, \cup, \cap, -)$ *is an implication algebra with respect to* V *and* $\Rightarrow$. *Moreover, the following equations hold*:

(1) For a detailed exposition of the theory of Boolean algebras see e.g. Sikorski [3]; see also [MM].

(i) $a \Rightarrow -b = b \Rightarrow -a$, (ii) $-(a \Rightarrow a) \Rightarrow b = V$, (iii) $a \cup b = (a \Rightarrow b) \Rightarrow b$, (iv) $a \cap b = -(-a \cup -b)$. *Conversely, if* $(A, V, \Rightarrow, -)$ *is an abstract algebra such that* $(A, V, \Rightarrow)$ *is an implication algebra and equations* (i), (ii) *hold, then* $(A, V, \Rightarrow, \cup, \cap, -)$, *where the operations* $\cup, \cap$ *are defined by* (iii) *and* (iv), *is a Boolean algebra.*

It follows from (b_1), IV 6 and IV 1.2 that

1.3. *Every Boolean algebra is a distributive lattice with unit element* V *and zero element* Λ.

Let $\mathfrak{A} = (A, \cup, \cap, V, \Lambda)$ be a distributive lattice with zero element Λ and unit element V. An element $a \in A$ is said to be *complemented* if there is an element $b \in A$ such that

$$(c_1) \qquad a \cup b = V,$$

$$(c_2) \qquad a \cap b = \Lambda.$$

If $a \in A$ is complemented, then there exists only one element $b \in A$ satisfying (c_1) and (c_2). The easy proof of this is omitted. If a is a complemented element in A, then the element b satisfying (c_1) and (c_2) will be called the *complement of a* and will be denoted by $-a$. Observe that if a is complemented, then $-a$ is also complemented, viz. a is the complement of $-a$. Thus, for every complemented element a,

$$(7) \qquad a \cup -a = V, \quad a \cap -a = \Lambda \quad \text{and} \quad a = --a.$$

Note that

1.4. *An abstract algebra* $(A, V, \Rightarrow, \cup, \cap, -)$ *is a Boolean algebra if and only if* $(A, V, \cup, \cap)$ *is a distributive lattice with unit element* V *and zero element* Λ *and every element in* A *is complemented. Moreover, the operation* $\Rightarrow$ *satisfies the condition* $a \Rightarrow b = -a \cup b$ *for all* a, b *in* A.

We state without proof that:

1.5. *The set of all complemented elements in any distributive lattice* $(A, \cup, \cap, V, \Lambda)$ *with unit element* V *and zero element* Λ *is a non-degenerate subalgebra of this lattice and is a Boolean algebra under the operations* $V, \Rightarrow, \cup, \cap, -$, *where the operation* $\Rightarrow$ *is defined by the equation* $a \Rightarrow b = -a \cup b$.

2. Subalgebras of Boolean algebras. The aim of this section is to list some theorems on subalgebras of Boolean algebras which will be useful in the sequel.

2.1. *The subalgebra of a Boolean algebra* $(A, V, \Rightarrow, \cup, \cap, -)$ *generated by a non-empty subset* $A_0 \subset A$ *is the set consisting of all elements* a *of the form*

$$(1) \qquad a = (a_{11} \cup \ldots \cup a_{1n_1}) \cap \ldots \cap (a_{m1} \cup \ldots \cap a_{mn_m}),$$

where either $a_{ij} \in A_0$ *or* $-a_{ij} \in A_0$.

The easy proof of this is omitted.

2.2. *If* $(A_0, \cup, \cap)$ *is a sublattice of a Boolean algebra* $\mathfrak{A} = (A, V, \Rightarrow, \cup, \cap, -)$ *and both the unit element* V *and the zero element* Λ *of* $\mathfrak{A}$ *belong to* A_0, *then the Boolean subalgebra of* $\mathfrak{A}$ *generated by* A_0 *is the set of all elements* a *of the form*

$$(2) \qquad a = (a_1 \Rightarrow b_1) \cap \ldots \cap (a_n \Rightarrow b_n)$$

where $a_1, \ldots, a_n, b_1, \ldots, b_n$ *are in* A_0.

By 2.1 it suffices to prove that if $a_1, \ldots, a_k, b_1, \ldots, b_m$ are in A_0, $(k, m \geqslant 0, k+m > 0)$, then there are elements a and b in A_0 such that $-a_1 \cup \ldots \cup -a_k \cup b_1 \cup \ldots \cup b_m = -a \cup b$. Clearly, the elements

$$a = \begin{cases} a_1 \cap \ldots \cap a_k & \text{if } k > 0, \\ V & \text{if } k = 0, \end{cases} \qquad b = \begin{cases} b_1 \cup \ldots \cup b_m & \text{if } m > 0, \\ \Lambda & \text{if } m = 0 \end{cases}$$

satisfy the above equation.

2.3. *A subalgebra of a Boolean algebra generated by a set consisting of* m *elements contains at most* 2^{2^m} *elements.*

This is a simple consequence of 2.1.

3. Filters and implicative filters in Boolean algebras. Any Boolean algebra $(A, V, \Rightarrow, \cup, \cap, -)$ is an implication algebra with respect to the operations $V, \Rightarrow$, but it is also a distributive lattice. Therefore, it is possible to consider both the notion of an implicative filter and the notion of a filter in Boolean algebras. However, because of 1 (b_1) and IV 2.1 the following statement holds:

3.1. *A subset of the set of all elements in a Boolean algebra is an implicative filter if and only if it is a filter.*

The following theorem follows from 1 (b_1), IV 6.3, 3.1 and the equational definability of the class of all Boolean algebras.

3.2. *Let* $\mathfrak{A} = (A, V, \Rightarrow, \cup, \cap, -)$ *be a Boolean algebra and let* ∇ *be an implicative filter in* $\mathfrak{A}$. *Then the relation* $\approx_\nabla$ *determined by* ∇ (see II 1 (2)) *is a congruence in* $\mathfrak{A}$. *The quotient algebra* $\mathfrak{A}/\nabla$ *is a Boolean algebra and the conditions* (i), (ii), (iii) *of* II 3.2 *hold.*

We complete this section by stating the following theorem.

3.3. *The following conditions are equivalent for every implicative filter* ∇ *in any Boolean algebra* $\mathfrak{A} = (A, V, \Rightarrow, \cup, \cap, -)$:

(1) ∇ *is irreducible,*

(2) ∇ *is prime,*

(3) ∇ *is maximal,*

(4) *for every* $a \in A$ *exactly one of the elements* $a, -a$ *belongs to* ∇,

(5) $\mathfrak{A}/\nabla$ *is the two-element Boolean algebra.*

Conditions (1), (2), (3) are equivalent by II 6.1, II 6.2 and 1.2. The equivalence of (3), (4) and (5) follows from 3.1, 1 (b_1), 3.2 and IV 6.5.

4. Representation theorem for Boolean algebras. Given a Boolean algebra $\mathfrak{A} = (A, V, \Rightarrow, \cup, \cap, -)$, let $X(\mathfrak{A})$ be the set of all prime filters in $\mathfrak{A}$. For each $a \in A$, let

(1) $$h(a) = \{\nabla \in X(\mathfrak{A}): a \in \nabla\}.$$

Consider $X(\mathfrak{A})$ as a topological space, the class $P(\mathfrak{A}) = \{h(a)\}_{a \in A}$ being taken as a subbase (see I 2.1). This topological space will be said to be the *Stone space of* $\mathfrak{A}$. Using the notation above, the following well-known representation theorem for Boolean algebras can be formulated:

4.1. *Each Boolean algebra* $\mathfrak{A} = (A, V, \Rightarrow, \cup, \cap, -)$ *is isomorphic to a field of sets. More exactly,* $(P(\mathfrak{A}), X(\mathfrak{A}), \Rightarrow, \cup, \cap, -)$ *is a field of subsets of* $X(\mathfrak{A})$ *and the mapping* h *defined by* (1) *is an isomorphism of* $\mathfrak{A}$ *onto this field of sets. The Stone space* $X(\mathfrak{A})$ *of* $\mathfrak{A}$ *is a compact, totally disconnected Hausdorff space and* $P(\mathfrak{A})$ *is the class of all both open and closed subsets of* $X(\mathfrak{A})$ [2].

For a proof of 4.1 see e.g. [MM] (II § 8).

[2] Stone [2], [3].

5. Topological Boolean algebras (3). By a *topological Boolean algebra* we shall mean an abstract algebra $(A, V, \Rightarrow, \cup, \cap, -, \mathbf{I})$ where $(A, V, \Rightarrow, \cup, \cap, -)$ is a Boolean algebra and, moreover, the following equations hold:

$$(\mathrm{i}_1) \qquad \mathbf{I}(a\cap b) = \mathbf{I}a\cap \mathbf{I}b$$

$$(\mathrm{i}_2) \qquad \mathbf{I}a\cap a = \mathbf{I}a,$$

$$(\mathrm{i}_3) \qquad \mathbf{II}a = \mathbf{I}a,$$

$$(\mathrm{i}_4) \qquad \mathbf{I}V = V.$$

The operation $\mathbf{I}$ is then called an *interior operation.* Note that the class of all topological Boolean algebras is equationally definable.

If X is a topological space with an interior operation $\mathbf{I}$, then the family $B(X)$ of all subsets of X is a topological Boolean algebra with respect to the operations $V = X$, the operation $\Rightarrow$ defined by the formula

$$Y\Rightarrow Z = (X-Y)\cup Z, \qquad \text{for all subsets } Y, Z \text{ of } X,$$

the set-theoretical operations of union, intersection and complementation and the interior operation $\mathbf{I}$. Clearly, every subalgebra of this algebra is also a topological Boolean algebra called a *topological field of sets* or, more precisely, a topological field of subsets of X. We shall prove in Section 7 that these are typical examples of topological Boolean algebras.

Since the axioms (i_1)–(i_4) correspond to the axioms for interior operations in topological spaces (see I 2), the interior operation in a topological Boolean algebra has properties analogous to these of an interior operation in a topological space. This explains introducing terminology analogous to that used in the case of topological spaces.

The element $\mathbf{I}a$ will be said to be the *interior of* a. The element $-\mathbf{I}-a$ will be called the *closure of* a and will be denoted by $\mathbf{C}a$. Thus

$$(1) \qquad \mathbf{C}a = -\mathbf{I}-a \quad \text{and} \quad \mathbf{I}a = -\mathbf{C}-a.$$

It is easy to prove, just as in I 2, that for any a, b in a topological Boolean algebra

$$(\mathrm{c}_1) \qquad \mathbf{C}(a\cup b) = \mathbf{C}a\cup \mathbf{C}b,$$

(3) For a comprehensive exposition of topological Boolean algebras see McKinsey and Tarski [1], Sikorski [1], Nöbeling [1]. See also [MM].

(c_2) $$a \cup \mathbf{C}a = \mathbf{C}a,$$

(c_3) $$\mathbf{CC}a = \mathbf{C}a,$$

(c_4) $$\mathbf{C}\Lambda = \Lambda.$$

The operation $\mathbf{C}$ is then called a *closure operation.* By (i_1) and (c_1) we immediately conclude that

(2) *if* $a \leqslant b$, *then* $\mathbf{I}a \leqslant \mathbf{I}b$ *and* $\mathbf{C}a \leqslant \mathbf{C}b$.

Note that (i_2) and (c_2) are equivalent to

(3) $$\mathbf{I}a \leqslant a \quad \text{and} \quad a \leqslant \mathbf{C}a,$$

respectively.

An element $a \in A$ is said to be *open* (*closed*) if $a = \mathbf{I}a$ (if $a = \mathbf{C}a$). It is easy to prove that an element a is open (closed) if and only if its complement $-a$ is closed (open). The meet (join) of a finite number of open (closed) elements is open (closed). For any set $\{a_t\}_{t \in T}$ of open (closed) elements a_t, $t \in T$, if there exists a least upper bound l.u.b. $\{a_t\}_{t \in T}$ (if there exists a greatest lower bound g.l.b. $\{a_t\}_{t \in T}$), then it is an open element (then it is a closed element).

By (i_2), (i_3), (2) we infer that if a is an open element, then for every element b

$$a \leqslant b \quad \text{if and only if} \quad a \leqslant \mathbf{I}b.$$

This means by (3) that $\mathbf{I}b$ is the greatest open element contained in b. Similarly, by (c_2), (c_3) and (2) we infer that if a is closed, then for every b

$$b \leqslant a \quad \text{if and only if} \quad \mathbf{C}b \leqslant a.$$

Thus by (3) $\mathbf{C}b$ is the least closed element containing b.

6. I-filters in topological Boolean algebras. Since every topological Boolean algebra is a Boolean algebra, we can consider the notion of a filter (see Sec. 3) which coincides with the notion of an implicative filter (see 3.1).

A filter ∇ in a topological Boolean algebra $(A, V, \Rightarrow, \cup, \cap, -, \mathbf{I})$ is said to be an **I**-*filter* provided

(1) $a \in \nabla$ implies that $\mathbf{I}a \in \nabla$, for every $a \in A$.

The set of all elements in a topological Boolean algebra is an **I**-filter. The set $\{V\}$ is also an **I**-filter, called the *unit* **I**-*filter*. For every fixed element $a \in A$, the set $\{x \in A: \mathbf{I}a \leqslant x\}$ is the least $\mathbf{I}$-filter containing a and is called the *principal* $\mathbf{I}$-*filter generated by a.*

6.1. *The* $\mathbf{I}$-*filter generated by a non-empty set* A_0 *of elements of a topological Boolean algebra* $(A, V, \Rightarrow, \cup, \cap, -, \mathbf{I})$ *is the set of all elements x in A for which there exist* $a_1, \ldots, a_n$ *in* A_0 *such that* $\mathbf{I}a_1 \cap \ldots \cap \mathbf{I}a_n \leqslant x$.

The easy proof by verification is left to the reader.

6.2. *The* $\mathbf{I}$-*filter generated by an* $\mathbf{I}$-*filter* ∇ *and an element* a_0 *in a topological Boolean algebra* $\mathfrak{A}$ *is the set of all elements x in* $\mathfrak{A}$ *for which there exists a* $c \in \nabla$ *such that* $\mathbf{I}a_0 \cap \mathbf{I}c \leqslant x$.

The easy proof is left to the reader.

An $\mathbf{I}$-filter ∇ in a topological Boolean algebra is said to be *proper* if there exists an element a in that algebra such that $a \notin \nabla$; this is equivalent to the condition that $\wedge \notin \nabla$.

An $\mathbf{I}$-filter ∇ is said to be *maximal* provided it is proper and is not a proper subset of any proper $\mathbf{I}$-filter.

An $\mathbf{I}$-filter ∇ is said to be *irreducible* provided it is proper and for any two proper $\mathbf{I}$-filters ∇_1, ∇_2 the condition $\nabla = \nabla_1 \cap \nabla_2$ implies that either $\nabla = \nabla_1$ or $\nabla = \nabla_2$.

An $\mathbf{I}$-filter ∇ is said to be $\mathbf{I}$-*prime* provided it is proper and the condition $\mathbf{I}a \cup \mathbf{I}b \in \nabla$ implies that either $\mathbf{I}a \in \nabla$ or $\mathbf{I}b \in \nabla$, for all elements a, b in the topological Boolean algebra under consideration.

We state without proof the following simple result.

6.3. *Every maximal* $\mathbf{I}$-*filter is an irreducible* $\mathbf{I}$-*filter.*

We shall prove that

6.4. *An* $\mathbf{I}$-*filter is irreducible if and only if it is* $\mathbf{I}$-*prime.*

Suppose that an $\mathbf{I}$-filter ∇ is not $\mathbf{I}$-prime. Then $\mathbf{I}a \cup \mathbf{I}b \in \nabla$ and $\mathbf{I}a \notin \nabla$, $\mathbf{I}b \notin \nabla$ for some elements a, b in the topological Boolean algebra under consideration. Let ∇_1 be the $\mathbf{I}$-filter generated by ∇ and a, and let ∇_2 be the $\mathbf{I}$-filter generated by ∇ and b. We show that

$$\nabla = \nabla_1 \cap \nabla_2. \tag{2}$$

Clearly, $\nabla \subset \nabla_1 \cap \nabla_2$. Suppose that $x \in \nabla_1 \cap \nabla_2$. Then by 6.2 $\mathbf{I}a \cap \mathbf{I}c_1 \leqslant x$ and $\mathbf{I}b \cap \mathbf{I}c_2 \leqslant x$ for some c_1, c_2 in ∇. Let $c = c_1 \cap c_2$. Then $\mathbf{I}c = \mathbf{I}c_1 \cap \mathbf{I}c_2$ and we have $\mathbf{I}a \cap \mathbf{I}c \leqslant x$, $\mathbf{I}b \cap \mathbf{I}c \leqslant x$. Consequently,

$(\mathbf{I}a \cup \mathbf{I}b) \cap \mathbf{I}c = (\mathbf{I}a \cap \mathbf{I}c) \cup (\mathbf{I}b \cap \mathbf{I}c) \leqslant x$. Since $\mathbf{I}a \cup \mathbf{I}b \in \nabla$ and $\mathbf{I}c \in \nabla$, we infer that $x \in \nabla$. Thus $\nabla_1 \cap \nabla_2 \subset \nabla$, which proves that (2) holds. Observe that ∇_1 and ∇_2 are proper. Otherwise we would have by (2), either $\nabla = \nabla_1$ or $\nabla = \nabla_2$, which contradicts our hypothesis that $\mathbf{I}a \notin \nabla$ and $\mathbf{I}b \notin \nabla$. Hence ∇ is not irreducible.

Conversely, suppose that ∇ is not an irreducible $\mathbf{I}$-filter. Then $\nabla = \nabla_1 \cap \nabla_2$, where $\nabla_1 \neq \nabla$ and $\nabla_2 \neq \nabla$. Thus $\nabla_1 \neq \nabla_2$. Then there exist $a \in \nabla_1$, $a \notin \nabla$ and $b \in \nabla_2$, $b \notin \nabla$ and $a \neq b$. Consequently, $\mathbf{I}a \in \nabla_1$ and $\mathbf{I}b \in \nabla_2$. Hence $\mathbf{I}a \cup \mathbf{I}b \in \nabla_1$ and $\mathbf{I}a \cup \mathbf{I}b \in \nabla_2$. Thus $\mathbf{I}a \cup \mathbf{I}b \in \nabla$. But neither $\mathbf{I}a$ belongs to ∇ nor $\mathbf{I}b$ belongs to ∇. Thus ∇ is not $\mathbf{I}$-prime.

It is easy to prove that

6.5. *The union of any chain of proper* $\mathbf{I}$*-filters in a topological Boolean algebra is a proper* $\mathbf{I}$*-filter.*

The proof is left to the reader.

6.6. *Each proper* $\mathbf{I}$*-filter in a topological Boolean algebra is contained in a maximal* $\mathbf{I}$*-filter.*

The proof follows from 6.5 by the application of I 3.1.

6.7. *For every element* a *in a topological Boolean algebra, if* $\mathbf{I}a \neq \wedge$, *then there exists a maximal* $\mathbf{I}$*-filter* ∇ *such that* $a \in \nabla$.

This follows from 6.6.

6.8. *If* ∇_0 *is an* $\mathbf{I}$*-filter in a topological Boolean algebra and* $\mathbf{I}a \notin \nabla_0$, *then there exists an irreducible* $\mathbf{I}$*-filter* ∇ *such that* $\nabla_0 \subset \nabla$ *and* $a \notin \nabla$.

The proof, similar to that of III 1.12, is left to the reader.

6.9. *If* $K(h)$ *is the kernel of an epimorphism* h *of a topological Boolean algebra* $\mathfrak{A} = (A, V, \Rightarrow, \cup, \cap, -, \mathbf{I})$ *onto a similar algebra* $\mathfrak{B} = (B, V', \Rightarrow, \cup, \cap, -, \mathbf{I})$, i.e. $K(h) = h^{-1}(V')$, *then* $K(h)$ *is an* $\mathbf{I}$*-filter. Moreover,* $h(a) = h(b)$ *is equivalent to* $a \Rightarrow b \in K(h)$ *and* $b \Rightarrow a \in K(h)$. *Consequently, the relation* $\approx_{K(h)}$ *determined by* $K(h)$ *is a congruence relation in* $\mathfrak{A}$. *The quotient algebra* $\mathfrak{A}/\approx_{K(h)}$ *is a topological Boolean algebra isomorphic to* $\mathfrak{B}$.

Since every topological Boolean algebra is a Boolean algebra and consequently, by 1 (b_1), a pseudo-Boolean algebra, we infer that it is a relatively pseudo-complemented lattice. By IV 2.3 $K(h)$ is a filter. Suppose that $a \in K(h)$, i.e. $h(a) = V'$. Hence $\mathbf{I}h(a) = h(\mathbf{I}a) = V'$. Thus

$\mathbf{I}a \in K(h)$. This proves that $K(h)$ is an $\mathbf{I}$-filter. The second part of 6.9 also follows from IV 2.3 and the equational definability of topological Boolean algebras.

6.10. *Let* $\mathfrak{A} = (A, V, \Rightarrow, \cup, \cap, -, \mathbf{I})$ *be a topological Boolean algebra and let* ∇ *be an* $\mathbf{I}$*-filter in* $\mathfrak{A}$. *Then the relation* $\approx_\nabla$ *determined by* ∇ *is a congruence relation in* $\mathfrak{A}$. *The quotient algebra* $\mathfrak{A}/\nabla$ *is a topological Boolean algebra and the conditions* (i), (ii), (iii) *of* II 3.2 *hold.*

Because of 3.2 it is sufficient to prove that the relation $\approx_\nabla$ preserves the operation $\mathbf{I}$. By 1 (b_1) and 1.1 (5) for each a, b in every topological Boolean algebra

(3) $$a \Rightarrow b = -a \cup b.$$

We shall show that

(4) $$\mathbf{I}(a \Rightarrow b) \leqslant \mathbf{I}a \Rightarrow \mathbf{I}b \quad \textit{for all } a, b \textit{ in } A.$$

Note that $(a \cap -a) \cup (a \cap b) = \wedge \cup (a \cap b) = a \cap b \leqslant b$. Hence, by the distributivity, $a \cap (-a \cup b) \leqslant b$. Applying 5 (2) and 5 ($i_1$), we get $\mathbf{I}a \cap \cap \mathbf{I}(-a \cup b) = \mathbf{I}(a \cap (-a \cup b)) \leqslant \mathbf{I}b$. Consequently, by 1 ($b_1$) and IV 1 (r), we get $\mathbf{I}(-a \cup b) \leqslant \mathbf{I}a \Rightarrow \mathbf{I}b$. Applying (3) we obtain (4).

If $a \Rightarrow b \in \nabla$ and $b \Rightarrow a \in \nabla$, then $\mathbf{I}(a \Rightarrow b) \in \nabla$ and $\mathbf{I}(b \Rightarrow a) \in \nabla$. Hence, by (4), we infer that $\mathbf{I}a \Rightarrow \mathbf{I}b \in \nabla$ and $\mathbf{I}b \Rightarrow \mathbf{I}a \in \nabla$. We have proved that the condition $a \approx_\nabla b$ implies that $\mathbf{I}a \approx_\nabla \mathbf{I}b$, i.e. the relation $\approx_\nabla$ preserves the operation $\mathbf{I}$.

6.11. *For every* $\mathbf{I}$*-filter* ∇ *in a topological Boolean algebra* $\mathfrak{A}$ *the following conditions are equivalent*:

(5) ∇ *is a maximal* $\mathbf{I}$*-filter*,

(6) *for every element* a *in* $\mathfrak{A}$ *exactly one of the elements* $\mathbf{I}a$, $-\mathbf{I}a$ *belongs to* ∇,

(7) *in the quotient algebra* $\mathfrak{A}/\nabla$ *every open element is closed and every closed element is open; all open (closed) elements form the two-element Boolean algebra.*

Conditions (5) and (6) are equivalent. Indeed, if ∇ is maximal, then it is proper and consequently at most one of the elements $\mathbf{I}a$, $-\mathbf{I}a$ belongs to ∇. On the other hand, by 6.3 and 6.4 ∇ is $\mathbf{I}$-prime. Since $\mathbf{I}a \cup -\mathbf{I}a$

$= V$ belongs to ∇, we infer that at least one of the elements $\mathbf{I}a$, $-\mathbf{I}a$ belongs to ∇. Thus (5) implies (6). Condition (6) clearly implies (5).

If (6) holds, then ∇ is proper and consequently $\mathfrak{A}/\nabla$ is not degenerate. If $||a||$ is open, then $||a|| = \mathbf{I}||a|| = ||\mathbf{I}a||$. By (6) and 6.10 either $||\mathbf{I}a|| = V$ or $||-\mathbf{I}a|| = -||\mathbf{I}a|| = V$, i.e. $||\mathbf{I}a|| = \Lambda$. Thus either $||a|| = V$ or $||a|| = \Lambda$. Consequently, there are only two open elements in $\mathfrak{A}/\nabla$, the zero element and the unit element. Since they are simultaneously closed, we infer that every open element in $\mathfrak{A}/\nabla$ is closed. If $||a||$ is closed, then $||-a||$ is open. Consequently, by the part just proved, $||-a||$ is closed, i.e. $||a||$ is open. Thus every closed element in $\mathfrak{A}/\nabla$ is open. We have proved that (6) implies (7). The condition (7) implies (6). In fact, if $\mathbf{I}a \notin \nabla$ for some a in $\mathfrak{A}$, then $||\mathbf{I}a|| \neq V$, i.e. $||\mathbf{I}a|| = \Lambda$. Consequently $||-\mathbf{I}a|| = -||\mathbf{I}a|| = V$. Thus $-\mathbf{I}a \in \nabla$. This completes the proof of 6.11.

7. Representation theorem for topological Boolean algebras. The aim of this section is to prove the following representation theorem.

7.1. *For every topological Boolean algebra* $\mathfrak{A} = (A, V, \Rightarrow, \cup, \cap, -, \mathbf{I})$ *there exists a monomorphism h from* $\mathfrak{A}$ *into a topological field of all subsets of a topological space X*[4].

By 4.1 there exists a monomorphism h from the Boolean algebra $(A, V, \Rightarrow, \cup, \cap, -)$ into the field of all subsets of a set $X \neq O$. We shall introduce a topology in X as follows. For each $Y \subset X$ we define $\mathbf{I}Y$ as the union of all sets $h(a)$ such that $h(a) \subset Y$ and $a = \mathbf{I}a$. The operation $\mathbf{I}$ thus defined satisfies the following conditions for all subsets Y, Z of X:

(1) $$\mathbf{I}Y \subset Y,$$

(2) $$\text{if} \quad Y \subset Z, \quad \text{then} \quad \mathbf{I}Y \subset \mathbf{I}Z,$$

(3) $$\mathbf{I}Y \subset \mathbf{I}\mathbf{I}Y,$$

(4) $$\mathbf{I}(Y \cap Z) = \mathbf{I}Y \cap \mathbf{I}Z,$$

(5) $$\mathbf{I}X = X.$$

Moreover, for each $a \in A$

(6) $$h(\mathbf{I}a) = \mathbf{I}h(a).$$

[4] McKinsey and Tarski [1].

Conditions (1) and (2) follow directly from the definition of the operation $\mathbf{I}$. To prove (3) observe that $\mathbf{II}Y$ is the union of all sets $h(b)$ such that $h(b) \subset \mathbf{I}Y$ and $b = \mathbf{I}b$. If $h(a) \subset Y$ and $a = \mathbf{I}a$, then by the definition of $\mathbf{I}Y$, we infer that $h(a) \subset \mathbf{I}Y$ and therefore is one of the sets whose union is the set $\mathbf{II}Y$. Thus (3) holds. By (1) and (2) $\mathbf{I}(Y \cap Z) \subset \mathbf{I}Y$ and $\mathbf{I}(Y \cap Z) \subset \mathbf{I}Z$. Hence $\mathbf{I}(Y \cap Z) \subset \mathbf{I}Y \cap \mathbf{I}Z$. On the other hand, if $x \in \mathbf{I}Y \cap \mathbf{I}Z$, then there exists an $h(a)$ such that $x \in h(a)$, $h(a) \subset Y$, $a = \mathbf{I}a$ and there exists an $h(b)$ such that $x \in h(b)$, $h(b) \subset Z$, $b = \mathbf{I}b$. Hence, $x \in h(a) \cap h(b) = h(a \cap b)$, where $h(a) \cap h(b) \subset Y \cap Z$ and, $\mathbf{I}(a \cap b) = \mathbf{I}a \cap \mathbf{I}b = a \cap b$. Thus $x \in \mathbf{I}(Y \cap Z)$. This proves that $\mathbf{I}Y \cap \mathbf{I}Z \subset \mathbf{I}(Y \cap Z)$ and consequently that (4) holds. To prove (5) observe that $h(V) \subset X$ and $V = \mathbf{I}V$; thus $X = h(V) \subset \mathbf{I}X$, i.e. $X = \mathbf{I}X$. To prove (6) note that $h(\mathbf{I}a) \subset \mathbf{I}h(a)$ because $h(\mathbf{I}a) \subset h(a)$ and $\mathbf{II}a = \mathbf{I}a$, i.e. $h(a)$ is one of the sets whose union is the set $\mathbf{I}h(a)$. On the other hand, if $h(b) \subset h(a)$ and $b = \mathbf{I}b$, then since h is a Boolean monomorphism, we get $b \subset a$. Consequently, by 5 (2), $\mathbf{I}b \subset \mathbf{I}a$. Hence, since $b = \mathbf{I}b$, we infer that $b \subset \mathbf{I}a$. Thus $h(b) \subset h(\mathbf{I}a)$. Consequently, the union of all sets $h(b)$ such that $h(b) \subset h(a)$ and $b = \mathbf{I}b$ is contained in $h(\mathbf{I}a)$. This proves that $\mathbf{I}h(a) \subset h(\mathbf{I}a)$. Thus (6) holds.

By (1), (3), (4) and (5) the operation $\mathbf{I}$ is an interior operation in X, i.e. X may be considered as a topological space with the interior operation $\mathbf{I}$. Condition (6) states that h preserves the interior operation. Since h preserves also all other operations in $\mathfrak{A}$ and is one-one, it is a monomorphism from $\mathfrak{A}$ into the topological field of all subsets of X.

8. Strongly compact topological spaces. A topological space $\mathbf{X}$ is said to be *strongly compact* if for every family $\{G_u\}_{u \in U}$ of open subsets G_u, $u \in U$ of X the condition $\bigcup_{u \in U} G_u = X$ implies that there exists a $G_{u_0} = X$ for some $u_0 \in U$. Passing to complements, we infer that X is strongly compact if the intersection of any class of non-empty closed subsets of X is not empty.

8.1. *Every topological space X is an open subspace of a strongly compact space X_0 such that $X_0 - X$ is a one-point set and the class of all open subsets of X_0 is composed of X_0 and of all open subsets of X.*

Let x_0 be an arbitrary element such that $x_0 \notin X$. Put $X_0 = X \cup \{x_0\}$ and let

$$\mathbf{I}_0 X_0 = X_0, \tag{1}$$

(2) $$\mathbf{I}_0(Y \cup \{x_0\}) = \mathbf{I}Y, \quad \text{for every } Y \subset X,$$

(3) $$\mathbf{I}_0 Y = \mathbf{I}Y, \quad \text{for every } Y \subset X,$$

where $\mathbf{I}$ is the interior operation in the space X.

It is easy to verify that $\mathbf{I}_0$ is an interior operation in X_0 and that the open sets in X_0 are the open subsets of X and the set X_0 itself. In particular, X is an open subset of X_0. Hence if the union of an arbitrary family of open sets in X_0 is equal to X_0, then x_0 belongs to this union and consequently x_0 belongs to at least one of the summands. But there is only one open set containing x_0, namely the whole space X_0. Thus at least one of summands must be equal to X_0. This proves that X_0 is strongly compact.

The space X_0 constructed in the proof of 8.1 will be called the *one-point strong compactification* of the space X.

9. A lemma on imbedding for topological Boolean algebras. The aim of this section is to prove the following theorem.

9.1. *Let* $\mathfrak{A} = (A, \vee, \Rightarrow, \cup, \cap, -, \mathbf{I})$ *be a topological Boolean algebra, let* A_0 *be a finite subset of* A *and let* $\mathfrak{A}_1 = (A_1, \vee, \Rightarrow, \cup, \cap, -)$ *be the Boolean subalgebra of* $(A, \vee, \Rightarrow, \cup, \cap, -)$ *generated by* A_0. *Then there exists an interior operation* $\mathbf{I}_1$ *in* $\mathfrak{A}_1$ *such that for every* $a \in A_1$

(1) $$\text{if} \quad \mathbf{I}a \in A_0, \quad \text{then} \quad \mathbf{I}_1 a = \mathbf{I}a.$$

The algebra $\mathfrak{A}_1$ *contains at most* 2^{2^r} *elements, where* r *is the number of elements in* A_0 [5].

The estimation of the number of elements in $\mathfrak{A}_1$ follows from 2.3. Let B_0 be the set of all open elements in $\mathfrak{A}$ which belong to A_0. Let $\mathbf{I}_1 \vee = \vee$ and for every $a \in A_1$, $a \neq \vee$ let $\mathbf{I}_1 a$ be the join of all meets $b_1 \cap \ldots \cap b_n \leqslant a$, where $b_1, \ldots, b_n \in B_0$ or $\mathbf{I}_1 a = \wedge$ if such meets do not exist. Then it is easy to verify that $\mathbf{I}_1$ is an interior operation in $\mathfrak{A}_1$. Note that $\mathbf{I}_1 a \leqslant \mathbf{I}a$ for every $a \in A_1$ by the definition of $\mathbf{I}_1$. On the other hand, if $\mathbf{I}a \in A_0$, then $\mathbf{I}a \in B_0$ and $\mathbf{I}a \leqslant a$. Thus $\mathbf{I}a \leqslant \mathbf{I}_1 a$ and so $\mathbf{I}a = \mathbf{I}_1 a$, which completes the proof of 9.1.

[5] McKinsey and Tarski [1].

10. Connections between topological Boolean algebras, pseudo-Boolean algebras, relatively pseudo-complemented lattices, contrapositionally complemented lattices and semi-complemented lattices. For any topological Boolean algebra $\mathfrak{A} = (A, \vee, \Rightarrow, \cup, \cap, -, \mathbf{I})$, let us denote by $G(\mathfrak{A})$ the set of all open elements in $\mathfrak{A}$. Since the join and the meet of two open elements is open and, moreover, $\wedge$, $\vee$ are open, $G(\mathfrak{A})$ is a sublattice of $\mathfrak{A}$ containing $\wedge$ and $\vee$. For every a, b in $G(\mathfrak{A})$, let us put

(1) $$a \Rightarrow_1 b = \mathbf{I}(a \Rightarrow b) = \mathbf{I}(-a \cup b), \qquad \urcorner a = \mathbf{I}-a.$$

10.1. *The algebra* $\mathfrak{G}(\mathfrak{A}) = (G(\mathfrak{A}), \vee, \Rightarrow_1, \cup, \cap, \urcorner)$ *is a pseudo-Boolean algebra.*

To begin with we shall prove that

(2) $a \cap x \leqslant b$ if and only if $x \leqslant a \Rightarrow_1 b$, for all x, a, b in $G(\mathfrak{A})$.

Suppose that $a \cap x \leqslant b$. Then $-a \cup (a \cap x) \leqslant -a \cup b = a \Rightarrow b$. Hence $(-a \cup a) \cap (-a \cup x) \leqslant a \Rightarrow b$, i.e. $x \leqslant -a \cup x \leqslant a \Rightarrow b$. Consequently, $x = \mathbf{I}x \leqslant \mathbf{I}(a \Rightarrow b) = a \Rightarrow_1 b$. Conversely, suppose that $x \leqslant a \Rightarrow_1 b$. Then $a \cap x \leqslant a \cap (a \Rightarrow_1 b) = a \cap \mathbf{I}(a \Rightarrow b) \leqslant a \cap (-a \cup b) = a \cap b \leqslant b$. Hence $a \cap x \leqslant b$.

It follows from (2) that $(G(\mathfrak{A}), \vee, \Rightarrow_1, \cup, \cap)$ is a relatively pseudo-complemented lattice containing the zero element (see IV 1 (r)). Hence by IV 6.6 and (1) the algebra $\mathfrak{G}(\mathfrak{A})$ is a pseudo-Boolean algebra.

We shall prove the following representation theorem for pseudo-Boolean algebras.

10.2. *For any pseudo-Boolean algebra* $\mathfrak{B} = (B, \vee, \Rightarrow_1, \cup, \cap, \urcorner)$ *there exists a topological Boolean algebra* $\mathfrak{A}$ *such that* $\mathfrak{B}$ *is isomorphic to* $\mathfrak{G}(\mathfrak{A})$ [6].

Every pseudo-Boolean algebra is a distributive lattice with zero element $\wedge$ and unit element $\vee$ (see IV 1.2 and IV 6). By III 2.3 there exists a monomorphism h from $(B, \vee, \cup, \cap)$ into the set lattice of all subsets of a set $X \neq O$. Consider this set lattice as a Boolean algebra with respect to the set-theoretical operations $\vee = X, \Rightarrow, \cup, \cap, -$, where

(3) $$Y \Rightarrow Z = (X-Y) \cup Z, \qquad \text{for all subsets } Y, Z \text{ of } X.$$

Clearly, $h(\mathfrak{B})$ is a sublattice of the Boolean algebra in question containing the zero element O and the unit element X. Let $(A, X, \Rightarrow, \cup, \cap, -)$

[6] McKinsey and Tarski [2]; cf. also [MM].

be the Boolean subalgebra of the Boolean algebra of all subsets of X generated by $h(\mathfrak{B})$. Then by 2.2 every element Y of A is of the form

$$(4) \qquad Y = \big(h(a_1) \Rightarrow h(b_1)\big) \cap \ldots \cap \big(h(a_n) \Rightarrow h(b_n)\big), \quad \text{where } a_1, \ldots, a_n, b_1, \ldots, b_n \in B.$$

For any a, b in B we have

$$(5) \qquad h(a \Rightarrow_1 b) \leqslant h(a) \Rightarrow h(b).$$

In fact, $a \cap (a \Rightarrow_1 b) \leqslant b$, by IV 1 (5) and IV 6. Hence $h(a) \cap h(a \Rightarrow_1 b) \leqslant h(b)$. Thus $-h(a) \cup \big(h(a) \cap h(a \Rightarrow_1 b)\big) \leqslant -h(a) \cup h(b)$. Hence $-h(a) \cup \cup h(a \Rightarrow_1 b) \leqslant h(a) \Rightarrow h(b)$ and consequently (5) holds.

Now we shall prove that for any $a, b, a_1, \ldots, a_n, b_1, \ldots, b_n$ in B the condition

$$(6) \qquad \big(h(a_1) \Rightarrow h(b_1)\big) \cap \ldots \cap \big(h(a_n) \Rightarrow h(b_n)\big) \leqslant h(a) \Rightarrow h(b)$$

implies that

$$(7) \qquad h(a_1 \Rightarrow_1 b_1) \cap \ldots \cap h(a_n \Rightarrow_1 b_n) \leqslant h(a \Rightarrow_1 b).$$

In fact, (6) is equivalent to

$$h(a) \cap \big(h(a_1) \Rightarrow h(b_1)\big) \cap \ldots \cap \big(h(a_n) \Rightarrow h(b_n)\big) \leqslant h(b).$$

Hence, by (5), we get

$$h(a) \cap \big(h(a_1 \Rightarrow_1 b_1) \cap \ldots \cap h(a_n \Rightarrow_1 b_n)\big) \leqslant h(b).$$

Consequently,

$$h\big(a \cap (a_1 \Rightarrow_1 b_1) \cap \ldots \cap (a_n \Rightarrow_1 b_n)\big) \leqslant h(b).$$

Since h is an isomorphism, we obtain

$$a \cap (a_1 \Rightarrow_1 b_1) \cap \ldots \cap (a_n \Rightarrow_1 b_n) \leqslant b.$$

By IV 1 (r) and the definition of a pseudo-Boolean algebra (see IV 6) it follows from the last inequality that

$$(a_1 \Rightarrow_1 b_1) \cap \ldots \cap (a_n \Rightarrow_1 b_n) \leqslant a \Rightarrow_1 b.$$

Hence we obtain (7).

Since (6) implies (7), we infer that the equation

$$\big(h(a_1) \Rightarrow h(b_1)\big) \cap \ldots \cap \big(h(a_n) \Rightarrow h(b_n)\big) = \big(h(a_1') \Rightarrow h(b_1')\big) \cap \ldots \cap \big(h(a_n') \Rightarrow h(b_n')\big)$$

implies

$$h(a_1 \Rightarrow_1 b_1) \cap \ldots \cap h(a_n \Rightarrow_1 b_n) = h(a'_1 \Rightarrow_1 b'_1) \cap \ldots \cap h(a'_n \Rightarrow_1 b'_n)$$

for all $a_1, \ldots, a_n, b_1, \ldots, b_n, a'_1, \ldots, a'_n, b'_1, \ldots, b'_n$ in B.

We define the interior operation $\mathbf{I}$ in the Boolean algebra $(A, X, \Rightarrow, \cup, \cap, -)$ as follows: if $Y \in A$ has the representation given by (4), then we put

$$\mathbf{I}Y = h(a_1 \Rightarrow_1 b_1) \cap \ldots \cap h(a_n \Rightarrow_1 b_n).$$

By the result above $\mathbf{I}Y$ does not depend on the representation of the element Y in the form (4). Clearly,

(8) $\quad \mathbf{I}Y \in h(B) \quad$ for every $Y \in A$,

(9) $\quad \mathbf{I}Y = Y \quad$ for every $Y \in h(B)$.

The operation $\mathbf{I}$ satisfies axioms (i_1)–(i_4) of Sec. 5. Indeed, (i_1) follows directly from the definition of $\mathbf{I}$; (i_2) follows from (5); (i_3) follows from (i_1), (8) and (9); (i_4) follows from (9). Let us put $\mathfrak{A} = (A, X, \Rightarrow, \cup, \cap, -, \mathbf{I})$. Then $\mathfrak{A}$ is a topological Boolean algebra and it follows immediately from (8) and (9) that $h(\mathfrak{B}) = \mathfrak{G}(\mathfrak{A})$. Thus $\mathfrak{B}$ is isomorphic to $\mathfrak{G}(\mathfrak{A})$.

An element a in a topological Boolean algebra $\mathfrak{A}$ is said to be *dense* in $\mathfrak{A}$ if $\mathbf{C}a = V$. It follows from this definition that a is dense if and only if $\mathbf{I}-a = -\mathbf{C}a = \Lambda$, i.e. if and only if there exists no open element $b \neq \Lambda$ such that $b \leqslant -a$. This condition is equivalent to the condition that the meet of a with each open element $b \neq \Lambda$ is different from Λ.

For every topological Boolean algebra $\mathfrak{A}$ the algebra $\mathfrak{G}'(\mathfrak{A}) = (G(\mathfrak{A}), V, \Rightarrow_1, \cup, \cap)$ is a relatively pseudo-complemented lattice. Every subalgebra of this algebra is also a relatively pseudo-complemented lattice. Note that

10.3. *The algebra* $\mathfrak{G}_0(\mathfrak{A}) = (G_0(\mathfrak{A}), V, \Rightarrow_1, \cup, \cap)$, *where* $G_0(\mathfrak{A})$ *is the set of all open dense elements in a topological Boolean algebra* $\mathfrak{A}$ *and* $\Rightarrow_1$ *is defined by* (1), *is a relatively pseudo-complemented lattice.*

It is sufficient to show that $\mathfrak{G}_0(\mathfrak{A})$ is a subalgebra of $\mathfrak{G}'(\mathfrak{A})$. The join of two elements in $G_0(\mathfrak{A})$ belongs to $G_0(\mathfrak{A})$ because of 5 (c_1). It is easy to see that if $a, b \in G_0(\mathfrak{A})$, then for every open element $c \neq \Lambda$ in $\mathfrak{A}$, $c \cap (a \cap b) = (c \cap a) \cap b \neq \Lambda$, i.e. $a \cap b$ is dense. Thus $a \cap b \in G_0(\mathfrak{A})$.

If $a, b \in G_0(\mathfrak{A})$, then the open element $a \Rightarrow_1 b$ is also dense since $b \leqslant a \Rightarrow_1 b$ (see IV 1.3 (4)) and consequently $V = \mathbf{C}b \leqslant \mathbf{C}(a \Rightarrow_1 b)$. Thus $a \Rightarrow_1 b \in G_0(\mathfrak{A})$.

10.4. *For every relatively pseudo-complemented lattice $\mathfrak{B}$ with a zero element, there exists a topological Boolean algebra $\mathfrak{A}$ such that $\mathfrak{B}$ is isomorphic to $\mathfrak{G}'(\mathfrak{A})$,* i.e. *to the lattice of all open elements in $\mathfrak{A}$. For every relatively pseudo-complemented lattice $\mathfrak{B}$ without a zero element there exists a topological Boolean algebra $\mathfrak{A}$ such that $\mathfrak{B}$ is isomorphic to $\mathfrak{G}_0(\mathfrak{A})$,* i.e. *to the lattice of all open dense elements in $\mathfrak{A}$. Moreover, $G_0(\mathfrak{A})$ is then the class of all open elements in $\mathfrak{A}$ different from $\wedge$*[7].

The first part of this theorem is obvious because of IV 6.6 and 10.2. Suppose that $\mathfrak{B}$ does not contain a zero element and let $\mathfrak{B}^*$ be the relatively pseudo-complemented lattice obtained from $\mathfrak{B}$ by adding one new element $\wedge$ (see IV 1.6). By the first part of 10.4 there exists a topological Boolean algebra $\mathfrak{A}$ such that $\mathfrak{B}^*$ is isomorphic to $\mathfrak{G}'(\mathfrak{A})$. Observe that the set of all open dense elements in $\mathfrak{A}$ is then composed of all open elements in $\mathfrak{A}$ different from $\wedge$. Indeed, since in $\mathfrak{B}^*$ the meet of any two elements different from $\wedge$ is different from $\wedge$, the same holds in the isomorphic algebra $\mathfrak{G}'(\mathfrak{A})$. Consequently, every element in $\mathfrak{G}'(\mathfrak{A})$ different from $\wedge$ is dense. Thus $\mathfrak{G}_0(\mathfrak{A})$ consists of all elements in $\mathfrak{G}'(\mathfrak{A})$ different from $\wedge$ and obviously $\mathfrak{B}$ is isomorphic to $\mathfrak{G}_0(\mathfrak{A})$.

If $\mathfrak{A}$ is an arbitrary topological Boolean algebra and c is any open element in $\mathfrak{A}$ then the algebras $\mathfrak{G}(\mathfrak{A}, c) = (G(\mathfrak{A}), V, \Rightarrow_1, \cup, \cap, \sim_c)$ and $\mathfrak{G}_0(\mathfrak{A}, c) = (G_0(\mathfrak{A}), V, \Rightarrow_1, \cup, \cap, \sim_c)$, where the operation $\Rightarrow_1$ is defined by (1) and the operation $\sim_c$ is defined by

(10) $$\sim_c a = a \Rightarrow_1 c,$$

are contrapositionally complemented lattices (see IV 4.1, IV 1.5 (13)).

It follows from 10.4, IV 4.1 and IV 1.5 (13) that

10.5. *For each contrapositionally complemented lattice $\mathfrak{B}$ with a zero element there exist a topological Boolean algebra $\mathfrak{A}$ and an open element c in $\mathfrak{A}$ such that $\mathfrak{B}$ is isomorphic to $\mathfrak{G}(\mathfrak{A}, c)$. If $\mathfrak{B}$ does not contain a zero*

[7] Rasiowa and Sikorski [3].

element, then there exist a topological Boolean algebra $\mathfrak{A}$ *and an open element* c *in* $\mathfrak{A}$ *such that* $\mathfrak{B}$ *is isomorphic to* $\mathfrak{G}_0(\mathfrak{A}, c)$ (8).

If $\mathfrak{A}$ is an arbitrary topological Boolean algebra and $\mathfrak{G}'(\mathfrak{A}) = (G(\mathfrak{A}), V, \Rightarrow_1, \cup, \cap)$ the relatively pseudo-complemented lattice of all open elements in $\mathfrak{A}$, then we can introduce in $G(\mathfrak{A})$ an operation $\dot{-}$ of semi-complementation (see IV 5), defining $\dot{-}$ on the set $G(\mathfrak{A})-\{V\}$ in an arbitrary way and setting $\dot{-}V = \Lambda$. Then $(G(\mathfrak{A}), V, \Rightarrow_1, \cup, \cap, \dot{-})$ is a semi-complemented lattice. It follows from 10.4 that

10.6. *For every semi-complemented lattice* $\mathfrak{B}$ *there exists a topological Boolean algebra* $\mathfrak{A}$ *and an operation* $\dot{-}$ *of semi-complementation on* $G(\mathfrak{A})$ *such that* $\mathfrak{B}$ *is isomorphic to* $(G(\mathfrak{A}), V, \Rightarrow_1, \cup, \cap, \dot{-})$ (9).

11. Lemmas on imbeddings for pseudo-Boolean algebras, relatively pseudo-complemented lattices, contrapositionally complemented lattices and semi-complemented lattices. Theorems 9.1 and 10.2 enable us to prove the following theorem for pseudo-Boolean algebras.

11.1. *Let* $\mathfrak{A}$ *be a pseudo-Boolean algebra and let* A_0 *be a finite set of* r *elements in* $\mathfrak{A}$. *Then there exists a finite pseudo-Boolean algebra* $\mathfrak{A}'$ *containing at most* 2^{2^r} *elements such that* $A_0 \cup \{V\}$ *is contained in* $\mathfrak{A}'$ *and for all* a, b, c *in* $A_0 \cup \{V\}$:

(1) *if* c *is the join of* a *and* b *in* $\mathfrak{A}$, *then* c *is the join of* a *and* b *in* $\mathfrak{A}'$,

(2) *if* c *is the meet of* a *and* b *in* $\mathfrak{A}$, *then* c *is the meet of* a *and* b *in* $\mathfrak{A}'$,

(3) *if* c *is the pseudo-complement of* a *relative to* b *in* $\mathfrak{A}$, *then* c *is the pseudo-complement of* a *relative to* b *in* $\mathfrak{A}'$,

(4) *if* c *is the pseudo-complement of* a *in* $\mathfrak{A}$, *then* c *is the pseudo-complement of* a *in* $\mathfrak{A}'$ (10).

By 10.2 we can identify $\mathfrak{A}$ with the pseudo-Boolean algebra $\mathfrak{G}(\mathfrak{B})$ of all open elements in a topological Boolean algebra $\mathfrak{B}$. Let $\mathfrak{B}_1$ be the Boolean subalgebra of $\mathfrak{B}$ generated by A_0. By 9.1 there exists an interior operation $\mathbf{I}_1$ in $\mathfrak{B}_1$ such that for each a in $\mathfrak{B}_1$

(5) $$\text{if} \quad \mathbf{I}a \in A_0, \quad \text{then} \quad \mathbf{I}_1 a = \mathbf{I}a.$$

(8) Rasiowa and Sikorski [3].

(9) Rasiowa and Sikorski [3].

(10) McKinsey and Tarski [2]; cf. also [MM].

We shall consider $\mathfrak{B}_1$ as a topological Boolean algebra with the interior operation $\mathbf{I}_1$. Let $\mathfrak{A}' = (G(\mathfrak{B}_1), V, \Rightarrow', \cup, \cap, \neg')$, where $G(\mathfrak{B}_1)$ is the set of all open elements in $\mathfrak{B}_1$ and for all a, b in $G(\mathfrak{B}_1)$

(6) $$a \Rightarrow' b = \mathbf{I}_1(-a \cup b), \qquad \neg' a = \mathbf{I}_1 - a.$$

For the pseudo-Boolean algebra $\mathfrak{A}'$ conditions (1), (2) are satisfied since $\mathfrak{G}(\mathfrak{B})$ and $\mathfrak{A}'$ are sublattices of $\mathfrak{B}$. If a, b, c are in $A_0 \cup \{V\}$, then the condition $c = a \Rightarrow_1 b = \mathbf{I}(-a \cup b)$ implies by (5) that $c = \mathbf{I}_1(-a \cup b) = a \Rightarrow' b$, which proves (3). The proof of (4) is analogous. Since by 2.3 the Boolean algebra $\mathfrak{B}_1$ contains at most 2^{2^r} elements, the algebra $\mathfrak{A}'$ satisfies the same condition.

From IV 1.6, IV 6.6 and 11.1 we get analogous theorems for relatively pseudo-complemented lattices.

11.2. *Let $\mathfrak{A}$ be a relatively pseudo-complemented lattice and let A_0 be a finite set of r elements in $\mathfrak{A}$. Then there exists a finite relatively pseudo-complemented lattice $\mathfrak{A}'$ containing at most 2^{2^r} elements such that $A_0 \cup \{V\}$ is contained in $\mathfrak{A}'$ and for all a, b, c in $A_0 \cup \{V\}$:*

(7) *if c is the join of a and b in $\mathfrak{A}$, then c is the join of a and b in $\mathfrak{A}'$,*

(8) *if c is the meet of a and b in $\mathfrak{A}$, then c is the meet of a and b in $\mathfrak{A}'$,*

(9) *if c is the pseudo-complement of a relative to b in $\mathfrak{A}$, then c is the pseudo-complement of a relative to b in $\mathfrak{A}'$.*

We shall prove that for contrapositionally complemented lattices the following theorem, analogous to 11.2, holds:

11.3. *Let $\mathfrak{A}$ be a contrapositionally complemented lattice and let A_0 be a finite set of elements in $\mathfrak{A}$ consisting of r elements. Then there exists a finite contrapositionally complemented lattice $\mathfrak{A}'$ consisting of at most $2^{2^{r+1}}$ elements such that $A_0 \cup \{V\}$ is contained in $\mathfrak{A}'$ and for all a, b, c in $A_0 \cup \{V\}$ conditions* (7), (8), (9) *hold and, moreover,*

(10) *if c is the contrapositional complement of a in $\mathfrak{A}$, then c is the contrapositional complement of a in $\mathfrak{A}'$.*

By IV 4.1 $\mathfrak{A}$ is a relatively pseudo-complemented lattice and the contrapositional complement is defined by means of the formula

(c*) $$\sim_c a = a \Rightarrow \sim_c V, \qquad \text{for all } a \text{ in } \mathfrak{A}.$$

By 11.2 there exists a relatively pseudo-complemented lattice $\mathfrak{A}'$ containing at most $2^{2^{r+1}}$ elements and such that $A_0 \cup \{\sim_c V\} \cup \{V\}$ is contained in $\mathfrak{A}'$ and (7), (8), (9) are satisfied for all a, b, c in $A_0 \cup \{\sim_c V\} \cup \cup \{V\}$. Let us define an operation of contrapositional complementation in $\mathfrak{A}'$ as follows:

(11) $$\sim_c' a = a \Rightarrow' \sim_c V, \quad \text{for all } a \text{ in } \mathfrak{A}',$$

where $\Rightarrow'$ denotes relative pseudo-complementation in $\mathfrak{A}'$. Then by IV 4.1 and IV 1.5 (13) we can treat $\mathfrak{A}'$ as a contrapositionally complemented lattice and, clearly, conditions (7), (8), (9), (10) are satisfied for all a, b, c in $A_0 \cup \{V\}$.

For semi-complemented lattices (see IV 5) an analogous theorem to 11.1 holds too. Namely

11.4. *Let $\mathfrak{A}$ be a semi-complemented lattice and let A_0 be a finite set of elements in $\mathfrak{A}$ consisting of r elements. Then there exists a finite semi-complemented lattice $\mathfrak{A}'$ consisting of at most 2^{2^r} elements, containing $A_0 \cup \{V\}$, and such that for all a, b, c in $A_0 \cup \{V\}$ conditions* (1), (2), (3) *are satisfied and, moreover,*

(12) *if c is the semi-complement of a in $\mathfrak{A}$, then c is the semi-complement of a in $\mathfrak{A}'$.*

Theorem 11.4 follows easily from the definition of semi-complemented lattices and from 11.2.

Exercises

1. Let $\mathfrak{A} = (A, V, \Rightarrow, \cup, \cap, -)$ be a Boolean algebra (cf. Sec. 5). Let $A_0 = \{a_t\}_{t\in T}$ and $B_0 = \{b_u\}_{u\in U}$ be any subsets of A. Since A is ordered by the lattice ordering relation $\leqslant$, l.u.b. A_0 and g.l.b.B_0 (see I 3) can exist. Instead of l.u.b.A_0 and g.l.b.B_0, we shall write $(\mathfrak{A}) \bigcup_{t\in T} a_t$ and $(\mathfrak{A}) \bigcap_{u\in U} b_u$, respectively. Let h be a homomorphism from $\mathfrak{A}$ into a Boolean algebra $\mathfrak{B} = (B, V, \Rightarrow, \cup, \cap, -)$. Prove that $(\mathfrak{B}) \bigcup_{t\in U} h(a_t) \leqslant h((\mathfrak{A}) \bigcup_{t\in T} a_t)$ and $h((\mathfrak{A}) \bigcap_{u\in U} b_u) \leqslant (\mathfrak{B}) \bigcap_{u\in U} h(b_u)$ whenever all the least upper bounds and greatest lower bounds appearing in the above formulas exist.

2. Let $\mathfrak{A} = (A, V, \Rightarrow, \cup, \cap, -)$ be a Boolean algebra and let h be the Stone monomorphism (i.e. that defined by (1) in 4) of $\mathfrak{A}$ into the Boolean algebra $(B(X), X, \Rightarrow, \cup, \cap, -)$ of all subsets of X, where X is the set of all maximal implicative filters in $\mathfrak{A}$. Consider X as a topological space, the topology being defined as in 4.1. Suppose that for a, a_t, b, b_u in A, $t \in T$, $u \in U$, $a = (\mathfrak{A}) \bigcup_{t \in T} a_t$ and $b = (\mathfrak{A}) \bigcap_{u \in U} b_u$. Prove that the sets

$$h(a) - \bigcup_{t \in T} h(a_t), \qquad \bigcap_{u \in U} h(b_u) - h(b),$$

where $\bigcup$ and $\bigcap$ denote set-theoretical union and intersection, respectively, are nowhere dense, i.e. the complement of the closure of each of these sets is dense in X(11).

3. Let $\mathfrak{A} = (A, \vee, \Rightarrow, \cup, \cap, -)$ be a Boolean algebra and let $a_n = (\mathfrak{A}) \bigcup_{u \in U_n} a_{nu}$, $b_n = (\mathfrak{A}) \bigcap_{v \in V_n} b_{nv}$, $n = 1, \ldots,$ where a_n, b_n, a_{nu}, b_{nv} are elements of A for $u \in U_n$, $v \in V_n$. Assume that $a \in A$ and $a \neq \wedge$. Prove that there exists a maximal implicative filter ∇ of $\mathfrak{A}$ for which $a \in \nabla$ and which satisfies the following conditions:

(i) if $a_n \in \nabla$, then $a_{nu} \in \nabla$ for some $u \in U_n$, $n = 1, 2, \ldots,$

(ii) if $b_{nv} \in \nabla$ for each $v \in V_n$, then $b_n \in \nabla$, $n = 1, 2, \ldots$ (12).

For the proof use Ex. 2 and the following modification of Baire's theorem due to Čech (13): the complement of any set of the first category (i.e. of the union of a sequence of nowhere dense sets) in a compact Hausdorff space is a dense set.

4. Prove that under the hypotheses adopted in Ex. 3, there exists a homomorphism h of $\mathfrak{A}$ onto the two-element Boolean algebra $\mathfrak{B}_0$ such that $h(a) = \vee$ and, moreover, $h(a_n) = (\mathfrak{B}_0) \bigcup_{u \in U_n} h(a_{nu})$, $h(b_n) = (\mathfrak{B}_0) \bigcap_{v \in V_n} h(b_{nv})$, $n = 1, 2, \ldots$ (14). For the proof use Ex. 3, 3.2 and 3.3.

5. Prove that under the hypotheses adopted in Ex. 3, there exists a monomorphism h_0 of $\mathfrak{A}$ into the Boolean algebra of all subsets of a set X_0 such that $h_0(a_n) = \bigcup_{u \in U_n} h_0(a_{nu})$, $h_0(b_n) = \bigcap_{v \in V_n} h_0(b_{nv})$, $n = 1, 2, \ldots,$ where $\bigcup$ and $\bigcap$ denote the set-theoretical union and intersection, respectively (15). To prove this, use Ex. 2. Define X_0 to be equal to the difference between X and the union of all sets $h(a_n) - \bigcup_{u \in U_n} h(a_{nu})$ and $\bigcap_{v \in V_n} h(b_{nv}) - h(b_n)$, $n = 1, 2, \ldots,$ and put $h_0(b) = h(b) \cap X_0$ for each $b \in A$. In order to show that h_0 is a one-one mapping apply Čech's modification of the Baire theorem as formulated in Ex. 3.

6. Prove an analogue of the theorem formulated in Ex. 5 for topological Boolean algebras (16). Apply Ex. 5 and the proof of 7.1.

7. Let $\mathfrak{B}$ be a topological Boolean algebra with an interior operation $\mathbf{I}$ and let $\mathfrak{A} = \mathfrak{G}(\mathfrak{B})$ be the pseudo-Boolean algebra of all open elements of $\mathfrak{B}$ (cf. Sec. 10).

(11) Rasiowa and Sikorski [1].
(12) Rasiowa and Sikorski [1].
(13) Čech [1].
(14) Rasiowa and Sikorski [1].
(15) Rasiowa and Sikorski [3], [MM] II 9.4.
(16) Rasiowa and Sikorski [3], [MM] III 4.3.

Let a_t, $t \in T$ be any elements of $\mathfrak{A}$. Prove that: (i) $(\mathfrak{A}) \bigcup_{t\in T} a_t$ exists if and only if $(\mathfrak{B}) \bigcup_{t\in T} a_t$ exists and $(\mathfrak{A}) \bigcup_{t\in T} a_t = (\mathfrak{B}) \bigcup_{t\in T} a_t$. (ii) If $(\mathfrak{B}) \bigcap_{t\in T} a_t$ exists, then $(\mathfrak{A}) \bigcap_{t\in T} a_t$ exists too, and $(\mathfrak{A}) \bigcap_{t\in T} a_t = \mathbf{I}(\mathfrak{B}) \bigcap_{t\in T} a_t$.

8. Prove an analogue of the theorem in Ex. 5 for pseudo-Boolean algebras [17]. Apply 10.2, Ex. 7 and Ex. 6.

9. Prove an analogue of the theorem in Ex. 5 for relatively pseudo-complemented lattices, contrapositionally complemented lattices and semi-complemented lattices [18]. Apply 10.4, 10.5, 10.6, respectively, and a similar argument to that used in Ex. 8.

[17] Rasiowa and Sikorski [3], [MM] IV 9.2.

[18] Rasiowa and Sikorski [3].

CHAPTER VII

POST ALGEBRAS

Introduction. The first axiom system for the algebras corresponding to the m-valued logics of Post was given by Rosenbloom [1], who called them Post algebras (of order $m = 2, 3, \ldots$). It was subsequently simplified by Epstein [1] and Traczyk [1], [3], [5], who proved the equational definability of the class of Post algebras (of each of the orders $m = 2, 3, \ldots$) and formulated the first equational system of axioms.

The theory of Post algebras has been developed in several papers by Traczyk ([1]-[8]) and also by Chang and Horn [1], Dwinger [1]-[3], Malcev [2], [3], Rousseau [2], [4] and other authors.

The definition of Post algebras as given here is based on the results due to Rousseau [2], who presented them at the Seminar on Foundations of Mathematics at the Polish Academy of Sciences Institute of Mathematics in October 1966. He was the first to observe that each Post algebra is a pseudo-Boolean algebra. This is especially important for metalogical applications.

Rousseau's system of axioms is inspired by the following idea. A Post algebra $\mathfrak{P}$ of order m is a coproduct of a chain $\mathfrak{C} = \{e_0, \ldots, e_{m-1}\}$, $\Lambda = e_0 \leqslant \ldots \leqslant e_{m-1} = V$, and a Boolean algebra $\mathfrak{B}_{\mathfrak{P}}$ with zero element Λ and unit element V. This means that their union generates $\mathfrak{P}$ and, for every distributive lattice $\mathfrak{A}$ with Λ and V, if h_1 is a homomorphism of $\mathfrak{C}$ into $\mathfrak{A}$ and h_2 is a homomorphism of $\mathfrak{B}_{\mathfrak{P}}$ into $\mathfrak{A}$, then there exists a homomorphism h of $\mathfrak{P}$ into $\mathfrak{A}$ which is a common extension of h_1 and h_2.

This chapter contains a brief exposition of the basic properties of Post algebras, the theory of filters and Traczyk's representation theorem. Also we prove the equivalence of Traczyk's and Rousseau's definitions of Post algebras.

1. Definition and elementary properties. An abstract algebra

(1) $\mathfrak{P} = (P, V, \Rightarrow, \cup, \cap, \neg, D_1, \ldots, D_{m-1}, e_0, \ldots, e_{m-1}), \quad m \geqslant 2,$

where $V, e_0, \ldots, e_{m-1}$ are zero-argument operations, $\neg, D_1, \ldots, D_{m-1}$ are one-argument operations and $\Rightarrow, \cup, \cap$ are two-argument operations, is said to be a *Post algebra of order m* if [1]

(p_0) $(P, V, \Rightarrow, \cup, \cap, \neg)$ is a pseudo-Boolean algebra,

and for all $a, b \in P$ the following equations hold:

(p_1) $D_i(a \cup b) = D_i(a) \cup D_i(b),\ i = 1, \ldots, m-1,$

(p_2) $D_i(a \cap b) = D_i(a) \cap D_i(b),\ i = 1, \ldots, m-1,$

(p_3) $D_i(a \Rightarrow b) = \big(D_1(a) \Rightarrow D_1(b)\big) \cap \ldots \cap \big(D_i(a) \Rightarrow D_i(b)\big),$
$i = 1, \ldots, m-1,$

(p_4) $D_i(\neg a) = \neg D_1(a),\ i = 1, \ldots, m-1,$

(p_5) $D_i\big(D_j(a)\big) = D_j(a),\ i, j = 1, \ldots, m-1,$

(p_6) $D_i(e_j) = \begin{cases} V & \text{if } i \leqslant j, \\ \neg V & \text{if } i > j, \end{cases} \quad i = 1, \ldots, m-1, \quad j = 0, \ldots, m-1,$

(p_7) $a = D_1(a) \cap e_1 \cup \ldots \cup D_{m-1}(a) \cap e_{m-1},$

(p_8) $D_1(a) \cup \neg D_1(a) = V.$

Observe that the class of all Post algebras of any order $m \geqslant 2$ is equationally definable since the class of all pseudo-Boolean algebras is equationally definable.

A Post algebra is said to be *non-degenerate* if it is a non-degenerate algebra, i.e. if it contains at least two different elements.

It follows from the above definition of Post algebras that if (1) is a Post algebra, then $(P, V, \Rightarrow, \cup, \cap)$ is a relatively pseudo-complemented lattice.

Let us note that the following holds:

1.1. *If* $\mathfrak{P} = (P, V, \Rightarrow, \cup, \cap, \neg, D_1, e_0, e_1)$ *is a Post algebra of order* 2, *then*

(2) $$e_1 = V, \quad e_0 = \neg V = \Lambda,$$

(3) $$D_1(a) = a \quad \textit{for all} \quad a \in P,$$

and $(P, V, \Rightarrow, \cup, \cap, \neg)$ *is a Boolean algebra.*

[1] See Rousseau [2].

By (p_7) for every $a \in P$

(4) $$a = D_1(a) \cap e_1.$$

Thus $a \leqslant e_1$ for every $a \in P$. Consequently, e_1 is the unit element in the pseudo-Boolean algebra $(A, V, \Rightarrow, \cup, \cap, \neg)$, i.e. $e_1 = V$. By (p_7), (p_6) and IV 6.1 (1) we obtain $e_0 = D_1(e_0) \cap e_1 = \neg V \cap e_1 = \Lambda \cap e_1 = \Lambda$. Thus (2) holds. By (4) and (2) we get $a = D_1(a) \cap V = D_1(a)$, which proves (3). It follows from (3) and (p_8) that $a \cup \neg a = V$ for each $a \in P$. Hence, by VI 1 ((b_1), (b_2)), $(P, V, \Rightarrow, \cup, \cap, \neg)$ is a Boolean algebra.

1.2. *In any Post algebra* (1) *the following conditions are satisfied*:

(5) $(P, V, \cup, \cap)$ *is a distributive lattice with unit element* V *and zero element* $\Lambda = \neg V$,

(6) $\Lambda = e_0 \leqslant e_1 \leqslant \ldots \leqslant e_{m-1} = V$,

(7) $D_i(a) \leqslant D_j(a)$ *for* $j \leqslant i$, $i, j = 1, \ldots, m-1$, $a \in P$,

(8) *if* $a \leqslant b$, *then* $D_i(a) \leqslant D_i(b)$, *for all* $i = 1, \ldots, m-1$, *and all* $a, b \in P$,

(9) $D_{m-1}(a) \leqslant a \leqslant D_1(a)$, *for each* $a \in P$,

(10) $e_0 \Rightarrow a = e_{m-1}$, $e_1 \Rightarrow a = D_1(a)$, $e_{m-1} \Rightarrow a = a$, *for each* $a \in P$,

(11) $e_i \Rightarrow a = D_1(a) \cap e_1 \cup \ldots \cup D_{i-1}(a) \cap e_{i-1} \cup D_i(a)$, *for* $1 < i < m-1$, $a \in P$,

(12) $a \Rightarrow e_0 = \neg a$, $a \Rightarrow e_{m-1} = e_{m-1}$, *for each* $a \in P$,

(13) $a \Rightarrow e_i = e_i \cup \neg D_{i+1}(a)$, *for* $0 \leqslant i < m-1$, $a \in P$,

(14) $$e_i \Rightarrow e_j = \begin{cases} e_{m-1} & \text{if } i \leqslant j, \\ e_j & \text{if } i > j,\ e_i \neq e_j, \end{cases}$$ *for* $i, j = 0, \ldots, m-1$,

(15) *in any non-degenerate Post algebra* (1), *if* $i \neq j$, *then* $e_i \neq e_j$, $i, j = 0, \ldots, m-1$.

Since $(A, V, \Rightarrow, \cup, \cap, \neg)$ is a pseudo-Boolean algebra, (5) follows from IV 6, IV 1.1, IV 1.2, IV 6.1.

Applying to e_i $(i = 0, \ldots, m-1)$ the axioms (p_7), (p_6) and (5), we get

$$e_0 = \wedge, \quad e_i = e_1 \cup \ldots \cup e_i, \quad i = 1, \ldots, m-1.$$

Hence $\wedge = e_0 \leqslant e_1 \leqslant \ldots \leqslant e_{m-1}$. Consequently,

$$e_i \cap e_j = e_i \quad \text{for} \quad i \leqslant j, \quad i, j = 0, \ldots, m-1.$$

By (p_7) and the last result we get $a \cap e_{m-1} = a$ for each $a \in P$. Hence $a \leqslant e_{m-1}$ for every $a \in P$, i.e. e_{m-1} is the unit element. Thus $e_{m-1} = \vee$ and (6) holds.

Operating on (p_7) with D_i $(i = 1, \ldots, m-1)$ and applying (p_1), (p_2), (p_5) and (p_6), we get

$$D_i(a) = D_i(a) \cup \ldots \cup D_{m-1}(a), \quad i = 1, \ldots, m-1, \; a \in P.$$

This proves that (7) holds.

Suppose that $a \leqslant b$ for some $a, b \in P$. Then $a \cap b = a$. Hence, by (p_2), $D_i(a) \cap D_i(b) = D_i(a \cap b) = D_i(a)$, which is equivalent to $D_i(a) \leqslant D_i(b)$. Thus (8) holds.

By (p_7) and (7), $a \cap D_1(a) = a$, i.e. $a \leqslant D_1(a)$ for each $a \in P$. By (p_7), (7), (5) and (6), $a \cap D_{m-1}(a) = D_{m-1}(a)$, i.e. $D_{m-1}(a) \leqslant a$, for all $a \in P$. Thus (9) holds.

By (6) and IV 1.3 (2), IV 1.5 (13), for every $a \in P$

$$(16) \qquad e_0 \Rightarrow a = \wedge \Rightarrow a = \vee = e_{m-1} \quad \text{and} \quad e_{m-1} \Rightarrow a = \vee \Rightarrow a = a.$$

By (p_7), (p_3), (p_6) and (6), (16) for every $a \in P$

$$\begin{aligned} e_1 \Rightarrow a &= D_1(e_1 \Rightarrow a) \cap e_1 \cup \ldots \cup D_{m-1}(e_1 \Rightarrow a) \cap e_{m-1} \\ &= (D_1(e_1) \Rightarrow D_1(a)) \cap e_1 \cup \ldots \cup \big((D_1(e_1) \Rightarrow D_1(a)) \cap \ldots \\ &\qquad \ldots \cap (D_{m-1}(e_1) \Rightarrow D_{m-1}(a))\big) \cap e_{m-1} \\ &= (e_{m-1} \Rightarrow D_1(a)) \cap e_1 \cup \ldots \cup \big((e_{m-1} \Rightarrow D_1(a)) \cap \ldots \\ &\qquad \ldots \cap (e_0 \Rightarrow D_{m-1}(a))\big) \cap e_{m-1} \\ &= D_1(a) \cap e_1 \cup \ldots \cup D_1(a) \cap e_{m-1} = D_1(a). \end{aligned}$$

This together with (16) proves that (10) holds.

The analogous proof of (11), based on (p_7), (p_3), (p_6), (6), (7), (16), is left to the reader.

Since $(P, \vee, \Rightarrow, \cup, \cap, \neg)$ is a pseudo-Boolean algebra, (12) follows from (6), IV 6.1 (2), IV 1.3 (7).

By (7) and IV 6.2 (5) for every $a \in P$

(17) $\neg D_i(a) \leqslant \neg D_j(a)$ if $i \leqslant j$, $i, j = 1, \ldots, m-1$.

Using the representation form (p_7) for $a \Rightarrow e_i$ ($0 \leqslant i < m-1$), and applying succesively (p_3), (p_6), (6), (12), (17), we get (13).

The condition $e_i \Rightarrow e_j = e_{m-1}$ if $i \leqslant j$ is satisfied because of (6) and IV 1.3 (2). If $i > j$ and $e_i \neq e_j$, then by (13), (p_6) and (6) we get

$$e_i \Rightarrow e_j = e_j \cup \neg D_{j+1}(e_i) = e_j \cup \neg e_{m-1} = e_j \cup \neg V = e_j \cup \Lambda = e_j.$$

Thus (14) holds. If $\mathfrak{P}$ is non-degenerate, then $V \neq \neg V$ and hence by (p_6), $e_i \neq e_j$ for $i \neq j$. Thus (15) holds.

1.3. *In any Post algebra* (1), *for any* $j = 1, \ldots, m-1$ *and any* $a \in P$, *the condition*

(18) $e_{i+1} \leqslant D_j(a) \cup e_i$ *for some* $i = 0, \ldots, m-2$

implies

(19) $$D_j(a) = e_{m-1}.$$

Suppose that (18) holds. Then $(D_j(a) \cup e_i) \cap e_{i+1} = e_{i+1}$. Applying ($p_1$), ($p_2$) and ($p_5$), we get $(D_j(a) \cup D_{i+1}(e_i)) \cap D_{i+1}(e_{i+1}) = D_{i+1}(e_{i+1})$. Hence, by ($p_6$), we obtain $(D_j(a) \cup e_0) \cap e_{m-1} = e_{m-1}$. Thus $D_j(a) = e_{m-1}$.

1.4. *In any Post algebra* (1) *the set* $B_{\mathfrak{P}}$ *of all elements of the form* $D_i(a)$, *where* $i = 1, \ldots, m-1$ *and* $a \in P$, *is the set of all complemented elements in the distributive lattice* (P, V, $\cup$, $\cap$). *Moreover, the set* $B_{\mathfrak{P}}$ *is closed under the operations* $V, \Rightarrow, \cup, \cap, \neg$ *and the algebra* $\mathfrak{B}_{\mathfrak{P}} = (B_{\mathfrak{P}}, V, \Rightarrow, \cup, \cap, \neg)$ *is a Boolean algebra.*

Since (P, $V, \Rightarrow, \cup, \cap, \neg$) is a pseudo-Boolean algebra, by IV 6.2 (7)

(20) $a \cap \neg a = \Lambda$ for every $a \in P$.

In particular,

(21) $D_i(a) \cap \neg D_i(a) = \Lambda$ for every $i = 1, \ldots, m-1$ and $a \in P$.

On the other hand, by (p_5) and (p_8),

$$D_i(a) \cup \neg D_i(a) = D_1(D_i(a)) \cup \neg D_1(D_i(a)) = V,$$

i.e.

(22) $D_i(a) \cup \neg D_i(a) = V$ for $i = 1, \ldots, m-1$ and every $a \in P$.

By (21) and (22) every element in $B_{\mathfrak{P}}$ is complemented (see VI p. 112), viz.

(23) *$\neg D_i(a)$ is the complement of $D_i(a)$, for $i = 1, \ldots, m-1$ and $a \in P$.*

Now let us suppose that a is an arbitrary complemented element in P. We shall prove first of all that the complement of a is equal to $\neg a$. Indeed, suppose that b is the complement of a. Then

(24) $$a \cap b = \Lambda,$$

(25) $$a \cup b = V.$$

By (24) and IV 6, IV 1 (r), IV 6.1 (2)

(26) $$b \leqslant a \Rightarrow \Lambda = \neg a.$$

Consequently, by (25),

$$V = a \cup b \leqslant a \cup \neg a,$$

i.e. $a \cup \neg a = V$. Thus, by (20), $\neg a$ is the complement of a. By (p_7) and (p_4)

$$\begin{aligned} \neg a &= D_1(\neg a) \cap e_1 \cup \ldots \cup D_{m-1}(\neg a) \cap e_{m-1} \\ &= \neg D_1(a) \cap (e_1 \cup \ldots \cup e_{m-1}) = \neg D_1(a); \end{aligned}$$

thus $\neg a = \neg D_1(a)$. Therefore, since a and $D_1(a)$ are complemented, we obtain by VI 1 (7)

$$a = \neg\neg a = \neg\neg D_1(a) = D_1(a).$$

Thus every complemented element a is of the form $D_1(a)$ and consequently belongs to $B_{\mathfrak{P}}$. Thus $B_{\mathfrak{P}}$ is the set of all complemented elements in the Post algebra (1).

By (23) and VI 1.5 the set $B_{\mathfrak{P}}$ is closed under the operations V, $\cup$, $\cap$, $\neg$. We shall prove that for all $a, b \in P$ and $i, j = 1, \ldots, m-1$

(27) $$D_i(a) \Rightarrow D_j(b) = \neg D_i(a) \cup D_j(b).$$

Since $(P, V, \Rightarrow, \cup, \cap, \neg)$ is a pseudo-Boolean algebra, by IV 6.2 (12)

(28) $$\neg D_i(a) \cup D_j(b) \leqslant D_i(a) \Rightarrow D_j(b).$$

Moreover, by IV 1.3 (5) and IV 6

$$D_i(a) \cap (D_i(a) \Rightarrow D_j(b)) \leqslant D_j(b).$$

Adding $\neg D_i(a)$ on both sides of the above inequality and applying the distributive law and (22), we get $\neg D_i(a) \cup (D_i(a) \Rightarrow D_j(b)) \leqslant \neg D_i(a) \cup D_j(b)$. Thus $D_i(a) \Rightarrow D_j(b) \leqslant \neg D_i(a) \cup D_j(b)$. Hence, by (28), we get (27). It follows from (27) that $B_{\mathfrak{P}}$ is also closed under the operation $\Rightarrow$. Hence, by VI 1.5, the algebra $\mathfrak{B}_{\mathfrak{P}} = (B_{\mathfrak{P}}, V, \Rightarrow, \cup, \cap, \neg)$ is a Boolean algebra.

The next two theorems state the equivalence of the definition of Post algebras adopted in this section to another definition of such algebras.

1.5. *Let an algebra* (1) *be a Post algebra of order* $m \geqslant 2$ *and let us introduce a new operation* C *by means of the equation*

(29) $$Ca = \neg D_1(a), \quad \textit{for each } a \in P.$$

Then the abstract algebra $\mathfrak{P}^* = (P, V, \cup, \cap, C, D_1, \ldots, D_{m-1}, e_0, \ldots \ldots, e_{m-1})$ *has the following properties* (2):

(p_1^*) $(P, V, \cup, \cap)$ *is a distributive lattice with unit element* $V = e_{m-1}$ *and zero element* $\Lambda = e_0$,

(p_2^*) $$e_0 \leqslant e_1 \leqslant \ldots \leqslant e_{m-1},$$

(p_3^*) $$D_1(a) \cap C(a) = \Lambda, \quad D_1(a) \cup C(a) = V,$$

(p_4^*) $$D_i(a) \cap D_j(a) = D_j(a) \quad \textit{if} \quad i \leqslant j, \quad i, j = 1, \ldots, m-1,$$

(p_5^*) $$D_i(a \cup b) = D_i(a) \cup D_i(b), \quad D_i(a \cap b) = D_i(a) \cap D_i(b), \quad i, j = 1, \ldots, m-1,$$

(p_6^*) $$D_i(D_j(a)) = D_j(a), \quad i, j = 1, \ldots, m-1,$$

(p_7^*) $$D_i(e_j) = \begin{cases} e_{m-1} & \textit{if } i \leqslant j, \\ e_0 & \textit{if } i > j, \end{cases} \quad i = 1, \ldots, m-1, \; j = 0, \ldots, m-1,$$

(p_8^*) $$a = D_1(a) \cap e_1 \cup \ldots \cup D_{m-1}(a) \cap e_{m-1},$$

where a, b *are arbitrary elements of* P.

(2) The conditions (p_1^*)-(p_8^*) form the system of axioms for Post algebras (of order m) given by Traczyk [5].

Indeed, (p_1^*) and (p_2^*) hold by 1.2 (5), (6); conditions (p_3^*) follow from (29), (21) and (p_8); (p_4^*) follows from 1.2 (7); conditions (p_5^*) are identical to (p_1) and (p_2); conditions (p_6^*), (p_7^*), (p_8^*) are identical to (p_5), (p_6) and (p_7), respectively.

1.6 [3]. *Let an abstract algebra* $\mathfrak{P}^* = (P, V, \cup, \cap, C, D_1, \ldots, D_{m-1}, e_0, \ldots, e_{m-1})$ *satisfy the conditions* (p_1^*)-(p_8^*) *formulated in Theorem* 1.5. *Let* $\Rightarrow$ *and* $\neg$ *be two new operations defined by means of the following equations*

$$(30)\quad a \Rightarrow b = \big(CD_1(a) \cup D_1(b)\big) \cap e_1 \cup \cup\Big(\big(CD_1(a) \cup D_1(b)\big) \cap \big(CD_2(a) \cup D_2(b)\big)\Big) \cap e_2 \cup \ldots \ldots \cup \Big(\big(CD_1(a) \cup D_1(b)\big) \cap \ldots \cap \big(CD_{m-1}(a) \cup D_{m-1}(b)\big)\Big) \cap e_{m-1},$$

$$(31)\quad \neg a = a \Rightarrow e_0,$$

for all a, b in P. Then the algebra $\mathfrak{P} = (P, V, \Rightarrow, \cup, \cap, \neg, D_1, \ldots \ldots, D_{m-1}, e_0, \ldots, e_{m-1})$ *is a Post algebra of order m.*

First of all let us observe that all elements of the form $D_i(a)$, where $i = 1, \ldots, m-1$, $a \in P$, are complemented, viz. for every $a \in P$

(32) $\quad CD_i(a)$ *is the complement of* $D_i(a)$, $\quad i = 1, \ldots, m-1$.

In fact, by (p_3^*) and (p_6^*) we have

$$D_1(a) \cup CD_1(a) = D_1\big(D_1(a)\big) \cup C\big(D_1(a)\big) = V$$

and

$$D_1(a) \cap CD_1(a) = D_1\big(D_1(a)\big) \cap C\big(D_1(a)\big) = \Lambda.$$

Hence, by (p_6^*),

$$D_i(a) \cup CD_i(a) = D_1\big(D_i(a)\big) \cup CD_1\big(D_i(a)\big) = V$$

and

$$D_i(a) \cap CD_i(a) = D_1\big(D_i(a)\big) \cap CD_1\big(D_i(a)\big) = \Lambda$$

for $i = 1, \ldots, m-1$ and every $a \in P$, which proves (32).

Consequently, by (p_4^*), VI 1.5 and VI 1 (3),

$$(33)\quad C\big(D_i(a) \cap D_j(a)\big) = CD_i(a) \cup CD_j(a) = CD_j(a) \quad \text{if} \quad i \leqslant j;$$

hence

[3] Theorems 1.5, 1.6 establish the equivalence of the definitions of Post algebras given by Rousseau and by Traczyk.

(34) $$CD_i(a) \leqslant CD_j(a) \quad \text{for} \quad i \leqslant j \text{ and every } a \in P.$$

Consequently, $D_j(a) \cap CD_i(a) \leqslant D_j(a) \cap CD_j(a) = e_0$, i.e.

(35) $$D_j(a) \cap CD_i(a) = e_0 \quad \text{if} \quad i \leqslant j, \text{ where } i, j = 1, \ldots, m-1, \ a \in P.$$

Note also that for all $a \in P$:

(36) $$D_j(CD_i(a)) = CD_i(a), \quad \text{for all } i, j = 1, \ldots, m-1.$$

Indeed, by (p_6^*), (p_5^*), (32), (p_1^*), (p_2^*) we get

$$D_i(a) \cup D_j(CD_i(a)) = D_j(D_i(a)) \cup D_j(CD_i(a)) = D_j(D_i(a) \cup CD_i(a)) = D_j(e_{m-1}) = V$$

and

$$D_i(a) \cap D_j(CD_i(a)) = D_j(D_i(a)) \cap D_j(CD_i(a)) = D_j(D_i(a) \cap CD_i(a)) = D_j(e_0) = e_0.$$

Thus $D_j(CD_i(a))$ is the complement of $D_i(a)$. Since for any complemented element there exists exactly one complement, we infer by (32) that (36) holds.

Now we are going to prove that for all $a, b \in P$

(37) $$a \cap (a \Rightarrow b) \leqslant b.$$

By (p_8^*) and (30)

$$a \cap (a \Rightarrow b) = (D_1(a) \cap e_1 \cup \ldots \cup D_{m-1}(a) \cap e_{m-1}) \cap \cap \big((CD_1(a) \cup D_1(b)) \cap e_1 \cup \ldots \ldots \cup (CD_1(a) \cup D_1(b)) \cap \ldots \cap (CD_{m-1}(a) \cup D_{m-1}(b)) \cap e_{m-1}\big).$$

Hence, by (p_1^*), $a \cap (a \Rightarrow b)$ is the join of all the following summands a_{ij} where $i, j = 1, \ldots, m-1$:

(38) $$a_{ij} = (D_i(a) \cap e_i) \cap (CD_1(a) \cup D_1(b)) \cap \ldots \cap (CD_j(a) \cup D_j(b)) \cap e_j.$$

If $i \leqslant j$, then $a_{ij} \leqslant (D_i(a) \cap e_i) \cap (CD_i(a) \cup D_i(b)) \cap e_j$. Applying ($p_2^*$), (32) and the distributive law, we get

(39) $$a_{ij} \leqslant (D_i(a) \cap CD_i(a)) \cap e_i \cup (D_i(a) \cap D_i(b)) \cap e_i = e_0 \cup (D_i(a) \cap D_i(b)) \cap e_i \leqslant D_i(b) \cap e_i.$$

If $i > j$, then by (35), $D_i(a) \cap CD_j(a) = e_0$. Hence, applying ($p_2^*$) and the distributive law, we obtain

$$a_{ij} \leqslant \big(D_i(a) \cap e_i\big) \cap \big(CD_j(a) \cup D_j(b)\big) \cap e_j \leqslant D_i(a) \cap D_j(b) \cap e_j$$
$$\leqslant D_j(b) \cap e_j.$$

Consequently, $a \cap (a \Rightarrow b) \leqslant D_1(b) \cap e_1 \cup \ldots \cup D_{m-1}(b) \cap e_{m-1} = b$.

In order to prove 1.6 we shall show that the operation $\Rightarrow$ defined by (30) satisfies the condition (r) in IV 1, i.e. that

(40) $\quad a \cap x \leqslant b \quad$ if and only if $\quad x \leqslant a \Rightarrow b \quad$ for all $x, a, b \in P$.

Suppose that $a \cap x \leqslant b$. Then by (p_1^*) and (p_5^*)

$$D_i(a) \cap D_i(x) \leqslant D_i(b) \quad \text{for} \quad i = 1, \ldots, m-1.$$

Hence $CD_i(a) \cup \big(D_i(a) \cap D_i(x)\big) \leqslant CD_i(a) \cup D_i(b)$ for $i = 1, \ldots, m-1$. Applying the distributive law, we get

$$\big(CD_i(a) \cup D_i(a)\big) \cap \big(CD_i(a) \cup D_i(x)\big) \leqslant CD_i(a) \cup D_i(b).$$

Hence, by (32), we obtain

(41) $$D_i(x) \leqslant CD_i(a) \cup D_i(b), \quad i = 1, \ldots, m-1.$$

By (p_4^*), (p_8^*), (41) and (30) we get

$$x = D_1(x) \cap e_1 \cup \ldots \cup D_{m-1}(x) \cap e_{m-1}$$
$$\leqslant D_1(x) \cap e_1 \cup D_1(x) \cap D_2(x) \cap e_2 \cup \ldots \cup D_1(x) \cap \ldots \cap D_{m-1}(x) \cap e_{m-1}$$
$$\leqslant \big(CD_1(a) \cup D_1(b)\big) \cap e_1 \cup \ldots$$
$$\ldots \cup \big(CD_1(a) \cup D_1(b)\big) \cap \ldots \cap \big(CD_{m-1}(a) \cup D_{m-1}(b)\big) \cap e_{m-1}$$
$$= a \Rightarrow b.$$

Thus $a \cap x \leqslant b$ implies $x \leqslant a \Rightarrow b$. On the other hand, if $x \leqslant a \Rightarrow b$, then by (37) we get $a \cap x \leqslant a \cap (a \Rightarrow b) \leqslant b$. This completes the proof of (40).

It follows from (40), IV 1 (r) and IV 6.6 that $(P, V, \Rightarrow, \cup, \cap, \neg)$ is a pseudo-Boolean algebra. The axioms (p_1), (p_2) are satisfied by (p_5^*). By (30) and (p_6^*) we easily get

(42) $$D_i(a) \Rightarrow D_i(b) = CD_i(a) \cup D_i(b) \quad \text{for} \quad i = 1, \ldots, m-1.$$

It follows from (30), (p_5^*), (p_7^*), (p_6^*) and (42) that (p_3) holds. Observe that by (31), (p_7^*) and (42)

$$\neg D_1(a) = D_1(a) \Rightarrow e_0 = D_1(a) \Rightarrow D_1(e_0) = CD_1(a) \cup D_1(e_0)$$
$$= CD_1(a),$$

i.e.

(43) $$\neg D_1(a) = CD_1(a), \quad \text{for each } a \in P.$$

By (31), (p_3), (42), (p_7^*), (34) and (43)

$$D_i(\neg a) = D_i(a \Rightarrow e_0) = \big(D_1(a) \Rightarrow D_1(e_0)\big) \cap \ldots \cap \big(D_i(a) \Rightarrow D_i(e_0)\big)$$
$$= CD_1(a) \cap \ldots \cap CD_i(a) = CD_1(a) = \neg D_1(a).$$

Thus the axiom (p_4) holds. The axioms (p_5), (p_6) and (p_7) hold because of (p_6^*), (p_7^*), (p_1^*) and (p_8^*). The axiom (p_8) holds by (43) and (32). This completes the proof of the theorem.

2. Examples of Post algebras. Let $P_m = \{e_0, \ldots, e_{m-1}\}$, where $m \geq 2$, $e_i \neq e_j$ for $i \neq j$. Consider the following algebra $\mathfrak{P}_m = (P_m, V, \Rightarrow, \cup, \cap, \neg, D_1, \ldots, D_{m-1}, e_0, \ldots, e_{m-1})$ whose operations are defined by means of the following equations

(1) $$V = e_{m-1},$$

(2) $$e_i \Rightarrow e_j = \begin{cases} V & \text{if} \quad i \leqslant j, \\ e_j & \text{if} \quad i > j, \end{cases}$$

(3) $$e_i \cup e_j = e_k, \quad \text{where} \quad k = \max(i, j),$$

(4) $$e_i \cap e_j = e_k, \quad \text{where} \quad k = \min(i, j),$$

(5) $$\neg e_i = e_i \Rightarrow e_0,$$

(6) $$D_i(e_j) = \begin{cases} V & \text{if} \quad i \leqslant j, \\ e_0 & \text{if} \quad i > j. \end{cases}$$

It is easy to verify that $\mathfrak{P}_m$ is a Post algebra of order m. It will be called the *m-element Post algebra of order m.* Note that any two m-element Post algebras of order m are isomorphic. Observe that if $m = 2$, then $(P_2, V, \Rightarrow, \cup, \cap, \neg)$ is the two-element Boolean algebra (VI 1). By (6) the operation D_1 is then the identity operation on P_2.

2.1. *The Post algebra $\mathfrak{P}_m$ is functionally complete* [(4)], i.e. *every map-*

[(4)] For the proof see Rosser and Turquette [1].

ping $f\colon P_m^k \to P_m$, $k = 0, 1, \ldots,$ *can be represented as a composition of the operations*: $\bigvee, \Rightarrow, \cup, \cap, \neg, D_1, \ldots, D_{m-1}, e_0, \ldots, e_{m-1}$.

For $k = 0$ the theorem holds, since $e_0, \ldots, e_{m-1}$ are all zero-argument operations on P_m. Suppose that for all $0 \leqslant n \leqslant k$ any n-argument mapping $g\colon P_m^n \to P_m$ can be represented by means of the above-mentioned operations in $\mathfrak{P}_m$. Let $f\colon P_m^{k+1} \to P_m$. Let us put

(7) $$J_i(x) = D_i(x) \cap \neg D_{i+1}(x), \quad i = 0, \ldots, m-1.$$

Thus $J_0(x) = \neg D_1(x)$, $J_1(x) = D_1(x) \cap \neg D_2(x), \ldots, J_{m-1}(x) = D_{m-1}(x)$. It is easy to verify that for every $i, j = 0, \ldots, m-1$

(8) $$J_i(e_j) = \begin{cases} e_{m-1} & \text{if } i = j, \\ e_0 & \text{if } i \neq j. \end{cases}$$

It follows from (8) that

(9) $$f(x_1, \ldots, x_k, x_{k+1}) = \big(f(x_1, \ldots, x_k, e_0) \cap J_0(x_{k+1})\big) \cup \ldots$$
$$\ldots \cup \big(f(x_1, \ldots, x_k, e_{m-1}) \cap J_{m-1}(x_{k+1})\big).$$

Indeed, for every $j = 0, \ldots, m-1$, if $x_{k+1} = e_j$, then the only summand on the right side of (9) different from e_0 is $f(x_1, \ldots, x_k, e_j) \cap J_j(e_j)$ $= f(x_1, \ldots, x_k, x_{k+1})$. This completes the proof of 2.1.

Every Post algebra $\mathfrak{P} = (P, \bigvee, \Rightarrow, \cup, \cap, \neg, D_1, \ldots, D_{m-1}, e_0, \ldots$ $\ldots, e_{m-1})$ of order m uniquely determines the Boolean algebra $\mathfrak{B}_{\mathfrak{P}}$ $= (B_{\mathfrak{P}}, \bigvee, \Rightarrow, \cup, \cap, \neg)$ of all complemented elements in the distributive lattice $(P, \bigvee, \cup, \cap)$ (see 1.4). Conversely, given a Boolean algebra $\mathfrak{B} = (B, \bigvee_{\mathfrak{B}}, \Rightarrow, \cup, \cap, -)$, it is possible to construct a Post algebra $\mathfrak{P}$ (of any order $m > 2$) in such a way that $\mathfrak{B}_{\mathfrak{P}}$ is isomorphic to $\mathfrak{B}$. For this purpose we use the following method. Let P be the set of all $(m-1)$-element sequences $(b_1, \ldots, b_{m-1})$, where $b_i \in B$ for $i = 1, \ldots, m-1$ and $b_{m-1} \leqslant \ldots \leqslant b_1$. Put for any elements $(b_1, \ldots, b_{m-1})$, $(c_1, \ldots, c_{m-1})$ in P

(10) $$\bigvee = (\bigvee_{\mathfrak{B}}, \ldots, \bigvee_{\mathfrak{B}}),$$

(11) $$(b_1, \ldots, b_{m-1}) \Rightarrow (c_1, \ldots, c_{m-1})$$
$$= \big(b_1 \Rightarrow c_1, (b_1 \Rightarrow c_1) \cap (b_2 \Rightarrow c_2), \ldots$$
$$\ldots, (b_1 \Rightarrow c_1) \cap \ldots \cap (b_{m-1} \Rightarrow c_{m-1})\big).$$

(12) $$(b_1, \ldots, b_{m-1}) \cup (c_1, \ldots, c_{m-1}) = (b_1 \cup c_1, \ldots, b_{m-1} \cup c_{m-1}),$$

(13) $(b_1, \ldots, b_{m-1}) \cap (c_1, \ldots, c_{m-1}) = (b_1 \cap c_1, \ldots, b_{m-1} \cap c_{m-1})$,

(14) $\neg(b_1, \ldots, b_{m-1}) = (-b_1, \ldots, -b_1)$,

(15) $D_i(b_1, \ldots, b_{m-1}) = (b_i, \ldots, b_i), \quad i = 1, \ldots, m-1$,

(16) $e_i = (\underbrace{\bigvee_{\mathfrak{B}}, \ldots, \bigvee_{\mathfrak{B}}}_{i\text{-times}}, \bigwedge_{\mathfrak{B}}, \ldots, \bigwedge_{\mathfrak{B}}), \quad i = 0, \ldots, m-1,$

where $\bigwedge_{\mathfrak{B}} = -\bigvee_{\mathfrak{B}}$.

It is easy to verify that the algebra $\mathfrak{P} = (P, \bigvee, \Rightarrow, \cup, \cap, \neg, D_1, \ldots$ $\ldots, D_{m-1}, e_0, \ldots, e_{m-1})$, whose operations are defined by (10)-(16), is a Post algebra of order m. Since the Boolean algebra $\mathfrak{B}_{\mathfrak{P}}$ determined by $\mathfrak{P}$ consists of all elements in P of the form $(b, \ldots, b)$ where $b \in B$, it is isomorphic to $\mathfrak{B}$.

3. Filters and D-filters in Post algebras. Since every Post algebra

(1) $\mathfrak{P} = (P, \bigvee, \Rightarrow, \cup, \cap, \neg, D_1, \ldots, D_{m-1}, e_0, \ldots, e_{m-1})$

is a relatively pseudo-complemented lattice with respect to the operations $\bigvee, \Rightarrow, \cup, \cap, \neg$, it follows from IV 2.1 that the notion of a filter coincides with the notion of an implicative filter in Post algebras.

A filter ∇ in a Post algebra $\mathfrak{P}$ will be said to be a *D-filter* [5] if $a \in \nabla$ implies $D_i(a) \in \nabla$ for $i = 1, \ldots, m-1$. A D-filter ∇ is said to be *proper* if $a \notin \nabla$ for some element $a \in P$. This condition is equivalent to the condition that $\bigwedge \notin \nabla$. A D-filter is said to be *irreducible* if it is proper and is not the intersection of any two proper D-filters different from it. A D-filter ∇ is said to be *prime* if it is proper and the condition $a \cup b \in \nabla$ implies that either $a \in \nabla$ or $b \in \nabla$. A D-filter is said to be *maximal* if it is proper and is not a proper subset of any proper D-filter.

If a is any element in a Post algebra (1), then the set of all elements x in this algebra such that $D_{m-1}(a) \leqslant x$ is the D-filter generated by a and will be called the *principal D-filter generated by a*. The easy proof, using 1.2 (8), 1.2 (9) and 1 (p_5), is left to the reader.

[5] Proper D-filters in Post algebras of order $m \geqslant 2$ are filters of order $m-1$ in the sense of Traczyk (see Traczyk [3]). Prime D-filters coincide with prime filters of order $m-1$ in each Post algebra of order $m \geqslant 2$.

3.1. *The D-filter generated by a set A_0 of elements in a Post algebra* (1) *is the set of all elements a in $\mathfrak{P}$ such that there exist $a_1, \ldots, a_n$ in A_0 for which $D_{m-1}(a_1) \cap \ldots \cap D_{m-1}(a_n) \leqslant a$.*

The easy proof by verification, using 1.2 (8), 1.2 (9), 1 (p_2), 1 (p_5), is left to the reader.

3.2. *The D-filter generated by a fixed element a_0 and a D-filter ∇ in a Post algebra* (1) *is the set of all elements a in $\mathfrak{P}$ such that $D_{m-1}(a_0) \cap \cap D_{m-1}(c) \leqslant a$ for an element $c \in \nabla$.*

The proof, similar to that of 3.1, is left to the reader.

Let us note that

3.3. *Every maximal D-filter is irreducible.*

The easy proof is omitted.

3.4. *For every D-filter ∇ in a Post algebra $\mathfrak{P}$ of order m, the following conditions are equivalent*:

(i) ∇ *is maximal,*

(ii) ∇ *is irreducible,*

(iii) ∇ *is prime,*

(iv) *for each element a of $\mathfrak{P}$ exactly one of the elements $D_{m-1}(a)$, $\neg D_{m-1}(a)$ is in ∇.*

By 3.3 (i) implies (ii). Suppose that a proper D-filter ∇ is not prime. Then there exist a, b in $\mathfrak{P}$ such that $a \cup b \in \nabla$, $a \notin \nabla$ and $b \notin \nabla$. Let ∇_1 be the D-filter generated by ∇ and a and let ∇_2 be the D-filter generated by ∇ and b. Clearly, $\nabla \subset \nabla_1 \cap \nabla_2$. Suppose that $x \in \nabla_1 \cap \nabla_2$. Then by 3.2, $D_{m-1}(a) \cap D_{m-1}(c_1) \leqslant x$ and $D_{m-1}(b) \cap D_{m-1}(c_2) \leqslant x$ for some $c_1, c_2 \in \nabla$. Let $c = c_1 \cap c_2$. By 1.2 (8), $D_{m-1}(c) \leqslant D_{m-1}(c_1)$ and $D_{m-1}(c) \leqslant D_{m-1}(c_2)$. Hence $D_{m-1}(a) \cap D_{m-1}(c) \leqslant x$ and $D_{m-1}(b) \cap D_{m-1}(c) \leqslant x$. Consequently, by one of the distributivity laws, $(D_{m-1}(a) \cup D_{m-1}(b)) \cap \cap D_{m-1}(c) \leqslant x$. Applying 1 ($p_1$), we get $D_{m-1}(a \cup b) \cap D_{m-1}(c) \leqslant x$. Since $a \cup b$, $c \in \nabla$, we infer that $D_{m-1}(a \cup b) \in \nabla$ and $D_{m-1}(c) \in \nabla$. Thus $D_{m-1}(a \cup b) \cap D_{m-1}(c) \in \nabla$ and consequently $x \in \nabla$. Hence $\nabla_1 \cap \cap \nabla_2 \subset \nabla$. We have just proved that $\nabla = \nabla_1 \cap \nabla_2$, i.e. ∇ is not irreducible. Thus (ii) implies (iii). Suppose that ∇ is a prime D-filter. By definition ∇ is proper. Hence, by 1 (21) for each a in $\mathfrak{P}$ at most one of the elements $D_{m-1}(a)$, $\neg D_{m-1}(a)$ is in ∇. On the other hand, by 1 (22), for each a in $\mathfrak{P}$, $D_{m-1}(a) \cup \neg D_{m-1}(a) \in \nabla$. Since ∇ is prime, at least one of the elements $D_{m-1}(a)$, $\neg D_{m-1}(a)$ is in ∇. Thus (iii) implies (iv). Suppose that (iv) holds. If $a \notin \nabla$, then by 1.2 (9), $D_{m-1}(a) \notin \nabla$. Hence $\neg D_{m-1}(a) \in \nabla$. This implies that the D-filter generated by ∇ and a is not proper. Thus ∇ is a maximal D-filter, i.e. (iv) implies (i).

3.5. *The union of any chain of proper D-filters is a proper D-filter.*

The proof by an easy verification is omitted.

3.6. *Each proper D-filter is contained in a maximal D-filter.*

The proof, based on 3.5 and I 3.1, is omitted.

3.7. *For any element a in a Post algebra $\mathfrak{P}$ of order m, if $D_{m-1}(a) \neq \wedge$, then there exists a maximal D-filter ∇ such that $a \in \nabla$.*

This follows from 3.6, since the principal D-filter generated by a is then proper.

3.8. *If ∇_0 is a D-filter in a Post algebra $\mathfrak{P}$ of order m and $D_{m-1}(a) \notin \nabla_0$, then there exists an irreducible D-filter ∇ such that $\nabla_0 \subset \nabla$ and $a \notin \nabla$.*

The proof, similar to that of III 1.12, is left to the reader.

Given an arbitrary Post algebra (1), we shall denote by $\mathfrak{B}_{\mathfrak{P}}$ the Boolean algebra $(B_{\mathfrak{P}}, \vee, \Rightarrow, \cup, \cap, \neg)$ of all complemented elements in $\mathfrak{P}$ (see 1.4). The following theorem states a connection between filters in a Post algebra $\mathfrak{P}$ and filters in the Boolean algebra $\mathfrak{B}_{\mathfrak{P}}$ of complemented elements in $\mathfrak{P}$.

3.9. *If ∇ is a filter in a Post algebra $\mathfrak{P}$, then $\nabla_0 = \nabla \cap B_{\mathfrak{P}}$ is a filter in $\mathfrak{B}_{\mathfrak{P}}$. Moreover, ∇ is proper if and only if ∇_0 is proper. If ∇ is prime, then ∇_0 is prime* (6).

If $a, b \in \nabla_0$, then clearly $a \cap b \in \nabla_0$. If $a \in \nabla_0$, $a \leqslant b \in B_{\mathfrak{P}}$, then $b \in \nabla$ and hence $b \in \nabla_0$. The filter ∇ is proper if and only if $\wedge = e_0 \notin \nabla$, which is equivalent to the condition that $\wedge = e_0 \notin \nabla_0$, i.e. to the condition that ∇_0 is proper. If ∇ is prime and, for some $a, b \in B_{\mathfrak{P}}$, $a \cup b \in \nabla_0$, then $a \cup b \in \nabla \cap B_{\mathfrak{P}}$. Hence either $a \in \nabla$ or $b \in \nabla$, i.e. either $a \in \nabla_0$ or $b \in \nabla_0$. Thus ∇_0 is prime.

3.10. *If ∇_0 is a filter in the Boolean algebra $\mathfrak{B}_{\mathfrak{P}}$ of all complemented elements in a Post algebra $\mathfrak{P}$, then*

(6) See Traczyk [3].

(i) *the set* ∇_i $(i = 1, \ldots, m-1)$ *defined by the equivalence*

$$a \in \nabla_i \quad \text{if and only if} \quad D_i(a) \in \nabla_0, \tag{2}$$

is a filter in $\mathfrak{P}$, *and* $\nabla_i \cap B_{\mathfrak{P}} = \nabla_0$,

(ii) ∇_{m-1} *is a* D-*filter in* $\mathfrak{P}$,

(iii) ∇_0 *is proper if and only if for each* $i = 1, \ldots, m-1$, *the filter* ∇_i *is proper* (7).

Suppose that $a, b \in \nabla_i$. Then by (2), $D_i(a), D_i(b) \in \nabla_0$. Hence, by 1 ($p_2$), $D_i(a \cap b) = D_i(a) \cap D_i(b) \in \nabla_0$, i.e. $a \cap b \in \nabla_i$. If $a \in \nabla_i$ and $a \leqslant b$, then $D_i(a) \in \nabla_0$ and by 1.2 (8), $D_i(a) \leqslant D_i(b)$. Hence $D_i(b) \in \nabla_0$, i.e. $b \in \nabla_i$. Thus ∇_i is a filter. By (2), 1.4, 1 (p_5) the second part of (i) holds. Suppose that $a \in \nabla_{m-1}$. Then $D_{m-1}(a) \in \nabla_0$. By 1.2 (7), $D_{m-1}(a) \leqslant D_i(a)$ for each $i = 1, \ldots, m-1$. Hence $D_i(a) \in \nabla_0$ and by 1 (p_5) $D_{m-1}(D_i(a)) \in \nabla_0$. Thus $D_i(a) \in \nabla_{m-1}$, i.e. ∇_{m-1} is a D-filter. The filter ∇_0 is proper if and only if $D_i(e_0) = e_0 = \wedge \notin \nabla_0$. By (2) this holds if and only if $e_0 = \wedge \notin \nabla_i$, i.e. if and only if ∇_i is a proper filter in $\mathfrak{P}$.

3.11. *If* ∇_0 *is a prime filter in the Boolean algebra* $\mathfrak{B}_{\mathfrak{P}}$ *of all complemented elements of a Post algebra* $\mathfrak{P}$, *then for* $i = 1, \ldots, m-1$, *the set* ∇_i *defined by* (2) *is a prime filter in* $\mathfrak{P}$. *Moreover,*

$$e_i \in \nabla_i \quad \text{and} \quad e_{i-1} \notin \nabla_i \quad \text{for} \quad i = 1, \ldots, m-1, \tag{3}$$

$$\nabla_1 \supset \nabla_2 \supset \ldots \supset \nabla_{m-1}, \quad \nabla_i \neq \nabla_{i-1} \quad \text{for} \quad i = 2, \ldots, m-1 \tag{4}$$ (8).

Suppose that ∇_0 is a prime filter in $\mathfrak{B}_{\mathfrak{P}}$. By 3.10 ∇_i $(i = 1, \ldots, m-1)$ is a filter in $\mathfrak{P}$. If $a \cup b \in \nabla_i$, then by (2) and 1 (p_1), $D_i(a \cup b) = D_i(a) \cup D_i(b) \in \nabla_0$. Hence either $D_i(a) \in \nabla_0$ or $D_i(b) \in \nabla_0$. Consequently, by (2) either $a \in \nabla_i$ or $b \in \nabla_i$. Thus ∇_i is a prime filter in $\mathfrak{P}$. Since $\vee = e_{m-1} \in \nabla_0$, we have by 1 ($p_6$), $D_i(e_i) = e_{m-1} \in \nabla_0$ $(i = 1, \ldots, m-1)$. Hence, by (2), $e_i \in \nabla_i$. On the other hand, by 1 (p_6), $D_i(e_{i-1}) = e_0 \notin \nabla_0$, since ∇_0 is a prime filter and therefore proper. Thus $e_{i-1} \notin \nabla_i$. Consequently, (3) holds. If $a \in \nabla_i$, $i = 2, \ldots, m-1$, then $D_i(a) \in \nabla_0$. By 1.2 (7), $D_i(a) \leqslant D_{i-1}(a)$. Thus $D_{i-1}(a) \in \nabla_0$ and consequently $a \in \nabla_{i-1}$. Hence $\nabla_1 \supset \nabla_2 \supset \ldots \supset \nabla_{m-1}$. By (3) $\nabla \neq \nabla_{i-1}$. Thus (4) holds.

(7) See Traczyk [3].

(8) See Traczyk [3].

3.12. *For every properly descending chain* $\nabla_1 \supset \ldots \supset \nabla_{m-1}$ *of prime filters in a Post algebra* $\mathfrak{P}$ *of order* $m \geqslant 2$ *the following condition is satisfied*:

(5) $\qquad e_i \in \nabla_i \quad$ *and* $\quad e_{i-1} \notin \nabla_i \quad$ *for* $\quad i = 1, \ldots, m-1$[9].

Clearly, $e_{m-1} = V \in \nabla_{m-1}$. Suppose that there exists an integer i, $1 \leqslant i \leqslant m-2$, such that $e_i \notin \nabla_i$. Let i_0 be the greatest integer $1 \leqslant i_0 \leqslant m-2$ satisfying this condition, i.e.

(6) $$e_{i_0} \notin \nabla_{i_0},$$

(7) $$e_{i_0+1} \in \nabla_{i_0+1} \subset \nabla_{i_0}.$$

Let $a = D_1(a) \cap e_1 \cup \ldots \cup D_{m-1}(a) \cap e_{m-1}$ be an element which belongs to ∇_{i_0} and does not belong to ∇_{i_0+1}. Such an element exists since $\nabla_1 \supset \ldots \supset \nabla_{m-1}$ is a properly descending chain. Since $a \notin \nabla_{i_0+1}$, we infer that $D_{i_0+1}(a) \cap e_{i_0+1} \notin \nabla_{i_0+1}$. Hence, by (7),

(8) $$D_{i_0+1}(a) \notin \nabla_{i_0+1}.$$

On the other hand, it follows from $D_1(a) \cap e_1 \cup \ldots \cup D_{i_0}(a) \cap e_{i_0} \leqslant e_{i_0} \notin \nabla_{i_0}$ that $D_1(a) \cap e_1 \cup \ldots \cup D_{i_0}(a) \cap e_{i_0} \notin \nabla_{i_0}$. Since ∇_{i_0} is a prime filter and $a \in \nabla_{i_0}$, we infer that

(9) $$D_{i_0+1}(a) \cap e_{i_0+1} \cup \ldots \cup D_{m-1}(a) \cap e_{m-1} \in \nabla_{i_0}.$$

Hence

(10) $$D_{i_0+1}(a) \in \nabla_{i_0}.$$

Indeed, otherwise we would have for every $i \geqslant i_0+1$, $D_i(a) \notin \nabla_{i_0}$ by 1.2 (7), and consequently (9) would not be satisfied.

By Theorems 3.9, VI 3.1, VI 3.3 and the inclusion $\nabla_{i_0} \supset \nabla_{i_0+1}$, the sets $\nabla_{i_0} \cap B_{\mathfrak{P}}$ and $\nabla_{i_0+1} \cap B_{\mathfrak{P}}$ are maximal filters in the Boolean algebra $\mathfrak{B}_{\mathfrak{P}}$ such that $\nabla_{i_0} \cap B_{\mathfrak{P}} \supset \nabla_{i_0+1} \cap B_{\mathfrak{P}}$. Thus

(11) $$\nabla_{i_0} \cap B_{\mathfrak{P}} = \nabla_{i_0+1} \cap B_{\mathfrak{P}}.$$

It follows from (8), (10) and 1.4 that

(12) $\qquad D_{i_0+1}(a) \in \nabla_{i_0} \cap B_{\mathfrak{P}} \quad$ and $\quad D_{i_0+1}(a) \notin \nabla_{i_0+1} \cap B_{\mathfrak{P}}$,

which contradicts (11).

[9] See Traczyk [3].

By a similar argument we get a contradiction if we assume that i_0 is the least integer $1 \leqslant i_0 \leqslant m-1$ such that $e_{i_0-1} \in \nabla_{i_0}$. The proof is left to the reader.

A filter ∇ in a Post algebra of order m which contains e_i and does not contain e_{i-1} $(i = 1, \ldots, m-1)$ will be called a *filter of order i*. Observe that, by 1 (p_6), every proper D-filter is a filter of order $m-1$.

It follows from 3.11 and the above definition that

3.13. *If ∇_0 is a prime filter in the Boolean algebra $\mathfrak{B}_{\mathfrak{P}}$ of all complemented elements in a Post algebra $\mathfrak{P}$ of order m, then the sets ∇_i $(i = 1, \ldots$ $\ldots, m-1)$ defined by the equivalence* (2) *are prime filters of order i in $\mathfrak{P}$ and $\nabla_1 \supset \ldots \supset \nabla_{m-1}$ is a properly descending chain* [10].

By 3.12 we get

3.14. *If $\nabla_1 \supset \ldots \supset \nabla_{m-1}$ is a properly descending chain of prime filters in a Post algebra $\mathfrak{P}$ of order $m > 2$, then for every $i = 1, \ldots, m-1$, ∇_i is a filter of order i* [11].

3.15. *In any Post algebra $\mathfrak{P}$ of order $m > 2$ every prime filter ∇ is a member of a properly descending chain of $m-1$ prime filters in $\mathfrak{P}$* [12].

Let i_0 be the least integer $1 \leqslant i_0 \leqslant m-1$ such that $e_{i_0} \in \nabla$ and let $\nabla_0 = \nabla \cap B_{\mathfrak{P}}$. By 3.9, ∇_0 is a prime filter in the Boolean algebra $\mathfrak{B}_{\mathfrak{P}}$ of all complemented elements in $\mathfrak{P}$. For every $i = 1, \ldots, m-1$, let

$$\nabla_i = \{a \in P\colon D_i(a) \in \nabla_0\}. \tag{13}$$

By 3.13, $\nabla_1 \supset \ldots \supset \nabla_{m-1}$ is a properly descending chain of prime filters in $\mathfrak{P}$ and for every $i = 1, \ldots, m-1$, ∇_i is a filter of order i. We shall prove that $\nabla = \nabla_{i_0}$. Let $a = D_1(a) \cap e_1 \cup \ldots \cup D_{m-1}(a) \cap e_{m-1}$ be an arbitrary element in $\mathfrak{P}$ (see 1 (p_7)). Since ∇ is a prime filter of order i_0, the condition $a \in \nabla$ implies that $D_{i_0}(a) \in \nabla \cap B_{\mathfrak{P}} = \nabla_0$. Indeed, by 1.4, $D_{i_0}(a) \in B_{\mathfrak{P}}$. On the other hand, we have $e_{i_0} \in \nabla$ and $e_{i_0-1} \notin \nabla$. By an analogous argument to that used in the proof of 3.12 we infer that $D_{i_0}(a) \in \nabla$. Thus $D_{i_0}(a) \in \nabla_0$. Hence, by (13), $a \in \nabla_{i_0}$, which proves that $\nabla \subset \nabla_{i_0}$. If $a \in \nabla_{i_0}$, then, by (13), $D_{i_0}(a) \in \nabla_0$ and consequently $D_{i_0}(a) \in \nabla$. Since $e_{i_0} \in \nabla$, we get $D_{i_0}(a) \cap e_{i_0} \in \nabla$ and hence $a \in \nabla$. Thus $\nabla_{i_0} \subset \nabla$.

[10] See Traczyk [3].
[11] See Traczyk [3].
[12] See Traczyk [3].

It follows from the proof of 3.15 and from 3.13 that the following theorem holds.

3.16. *A subset* ∇ *of the set of all elements in a Post algebra* $\mathfrak{P}$ *of order* $m > 2$ *is a prime filter of order* i, $i = 1, \ldots, m-1$, *if and only if there exists a prime filter* ∇_0 *in the Boolean algebra of all complemented elements in* $\mathfrak{P}$ *such that for every* a *in* $\mathfrak{P}$

(14) $$a \in \nabla \quad \textit{if and only if} \quad D_i(a) \in \nabla_0$$ [13].

Let us note that

3.17. *If* ∇_1 *and* ∇_2 *are two prime filters of the same order* i, $i = 1, \ldots, m-1$, *in a Post algebra* $\mathfrak{P}$ *of order* $m > 2$ *and* $\nabla_1 \subset \nabla_2$, *then* $\nabla_1 = \nabla_2$ [14].

If the hypotheses of 3.17 are satisfied, then $\nabla_0' = \nabla_1 \cap B_{\mathfrak{P}}$ and $\nabla_0'' = \nabla_2 \cap B_{\mathfrak{P}}$ are prime filters in the Boolean algebra $\mathfrak{B}_{\mathfrak{P}}$ of all complemented elements in $\mathfrak{P}$ such that

(15) $$a \in \nabla_1 \quad \textit{if and only if} \quad D_i(a) \in \nabla_0'$$

and

(16) $$a \in \nabla_2 \quad \textit{if and only if} \quad D_i(a) \in \nabla_0''.$$

This follows from the proof of 3.15. Since $\nabla_1 \subset \nabla_2$, we get $\nabla_0' \subset \nabla_0''$. By VI 3.1, VI 3.3, the filters ∇_0', ∇_0'' are maximal. Thus $\nabla_0' = \nabla_0''$. Hence, by (15) and (16), we get $\nabla_1 = \nabla_2$.

4. Post homomorphisms[15]. The following theorems state a connection between epimorphisms of Post algebras onto similar algebras and D-filters.

4.1. *If* $K(h)$ *is the kernel of an epimorphism* h *of a Post algebra* $\mathfrak{P} = (P, V, \Rightarrow, \cup, \cap, \neg, D_1, \ldots, D_{m-1}, e_0, \ldots, e_{m-1})$ *onto a similar algebra* $\mathfrak{A} = (A, V, \Rightarrow, \cup, \cap, \neg, D_1, \ldots, D_{m-1}, e_0, \ldots, e_{m-1})$, i.e. $K(h) = h^{-1}(V)$, *then* $K(h)$ *is a* D*-filter. Moreover,* $h(a) = h(b)$ *is equivalent to* $a \Rightarrow b \in K(h)$ *and* $b \Rightarrow a \in K(h)$. *Consequently, the relation* $\approx_{K(h)}$ *determined by* $K(h)$ *is a congruence in* $\mathfrak{P}$. *The quotient algebra* $\mathfrak{P}/\approx_{K(h)}$ *is a Post algebra isomorphic to* $\mathfrak{A}$.

(13) See Traczyk [3].

(14) See Traczyk [3].

(15) Post homomorphisms in the sense adopted here differ from those considered by Traczyk [3].

Since the class of all Post algebras is equationally definable, it is closed under epimorphisms and therefore the epimorphic image $\mathfrak{A} = h(\mathfrak{P})$ of a Post algebra $\mathfrak{P}$ is a Post algebra.

The algebra $(P, V, \Rightarrow, \cup, \cap)$ is a relatively pseudo-complemented lattice by 1 (p_0). Hence, by IV 2.3, $K(h)$ is a filter in $\mathfrak{P}$. Suppose that $a \in K(h)$. Then $h(a) = V$. Hence (see 1.2 (6), 1 (p_6)), for each $i = 1, \ldots, m-1$, $h(D_i(a)) = D_i(h(a)) = D_i(V) = D_i(e_{m-1}) = V$. Thus $D_i(a) \in K(h)$. This proves that $K(h)$ is a D-filter in $\mathfrak{P}$.

By IV 2.3, $h(a) = h(b)$ is equivalent to $a \Rightarrow b \in K(h)$ and $b \Rightarrow a \in K(h)$. Hence, by I 4.7, the relation $a \approx_{K(h)} b$ which holds between a and b if and only if $a \Rightarrow b \in K(h)$ and $b \Rightarrow a \in K(h)$ is a congruence in $\mathfrak{P}$ and $\mathfrak{P}/\approx_{K(h)}$ is isomorphic to $\mathfrak{A}$.

4.2. *Let* $\mathfrak{P} = (P, V, \Rightarrow, \cup, \cap, \neg, D_1, \ldots, D_{m-1}, e_0, \ldots, e_{m-1})$ *be a non-degenerate Post algebra of order m and let* ∇ *be a D-filter in* $\mathfrak{P}$. *Then the relation* $\approx_\nabla$ *determined by* ∇ *is a congruence in* $\mathfrak{P}$. *The quotient algebra* $\mathfrak{P}/\approx_\nabla$ *denoted by* $\mathfrak{P}/\nabla$ *is a Post algebra of order m. The mapping* $h(a) = ||a||$ *is an epimorphism of* $\mathfrak{P}$ *onto* $\mathfrak{P}/\nabla$ *and* ∇ *is the kernel of h. For every* $a \in P$, $a \in \nabla$ *if and only if* $a \approx_\nabla V$. *The algebra* $\mathfrak{P}/\nabla$ *is degenerate if and only if* ∇ *is not proper.*

Since any D-filter is a filter and any Post algebra is a pseudo-Boolean algebra, by IV 6.3 we conclude that the relation $\approx_\nabla$ is an equivalence relation preserving the operations $V, \Rightarrow, \cup, \cap, \neg$. Suppose that $a \approx_\nabla b$. By 1 (p_3) and the definition of D-filters we get

$$D_{m-1}(a \Rightarrow b) = (D_1(a) \Rightarrow D_1(b)) \cap \ldots \cap (D_{m-1}(a) \Rightarrow D_{m-1}(b)) \in \nabla.$$

Hence, for each $i = 1, \ldots, m-1$, $D_i(a) \Rightarrow D_i(b) \in \nabla$. By a similar argument $D_i(b) \Rightarrow D_i(a) \in \nabla$. Thus $D_i(a) \approx_\nabla D_i(b)$ for all $i = 1, \ldots, m-1$. Consequently, $\approx_\nabla$ is a congruence relation in $\mathfrak{P}$. Hence the mapping h given by $h(a) = ||a|| \in P/\nabla$ is an epimorphism of $\mathfrak{P}$ onto the quotient algebra $\mathfrak{P}/\nabla$. Since the class of all Post algebras is equationally definable, the algebra $\mathfrak{P}/\nabla$ is a Post algebra (see I 4.11). Clearly, ∇ is the kernel of h, which is equivalent to the condition that for each $a \in P$, $a \in \nabla$ if and only if $a \approx_\nabla V$. The condition that $\mathfrak{P}/\nabla$ is degenerate if and only if ∇ is not proper is obvious.

The next theorems concern epimorphisms of non-degenerate Post algebras of order m onto m-element Post algebras of order m.

4.3. *If h is an epimorphism of a non-degenerate Post algebra* $\mathfrak{P} = (P, V, \Rightarrow, \cup, \cap, \neg, D_1, \ldots, D_{m-1}, e_0, \ldots, e_{m-1})$ *of order m onto the m-element Post algebra* $\mathfrak{P}_m$ (see Sec. 2), *then the set* $K(h) \cap B_{\mathfrak{P}} = \{a \in B_{\mathfrak{P}}: h(a) = e_{m-1}\}$ *is a prime filter in the Boolean algebra* $\mathfrak{B}_{\mathfrak{P}}$ *of all complemented elements in* $\mathfrak{P}$ (see 1.4).

By 1.2 (6), $K(h) \cap B_{\mathfrak{P}} = \{a \in B_{\mathfrak{P}}: h(a) = e_{m-1}\}$. By 4.1 and 3.9, $K(h) \cap B_{\mathfrak{P}}$ is a filter in $\mathfrak{B}_{\mathfrak{P}}$. If $a \in B_{\mathfrak{P}}$, then by 1.4, $a = D_j(b)$ for some $b \in P$ and some $j = 1, \ldots, m-1$. Hence $h(a) = h(D_j(b)) = D_j(h(b)) \in \{e_0, e_{m-1}\}$ (see sec 2). Suppose that $a \cup b \in K(h) \cap B_{\mathfrak{P}}$ for some $a, b \in B_{\mathfrak{P}}$. Then $h(a \cup b) = e_{m-1}$. Consequently, $h(a) \cup h(b) = e_{m-1}$. Since $h(a), h(b) \in \{e_0, e_{m-1}\}$, we infer that either $h(a) = e_{m-1}$ or $h(b) = e_{m-1}$, i.e. either $a \in K(h) \cap B_{\mathfrak{P}}$ or $b \in K(h) \cap B_{\mathfrak{P}}$. Thus $K(h) \cap B_{\mathfrak{P}}$ is a prime filter in $\mathfrak{B}_{\mathfrak{P}}$.

4.4. *If $K(h)$ is the kernel of an epimorphism h of a non-degenerate Post algebra* $\mathfrak{P} = (P, V, \Rightarrow, \cup, \cap, \neg, D_1, \ldots, D_{m-1}, e_0, \ldots, e_{m-1})$ *of order $m > 2$ onto the m-element Post algebra* $\mathfrak{P}_m$ (see Sec. 2), *then $K(h)$ is a prime D-filter.*

Observe that by 4.3, $K(h) \cap B_{\mathfrak{P}}$ is a prime filter in the Boolean algebra $\mathfrak{B}_{\mathfrak{P}}$ of all complemented elements in $\mathfrak{P}$ (see 1.4). Hence, by 3.10 and 3.11, the set

(2) $$\nabla_{m-1} = \{a \in P: D_{m-1}(a) \in K(h) \cap B_{\mathfrak{P}}\}$$

is a prime D-filter in $\mathfrak{P}$. We shall prove that

(3) $$\nabla_{m-1} = K(h).$$

If $a \in K(h)$, then $h(a) = e_{m-1}$. Hence $h(D_{m-1}(a)) = D_{m-1}(h(a)) = D_{m-1}(e_{m-1}) = e_{m-1}$. Thus $D_{m-1}(a) \in K(h) \cap B_{\mathfrak{P}}$, i.e. $a \in \nabla_{m-1}$. Conversely, if $a \in \nabla_{m-1}$, then $D_{m-1}(a) \in K(h)$. Hence $h(D_{m-1}(a)) = D_{m-1}(h(a)) = e_{m-1}$. But $h(a) \in \{e_0, \ldots, e_{m-1}\}$. This implies by 1 ($p_6$) that $h(a) = e_{m-1}$, i.e. $a \in K(h)$. Thus $K(h) = \nabla_{m-1}$, which completes the proof of 4.4.

4.5. *For any D-filter ∇ in a non-degenerate Post algebra $\mathfrak{P}$ of order m the following conditions are equivalent*:

(4) *∇ is a prime D-filter,*

(5) *$\mathfrak{P}/\nabla$ is the m-element Post algebra $\mathfrak{P}_m$ of order m* [16].

[16] See Traczyk [3].

By 4.2 and 4.4 condition (5) implies (4). Suppose that ∇ is a prime D-filter in $\mathfrak{P}$. Since $D_{m-1}(e_j) = \Lambda$ for $j < m-1$, we infer that $e_j \notin \nabla$, for $j < m-1$. Hence, for each $i, j = 0, \ldots, m-1$, the condition $i \neq j$ implies $||e_i|| \neq ||e_j||$. Indeed, if $i < j$, then by 1.2 (14), $e_j \Rightarrow e_i = e_i \notin \nabla$, i.e. e_i is not equivalent to e_j, which implies that $||e_i|| \neq ||e_j||$. Thus the algebra $\mathfrak{P}/\nabla$ contains at least m different elements. Let a be an arbitrary element in the Post algebra $\mathfrak{P}$. Since by 3.9, $\nabla_0 = \nabla \cap B_{\mathfrak{P}}$ is a prime filter in the Boolean algebra $\mathfrak{B}_{\mathfrak{P}}$ of all complemented elements in $\mathfrak{P}$ (see 1.4), for each $i = 1, \ldots, m-1$, either $D_i(a) \in \nabla_0$ or $\neg D_i(a) \in \nabla_0$ (see VI 3.3). Hence either $D_i(a) \in \nabla$ or $\neg D_i(a) \in \nabla$. By 1.2 (7) if $D_i(a) \in \nabla$, then for every $1 \leqslant j \leqslant i$, $D_j(a) \in \nabla$. By 1.4, 1.2 (7) and VI 1.1 (14) the condition $i \leqslant j$ implies $\neg D_i(a) \leqslant \neg D_j(a)$. Hence, if $\neg D_i(a) \in \nabla$, then for each $j \geqslant i$, $\neg D_j(a) \in \nabla$. Consequently, for every element a in the Post algebra $\mathfrak{P}$ there exists an $i \in \{0, \ldots, m-1\}$ such that for every $0 < j \leqslant i$, $D_j(a) \in \nabla$ and for every $i < j \leqslant m-1$, $\neg D_j(a) \in \nabla$. We shall prove that $||a|| = ||e_i||$. If $i = 0$, then by 1.2 (12), 1 (p_4) and 1.2 (10), we have $D_{m-1}(a \Rightarrow e_0) = D_{m-1}(\neg a) = \neg D_1(a) \in \nabla$ and $D_{m-1}(e_0 \Rightarrow a) = D_{m-1}(e_{m-1}) = V \in \nabla$. Hence, by 1.2 (9), $a \Rightarrow e_0 \in \nabla$ and $e_0 \Rightarrow a \in \nabla$, i.e. $||a|| = ||e_0||$. Suppose that $0 < i \leqslant m-1$. Then by 1 (p_3), 1 (p_6), 1.2,

$$
\begin{aligned}
D_{m-1}(a \Rightarrow e_i) &= (D_1(a) \Rightarrow D_1(e_i)) \cap \ldots \cap (D_{m-1}(a) \Rightarrow D_{m-1}(e_i)) \\
&= (D_1(a) \Rightarrow e_{m-1}) \cap \ldots \cap (D_i(a) \Rightarrow e_{m-1}) \cap \\
&\qquad \cap (D_{i+1}(a) \Rightarrow e_0) \cap \ldots \cap (D_{m-1}(a) \Rightarrow e_0) \\
&= \neg D_{i+1}(a) \cap \ldots \cap \neg D_{m-1}(a).
\end{aligned}
$$

Since $\neg D_{i+1}(a), \ldots, \neg D_{m-1}(a) \in \nabla$, we get $D_{m-1}(a \Rightarrow e_i) \in \nabla$. On the other hand, by 1 (p_3), 1 (p_6), 1.2,

$$
\begin{aligned}
D_{m-1}(e_i \Rightarrow a) &= (D_1(e_i) \Rightarrow D_1(a)) \cap \ldots \cap (D_{m-1}(e_i) \Rightarrow D_{m-1}(a)) \\
&= (e_{m-1} \Rightarrow D_1(a)) \cap \ldots \cap (e_{m-1} \Rightarrow D_i(a)) \cap \\
&\qquad \cap (e_0 \Rightarrow D_{i+1}(a)) \cap \ldots \cap (e_0 \Rightarrow D_{m-1}(a)) \\
&= D_1(a) \cap \ldots \cap D_i(a).
\end{aligned}
$$

Since $D_1(a), \ldots, D_i(a) \in \nabla$, we infer that $D_{m-1}(e_i \Rightarrow a) \in \nabla$ and hence by 1.2 (9), $e_i \Rightarrow a \in \nabla$. Thus $a \approx_\nabla e_i$, i.e. $||a|| = ||e_i||$. This proves that condition (4) implies (5), which completes the proof of 4.5.

4.6. *Let* $\mathfrak{P} = (P, V, \Rightarrow, \cup, \cap, \neg, D_1, \ldots, D_{m-1}, e_0, \ldots, e_{m-1})$ *and* $\mathfrak{P}' = (P', V, \Rightarrow, \cup, \cap, \neg, D_1', \ldots, D_{m-1}', e_0', \ldots, e_{m-1}')$ *be Post algebras and let* h_0 *be a homomorphism of the Boolean algebra* $\mathfrak{B}_{\mathfrak{P}}$ *of all complemented elements in* $\mathfrak{P}$ *into the Boolean algebra* $\mathfrak{B}_{\mathfrak{P}'}$ *of all complemented elements in* $\mathfrak{P}'$. *Then* h_0 *can be extended to a homomorphism* h *from* $\mathfrak{P}$ *into* $\mathfrak{P}'$. *Moreover, if* h_0 *is an epimorphism of* $\mathfrak{B}_{\mathfrak{P}}$ *onto* $\mathfrak{B}_{\mathfrak{P}'}$, *then* h *is an epimorphism of* $\mathfrak{P}$ *onto* $\mathfrak{P}'$ [(17)].

Let us put

(6) $$h(e_i) = e_i', \quad i = 1, \ldots, m-2,$$

(7) $$h(x) = h_0(D_1(x)) \cap e_1' \cup \ldots \cup h_0(D_{m-1}(x)) \cap e_{m-1}'.$$

Then by (7), 1 (p_5), the distributive law and 1.2 (6) we have

$$h(D_i(x)) = h_0(D_i(x)) \cap e_1' \cup \ldots \cup h_0(D_i(x)) \cap e_{m-1}' = h_0(D_i(x)).$$

Thus

(8) $h(D_i(x)) = h_0(D_i(x))$ for every $i = 1, \ldots, m-1$ and $x \in P$,

i.e. h is an extension of h_0 since the elements $D_i(x)$, $i = 1, \ldots, m-1$, $x \in P$, by 1.4 are all elements in $\mathfrak{B}_{\mathfrak{P}}$. It follows from (8) that

(9) $$h(e_0) = e_0' \quad \text{and} \quad h(e_{m-1}) = e_{m-1}'.$$

Moreover, by (7), 1 (p_1), 1 (p_2) and the distributive law in $\mathfrak{P}'$, we get

(10) $h(x \cup y) = h(x) \cup h(y)$, $h(x \cap y) = h(x) \cap h(y)$ for any x, y in $\mathfrak{P}$.

We shall prove that for any $i = 1, \ldots, m-1$ and each x in $\mathfrak{P}$

(11) $$D_i'(h(x)) = h(D_i(x)).$$

By (7), 1 (p_1), 1 (p_2), 1 (p_6), 1.2 (5), 1.2 (7) we get

$$\begin{aligned} D_i'(h(x)) &= D_i'\big(h_0(D_1(x))\big) \cap D_i'(e_1') \cup \ldots \cup D_i'\big(h_0(D_{m-1}(x))\big) \cap D_i'(e_{m-1}') \\ &= D_i'\big(h_0(D_i(x))\big) \cup \ldots \cup D_i'\big(h_0(D_{m-1}(x))\big) \\ &= D_i'\big(h_0(D_i(x) \cup \ldots \cup D_{m-1}(x))\big) = D_i'\big(h_0(D_i(x))\big). \end{aligned}$$

Since $h_0(D_i(x))$ is a complemented element in $\mathfrak{P}'$, by 1.4 it is of the

(17) See Traczyk [3].

form $D'_j(y)$ for some $j = 1, \ldots, m-1$ and $y \in P'$. Hence, by 1 (p_5) and (8), we get $D'_i\big(h_0(D_i(x))\big) = h_0(D_i(x)) = h(D_i(x))$, which proves (11).

Let us put

$$(12) \qquad C(x) = \neg D_1(x) \quad \text{for each} \quad x \in P,$$

$$(13) \qquad C'(y) = \neg D'_1(y) \quad \text{for each} \quad y \in P'.$$

By 1 (23), $\neg D_1(x)$ $(\neg D'_1(y))$ is the complement of $D_1(x)$ (of $D'_1(y)$) in the Boolean algebra $\mathfrak{B}_{\mathfrak{P}}$ (in the Boolean algebra $\mathfrak{B}_{\mathfrak{P}'}$). Since h_0 is a Boolean homomorphism from $\mathfrak{B}_{\mathfrak{P}}$ into $\mathfrak{B}_{\mathfrak{P}'}$ it preserves the complement. On the other hand, for any $a \in B_{\mathfrak{P}}$, we have $h(a) = h_0(a)$. Hence, by (11), (12) and (13), we get

$$h(C(x)) = h(\neg D_1(x)) = h_0(\neg D_1(x)) = \neg h_0(D_1(x)) = \neg h(D_1(x))$$
$$= \neg D'_1(h(x)) = C'(h(x)).$$

Thus

$$(14) \qquad h(C(x)) = C'(h(x)).$$

Observe that in any lattice $(A, \cup, \cap)$ there exists at most one operation $\Rightarrow$ of relative pseudo-complementation. Indeed, by IV 1 (r) and IV 1.3 (5),

$$a \Rightarrow b = \text{l.u.b.}\{x \in A\colon a \cap x \leqslant b\}, \qquad \text{for all } a, b \text{ in } A.$$

It follows from this remark and from 1.5, 1.6 that the operation $\Rightarrow$ of relative pseudo-complementation in $\mathfrak{P}$ (in $\mathfrak{P}'$) is equationally definable by means of the operations $\cup, \cap, C, D_1, \ldots, D_{m-1}, e_0, \ldots \ldots, e_{m-1}$ (by means of the operations $\cup, \cap, C', D'_1, \ldots, D'_{m-1}, e'_0, \ldots \ldots, e'_{m-1}$). Hence, by (10), (14), (11) and (6) we get

$$(15) \qquad h(x \Rightarrow y) = h(x) \Rightarrow h(y), \qquad \text{for all } x, y \text{ in } \mathfrak{P}.$$

Consequently, by 1.2 (12) and (6), we get

$$(16) \quad h(\neg x) = h(x \Rightarrow e_0) = h(x) \Rightarrow e'_0 = \neg h(x), \qquad \text{for any } x, \text{ in } \mathfrak{P}.$$

Thus h is a homomorphism of $\mathfrak{P}$ into $\mathfrak{P}'$. The second part of 4.6 is easy to prove. The proof is left to the reader.

4.7. *If h_0 is an isomorphism of the Boolean algebra $\mathfrak{B}_{\mathfrak{P}}$ of all complemented elements in a Post algebra $\mathfrak{P}$ onto the Boolean algebra $\mathfrak{B}_{\mathfrak{P}'}$*

of all complemented elements in a Post algebra $\mathfrak{P}'$ *of the same order as* $\mathfrak{P}$, *then* h_0 *can be extended to an isomorphism* h *of* $\mathfrak{P}$ *onto* $\mathfrak{P}'$.

Let h_0 be an isomorphism of $\mathfrak{B}_{\mathfrak{P}}$ onto $\mathfrak{B}_{\mathfrak{P}'}$. By 4.6, h_0 can be extended to an epimorphism h of $\mathfrak{P}$ onto $\mathfrak{P}'$. It remains to show that h is one-one. Suppose that $x \neq y$ for some x, y in $\mathfrak{P}$. Consequently, by 1 (p_7), there exists an index i_0 such that $D_{i_0}(x) \neq D_{i_0}(y)$. Hence, $h(D_{i_0}(x)) = h_0(D_{i_0}(x)) \neq h_0(D_{i_0}(y)) = h(D_{i_0}(y))$. Hence, by (11), $D'_{i_0}h(x)) \neq D'_{i_0}(h(y))$. This implies that $h(x) \neq h(y)$.

5. Post fields of sets. A topological space X will be said to be a *Post space of order* $m, m \geqslant 2$ [18] if

(1) X is the set-theoretic union of disjoint topological spaces X_i, $1 \leqslant i \leqslant m-1$,

(2) there exists a compact totally disconnected Hausdorff space X_0 and homeomorphisms g_i: $X_i \to X_0$ of X_i onto X_0, $1 \leqslant i \leqslant m-1$,

(3) the family of sets $B(X) = \{\bigcup_{i=1}^{m-1} g_i^{-1}(U)$: *U is an open and closed subset of* $X_0\}$ is a base for the open sets in X.

It is easy to see that *any Post space is compact.*

If X is a Post space of order m and X_i, g_i, $i = 1, \ldots, m-1$, X_0 and $B(X)$ have the meaning as indicated in (1), (2), (3), then we shall often write $X = (\{X_i, g_i\}_{i=1,\ldots,m-1}, X_0, B(X))$.

5.1. *For any Post space* $X = (\{X_i, g_i\}_{i=1,\ldots,m-1}, X_0, B(X))$, $\mathfrak{B}(X) = (B(X), X, \Rightarrow, \cup, \cap, -)$, *where for all sets* Y, Z *in* $B(X)$

(4) $$Y \Rightarrow Z = (X-Y) \cup Z \quad \textit{and} \quad -Y = X-Y,$$

is the field (see VI 1) *of all simultaneously open and closed subsets of* X.

Clearly, $X \in B(X)$ and the union of any two sets in $B(X)$ also belongs to $B(X)$. It is easy to verify that if $Y \in B(X)$, i.e. $Y = \bigcup_{i=1}^{m-1} g_i^{-1}(U)$ for some open and closed subset U of X_0, then by (1), (2), (3) we have the complement $-Y = \bigcup_{i=1}^{m-1} g_i^{-1}(-U) \in B(X)$. Thus $B(X)$ is closed under

[18] See Traczyk [3], [4]; Dwinger [1].

the operations $\cup$, $-$ and contains X, which proves that $\mathfrak{B}(X)$ is a field of some both open and closed subsets of X. Suppose that Y is an open and closed subset of X. Then $Y = \bigcup_{t\in T} G_t$, where $G_t = \bigcup_{i=1}^{m-1} g_i^{-1}(U_t)$ for $t \in T$ and U_t is an open and closed subset of X_0. Thus

$$(5) \qquad Y = \bigcup_{t\in T} \bigcup_{i=1}^{m-1} g_i^{-1}(U_t).$$

Since Y is a closed subset of a compact topological space, it follows from (5) that there exist $t_1, \ldots, t_n \in T$ such that

$$Y = \bigcup_{j=1}^{n} \bigcup_{i=1}^{m-1} g_i^{-1}(U_{t_j}) = \bigcup_{i=1}^{m-1} g_i^{-1}(U_{t_1} \cup \ldots \cup U_{t_n}).$$

The set $U_{t_1} \cup \ldots \cup U_{t_n}$ is an open and closed subset of X_0; hence $Y \in B(X)$.

Given any Post space $X = (\{X_i, g_i\}_{i=1,\ldots,m-1}, X_0, B(X))$, let us put

$$(6) \; E_0 = O, \; E_1 = X_1, \; E_2 = X_1 \cup X_2, \; \ldots, \; E_{m-1} = X_1 \cup \ldots \cup X_{m-1} = X.$$

Clearly,

$$(7) \qquad O = E_0 \subset E_1 \subset E_2 \subset \ldots \subset E_{m-1} = X.$$

With the above notation the following result holds:

5.2. *For any* $Y, Z \in B(X)$,

(i) *if* $Y \cap E_j \subset E_{j-1}$, *for some* $j = 1, \ldots, m-1$, *then* $Y = O$,

(ii) *if* $Y \cap E_j = Z \cap E_j$, *for some* $j = 1, \ldots, m-1$, *then* $Y = Z$.

To prove (i) observe that $Y = \bigcup_{i=1}^{m-1} g_i^{-1}(U)$ for some open and closed subset U of X_0. Hence, by (1), (2), (3) and (6), we get

$$(8) \qquad Y \cap E_j = \bigcup_{i=1}^{j} g_i^{-1}(U), \quad j = 1, \ldots, m-1.$$

If $Y \cap E_j \subset E_{j-1}$, then by (8), (1) and (2) we infer that $g_j^{-1}(U) = O$ and consequently $U = O$. Thus $Y = O$.

To prove (ii) suppose that $Y \cap E_j \subset Z \cap E_j$ for some $j = 1, \ldots, m-1$. Hence $-Z \cap Y \cap E_j = -Z \cap Z \cap E_j = O \subset E_{j-1}$. Since $-Z \cap Y = -(-Y \cup Z) = -(Y \Rightarrow Z) \in B(X)$, by (i) we get $-Z \cap Y = O$. By an analogous argument $-Y \cap Z = O$. Thus $Y = Z$.

Let $X = (\{X_i, g_i\}_{i=1,\ldots,m-1}, X_0, B(X))$ be an arbitrary Post space, let $E_0, \ldots, E_{m-1}$ be subsets of X defined by (6) and let $P(X)$ be the class of all subsets Y of X of the form

(9) $$Y = Y_1 \cap E_1 \cup \ldots \cup Y_{m-1} \cap E_{m-1}, \quad \text{where} \quad Y_i \in B(X), \quad i = 1, \ldots, m-1.$$

Clearly,

(10) $$B(X) \subset P(X) \quad \text{and} \quad E_i \in P(X), \quad i = 1, \ldots, m-1.$$

We shall prove that under the above hypotheses

5.3. *For each set Y in $P(X)$ there exists exactly one representation*

(11) $$Y = D_1 \cap E_1 \cup \ldots \cup D_{m-1} \cap E_{m-1},$$

where $D_i \in B(X)$ for $i = 1, \ldots, m-1$ and

(12) $$D_1 \supset D_2 \supset \ldots \supset D_{m-1}.$$

This representation will be called *monotonic.*

If $Y \in P(X)$, then Y can be represented in the form (9). Assuming $D_i = Y_i \cup \ldots \cup Y_{m-1}$, we get the required representation (11). Suppose that

(13) $$Y = D_1 \cap E_1 \cup \ldots \cup D_{m-1} \cap E_{m-1} = D'_1 \cap E_1 \cup \ldots \cup D'_{m-1} \cap E_{m-1},$$

where $D_i, D'_i \in B(X)$, $i = 1, \ldots, m-1$, $D_1 \supset \ldots \supset D_{m-1}$, $D'_1 \supset \ldots \supset D'_{m-1}$. Then by (6), $Y \cap E_1 = D_1 \cap E_1 = D'_1 \cap E_1$. Hence, by 5.2 (ii), $D_1 = D'_1$. Now, let us assume that $D_j = D'_j$ for $0 < j < k \leqslant m-1$ and consider the meets of E_k and of each of the two sides of equation (13). Then we get

$$D_1 \cap E_1 \cup \ldots \cup D_k \cap E_k = D'_1 \cap E_1 \cup \ldots \cup D'_k \cap E_k.$$

Hence

$$-D'_k \cap D_k \cap E_k \subset -D'_k \cap D_1 \cap E_1 \cup \ldots \cup -D'_k \cap D_k \cap E_k \subset E_{k-1}.$$

Applying 5.2 (i), we get $-D'_k \cap D_k = O$. By a similar argument $-D_k \cap D'_k = O$. Thus $D_k = D'_k$. Consequently, using an inductive argument, we get $D_i = D'_i$ for any $i = 1, \ldots, m-1$.

It follows from 5.3 that there exist operations D_j: $P(X) \to B(X)$, $j = 1, \ldots, m-1$, on $P(X)$, which with every $Y \in P(X)$ associate uniquely determined coefficients $D_j(Y) \in B(X)$ of the monotonic representation of Y. Thus, for any $Y \in B(X)$, the following equation holds:

(14) $$Y = D_1(Y) \cap E_1 \cup \ldots \cup D_{m-1}(Y) \cap E_{m-1},$$

where

(15) $$D_j(Y) \in B(X), \quad j = 1, \ldots, m-1,$$

(16) $$D_1(Y) \supset \ldots \supset D_{m-1}(Y).$$

Consequently,

(17) $$D_i(Y) \cap D_j(Y) = D_j(Y) \quad \text{if} \quad i \leqslant j, \quad i, j \in \{1, \ldots, m-1\}.$$

If $Y \in B(X)$, then $Y = Y \cap E_1 \cup \ldots \cup Y \cap E_{m-1}$ is the unique monotonic representation of Y. Hence

(18) $$D_j(Y) = Y \quad \text{for any } Y \in B(X) \text{ and } j = 1, \ldots, m-1.$$

By (15) and (18) we get

(19) $$D_i(D_j(Y)) = D_j(Y) \quad \text{for any } Y \in P(X) \text{ and } i, j \in \{1, \ldots, m-1\}.$$

Observe that $E_i = X \cap E_1 \cup \ldots \cup X \cap E_i \cup O \cap E_{i+1} \cup \ldots \cup O \cap E_{m-1}$ is the unique monotonic representation of E_i, $i = 1, \ldots, m-1$. Hence, by (14) and (6), we get

(20) $$D_i(E_j) = \begin{cases} E_{m-1} & \text{if } i \leqslant j \\ E_0 & \text{if } i > j \end{cases} \quad \text{for all } i \in \{1, \ldots, m-1\},\ j \in \{0, \ldots, m-1\}.$$

Note also that $P(X)$ *is closed under the operations of set-theoretic union and intersection*, i.e. $(P(X), \cup, \cap)$ *is a set lattice*. In fact, if $Y, Z \in P(X)$, then

$$Y = D_1(Y) \cap E_1 \cup \ldots \cup D_{m-1}(Y) \cap E_{m-1},$$
$$Z = D_1(Z) \cap E_1 \cup \ldots \cup D_{m-1}(Z) \cap E_{m-1}.$$

Hence, by distributivity and (16), we get

(21) $$Y \cup Z = (D_1(Y) \cup D_1(Z)) \cap E_1 \cup \ldots \cup (D_{m-1}(Y) \cup D_{m-1}(Z)) \cap E_{m-1},$$

(22) $$Y \cap Z = (D_1(Y) \cap D_1(Z)) \cap E_1 \cup \ldots \cup (D_{m-1}(Y) \cap D_{m-1}(Z)) \cap E_{m-1}.$$

Clearly, $D_i(Y) \cup D_i(Z)$, $D_i(Y) \cap D_i(Z) \in B(X)$. Hence $Y \cup Z$, $Y \cap Z \in P(X)$. Moreover, the representations (21), (22) are monotonic. On the other hand, by (14) we get

$$(23) \qquad Y \cup Z = D_1(Y \cup Z) \cap E_1 \cup \ldots \cup D_{m-1}(Y \cup Z) \cap E_{m-1},$$

$$(24) \qquad Y \cap Z = D_1(Y \cap Z) \cap E_1 \cup \ldots \cup D_{m-1}(Y \cap Z) \cap E_{m-1}.$$

It follows from the uniqueness of the monotonic representations and from (21), (22), (23), (24) that

$$(25) \quad D_i(Y \cup Z) = D_i(Y) \cup D_i(Z), \qquad D_i(Y \cap Z) = D_i(Y) \cap D_i(Z), \qquad i = 1, \ldots, m-1.$$

Let us introduce in $P(X)$ one more operation C by means of the following equation:

$$(26) \qquad C(Y) = -D_1(Y), \quad \text{for any } Y \in P(X).$$

Then we obtain

$$(27) \qquad D_1(Y) \cap C(Y) = O, \quad D_1(Y) \cup C(Y) = X.$$

Consider the abstract algebra $\mathfrak{P}^*(X) = (P(X), X, \cup, \cap, C, D_1, \ldots \ldots, D_{m-1}, E_0, \ldots, E_{m-1})$. We have proved before that $(P(X), X, \cup, \cap)$ is a set lattice, i.e. a distributive lattice with unit element $X = E_{m-1}$ and zero element $O = E_0$. It follows from this remark and from (7), (27), (17), (19), (25), (20), (14) that the axioms (p_1^*)–(p_8^*), formulated in 1.5, are satisfied. Applying 1.6, we infer that the abstract algebra $\mathfrak{P}(X) = (P(X), X, \Rightarrow, \cup, \cap, \neg, D_1, \ldots, D_{m-1}, E_0, \ldots, E_{m-1})$, where $\Rightarrow, \neg$ are operations defined by means of the equations

$$(28) \quad Y \Rightarrow Z = (CD_1(Y) \cup D_1(Z)) \cap E_1 \cup ((CD_1(Y) \cup D_1(Z)) \cap \cap (CD_2(Y) \cup D_2(Z))) \cap E_2 \cup \ldots \ldots \cup ((CD_1(Y) \cup D_1(Z)) \cap \ldots \cap (CD_{m-1}(Y) \cup D_{m-1}(Z))) \cap E_{m-1},$$

$$(29) \qquad \neg Y = Y \Rightarrow E_0,$$

is a Post algebra of order m. It will be said to be a *Post field of sets* (*of subsets of a Post space* X)[19]. We shall prove in Sec. 6 that Post fields of sets are typical examples of Post algebras.

[19] See Traczyk [3], [4]; Dwinger [1].

6. Representation theorem for Post algebras. The aim of this section is to prove the following representation theorem for Post algebras:

6.1. *Every Post algebra* $\mathfrak{P} = (P, V, \Rightarrow, \cup, \cap, \neg, D_1, \ldots, D_{m-1}, E_0, \ldots, E_{m-1})$ *is isomorphic to a Post field* $\mathfrak{P}(X) = (P(X), X, \Rightarrow, \cup, \cap, \neg, D_1, \ldots, D_{m-1}, E_0, \ldots, E_{m-1})$ *of subsets of a Post space* $X = (\{X_i, g_i\}_{i=1,\ldots,m-1}, X_0, B(X))$[20].

Let $\mathfrak{B}_{\mathfrak{P}} = (B_{\mathfrak{P}}, V, \Rightarrow, \cup, \cap, \neg)$ be the Boolean algebra of all complemented elements in $(P, V, \cup, \cap)$ (see 1.4). For any $a \in B_{\mathfrak{P}}$ let

(1) $$h_0(a) = \{\nabla_0 \in X_0 \colon a \in \nabla_0\},$$

where X_0 is the set of all prime filters of $\mathfrak{B}_{\mathfrak{P}}$. Moreover, let us put

(2) $$B_0(X_0) = \{h_0(a)\}_{a \in B_{\mathfrak{P}}}.$$

By VI 4.1, h_0 is a Boolean isomorphism of $\mathfrak{B}_{\mathfrak{P}}$ onto the field $\mathfrak{B}_0(X_0) = (B_0(X_0), X_0, \Rightarrow, \cup, \cap, -)$ of subsets of X_0 and X_0 becomes a compact totally disconnected Hausdorff space if we take $B_0(X_0)$ as a base for the topology of X_0. Moreover, $B_0(X_0)$ is the class of all both open and closed subsets of X_0.

For each $i = 1, \ldots, m-1$, let X_i be the set of all prime filters of order i of the Post algebra $\mathfrak{P}$ (see Sec. 3, p. 149) and let $g_i \colon X_i \to X_0$ be the mapping defined as follows:

(3) $$g_i(\nabla_i) = \nabla_i \cap B_{\mathfrak{P}} \quad \text{for every } \nabla_i \in X_i.$$

It follows from 3.10, 3.11, 3.16 that

(4) $g_i \colon X_i \to X_0$ is one-one and maps X_i onto X_0, $i = 1, \ldots, m-1$.

Thus every set X_i becomes a topological space homeomorphic with X_0 if we take $B_i(X_i) = \{g_i^{-1}(h_0(a))\}_{a \in B_{\mathfrak{P}}}$ as the base for the topology of X_i. Then g_i is a homeomorphism of X_i onto X_0, $i = 1, \ldots, m-1$.

Observe that

(5) $$X_i \cap X_j = O \quad \text{if} \quad i \neq j, \quad i, j \in \{1, \ldots, m-1\}.$$

Indeed, if $i < j$, then by the definition of a prime filter of order $n = 1, \ldots, m-1$, (see Sec. 3, p. 149), for any $\nabla_i \in X_i$ and any $\nabla_j \in X_j$

[20] Traczyk [3], [4]. For a modification of the representation theorem see Dwinger [1].

we have $e_i \in \nabla_i$ and $e_{j-1} \notin \nabla_j$. Since $i \leqslant j-1$, by 1.2 (6), $e_i \leqslant e_{j-1}$, and consequently by III 1 (f_2), $e_i \notin \nabla_j$. Thus $\nabla_i \neq \nabla_j$.

Let us put

$$X = X_1 \cup \ldots \cup X_{m-1}, \tag{6}$$

and

$$B(X) = \Big\{\bigcup_{i=1}^{m-1} g_i^{-1} h_0(a)\Big\}_{a \in B_{\mathfrak{P}}}. \tag{7}$$

We shall consider X as a topological space with $B(X)$ as the base of the topology. Since the space X satisfies conditions (1), (2), (3) in Sec. 5, it is a Post space of order m. By 5.1 $\mathfrak{B}(X) = \big(B(X), X, \Rightarrow, \cup, \cap, -\big)$ is the field of all both open and closed subsets of X.

Let $\mathfrak{P}(X) = \big(P(X), X, \Rightarrow, \cup, \cap, \neg, D_1, \ldots, D_{m-1}, E_0, \ldots, E_{m-1}\big)$ be the Post field of subsets of X constructed as in Sec. 5. By 5 (18) and 1.4 $\mathfrak{B}(X)$ is the Boolean algebra of all complemented elements in $\big(P(X), X, \cup, \cap\big)$.

Consider the mapping $h\colon B_{\mathfrak{P}} \to B(X)$, given by

$$h(a) = \bigcup_{i=1}^{m-1} g_i^{-1} h_0(a), \qquad \text{for every } a \in B_{\mathfrak{P}}. \tag{8}$$

We shall prove that h is a Boolean isomorphism of $\mathfrak{B}_{\mathfrak{P}}$ onto $\mathfrak{B}(X)$. Observe that h_0 is a Boolean isomorphism of $\mathfrak{B}_{\mathfrak{P}}$ onto $\mathfrak{B}_0(X_0)$ and it is easy to prove by applying (4) that $g_i^{-1}\colon 2^{X_0} \to 2^{X_i}$ is a Boolean isomorphism of the field of all subset of X_0 onto the field of all subsets of X_i, $i = 1, \ldots, m-1$. The proof is left to the reader. Thus $g_i^{-1} h_0$ is a Boolean isomorphism of $\mathfrak{B}_{\mathfrak{P}}$ onto the field $\mathfrak{B}_i(X_i) = \big(B_i(X_i), X_i, \Rightarrow, \cup, \cap, -\big)$ of subsets of X_i, $i = 1, \ldots, m-1$. Hence we have for any $i = 1, \ldots, m-1$

$$g_i^{-1} h_0(V) = X_i, \tag{9}$$

$$g_i^{-1} h_0(a \Rightarrow b) = \big(X_i - g_i^{-1} h_0(a)\big) \cup g_i^{-1} h_0(b) = g_i^{-1} h_0(a) \Rightarrow g_i^{-1} h_0(b), \tag{10}$$

$$g_i^{-1} h_0(a \cup b) = g_i^{-1} h_0(a) \cup g_i^{-1} h_0(b), \tag{11}$$

$$g_i^{-1} h_0(a \cap b) = g_i^{-1} h_0(a) \cap g_i^{-1} h_0(b), \tag{12}$$

$$g_i^{-1} h_0(\neg a) = X - g_i^{-1} h_0(a) = -g_i^{-1} h_0(a). \tag{13}$$

Applying (5)-(13) and the fact that $g_i^{-1}h_0$ for $i = 1, \ldots, m-1$ is a one-one mapping of $B_{\mathfrak{P}}$ onto $B_i(X_i)$, it is easy to show that h is a one-one mapping from $B_{\mathfrak{P}}$ onto $B(X)$ and that the following equations hold:

$$h(\vee) = X, \quad h(a \Rightarrow b) = (X - h(a)) \cup h(b), \quad h(a \cup b) = h(a) \cup h(b),$$
$$h(a \cap b) = h(a) \cap h(b), \quad h(-a) = X - h(a) = -h(a).$$

The proof is left to the reader. Thus h is a Boolean isomorphism of $\mathfrak{B}_{\mathfrak{P}}$ onto $\mathfrak{B}(X)$.

By 4.7, h can be extended to an isomorphism of $\mathfrak{P}$ onto $\mathfrak{P}(X)$. This completes the proof of Theorem 6.1.

Exercises

1. Let $\mathbf{L}_d$ be the class of all non-degenerate distributive lattices with a unit element and a zero element. A distributive lattice $\mathfrak{A} = (A, \vee, \cup, \cap, \wedge)$ in $\mathbf{L}_d$ will be said to be the *$\mathbf{L}_d$-product* (the *coproduct*) of the distributive lattices $\mathfrak{E}_1 = (E_1, \vee, \cup, \cap, \wedge)$, $\mathfrak{E}_2 = (E_2, \vee, \cup, \cap, \wedge)$ in $\mathbf{L}_d$ provided there exist two subalgebras $\mathfrak{A}_1 = (A_1, \vee, \cup, \cap, \wedge)$ and $\mathfrak{A}_2 = (A_2, \vee, \cup, \cap, \wedge)$ of $\mathfrak{A}$ such that: (i) the union $A_1 \cup A_2$ generates $\mathfrak{A}$; (ii) $\mathfrak{A}_i$ is isomorphic to $\mathfrak{E}_i$, $i = 1, 2$; (iii) if for each $i = 1, 2$, h_i is a homomorphism of $\mathfrak{A}_i$ into an algebra $\mathfrak{C} \in \mathbf{L}_d$, then there exists a homomorphism h of $\mathfrak{A}$ into $\mathfrak{C}$, which is a common extension of h_1 and h_2. Prove that any two algebras $\mathfrak{E}_1$, $\mathfrak{E}_2$ in $\mathbf{L}_d$ have a coproduct defined up to the isomorphism type (21).

2. Prove that each non-degenerate Post algebra of order $m > 2$ is the coproduct of a non-degenerate Boolean algebra and a chain $\wedge = e_0 \leqslant \ldots \leqslant e_{m-1} = \vee$, and conversely (22).

3. Omitting in the system of axioms (p_0)–(p_8) Sec. 1 for Post algebras the axiom (p_8), we obtain the system of axioms for pseudo-Post algebras. Prove that each non-degenerate pseudo-Post algebra of order $m \geqslant 2$ is the coproduct of a non-degenerate pseudo-Boolean algebra and a chain $\wedge = e_0 \leqslant \ldots \leqslant e_{m-1} = \vee$, and conversely (23).

4. Prove that for any elements $a, a_t \, (t \in T)$ in a Post algebra $\mathfrak{P}$: (i) if $a = (\mathfrak{P}) \bigcup_{t \in T} a_t$, then $D_i(a) = (\mathfrak{B}_{\mathfrak{P}}) \bigcup_{t \in T} D_i(a_t)$, $i = 1, 2, \ldots$, (ii) if $a = (\mathfrak{P}) \bigcap_{t \in T} a_t$, then $D_i(a) = (\mathfrak{B}_{\mathfrak{P}}) \bigcap_{t \in T} D_i(a_t)$, $i = 1, 2, \ldots$, where $(\mathfrak{P}) \bigcup_{t \in T} a_t$ $((\mathfrak{P}) \bigcap_{t \in T} a_t)$ and $(\mathfrak{B}_{\mathfrak{P}}) \bigcup_{t \in T} D_i(a_t)$ $((\mathfrak{B}_{\mathfrak{P}}) \bigcap_{t \in T} D_i(a_t))$ denote the least upper bound (the

(21) See Sikorski [2].
(22) See Rousseau [2], [4].
(23) See Rousseau [2], [4].

greatest lower bound) of $\{a_t\}_{t\in T}$ in $\mathfrak{P}$ and of $\{D_i(a_t)\}_{t\in T}$ in the Boolean algebra $\mathfrak{B}_{\mathfrak{P}}$ of all complemented elements in $\mathfrak{P}$, respectively [24].

5. Let $a_0, a_n, b_n, a_{nu}, b_{nv}$ ($u \in U_n, v \in V_n$ where U_n, V_n are arbitrary sets, $n = 1, 2, \ldots$) be elements of a non-degenerate Post algebra $\mathfrak{P}$ of order $m \geqslant 2$ such that (1) $a_n = (\mathfrak{P}) \bigcup_{u\in U_n} a_{nu}$, (2) $b_n = (\mathfrak{P}) \bigcap_{v\in V_n} b_{nv}$, (3) $D_{m-1}(a_0) \neq e_0 = \wedge$. Prove that there exists a prime D-filter ∇ in $\mathfrak{P}$ such that (i) $||a_n|| = (\mathfrak{P}/\nabla) \bigcup_{u\in U_n} ||a_{nu}||$, (ii) $||b_n|| = (\mathfrak{P}/\nabla) \bigcap_{v\in V_n} ||b_{nv}||$, $n = 1, 2, \ldots$, (iii) $a_0 \in \nabla$ [25]. Apply Ex. 4, Ex. 3 in Chap. VI, 3.10, 3.11 and 4.5.

6. Generalize the theorem in Ex. 5 of Chap. VI to Post algebras [26]. More exactly, prove that under the hypotheses adopted in Ex. 5 there is a monomorphism h of $\mathfrak{P}$ into a Post field $\mathfrak{P}(X)$ such that $h(a_n) = \bigcup_{u\in U_n} h(a_{nu})$ and $h(b_n) = \bigcap_{v\in V_n} h(b_{nv})$, $n = 1, 2, \ldots$, where $\bigcup$, $\bigcap$ denote the set-theoretic union and intersection, respectively.

(24) Epstein [1].
(25) Rasiowa [9].
(26) Włodarska [1], cf. Rasiowa [12].

PART TWO

NON-CLASSICAL LOGICS

CHAPTER VIII

IMPLICATIVE EXTENSIONAL PROPOSITIONAL CALCULI

Introduction. The algebraic approach to logic, begun by Boole and his followers (W. Stanley Jevons, Peirce, Schröder), has recently been taken up and strongly developed in connection with the formulation of non-classical logics: many-valued logics of Łukasiewicz and Post, intuitionistic logic, modal logic, positive logic, constructive logic with strong negation and others. These logics were inspired by purely philosophical considerations. It seems surprising that they have all led to interesting mathematical concepts and theories.

The main aim of this chapter is to give a tentative formulation of a general algebraic approach to a class of logics or, more precisely, to the propositional calculi of these logics. Hopefully the logics suitable for this approach will turn out to be broad, containing many representative cases (1). Since implication seems to be the most important connective in any logical system, we will examine logics with a connective which has certain natural properties associated with the concept of (possibly weak) implication. This connective will be called implication. The propositional calculi we consider are therefore called implicative calculi.

Each formalized system $\mathscr{S}$ of a propositional calculus is an ordered pair $\mathscr{S} = (\mathscr{L}, C_{\mathscr{L}})$, where $\mathscr{L}$ is a formalized language and $C_{\mathscr{L}}$ is a consequence operation in $\mathscr{L}$. The formalized language $\mathscr{L}$ of $\mathscr{S}$ is—roughly

(1) The definition of this class of logics and theorems relating to them, as given here, were presented at the Illinois Institute of Technology in November 1967 and at the Colloquium on Mathematics in the University of Illinois, Urbana, on December 21, 1967. The main idea is a modification of that presented in Rasiowa and Sikorski [3] and also of the one expounded in [MM]. Similar results have independently been obtained and presented in an elegant form by von Bummert [1].

speaking—a set of certain finite sequences of elements of the alphabet of $\mathscr{L}$, termed formulas. Among the elements of the alphabet of $\mathscr{L}$ there are propositional variables and connectives. The set of all formulas containing the connectives regarded as algebraic operations is a free algebra in the class of all similar algebras. Accordingly, the first step in defining a propositional calculus is to determine this free algebra called the algebra of formulas.

Given a formalized language $\mathscr{L}$, we interpret its formulas as polynomials (algebraic functions) in algebras associated with $\mathscr{L}$, i.e., algebras similar to the algebra of formulas of $\mathscr{L}$ with a distinguished element V, called the unit element. We regard elements of any algebra associated with $\mathscr{L}$ as truth values, in particular the element V as the value truth. If all propositional variables which occur in a formula α of $\mathscr{L}$ are interpreted as variables ranging over an algebra $\mathfrak{A}$ associated with $\mathscr{L}$ and all connectives in that formula as the corresponding algebraic operations in this algebra, then we obtain the interpretation of α as a polynomial determined by α in $\mathfrak{A}$.

The consequence operation $C_{\mathscr{L}}$ in any formalized language $\mathscr{L}$ will be introduced in the standard manner. This means that a set $\mathscr{A}_l$ of formulas of $\mathscr{L}$ will be distinguished as the set of logical axioms and certain rules of inference will be established which to some finite sequences of formulas called premisses assign a formula called the conclusion. Then the consequence operation $C_{\mathscr{L}}$ assigns to each set $\mathscr{A}$ of formulas of $\mathscr{L}$ the set $C_{\mathscr{L}}(\mathscr{A})$ of all its logical consequences, $C_{\mathscr{L}}(\mathscr{A})$ being defined as the least set of formulas of $\mathscr{L}$ containing the union of the sets $\mathscr{A}_l$ and $\mathscr{A}$ and closed under the rules of inference.

The class **S** of standard systems of implicative extensional propositional calculi as defined in Section 5 consists of propositional calculi $\mathscr{S} = (\mathscr{L}, C_{\mathscr{L}})$ which satisfy certain natural conditions concerning the consequence operation.

Each propositional calculus $\mathscr{S} = (\mathscr{L}, C_{\mathscr{L}})$ in the class **S** determines in a precise way a class of abstract algebras called $\mathscr{S}$-algebras. Roughly speaking, they are implicative algebras associated with $\mathscr{L}$ and have the following property: for every formula α derivable in $\mathscr{S}$, i.e. $\alpha \in C_{\mathscr{L}}(O)$, its interpretation as a polynomial in such an algebra is identically equal to the unit element V. We then say that α is valid in the algebra under consideration.

If a propositional calculus $\mathcal{S} = (\mathcal{L}, C_{\mathcal{L}})$ in $\mathbf{S}$ is consistent, i.e. if there exists a formula in $\mathcal{L}$ which is not derivable in $\mathcal{S}$, then by identifying any two formulas α, β of $\mathcal{L}$ for which both implications $(\alpha \Rightarrow \beta)$ and $(\beta \Rightarrow \alpha)$ are derivable in $\mathcal{S}$, we obtain from the algebra of formulas of $\mathcal{L}$ the quotient algebra $\mathfrak{A}(\mathcal{S})$ which is a non-degenerate free algebra in the class of all $\mathcal{S}$-algebras. It is called the algebra of the propositional calculus $\mathcal{S}$.

The completeness theorem (Sec. 7) for consistent propositional calculi in the class $\mathbf{S}$ states that the set $C_{\mathcal{L}}(O)$ of all derivable formulas in $\mathcal{S} = (\mathcal{L}, C_{\mathcal{L}})$ coincides with the set of all formulas valid in each $\mathcal{S}$-algebra.

The consistent propositional calculi in the class $\mathbf{S}$ are said to be logically equivalent (Sec. 8) if they determine the same class of $\mathcal{S}$-algebras. Each class $\mathbf{L} \subset \mathbf{S}$ of consistent logically equivalent propositional calculi is here called a logic. Observe that by the concept of logic adopted above, classical logic with the two connectives, negation and implication, is different from classical logic with the four connectives: negation, implication, disjunction and conjunction. We adopt the definition to keep things simple. Similarly, our restriction to standard systems of propositional calculi can easily be eliminated and is introduced here for the same reason.

The concept of an $\mathbf{L}$-theory of order zero is introduced in Section 9. With each $\mathbf{L}$-theory is connected the notion of an $\mathbf{L}$-model. We prove that for every consistent $\mathbf{L}$-theory there exists an $\mathbf{L}$-model, and a characterization of all theorems of consistent $\mathbf{L}$-theories by means of $\mathbf{L}$-models is established. The construction of the algebra $\mathfrak{A}(\mathcal{S}(\mathcal{A}))$ for a given consistent $\mathbf{L}$-theory $\mathcal{S}(\mathcal{A})$, similar to that of the algebra $\mathfrak{A}(\mathcal{S})$, is used in the proofs of these theorems. Moreover, if a certain condition concerning a logic $\mathbf{L}$ is satisfied, the existence of an $\mathbf{L}$-model for an $\mathbf{L}$-theory is equivalent to its consistency (Sec. 10).

The concept of a deductive filter in $\mathcal{S}$-algebras is introduced in Section 11. Deductive filters are kernels of epimorphisms from $\mathcal{S}$-algebras onto $\mathcal{S}$-algebras, and conversely (Sections 11, 12). On the other hand, there is a one-one correspondence betwen deductive filters in the algebra $\mathfrak{A}(\mathscr{S})$ of a propositional calculus $\mathscr{S}$ of a given logic **L** and **L**-theories based on $\mathscr{S}$ if any two theories which have the same set of theorems are identified. The quotient algebra $\mathfrak{A}(\mathscr{S})/\nabla_{\mathscr{A}}$, where $\nabla_{\mathscr{A}}$ is the filter corresponding to an **L**-theory $\mathscr{S}(\mathscr{A})$, is isomorphic with the algebra $\mathfrak{A}(\mathscr{S}(\mathscr{A}))$ of that theory. Accordingly, the study of **L**-theories is closely connected with that of deductive filters in the algebra $\mathfrak{A}(\mathscr{S})$ of the propositional calculus $\mathscr{S}$ in the **L** in question. This close relationship between **L**-theories and deductive filters suggests that we should distinguish certain kinds of **L**-theories with respect to the types of the corresponding deductive filters. For instance, consistent, irreducible and maximal **L**-theories are analogues of proper, irreducible and maximal deductive filters, respectively. Various metalogical theorems result immediately from corresponding theorems concerning deductive filters.

For example, the theorem which states that each consistent **L**-theory can be extended to an irreducible **L**-theory is a corollary if its algebraic analogue that every proper deductive filter can be extended to an irreducible deductive filter.

The results summarized in this chapter show a far-reaching parallelism between the structure of propositional calculi and **L**-theories of a given logic **L** on the one hand, and a well-defined class of abstract algebras and deductive filters on the other.

1. Formalized languages of zero order. By an *alphabet of a formalized language of zero order* in this book is meant any ordered system $A^0 = (V, L_0, L_1, L_2, U)$, where

(1) V, L_0, L_1, L_2, U are disjoint sets,

(2) the set V is infinite,

(3) the sets L_0, L_1 are finite (possibly empty),

(4) the set L_2 is finite and always contains an element denoted by $\Rightarrow$, called the *implication sign.*

Elements in V will be called *propositional variables* and will be denoted by $p, q, r, \ldots$, with indices if necessary. Elements in L_0, L_1, L_2 will be called *propositional constants, unary propositional connectives* and *binary propositional connectives*, respectively. Elements in U will be called *auxiliary signs*. We shall always assume that U contains two elements denoted by *(,)* and called the *parentheses* in A^0.

Elements in the union $V \cup L_0 \cup L_1 \cup L_2 \cup U$ will be called *signs of the alphabet* A^0.

The set F of all *formulas over the alphabet* A^0 is the least set of finite sequences of signs in A^0 such that

(f_1) all propositional variables (considered as one-element sequences) are in F,

(f_2) all propositional constants are in F,

(f_3) if α is in F and o is a unary propositional connective in A^0, then $o\alpha$ is in F,

(f_4) if α, β are in F and o is a binary connective in A^0, then $(\alpha o \beta)$ is in F.

The ordered pair

$$\mathscr{L} = (A^0, F), \tag{5}$$

where A^0 is an alphabet of a formalized language of zero order and F is the set of all formulas over the alphabet A^0, will be said to be a *formalized implicative language of zero order*, or briefly, a *formalized language of zero order*.

Let $\mathscr{L} = (A^0, F)$ and $\mathscr{L}' = (A^{0\prime}, F')$ be two formalized languages of zero order, where $A^0 = (V, L_0, L_1, L_2, U)$ and $A^{0\prime} = (V', L_0', L_1', L_2', U)$. The languages $\mathscr{L}$ and $\mathscr{L}'$ will be said to be *similar* provided $L_i = L_i'$, $i = 0, 1, 2$. The language $\mathscr{L}'$ is an *extension* of the language $\mathscr{L}$ provided $V \subset V'$ and $L_i \subset L_i'$, for $i = 0, 1, 2$.

It follows from the above definitions that

1.1. *If $\mathscr{L}$ and $\mathscr{L}'$ are similar formalized languages of zero order, then $\mathscr{L}'$ is an extension of $\mathscr{L}$ if and only if $V \subset V'$.*

Let $\mathscr{L} = (A^0, F)$ be a formalized language of zero order. A formula β in F is said to be a *direct subformula* of a formula α in F provided α has one of the forms:

(6) $(\beta o \gamma)$, $(\gamma o \beta)$, where o is a binary connective of A^0 and γ is a formula in F,

(7) $o\beta$, where o is a unary connective in A^0.

A formula β in F is said to be a *subformula* of a formula α in F provided there exists a sequence $\beta_1, \dots, \beta_m$ ($m \geqslant 1$) of formulas of F such that β_1 is β, β_m is α and, for $i = 2, \dots, m$, β_{i-1} is a direct subformula of β_i.

2. The algebra of formulas (2). In this section formalized languages of zero order will be investigated from the algebraic view-point. First of all let us note the following result.

2.1. *Let* $A^0 = (V, L_0, L_1, L_2, U)$ *be an alphabet of a formalized language of zero order, where* $L_0 = \{e_0, \dots, e_{m-1}\}$, $L_1 = \{o^1, \dots, o^s\}$, $L_2 = \{\Rightarrow, o_1, \dots, o_t\}$, $m, s, t = 0, 1, 2, \dots$ *and let* F *be the set of all formulas over* A^0. *Then* $\mathfrak{F} = (F, \Rightarrow, o_1, \dots, o_t, o^1, \dots, o^s, e_0, \dots, e_{m-1})$ *is an abstract algebra with* $t+1$ *binary operations* $\Rightarrow, o_1, \dots, o_t$, s *unary operations* $o^1, \dots, o^s$ *and* m *zero-argument operations* $e_0, \dots, e_{m-1}$, *defined as follows*:

(1) *the formulas* $(\alpha \Rightarrow \beta)$, $(\alpha o_i \beta)$ *are the results of the operations* $\Rightarrow, o_i$, $i = 1, \dots, t$, *respectively, performed on formulas* α, β;

(2) *the formulas* $o^i\alpha$ *are the results of the operations* o^i, $i = 1, \dots, s$, *respectively, performed on a formula* α;

(3) *the formulas* $e_0, \dots, e_{m-1}$ *are treated as zero-argument operations in* $\mathfrak{F}$.

Moreover, the set V *is the set of generators of* $\mathfrak{F}$.

This result follows directly from the definition of the set F of all formulas over A^0 (see Sec. 1). The algebra $\mathfrak{F}$ will be said to be the *algebra of formulas of the formalized language* $\mathscr{L} = (A^0, F)$.

2.2. *If* $\mathscr{L} = (A^0, F)$ *and* $\mathscr{L}' = (A^{0\prime}, F')$ *are two similar languages of zero order such that* $\mathscr{L}'$ *is an extension of* $\mathscr{L}$, *then the algebra of formulas of* $\mathscr{L}$ *is a subalgebra of the algebra of formulas of* $\mathscr{L}'$.

2.3. *The algebra* $\mathfrak{F}$ *of formulas of a formalized language* $\mathscr{L}$ *of zero order is a free algebra in the class of all similar algebras, the set* V *of all propositional variables in* $\mathscr{L}$ *being the set of free generators in* $\mathfrak{F}$.

Let $\mathfrak{F} = (F, \Rightarrow, o_1, \dots, o_t, o^1, \dots, o^s, e_0, \dots, e_{m-1})$ be the algebra of formulas of a formalized language $\mathscr{L} = (A^0, F)$ of zero order and

(2) The exposition in Sections 1–4 is a slight modification of that given in [MM]. The idea of treating the set of formulas of a formalized language as an abstract algebra is due to Lindenbaum and Tarski.

let V be the set of all propositional variables in $\mathscr{L}$. Let $\mathfrak{B} = (B, \rightarrow, \mathrm{o}_1, \dots, \mathrm{o}_t, \mathrm{o}^1, \dots, \mathrm{o}^s, \mathrm{e}_0, \dots, \mathrm{e}_{m-1})$ be an arbitrary similar algebra. Each mapping $f\colon V \rightarrow B$ can be extended to a mapping $h\colon F \rightarrow B$ defined by induction on the length of the formulas as follows:

(4) for any propositional variable p, let $hp = fp$;

(5) for any propositional constant e_i, $i = 0, \dots, m-1$, let $he_i = \mathrm{e}_i$;

(6) suppose that for some formulas α, β in F the values of $h\alpha$ and $h\beta$ are defined; we then define $h(\alpha \Rightarrow \beta)$, $h(\alpha o_i \beta)$, $i = 1, \dots, t$, $ho^i\alpha$, $i = 1, \dots, s$ by means of the equations

$$h(\alpha \Rightarrow \beta) = h\alpha \rightarrow h\beta, \quad h(\alpha o_i \beta) = h\alpha \mathrm{o}_i h\beta, \quad i = 1, \dots, t,$$

$$ho^i\alpha = \mathrm{o}^i h\alpha, \quad i = 1, \dots, s.$$

It follows from (4) that h is an extension of f. By (5) and (6) h is a homomorphism. This completes the proof of 2.3.

By a *substitution* from a formalized language $\mathscr{L} = (A^0, F)$ of zero order into a similar formalized language $\mathscr{L}' = (A^{0\prime}, F')$ of zero order we shall understand any mapping

(7) $$\mathfrak{s}\colon V \rightarrow F'$$

from the set V of all propositional variables in $\mathscr{L}$ into the set F' of all formulas in $\mathscr{L}'$.

Since any substitution $\mathfrak{s}\colon V \rightarrow F'$ from a formalized language $\mathscr{L}$ of zero order into a similar formalized language $\mathscr{L}'$ of zero order can be regarded as a mapping from the set of generators of the algebra $\mathfrak{F}$ of formulas of $\mathscr{L}$ into the set of all elements of the algebra $\mathfrak{F}'$ of formulas of $\mathscr{L}'$, we infer by 2.3 and I 4.4 that the following statement holds:

2.4. *Any substitution $\mathfrak{s}$ from a formalized language $\mathscr{L}$ of zero order into a similar formalized language $\mathscr{L}'$ of zero order can be extended in the unique way to a homomorphism from the algebra $\mathfrak{F}$ of formulas of $\mathscr{L}$ into the algebra $\mathfrak{F}'$ of formulas of $\mathscr{L}'$, which therefore will be denoted by the same letter $\mathfrak{s}$,* i.e. *we have*

(8) $\mathfrak{s}e_i = e_i$, *for each propositional constant* e_i,

(9) $\mathfrak{s}(\alpha \Rightarrow \beta) = (\mathfrak{s}\alpha \Rightarrow \mathfrak{s}\beta)$, $\mathfrak{s}(\alpha o_i \beta) = (\mathfrak{s}\alpha o_i \mathfrak{s}\beta)$, *for all formulas* α, β *in* $\mathscr{L}$, *and for each binary connective* o_i,

(10) $\mathfrak{s}o^i\alpha = o^i\mathfrak{s}\alpha$, *for each formula* α *in* $\mathscr{L}$ *and for each unary connective* o^i.

2.5. *Let* $\mathfrak{s}$ *be a substitution from a formalized language* $\mathscr{L}$ *of zero order into a similar language* $\mathscr{L}'$ *and let* $\mathfrak{s}'$ *be a substitution from* $\mathscr{L}'$ *into a similar language* $\mathscr{L}''$. *Then there exists a substitution* $\mathfrak{s}_1$ *from* $\mathscr{L}$ *into* $\mathscr{L}''$ *such that for every formula* α *of* $\mathscr{L}$

(11) $$\mathfrak{s}_1\alpha = \mathfrak{s}'(\mathfrak{s}\alpha).$$

Let us put

(12) $\mathfrak{s}_1 p = \mathfrak{s}'(\mathfrak{s}p)$ for every propositional variable p of $\mathscr{L}$.

By 2.4 both sides of (11) can be considered as homomorphisms from the algebra $\mathfrak{F}$ of formulas of $\mathscr{L}$ into the algebra $\mathfrak{F}''$ of formulas of $\mathscr{L}''$. Since by (12) they coincide on the set of generators of $\mathfrak{F}$, by I 4.4 they are equal.

3. Interpretation of formulas as mappings [3]. Let $\mathscr{L} = (A^0, F)$ be a formalized language of zero order where $A^0 = \{V, L_0, L_1, L_2, U\}$ and $L_0 = \{e_0, \ldots, e_{m-1}\}$, $L_1 = \{o^1, \ldots, o^s\}$, $L_2 = \{\Rightarrow, o_1, \ldots, o_t\}$, $m, s, t \in \{0, 1, 2, \ldots\}$.

An abstract algebra $\mathfrak{A} = (A, V_{\mathfrak{A}}, \rightarrow, \mathrm{o}_1, \ldots, \mathrm{o}_t, \mathrm{o}^1, \ldots, \mathrm{o}^s, \mathrm{e}_0, \ldots, \mathrm{e}_{m-1})$ will be said to be *associated* with the language $\mathscr{L}$ if the algebra $\mathfrak{A}^0 = (A, \rightarrow, \mathrm{o}_1, \ldots, \mathrm{o}_t, \mathrm{o}^1, \ldots, \mathrm{o}^s, \mathrm{e}_0, \ldots, \mathrm{e}_{m-1})$ is similar to the algebra $\mathfrak{F}$ of formulas of the language $\mathscr{L}$ (see Sec. 2) and $V_{\mathfrak{A}}$ is a zero-argument operation in $\mathfrak{A}$, i.e. a distinguished element in A. Sometimes instead of $V_{\mathfrak{A}}$ we shall also write V. In order to simplify the notation in the next sections we shall denote the operations in algebras associated with a formalized language $\mathscr{L}$ by the same signs as the corresponding operations in the algebra $\mathfrak{F}$ of formulas of $\mathscr{L}$.

By a *valuation* of $\mathscr{L}$ in an algebra $\mathfrak{A}$ associated with $\mathscr{L}$ we shall understand any mapping

(1) $$v\colon V \rightarrow A,$$

i.e. any point $v = \{v_p\}_{p \in V}$ of the Cartesian product A^V.

[3] Cf. [MM] Chap. VI. These mappings were first defined by Łukasiewicz and Tarski [1].

Every formula α in $\mathscr{L}$ uniquely determines a mapping

(2) $$\alpha_{\mathfrak{A}}\colon A^V \to A,$$

defined by induction on the length of α as follows:

(3)
$$\begin{aligned} p_{\mathfrak{A}}(v) &= v(p) \quad \text{for each propositional variable } p \text{ in } V,\\ e_{j\mathfrak{A}}(v) &= \mathrm{e}_j \quad \text{for} \quad j = 0, \ldots, m-1,\\ o^i\alpha_{\mathfrak{A}}(v) &= \mathrm{o}^i\alpha_{\mathfrak{A}}(v) \quad \text{for} \quad i = 1, \ldots, s,\\ (\alpha o_i \beta)_{\mathfrak{A}}(v) &= \alpha_{\mathfrak{A}}(v)\mathrm{o}_i\beta_{\mathfrak{A}}(v), \quad i = 1, \ldots, t,\\ (\alpha \Rightarrow \beta)_{\mathfrak{A}}(v) &= \alpha_{\mathfrak{A}}(v) \to \beta_{\mathfrak{A}}(v). \end{aligned}$$

Another definition of the mapping $\alpha_{\mathfrak{A}}(v)$ is by means of Theorems 2.3 and I 4.4. By these theorems every valuation $v\colon V \to A$ can be extended in a unique way to a homomorphism $v_{\mathfrak{A}}$ from the algebra $\mathfrak{F}$ of formulas of the formalized language $\mathscr{L}$ into $\mathfrak{A}^0$. Consequently, for every formula α in $\mathscr{L}$, $v_{\mathfrak{A}}(\alpha)$ is a uniquely determined element in $\mathfrak{A}$. If α is a fixed formula and v is considered to be variable, then $v_{\mathfrak{A}}(\alpha)$ is a function of the valuation v. Denoting this function by $\alpha_{\mathfrak{A}}$, we have by definition

(4) $$\alpha_{\mathfrak{A}}(v) = v_{\mathfrak{A}}(\alpha), \quad \text{for any valuation } v\colon V \to A.$$

The mapping $\alpha_{\mathfrak{A}}$ defined by (4) satisfies the equations (3) and therefore coincides with the mapping (2) defined by means of (3).

Let us note the following remark.

3.1. *For every valuation v of $\mathscr{L}$ in an algebra $\mathfrak{A}$ associated with $\mathscr{L}$, $\alpha_{\mathfrak{A}}(v)$ considered as a function of α, is a homomorphism from the algebra $\mathfrak{F}$ of formulas of $\mathscr{L}$ into the algebra $\mathfrak{A}^0$, obtained from $\mathfrak{A}$ by dropping the operation $\vee_{\mathfrak{A}}$ and such that $p_{\mathfrak{A}}(v) = v(p)$ for each propositional variable p in $\mathscr{L}$. The value of $\alpha_{\mathfrak{A}}(v)$ depends only on $v(p_1), \ldots, v(p_n)$, where $p_1, \ldots, p_n$ are all propositional variables appearing in α.*

Let $\mathfrak{B} = (B, \vee_{\mathfrak{B}}, \to, \mathrm{o}_1, \ldots, \mathrm{o}_t, \mathrm{o}^1, \ldots, \mathrm{o}^s, \mathrm{e}_0, \ldots, \mathrm{e}_{m-1})$ be an abstract algebra similar to an abstract algebra $\mathfrak{A}$ associated with the formalized language $\mathscr{L}$ of zero order. Let h be a homomorphism from $\mathfrak{A}$ into $\mathfrak{B}$. For every valuation v of $\mathscr{L}$ in $\mathfrak{A}$ let hv be the valuation of $\mathscr{L}$ in $\mathfrak{B}$ defined by the following equation

(5) $$hv(p) = h\big(v(p)\big) \quad \text{for any propositional variable } p \text{ in } \mathscr{L}.$$

Under the above hypotheses the following theorem holds.

3.2. *For every valuation v of $\mathscr{L}$ in $\mathfrak{A}$ and for every formula α in $\mathscr{L}$*

(6) $$\alpha_{\mathfrak{B}}(hv) = h\alpha_{\mathfrak{A}}(v),$$

and

(7) *the condition* $\alpha_{\mathfrak{A}}(v) = V_{\mathfrak{A}}$ *implies* $\alpha_{\mathfrak{B}}(hv) = V_{\mathfrak{B}}$.

By 3.1, for every fixed valuation v in $\mathfrak{A}$, both sides of (6) considered as functions of α are homomorphisms from the algebra $\mathfrak{F}$ of formulas in $\mathscr{L}$ into $\mathfrak{B}^0$. Since they coincide on the set of generators of $\mathfrak{F}$, by I 4.4 they are equal. Since h is a homomorphism from $\mathfrak{A}$ into $\mathfrak{B}$, we have $hV_{\mathfrak{A}} = V_{\mathfrak{B}}$. Hence, from (6), we get (7).

Let $\mathscr{L}$ and $\mathscr{L}'$ be similar formalized languages of zero order and let $\mathfrak{s}$ be an arbitrary substitution from $\mathscr{L}$ into $\mathscr{L}'$. For every algebra $\mathfrak{A}$ associated with $\mathscr{L}$ and $\mathscr{L}'$ the substitution $\mathfrak{s}$ determines a mapping $\mathfrak{s}_{\mathfrak{A}}$ which to every valuation v' of $\mathscr{L}'$ in $\mathfrak{A}$ assigns a valuation $v = \mathfrak{s}_{\mathfrak{A}} v'$ of $\mathscr{L}$ in $\mathfrak{A}$ defined as follows

(8) $v(p) = \mathfrak{s}_{\mathfrak{A}} v'(p) = \mathfrak{s}p_{\mathfrak{A}}(v')$ for each propositional variable p of $\mathscr{L}$.

Under the above hypotheses the following holds:

3.3. *For each formula α of $\mathscr{L}$*

(9) $$\mathfrak{s}\alpha_{\mathfrak{A}}(v') = \alpha_{\mathfrak{A}}(\mathfrak{s}_{\mathfrak{A}} v') = \alpha_{\mathfrak{A}}(v).$$

The proof is analogous to that of 3.2.

Let $\approx$ be a congruence relation in the algebra $\mathfrak{F}$ of formulas in a formalized language $\mathscr{L}$ of zero order. Consider an abstract algebra

(10) $$\mathfrak{F}/{\approx}(V) = (F/{\approx}, V, \Rightarrow, o_1, \ldots, o_t, o^1, \ldots, o^s, ||e_0||, \ldots, ||e_{m-1}||)$$

where $(F/{\approx}, \Rightarrow, o_1, \ldots, o_t, o^1, \ldots, o^s, ||e_0||, \ldots, ||e_{m-1}||)$ is the quotient algebra $\mathfrak{F}/{\approx}$ (see I 4, p. 12) and V is a distinguished element in $F/{\approx}$. Since $\mathfrak{F}/{\approx}(V)$ is an algebra associated with $\mathscr{L}$, we can interpret formulas of $\mathscr{L}$ as mappings from $\mathfrak{F}/{\approx}(V)^V$ into $\mathfrak{F}/{\approx}(V)$. Observe that every substitution $\mathfrak{s}$ from $\mathscr{L}$ into $\mathscr{L}$ determines a corresponding valuation v in $\mathfrak{F}/{\approx}(V)$ defined as follows

(11) $v(p) = ||\mathfrak{s}p||$ for every propositional variable p of $\mathscr{L}$.

In particular, if $\mathfrak{s}$ is the identity mapping, then the corresponding valuation v^0 is defined by the equation

(12) $v^0(p) = ||p||$ for every propositional variable p of $\mathscr{L}$,

and will be called the *canonical valuation in* $\mathfrak{F}/\approx(V)$. Under the above hypotheses and notation the following theorem holds.

3.4. *If* $\mathfrak{A} = \mathfrak{F}/\approx(V)$, *then for every substitution* $\mathfrak{s}$ *from* $\mathscr{L}$ *into* $\mathscr{L}$ *and for every formula* α *of* $\mathscr{L}$

(13) $$\alpha_{\mathfrak{A}}(v) = ||\mathfrak{s}\alpha||,$$

where v *is the valuation defined by* (11). *In particular,*

(14) $$\alpha_{\mathfrak{A}}(v^0) = ||\alpha||$$

for the canonical valuation v^0.

By 2.4 and 3.1, both sides of (13), considered as functions of α, are homomorphisms from the algebra $\mathfrak{F}$ of formulas of $\mathscr{L}$ into $\mathfrak{F}/\approx$ and they coincide on the set of propositional variables in $\mathscr{L}$, i.e. on the set of generators of $\mathfrak{F}$. Hence, by I 4.4, they are equal. Thus (13) holds and in consequence (14) holds too.

4. Consequence operations in formalized languages of zero order. Let $\mathscr{L} = (A^0, F)$ be a formalized language of zero order (see Sec. 1). By a *consequence operation in* $\mathscr{L}$ we mean any mapping $C_{\mathscr{L}}: 2^F \to 2^F$ satisfying the following conditions [4]:

(c_1) $\mathscr{A} \subset C_{\mathscr{L}}(\mathscr{A})$,

(c_2) if $\mathscr{A}_1 \subset \mathscr{A}_2$, then $C_{\mathscr{L}}(\mathscr{A}_1) \subset C_{\mathscr{L}}(\mathscr{A}_2)$,

(c_3) $C_{\mathscr{L}}(C_{\mathscr{L}}(\mathscr{A})) = C_{\mathscr{L}}(\mathscr{A})$,

for all subsets $\mathscr{A}, \mathscr{A}_1, \mathscr{A}_2$ of the set F of all formulas in $\mathscr{L}$. If a formalized language $\mathscr{L}$ is fixed, then instead of $C_{\mathscr{L}}$ we shall also write C.

A consequence operation C in a formalized language $\mathscr{L} = (A^0, F)$ of zero order is said to have a *finite character* if the following condition is satisfied: for every $\alpha \in F$ and every $\mathscr{A} \subset F$

(c_4) if $\alpha \in C(\mathscr{A})$, then there exists a finite subset $\mathscr{A}_0$ of $\mathscr{A}$ such that $\alpha \in C(\mathscr{A}_0)$.

A consequence operation C in a formalized language $\mathscr{L} = (A^0, F)$ can be introduced by the following method. We choose

(1) a set $\mathscr{A}_l \subset F$ of formulas of $\mathscr{L}$ called a *set of logical axioms*,

(2) a finite set $\{(r_1), \ldots, (r_k)\}$ *of rules of inference*,

[4] Tarski [2].

which (roughly speaking) are mappings assigning to some finite sequences $\alpha_1, \ldots, \alpha_n$ of formulas called *premises* a formula β called the *conclusion*.

More exactly, any rule of inference (r) is a mapping

(3) $(r)\colon P \to F, \quad \text{where} \quad P \subset F^n \quad \text{for some } n = 1, 2, \ldots.$

Instead of $(r)(\alpha_1, \ldots, \alpha_n) = \beta$ we shall usually write

$$(r) \ \frac{\alpha_1, \ldots, \alpha_n}{\beta}.$$

A subset $F_0 \subset F$ will be said to be *closed under a rule of inference* $(r)\colon P \to F$, $P \subset F^n$, provided the condition $(\alpha_1, \ldots, \alpha_n) \in F_0^n \cap P$ implies that $(r)(\alpha_1, \ldots, \alpha_n) \in F_0$.

Now we define the consequence operation C in $\mathscr{L}$ as follows: *for each set* $\mathscr{A} \subset F$, *the set* $C(\mathscr{A})$ *is the least set of formulas of* $\mathscr{L}$ *containing the union* $\mathscr{A}_l \cup \mathscr{A}$ *and closed under the rules of inference* $(r_1), \ldots, (r_k)$. The operation C satisfies the conditions (c_1), (c_2), (c_3) and will be said to be the *consequence operation in* $\mathscr{L}$ *determined by* $\mathscr{A}_l$ *and* $\{(r_1), \ldots, (r_k)\}$.

By a *formal proof of a formula* α *from a set* $\mathscr{A} \subset F$ *with respect to a set* $\mathscr{A}_l$ *of logical axioms and rules of inference* $(r_1), \ldots, (r_k)$ we mean any finite sequence $\alpha_1, \ldots, \alpha_n$ of formulas of $\mathscr{L}$ such that

(p_1) $\alpha_1 \in \mathscr{A}_l \cup \mathscr{A}$,

(p_2) for every $1 < i \leqslant n$, either $\alpha_i \in \mathscr{A}_l \cup \mathscr{A}$ or α_i is the conclusion of one of the rules of inference (r_j), $j = 1, \ldots, k$, of which the premises are some of the formulas $\alpha_1, \ldots, \alpha_{i-1}$,

(p_3) $\alpha_n = \alpha$.

Suppose that a set $\mathscr{A}_l$ of logical axioms and a set $\{(r_1), \ldots, (r_k)\}$ of rules of inference are fixed. If there exists a formal proof of a formula α from a set $\mathscr{A} \subset F$ with respect to $\mathscr{A}_l$ and $(r_1), \ldots, (r_k)$, then we shall write

(4) $$\mathscr{A} \vdash \alpha.$$

In particular, if $\mathscr{A} = O$, we write instead of (4)

(5) $$\vdash \alpha.$$

For any set $\mathscr{A} \subset F$ let

(6) $$D(\mathscr{A}) = \{\alpha \in F\colon \mathscr{A} \vdash \alpha\}.$$

It is easy to prove that

(7) $$D(\mathscr{A}) = C(\mathscr{A}) \quad \text{for each } \mathscr{A} \subset F,$$

where C is the consequence operation determined by the same set $\mathscr{A}_l$ of logical axioms and by the same set $\{(r_1), \ldots, (r_k)\}$ of rules of inference.

It follows from (6) and (7) that

(8) $$C(\mathscr{A}) = \{\alpha \in F\colon \mathscr{A} \vdash \alpha\}.$$

The equation (8) is another, equivalent definition of the *consequence operation C in $\mathscr{L}$ determined by a set $\mathscr{A}_l$ of logical axioms and by a set $\{(r_1), \ldots, (r_k)\}$ of rules of inference.*

It is also easy to prove applying (8) that each consequence operation defined as above has the property (c_4), i.e. has a finite character.

5. The class S of standard systems of implicative extensional propositional calculi. Let $\mathscr{L} = (A^0, F)$ be a formalized language of zero order (see Sec. 1) and let $C_{\mathscr{L}}$ be the consequence operation in $\mathscr{L}$ determined by a set $\mathscr{A}_l$ of logical axioms and by a set $\{(r_1), \ldots, (r_k)\}$ of rules of inference (see Sec. 4). Then the system

$$\mathscr{S} = (\mathscr{L}, C_{\mathscr{L}})$$

will be said to be a *system of propositional calculus* and the set $C_{\mathscr{L}}(O)$ will be said to be the *set of all derivable formulas in $\mathscr{S}$*.

In our book we shall investigate a certain class **S** of systems of propositional calculi called the *class of standard systems of implicative extensional propositional calculi.* The class **S** is defined as follows: a system $\mathscr{S} = (\mathscr{L}, C_{\mathscr{L}})$ of propositional calculus, where $C_{\mathscr{L}}$ is determined by a set $\mathscr{A}_l$ of logical axioms and by a set $\{(r_1), \ldots, (r_k)\}$ of rules of inference, is in **S** provided that the following conditions are satisfied:

(s_1) the set $\mathscr{A}_l$ of logical axioms is closed under substitutions, i.e. for every substitution $\mathfrak{s}$ from $\mathscr{L}$ into $\mathscr{L}$ (see 2 (7)) and every formula α of $\mathscr{L}$, if $\alpha \in \mathscr{A}_l$, then $\mathfrak{s}\alpha \in \mathscr{A}_l$;

(s_2) the rules of inference (r_i), $i = 1, \ldots, k$, are invariant under substitutions, i.e. if the rule (r_i), $i = 1, \ldots, k$, assigns to premises $\alpha_1, \ldots, \alpha_n$ the conclusion β, then for every substitution $\mathfrak{s}$ from $\mathscr{L}$ into $\mathscr{L}$ the same rule assigns to the formulas $\mathfrak{s}\alpha_1, \ldots, \mathfrak{s}\alpha_n$ the conclusion $\mathfrak{s}\beta$;

(s_3) for every formula α of $\mathscr{L}$, $(\alpha \Rightarrow \alpha) \in C_{\mathscr{L}}(O)$;

(s_4) for all formulas α, β of $\mathscr{L}$ and for every set $\mathscr{A}$ of formulas of $\mathscr{L}$, if α, $(\alpha \Rightarrow \beta) \in C_{\mathscr{L}}(\mathscr{A})$, then $\beta \in C_{\mathscr{L}}(\mathscr{A})$;

(s_5) for all formulas α, β, γ of $\mathscr{L}$ and for every set $\mathscr{A}$ of formulas of $\mathscr{L}$, if $(\alpha \Rightarrow \beta)$, $(\beta \Rightarrow \gamma) \in C_{\mathscr{L}}(\mathscr{A})$, then $(\alpha \Rightarrow \gamma) \in C_{\mathscr{L}}(\mathscr{A})$;

(s_6) for every formula α of $\mathscr{L}$ and for every set $\mathscr{A}$ of formulas of $\mathscr{L}$ the condition $\alpha \in C_{\mathscr{L}}(\mathscr{A})$ implies that for every formula β of $\mathscr{L}$, $(\beta \Rightarrow \alpha) \in C_{\mathscr{L}}(\mathscr{A})$;

(s_7) for all formulas α, β of $\mathscr{L}$ and for every set $\mathscr{A}$ of formulas of $\mathscr{L}$ the condition $(\alpha \Rightarrow \beta)$, $(\beta \Rightarrow \alpha) \in C_{\mathscr{L}}(\mathscr{A})$ implies that for each unary connective o of $\mathscr{L}$, the formula $(o\alpha \Rightarrow o\beta) \in C_{\mathscr{L}}(\mathscr{A})$;

(s_8) for all formulas α, β, γ, δ of $\mathscr{L}$ and for every set $\mathscr{A}$ of formulas of $\mathscr{L}$ the condition $(\alpha \Rightarrow \beta)$, $(\beta \Rightarrow \alpha)$, $(\gamma \Rightarrow \delta)$, $(\delta \Rightarrow \gamma) \in C_{\mathscr{L}}(\mathscr{A})$ implies that for each binary connective o of $\mathscr{L}$, $((\alpha o \gamma) \Rightarrow (\beta o \delta)) \in C_{\mathscr{L}}(\mathscr{A})$.

Let $\mathscr{S} = (\mathscr{L}, C_{\mathscr{L}})$ be a system in $\mathbf{S}$. According to the definition adopted at the beginning of this section, $C_{\mathscr{L}}(O)$ is the set of all derivable formulas in $\mathscr{S}$. By 4 ((8), (5)), we have

(1) $$\alpha \in C_{\mathscr{L}}(O) \quad \text{if and only if} \quad \vdash \alpha,$$

i.e. if and only if there exists a formal proof of α from the empty set of formulas in $\mathscr{L}$ with respect to the set $\mathscr{A}_l$ of logical axioms of $\mathscr{S}$ and the set $\{(r_1), \ldots, (r_k)\}$ of all rules of inference in $\mathscr{S}$. Such a formal proof will be said to be a *formal* proof in $\mathscr{S}$.

A system $\mathscr{S} = (\mathscr{L}, C_{\mathscr{L}})$ of propositional calculus in $\mathbf{S}$ will be said to be *consistent* provided there exists a formula α of $\mathscr{L}$ such that $\alpha \notin C_{\mathscr{L}}(O)$.

The following theorem completes this section.

5.1. *For every system* $\mathscr{S} = (\mathscr{L}, C_{\mathscr{L}}) \in \mathbf{S}$, *the set* $C_{\mathscr{L}}(O)$ *is closed under substitutions*, i.e. *for every formula* α *of* $\mathscr{L}$ *and for every substitution* $\mathfrak{s}$ *from* $\mathscr{L}$ *into* $\mathscr{L}$ *the condition* $\alpha \in C_{\mathscr{L}}(O)$ *implies that* $\mathfrak{s}\alpha \in C_{\mathscr{L}}(O)$.

The easy proof is omitted.

6. $\mathscr{S}$-algebras [5]**.** Any system $\mathscr{S} = (\mathscr{L}, C_{\mathscr{L}})$ of propositional calculus in **S** determines in the following way a class of abstract algebras called *$\mathscr{S}$-algebras*: an abstract algebra

$$(1) \qquad \mathfrak{A} = (A, V, \Rightarrow, o_1, \ldots, o_t, o^1, \ldots, o^s, e_0, \ldots, e_{m-1})$$

associated with the formalized language $\mathscr{L}$ is said to be an $\mathscr{S}$-algebra provided that

(a_1) if a formula α of $\mathscr{L}$ belongs to the set $\mathscr{A}_l$ of logical axioms of $\mathscr{S}$, then $\alpha_{\mathfrak{A}}(v) = V$ for every valuation v of $\mathscr{L}$ in $\mathfrak{A}$;

(a_2) if a rule of inference (r) in $\mathscr{S}$ assigns to the premises $\alpha_1, \ldots, \alpha_n$ the conclusion β, then for every valuation v of $\mathscr{L}$ in $\mathfrak{A}$ the condition $\alpha_{1\mathfrak{A}}(v) = V, \ldots, \alpha_{n\mathfrak{A}}(v) = V$ implies $\beta_{\mathfrak{A}}(v) = V$;

(a_3) for all $a, b, c \in A$, if $a \Rightarrow b = V$ and $b \Rightarrow c = V$, then $a \Rightarrow c = V$;

(a_4) for all $a, b \in A$, if $a \Rightarrow b = V$ and $b \Rightarrow a = V$, then $a = b$.

Let us note the following

6.1. *If α is a formula derivable in $\mathscr{S} = (\mathscr{L}, C_{\mathscr{L}})$, then*

$$(2) \qquad \alpha_{\mathfrak{A}}(v) = V$$

for every valuation v of $\mathscr{L}$ in every $\mathscr{S}$-algebra $\mathfrak{A}$.

Let $\mathscr{A}$ be the set of all formulas of $\mathscr{L}$ satisfying (2). By (a_1), the set $\mathscr{A}_l$ of logical axioms of $\mathscr{S}$ is contained in $\mathscr{A}$. By (a_2) the set $\mathscr{A}$ is closed under all rules of inference of $\mathscr{S}$. Thus, by the definition of $C_{\mathscr{L}}$ (see Sec. 4), we get $C_{\mathscr{L}}(O) \subset \mathscr{A}$, which by Sec. 5 proves that every derivable formula in $\mathscr{S}$ belongs to $\mathscr{A}$.

The next theorem states a relationship between $\mathscr{S}$-algebras and implicative algebras.

6.2. *For any system $\mathscr{S} = (\mathscr{L}, C_{\mathscr{L}})$ of propositional calculus in* **S**, *if an abstract algebra* (1) *is an $\mathscr{S}$-algebra, then* $(A, V, \Rightarrow)$ *is an implicative algebra.*

By 5 (s_3) and 5 (1), for every propositional variable p of $\mathscr{L}$, $\vdash (p \Rightarrow p)$.

[5] The concept of an $\mathscr{S}$-algebra, as introduced here, is a modification of that in Rasiowa and Sikorski [3]. Analogous notions have been applied by many authors. The idea is due to Łukasiewicz and Tarski [1], who instead of $\mathscr{S}$-algebra used the term *matrix*. The exposition in Sec. 5 is similar to that in [MM].

Hence, by 6.1 and 3 (3), $V = (p \Rightarrow p)_{\mathfrak{A}}(v) = p_{\mathfrak{A}}(v) \Rightarrow p_{\mathfrak{A}}(v) = v(p) \Rightarrow v(p)$ for every valuation v of $\mathscr{L}$ in any $\mathscr{S}$-algebra $\mathfrak{A}$. Thus

$$(3) \qquad a \Rightarrow a = V \quad \text{for each } a \text{ in } \mathfrak{A}.$$

On the other hand, by 5 (s_3), 5 (s_6) and 5 (1), for all propositional variables p, q of $\mathscr{L}$, $\vdash (q \Rightarrow (p \Rightarrow p))$. Hence, by 6.1, 3 (3) and (3), we get $V = (q \Rightarrow (p \Rightarrow p))_{\mathfrak{A}}(v) = v(q) \Rightarrow (v(p) \Rightarrow v(p)) = v(q) \Rightarrow V$ for every valuation v of $\mathscr{L}$ in any $\mathscr{S}$-algebra $\mathfrak{A}$. Thus

$$(4) \qquad a \Rightarrow V = V \quad \text{for each } a \text{ in } \mathfrak{A}.$$

On account of (3), (4), (a_3), (a_4) the conditions (i_1), (i_2), (i_3), (i_4) formulated in II 1 are satisfied, which proves 6.2.

Let $\mathfrak{F} = (F, \Rightarrow, o_1, \ldots, o_t, o^1, \ldots, o^s, e_0, \ldots, e_{m-1})$, where $t, s, m = 0, 1, 2, \ldots$, be the algebra of formulas of a formalized language $\mathscr{L} = (A^0, F)$ of zero order and let $\mathscr{S} = (\mathscr{L}, C_{\mathscr{L}})$ be a system of propositional calculus in the class **S**. Under the above hypotheses the following holds.

6.3. *The binary relation* $\leqslant$ *in* F *defined by the equivalence*

$$(5) \qquad \alpha \leqslant \beta \quad \text{if and only if} \quad \vdash (\alpha \Rightarrow \beta)$$

(i.e. *$(\alpha \Rightarrow \beta)$ is derivable in* $\mathscr{S}$) *is a quasi-ordering in* F.

By 5 (s_3) and 5 (1) this relation is reflexive. By 5 (s_5) and 5 (1), it is transitive. Thus 6.3 holds.

It follows from 6.3, I 3.2, 5 (s_7), 5 (s_8), 5 (1) that

6.4. *The binary relation* $\approx$ *in* F *defined as follows*

$$(6) \quad \alpha \approx \beta \quad \text{if and only if} \quad \vdash (\alpha \Rightarrow \beta) \text{ and } \vdash (\beta \Rightarrow \alpha) \text{ in } \mathscr{S},$$

is a congruence relation in the algebra $\mathfrak{F}$. *Moreover, the relation* $\leqslant$ *on* $F/\approx$ *defined by the equivalence*

$$(7) \qquad ||\alpha|| \leqslant ||\beta|| \quad \text{if and only if} \quad \vdash (\alpha \Rightarrow \beta) \text{ in } \mathscr{S},$$

is an ordering on $F/\approx$.

Observe that for any two formulas α, β in $\mathscr{L}$, if both formulas are derivable in $\mathscr{S} = (\mathscr{L}, C_{\mathscr{L}}) \in \mathbf{S}$, then $\alpha \approx \beta$ by (6) and 5 (s_6), 5 (1). On the other hand, if α is derivable in $\mathscr{S}$ and β is not derivable in $\mathscr{S}$, then the relation $\alpha \approx \beta$ does not hold, because the formula $(\alpha \Rightarrow \beta)$ is not derivable by 5 (s_4). Hence, any two derivable formulas in $\mathscr{S}$

determine the same equivalence class, which will be denoted by V. Thus, by definition,

(8) $V = ||\alpha||$ if and only if α is a derivable formula in $\mathscr{S}$.

For any system $\mathscr{S} = (\mathscr{L}, C_{\mathscr{L}})$ in **S**, the algebra

(9) $\mathfrak{A}(\mathscr{S}) = (F/\approx, V, \Rightarrow, o_1, \ldots, o_t, o^1, \ldots, o^s, ||e_0||, \ldots, ||e_{m-1}||)$,

such that $(F/\approx, \Rightarrow, o_1, \ldots, o_t, o^1, \ldots, o^s, ||e_0||, \ldots, ||e_{m-1}||)$ is the quotient algebra $\mathfrak{F}/\approx$ (where $\mathfrak{F}$ is the algebra of formulas of $\mathscr{L}$ and $\approx$ is the congruence relation defined by (6)) and V is the element of $F/\approx$ defined by (8), will be said to be the *algebra of the system* $\mathscr{S}$ [(6)].

By definition, the following equations hold in $\mathfrak{A}(\mathscr{S})$:

(10)
$$||\alpha|| \Rightarrow ||\beta|| = ||(\alpha \Rightarrow \beta)||,$$
$$||\alpha|| o_i ||\beta|| = ||(\alpha o_i \beta)||, \quad i = 1, \ldots, t,$$
$$o^i ||\alpha|| = ||o^i \alpha||, \quad i = 1, \ldots, s,$$

for all formulas α, β of $\mathscr{L}$.

By (7), (8) and (10) we get

(11) $||\alpha|| \leqslant ||\beta||$ if and only if $||\alpha|| \Rightarrow ||\beta|| = V$,

for all formulas α, β of $\mathscr{L}$.

By (8), (7) and 5 (s_6), we get

(12) $||\beta|| \leqslant V$ for every formula β in $\mathscr{L}$.

It follows from 6.4 and (12) that V is the greatest element in the ordered set $(F/\approx, \leqslant)$, where $\leqslant$ is defined by (7).

We shall prove that

6.5. *For any system $\mathscr{S} = (\mathscr{L}, C_{\mathscr{L}})$ in* **S**, *the algebra $\mathfrak{A}(\mathscr{S})$ is an $\mathscr{S}$-algebra. Moreover, $\mathfrak{A}(\mathscr{S})$ is non-degenerate if and only if the system $\mathscr{S}$ is consistent.*

Let v be a valuation of $\mathscr{L}$ in $\mathfrak{A}(\mathscr{S})$ and let $\mathfrak{s}$ be a substitution from $\mathscr{L}$ into $\mathscr{L}$ such that

(13) $v(p) = ||\mathfrak{s}p||$ for every propositional variable p in $\mathscr{L}$.

By 3.4 we get

(14) $\alpha_{\mathfrak{A}(\mathscr{S})}(v) = ||\mathfrak{s}\alpha||$ for every formula α of $\mathscr{L}$.

(6) The idea of this construction is due to Lindenbaum and Tarski.

In particular, for the canonical valuation v^0 (see 3 (12)),

$$(15) \qquad \alpha_{\mathfrak{A}(\mathscr{S})}(v^0) = ||\alpha||.$$

If $\alpha \in \mathscr{A}_l$, then by 5 (s_1) for every substitution $\mathfrak{s}$ from $\mathscr{L}$ into $\mathscr{L}$, $\mathfrak{s}\alpha \in \mathscr{A}_l$ and consequently by (8), $||\mathfrak{s}\alpha|| = V$. Thus $\alpha_{\mathfrak{A}(\mathscr{S})}(v) = V$, i.e. the condition ($a_1$) holds.

Suppose that a rule of inference (r) assigns to premises $\alpha_1, \ldots, \alpha_n$ a formula β as the conclusion and let $\alpha_{i\mathfrak{A}(\mathscr{S})}(v) = V$, $i = 1, \ldots, n$, for a valuation v of $\mathscr{L}$ in $\mathfrak{A}(\mathscr{S})$. Thus, by (14), for a substitution $\mathfrak{s}$ satisfying (13) we get $||\mathfrak{s}\alpha_i|| = V$, $i = 1, \ldots, n$. Hence, by (8), the formulas $\mathfrak{s}\alpha_1, \ldots, \mathfrak{s}\alpha_n$ are derivable in $\mathscr{S}$. Consequently, by 5 (s_2), the formula $\mathfrak{s}\beta$ is derivable in $\mathscr{S}$. Thus, by (8), $||\mathfrak{s}\beta|| = V$. Applying (14) we get $\beta_{\mathfrak{A}(\mathscr{S})}(v) = V$, which proves that ($a_2$) holds.

The conditions (a_3) and (a_4) are satisfied by (11) and 6.4. Thus $\mathfrak{A}(\mathscr{S})$ is an $\mathscr{S}$-algebra. The second part of 6.5 follows from (8).

6.6. *The algebra $\mathfrak{A}(\mathscr{S})$ of a consistent system $\mathscr{S} = (\mathscr{L}, C_{\mathscr{L}})$ of a propositional calculus belonging to* **S** *is a free algebra in the class of all $\mathscr{S}$-algebras, the set $\{||p||\}_{p \in V}$ (where V is the set of all propositional variables in $\mathscr{L}$) being the set of free generators in $\mathfrak{A}(\mathscr{S})$. If $p_1 \neq p_2$, then $||p_1|| \neq ||p_2||$.*

Clearly, the set $\{||p||\}_{p \in V}$ generates $\mathfrak{A}(\mathscr{S})$. Let f be an arbitrary mapping from $\{||p||\}_{p \in V}$ into the set A of all elements of an $\mathscr{S}$-algebra $\mathfrak{A} = (A, V_{\mathfrak{A}}, \Rightarrow, o_1, \ldots, o_t, o^1, \ldots, o^s, e_0, \ldots, e_{m-1})$. The mapping f induces a mapping $v\colon V \to A$, defined as follows

$$(16) \qquad v(p) = f||p|| \quad \text{for any } p \in V.$$

The mapping v is a valuation in the $\mathscr{S}$-algebra $\mathfrak{A}$. By 2.3 and I 4.4 v can be extended in a unique way to a homomorphism $v_{\mathfrak{A}}$ from the algebra $\mathfrak{F}$ of formulas of $\mathscr{L}$ into $\mathfrak{A}^0 = (A, \Rightarrow, o_1, \ldots, o_t, o^1, \ldots, o^s, e_0, \ldots, e_{m-1})$. Moreover, by 3 (4)

$$(17) \qquad \gamma_{\mathfrak{A}}(v) = v_{\mathfrak{A}}(\gamma) \quad \text{for every formula } \gamma \text{ of } \mathscr{L}.$$

Let α, β be arbitrary formulas in $\mathscr{L}$ such that $||\alpha|| = ||\beta||$. Then $\alpha \approx \beta$ and by (6) the formulas $(\alpha \Rightarrow \beta)$, $(\beta \Rightarrow \alpha)$ are derivable in $\mathscr{S}$. Applying 6.1, we get $(\alpha \Rightarrow \beta)_{\mathfrak{A}}(v) = V_{\mathfrak{A}}$ and $(\beta \Rightarrow \alpha)_{\mathfrak{A}}(v) = V_{\mathfrak{A}}$. Hence, by 3 (3) and ($a_4$), $\alpha_{\mathfrak{A}}(v) = \beta_{\mathfrak{A}}(v)$. Thus by (17) $v_{\mathfrak{A}}(\alpha) = v_{\mathfrak{A}}(\beta)$. We have proved that the condition $||\alpha|| = ||\beta||$ implies that $v_{\mathfrak{A}}(\alpha) = v_{\mathfrak{A}}(\beta)$. Consequently, the equation

$$h(||\alpha||) = v_{\mathfrak{A}}(\alpha) \tag{18}$$

defines a mapping $h\colon F/\approx \to A$. Observe that h is an extension of f, since $h(||p||) = v_{\mathfrak{A}}(p) = f||p||$ for every $p \in V$. Moreover, by (8) we have $h(V) = h(||\alpha||)$, where α is a formula derivable in $\mathcal{S}$. Hence, applying (18), (17) and 6.1, we get

$$h(V) = V_{\mathfrak{A}}. \tag{19}$$

Since $v_{\mathfrak{A}}$ is a homomorphism from the algebra $\mathfrak{F}$ of formulas of $\mathcal{L}$ into $\mathfrak{A}^0$, we infer by (18), (10), (19) that h is a homomorphism from $\mathfrak{A}(\mathcal{S})$ into $\mathfrak{A}$. This completes the proof of the first part of 6.6. If $p_1 \neq p_2$ for some propositional variables p_1, p_2 of $\mathcal{L}$, then $(p_1 \Rightarrow p_2)$ is not derivable in $\mathcal{S}$. Otherwise, by 5.1, for every formula β of $\mathcal{L}$ the formula $((p_1 \Rightarrow p_1) \Rightarrow \beta)$ would be derivable in $\mathcal{S}$ and consequently, by 5 (s_3), 5 (s_4), β would be derivable, which contradicts the hypothesis that $\mathcal{S}$ is consistent. Thus $||p_1|| \neq ||p_2||$.

7. Completeness theorem. Let $\mathcal{S} = (\mathcal{L}, C_{\mathcal{L}})$ be a consistent system in the class **S** of standard systems of implicative extensional propositional calculi (see Sec. 5). A formula α of $\mathcal{L}$ will be said to be *valid in an algebra* $\mathfrak{A}$ associated with $\mathcal{L}$ provided that

$$\alpha_{\mathfrak{A}}(v) = V \quad \text{for every valuation } v \text{ of } \mathcal{L} \text{ in } \mathfrak{A}. \tag{1}$$

A formula α of $\mathcal{L}$ will be said to be $\mathcal{S}$*-valid* if α is valid in every $\mathcal{S}$-algebra (see Sec. 6).

By 6.1 every formula α derivable in $\mathcal{S}$ is $\mathcal{S}$-valid. The converse statement is also true. The following theorem is called the *completeness theorem for propositional calculi in the class* **S**.

7.1. *A formula α of a consistent system $\mathcal{S} = (\mathcal{L}, C_{\mathcal{L}})$ in* **S** *is derivable in $\mathcal{S}$ if and only if it is $\mathcal{S}$-valid.*

Theorem 7.1 is a part of the following theorem:

7.2. *For every formula α of a consistent system $\mathcal{S} = (\mathcal{L}, C_{\mathcal{L}})$ in* **S** *the following conditions are equivalent:*

(i) *α is derivable in $\mathcal{S}$,*

(ii) *α is $\mathcal{S}$-valid,*

(iii) $\alpha_{\mathfrak{A}(\mathcal{S})}(v^0) = V$, *where* v^0 *is the canonical valuation in the algebra* $\mathfrak{A}(\mathcal{S})$ *of the system* $\mathcal{S}$,

(iv) $||\alpha|| = V$ *in the algebra* $\mathfrak{A}(\mathcal{S})$ *of the system* $\mathcal{S}$.

Observe that (i) implies (ii) by 6.1. Clearly, (ii) implies (iii). By 6 (15), (iii) implies (iv), and by 6 (8), (iv) implies (i), which completes the proof.

8. Logically equivalent systems. Two consistent systems $\mathcal{S} = (\mathcal{L}, C_{\mathcal{L}})$ and $\mathcal{S}' = (\mathcal{L}', C_{\mathcal{L}'})$ in **S** where $\mathcal{L}$ and $\mathcal{L}'$ are similar languages of zero order will be said to be *logically equivalent*, or to be *of the same logic*, provided the class of all $\mathcal{S}$-algebras coincides with the class of all $\mathcal{S}'$-algebras.

Any class **L** of consistent logically equivalent systems in **S** determines a *logic*. For instance, the class of all logically equivalent systems $\mathcal{S}$ in **S** for which the class of $\mathcal{S}$-algebras coincides with the class of pseudo-Boolean algebras (see IV 6) determines intuitionistic logic.

Let **L** be a class of consistent systems in **S** of the same logic. Let **K** be the class of $\mathcal{S}$-algebras for any $\mathcal{S} \in \mathbf{L}$. A formula α of an arbitrary $\mathcal{S} \in \mathbf{L}$ will be said to be **L**-*valid* provided for every valuation v in any $\mathfrak{A} \in \mathbf{K}$, $\alpha_{\mathfrak{A}}(v) = V$. It follows from 7.1 that

8.1. *For any formula* α *in any system* $\mathcal{S} \in \mathbf{L}$ *the following conditions are equivalent*:

(i) α *is derivable in* $\mathcal{S}$,

(ii) α *is* **L**-*valid.*

This theorem completely characterizes derivable formulas in any system $\mathcal{S} \in \mathbf{L}$, i.e. in any system of the same logic.

8.2. *Let* $\mathcal{S} = (\mathcal{L}, C_{\mathcal{L}})$ *be a consistent system in* **S** *and let* $\mathcal{L}'$ *be a formalized language of zero order similar to* $\mathcal{L}$ *such that the cardinal of the set of propositional variables in* $\mathcal{L}'$ *is greater than the cardinal of the set of propositional variables in* $\mathcal{L}$. *Then there exists a system* $\mathcal{S}' = (\mathcal{L}', C_{\mathcal{L}'})$ *in* **S** *logically equivalent to* $\mathcal{S}$.

The set $\mathcal{A}_l$ of logical axioms of $\mathcal{S}$ induces the set $\mathcal{A}'_l$ of logical axioms of $\mathcal{S}'$ as follows: for any formula β in $\mathcal{L}'$

(1) $\beta \in \mathcal{A}'_l$ if and only if there exists a substitution $\mathfrak{s}$ from $\mathcal{L}$ into $\mathcal{L}'$ such that $\beta = \mathfrak{s}\alpha$ and $\alpha \in \mathcal{A}_l$.

Similarly, any rule of inference (r) of $\mathscr{S}$ induces a corresponding rule of inference (r') in $\mathscr{S}'$ as follows: the rule (r') associates with the premises $\gamma_1, \ldots, \gamma_n$ the conclusion δ if and only if there exists a substitution $\mathfrak{s}$ from $\mathscr{L}$ into $\mathscr{L}'$ such that $\gamma_i = \mathfrak{s}\alpha_i$ for $i = 1, \ldots, n$, $\delta = \mathfrak{s}\beta$ and the rule of inference (r) associates with the premises $\alpha_1, \ldots, \alpha_n$ the conclusion β.

Let us introduce, moreover, in $\mathscr{S}'$ the following rules of inference:

$$(\mathrm{r}_1'^*)\ \frac{\alpha, (\alpha \Rightarrow \beta)}{\beta}, \qquad (\mathrm{r}_2'^*)\ \frac{(\alpha \Rightarrow \beta), (\beta \Rightarrow \gamma)}{(\alpha \Rightarrow \gamma)}, \qquad (\mathrm{r}_3'^*)\ \frac{\alpha}{\beta \Rightarrow \alpha},$$

$$(\mathrm{r}_4'^*)\ \frac{(\alpha \Rightarrow \beta), (\beta \Rightarrow \alpha)}{(o\alpha \Rightarrow o\beta)} \quad \text{for every unary connective } o \text{ in } \mathscr{L}',$$

$$(\mathrm{r}_5'^*)\ \frac{(\alpha \Rightarrow \beta), (\beta \Rightarrow \alpha), (\gamma \Rightarrow \delta), (\delta \Rightarrow \gamma)}{((\alpha o \gamma) \Rightarrow (\beta o \delta))}$$

for every binary connective o in $\mathscr{L}'$,

where α, β, γ, δ are arbitrary formulas of $\mathscr{L}'$. The consequence operation $C_{\mathscr{L}'}$ determined in $\mathscr{L}'$ by $\mathscr{A}_l'$, by all rules of inference corresponding to rules of inference in $\mathscr{S}$ and by the rules of inference $(\mathrm{r}_1'^*)$-$(\mathrm{r}_5'^*)$, is the required one, i.e. $\mathscr{S}' = (\mathscr{L}', C_{\mathscr{L}'}) \in \mathbf{S}$ and $\mathscr{S}'$ is logically equivalent to $\mathscr{S}$.

By 2.5 and the definition of $\mathscr{A}_l'$, the condition 5 (s_1) is satisfied. Similarly, by 2.5, it follows from the definition of the rules of inference corresponding to the rules of inference in $\mathscr{S}$ that they satisfy the condition 5 (s_2). Clearly, the rules $(\mathrm{r}_1'^*)$–$(\mathrm{r}_5'^*)$ satisfy the condition 5 (s_2) too. Observe that if $\alpha \in C_{\mathscr{L}}(O)$, then for every substitution $\mathfrak{s}$ from $\mathscr{L}$ into $\mathscr{L}'$ we get $\mathfrak{s}\alpha \in C_{\mathscr{L}'}(O)$. Indeed, if $\alpha \in C_{\mathscr{L}}(O)$, then there exists a formal proof $\alpha_1, \ldots, \alpha_n$ of α in $\mathscr{S}$. It is easy to verify that then $\mathfrak{s}\alpha_1, \ldots, \mathfrak{s}\alpha_n$ is a formal proof of $\mathfrak{s}\alpha$ in $\mathscr{S}'$. Thus $\mathfrak{s}\alpha \in C_{\mathscr{L}'}(O)$. It follows from this remark that the condition 5 (s_3) is also satisfied. The conditions 5 (s_4)–5 (s_8) are satisfied because of rules of inference $(\mathrm{r}_1'^*)$–$(\mathrm{r}_5'^*)$. Thus $\mathscr{S}' \in \mathbf{S}$.

We shall prove that every $\mathscr{S}$-algebra is an $\mathscr{S}'$-algebra. Let $\mathfrak{A}$ be an $\mathscr{S}$-algebra and suppose $\beta \in \mathscr{A}_l'$. By (1), $\beta = \mathfrak{s}\alpha$ for some substitution $\mathfrak{s}$ from $\mathscr{L}$ into $\mathscr{L}'$ and some formula $\alpha \in \mathscr{A}_l$. Let v' be an arbitrary valuation of $\mathscr{L}'$ in $\mathfrak{A}$. Let us define a valuation v of $\mathscr{L}$ in $\mathfrak{A}$ as follows:

(2) $v(p) = (\mathfrak{s}p)_{\mathfrak{A}}(v')$ for every propositional variable p of $\mathscr{L}$.

Then, by 3.3, for every formula γ of $\mathscr{L}$ we have

(3) $$\gamma_{\mathfrak{A}}(v) = (\mathfrak{s}\gamma)_{\mathfrak{A}}(v').$$

By (3) we get $\beta_{\mathfrak{A}}(v') = (\mathfrak{s}\alpha)_{\mathfrak{A}}(v') = \alpha_{\mathfrak{A}}(v)$. Since $\alpha \in \mathscr{A}_l$, by 6 (a_1) we have $\alpha_{\mathfrak{A}}(v) = V$. Hence $\beta_{\mathfrak{A}}(v') = V$, i.e. $\mathfrak{A}$ satisfies the condition 6 (a_1) for the system $\mathscr{S}'$. Suppose that (r') is a rule of inference in $\mathscr{S}'$ corresponding to a rule of inference (r) in $\mathscr{S}$. Suppose that (r') assigns to the formulas $\gamma_1, \ldots, \gamma_n$ of $\mathscr{L}'$ a formula δ of $\mathscr{L}'$. Then there exists a substitution $\mathfrak{s}$ from $\mathscr{L}$ into $\mathscr{L}'$ such that

(4) $$\gamma_i = \mathfrak{s}\alpha_i, \quad i = 1, \ldots, n, \quad \delta = \mathfrak{s}\beta$$

and (r) assigns to the premises $\alpha_1, \ldots, \alpha_n$ the conclusion β. Let v' be a valuation of $\mathscr{L}'$ in $\mathfrak{A}$ such that

(5) $$\gamma_{i\mathfrak{A}}(v') = V \quad \text{for every } i = 1, \ldots, n.$$

Let v be the valuation of $\mathscr{L}$ in $\mathfrak{A}$ defined by (2). Then, by (3), (4) and (5), we get $\alpha_{i\mathfrak{A}}(v) = (\mathfrak{s}\alpha_i)_{\mathfrak{A}}(v') = \gamma_{i\mathfrak{A}}(v') = V$, $i = 1, \ldots, n$. Hence, by 6 (a_2), $\beta_{\mathfrak{A}}(v) = V$. Applying (3) and (4), we get $\delta_{\mathfrak{A}}(v') = V$. Thus the condition 6 ($a_2$) holds for the rule (r') in the algebra $\mathfrak{A}$. By 3 (3), 6.2, II 1.2, II 1 (i_2) (i_3) (i_4), the condition 6 (a_2) also holds for the rules ($r_1'^*$)-($r_5'^*$) in the algebra $\mathfrak{A}$. Since the conditions 6 (a_3), 6 (a_4) are satisfied in $\mathfrak{A}$, we have proved that $\mathfrak{A}$ is an $\mathscr{S}'$-algebra.

Now let us assume that $\mathfrak{A}$ is an $\mathscr{S}'$-algebra. Suppose that $\alpha \in A_l$ and let v be an arbitrary valuation of $\mathscr{L}$ in $\mathfrak{A}$. Let $\mathfrak{s}$ be a one-one mapping from the set V of all propositional variables of $\mathscr{L}$ into the set V' of all propositional variables of $\mathscr{L}'$. Let v' be a valuation of $\mathscr{L}'$ in $\mathfrak{A}$ satisfying the following condition:

(6) $$v'(\mathfrak{s}p) = v(p) \quad \text{for every } p \in V.$$

Then, by 3.3, for every formula γ of $\mathscr{L}$

(7) $$\mathfrak{s}\gamma_{\mathfrak{A}}(v') = \gamma_{\mathfrak{A}}(v).$$

Since $\mathfrak{s}\alpha \in \mathscr{A}'$ and $\mathfrak{A}$ is an $\mathscr{S}'$-algebra, by 6 (a_1) we get $\mathfrak{s}\alpha_{\mathfrak{A}}(v') = V$. Hence, by (7), $\alpha_{\mathfrak{A}}(v) = V$. Thus $\mathfrak{A}$ satisfies the condition 6 (a_1) for the system $\mathscr{S}$.

Suppose that (r) is a rule of inference in $\mathscr{S}$ which assigns to the premises $\alpha_1, \ldots, \alpha_n$ the conclusion β. Let v be a valuation of $\mathscr{L}$ in $\mathfrak{A}$ such that

(8) $$\alpha_{i\mathfrak{A}}(v) = V, \quad i = 1, \ldots, n.$$

Let $\mathfrak{s}$ be a one-one mapping from V into V' and let v' be a valuation of $\mathscr{L}'$ in $\mathfrak{A}$ satisfying (6). Then by (7) and (8) we get

(9) $$\mathfrak{s}\alpha_{i\mathfrak{A}}(v') = V, \quad i = 1, \ldots, n.$$

Since the corresponding rule of inference (r') in $\mathscr{S}'$ assigns to the premises $\mathfrak{s}\alpha_1, \ldots, \mathfrak{s}\alpha_n$ the conclusion $\mathfrak{s}\beta$, it follows from (9) and 6 (a_2) that $\mathfrak{s}\beta_{\mathfrak{A}}(v') = V$. Hence, by (7), $\beta_{\mathfrak{A}}(v) = V$. Thus the condition 6 ($a_2$) is satisfied in $\mathfrak{A}$ for the system $\mathscr{S}$. Since the conditions 6 (a_3), 6 (a_4) are also satisfied in $\mathfrak{A}$, we have just proved that $\mathfrak{A}$ is an $\mathscr{S}$-algebra. This completes the proof of 8.2.

9. L-theories of zero order. Let **L** be a class of consistent logically equivalent systems in **S** and let $\mathscr{S} = (\mathscr{L}, C_{\mathscr{L}}) \in \mathbf{L}$. For every set $\mathscr{A}$ of formulas in $\mathscr{L}$ the system $\mathscr{S}(\mathscr{A}) = (\mathscr{L}, C_{\mathscr{L}}, \mathscr{A})$ will be said to be an **L**-*theory of zero order based on* $\mathscr{S}$ and the set $\mathscr{A}$ will be called the *set of specific axioms of* $\mathscr{S}(\mathscr{A})$. The set $C_{\mathscr{L}}(\mathscr{A})$ is considered to be the *set of all theorems of* $\mathscr{S}(\mathscr{A})$.

The theory $\mathscr{S}(O) = (\mathscr{L}, C_{\mathscr{L}}, O)$ will be identified with the system $\mathscr{S} = (\mathscr{L}, C_{\mathscr{L}})$.

By a *formal proof of a formula* α *in* $\mathscr{S}(\mathscr{A})$ we shall understand a formal proof of α from the set $\mathscr{A}$ of specific axioms of $\mathscr{S}(\mathscr{A})$ with respect to the set $\mathscr{A}_l$ of logical axioms of $\mathscr{S}$ and the rules of inference adopted in $\mathscr{S}$ (see Sec. 4). If there exists a formal proof of α in $\mathscr{S}(\mathscr{A})$, we shall write

(1) $$\mathscr{A} \vdash \alpha.$$

By 4 (8), for every formula α of $\mathscr{L}$

(2) $$\mathscr{A} \vdash \alpha \quad \text{if and only if} \quad \alpha \in C_{\mathscr{L}}(\mathscr{A}).$$

A theory $\mathscr{S}(\mathscr{A})$ is said to be *consistent* if there exists a formula α in $\mathscr{L}$ such that $\alpha \notin C_{\mathscr{L}}(\mathscr{A})$.

A valuation v of $\mathscr{L}$ in a non-degenerate $\mathscr{S}$-algebra $\mathfrak{A}$ will be said to be an **L**-*model of a formula* α of $\mathscr{L}$, if $\alpha_{\mathfrak{A}}(v) = V$. By an **L**-*model of an* **L**-*theory* $\mathscr{S}(\mathscr{A})$ we shall understand any valuation v of $\mathscr{L}$ in a non-degenerate $\mathscr{S}$-algebra $\mathfrak{A}$ such that for each $\alpha \in \mathscr{A}$, $\alpha_{\mathfrak{A}}(v) = V$. An **L**-model v of an **L**-theory $\mathscr{S}(\mathscr{A})$ will be said to be *adequate* provided that, for any formula α of $\mathscr{L}$, α is a theorem of this theory if and only if v is an **L**-model of α.

9.1. *If α is a theorem of an* **L**-*theory $\mathscr{S}(\mathscr{A}) = (\mathscr{L}, C_{\mathscr{L}}, \mathscr{A})$, then every* **L**-*model v of $\mathscr{S}(\mathscr{A})$ in any non-degenerate $\mathscr{S}$-algebra $\mathfrak{A}$ is an* **L**-*model of* α, i.e.

$$\alpha_{\mathfrak{A}}(v) = V. \tag{3}$$

The easy proof based on 6 (a_1) and 6 (a_2) is omitted.

Now we are going to give a method of constructing an **L**-model for every consistent **L**-theory $\mathscr{S}(\mathscr{A}) = (\mathscr{L}, C_{\mathscr{L}}, \mathscr{A})$.

Let $\mathfrak{F} = (F, \Rightarrow, o_1, \ldots, o_t, o^1, \ldots, o^s, e_0, \ldots, e_{m-1})$ be the algebra of formulas of $\mathscr{L}$ and let $\leqslant_{\mathscr{A}}$ be the binary relation on F defined as follows: for all $\alpha, \beta \in F$

$$\alpha \leqslant_{\mathscr{A}} \beta \quad \text{if and only if} \quad \mathscr{A} \vdash (\alpha \Rightarrow \beta). \tag{4}$$

By 4 (c_2), 5 (s_3) and 5 (s_5) this relation is reflexive and transitive, i.e. $\leqslant_{\mathscr{A}}$ is a quasi-ordering on F.

Let us define the binary relation $\approx_{\mathscr{A}}$ on F as follows

$$\alpha \approx_{\mathscr{A}} \beta \text{ if and only if } \alpha \leqslant_{\mathscr{A}} \beta \text{ and } \beta \leqslant_{\mathscr{A}} \alpha. \tag{5}$$

By (3) and (4) we have

$$\alpha \approx_{\mathscr{A}} \beta \quad \text{if and only if} \quad \mathscr{A} \vdash (\alpha \Rightarrow \beta) \text{ and } \mathscr{A} \vdash (\beta \Rightarrow \alpha). \tag{6}$$

By I 3.2 the relation $\approx_{\mathscr{A}}$ is an equivalence relation on F and by 5 (s_7), 5 (s_8) it is a congruence in the algebra $\mathfrak{F}$. Consider the quotient algebra

$$\mathfrak{F}/{\approx_{\mathscr{A}}} = (F/{\approx_{\mathscr{A}}}, \Rightarrow, o_1, \ldots, o_t, o^1, \ldots, o^s, ||e_0||_{\mathscr{A}}, \ldots, ||e_{m-1}||_{\mathscr{A}})$$

and for any $\alpha \in F$, let $||\alpha||_{\mathscr{A}}$ be the equivalence class determined by α.

By I 3.2 the following relation on $F/{\approx_{\mathscr{A}}}$ is an ordering on $F/{\approx_{\mathscr{A}}}$:

$$||\alpha||_{\mathscr{A}} \leqslant_{\mathscr{A}} ||\beta||_{\mathscr{A}} \quad \text{if and only if} \quad \alpha \leqslant_{\mathscr{A}} \beta \quad (\text{i.e. } \mathscr{A} \vdash (\alpha \Rightarrow \beta)). \tag{7}$$

Observe that for any α, β in F

$$\text{if} \quad \mathscr{A} \vdash \alpha, \quad \text{then} \quad \alpha \approx_{\mathscr{A}} \beta \quad \text{if and only if} \quad \mathscr{A} \vdash \beta. \tag{8}$$

Indeed, if α, β are theorems of $\mathscr{S}(\mathscr{A})$, then by 5 (s_6) and (6) $\alpha \approx_{\mathscr{A}} \beta$. If $\mathscr{A} \vdash \alpha$ and not $\mathscr{A} \vdash \beta$, then $(\alpha \Rightarrow \beta)$ is not a theorem of $\mathscr{S}(\mathscr{A})$ because of 5 (s_4). Thus $\alpha \approx_{\mathscr{A}} \beta$ does not hold.

By (8) any theorem of $\mathscr{S}(\mathscr{A})$ determines the same equivalence class in $F/\approx_{\mathscr{A}}$, which will be denoted by V. By definition we have

(9) $$V = ||\alpha||_{\mathscr{A}} \quad \text{if and only if} \quad \mathscr{A} \vdash \alpha.$$

Observe that by 5 (s_6), if $\mathscr{A} \vdash \alpha$, then for each β, $\mathscr{A} \vdash (\beta \Rightarrow \alpha)$. Hence by (7) and (9), for each $\beta \in F$

(10) $$||\beta||_{\mathscr{A}} \leqslant_{\mathscr{A}} V,$$

i.e. V is the greatest element in the ordered set $(F/\approx_{\mathscr{A}}, \leqslant_{\mathscr{A}})$.

The abstract algebra

(11) $$\mathfrak{A}(\mathscr{S}(\mathscr{A})) = (F/\approx_{\mathscr{A}}, V, \Rightarrow, o_1, \ldots, o_t, o^1, \ldots, o^s, ||e_0||_{\mathscr{A}}, \ldots, ||e_{m-1}||_{\mathscr{A}})$$

will be called the *algebra of the theory* $\mathscr{S}(\mathscr{A})$ [7].

We have

9.2. *For every* **L**-*theory* $\mathscr{S}(\mathscr{A}) = (\mathscr{L}, C_{\mathscr{L}}, \mathscr{A})$, *the algebra* $\mathfrak{A}(\mathscr{S}(\mathscr{A}))$ *is an* $\mathscr{S}$-*algebra. Moreover,* $\mathfrak{A}(\mathscr{S}(\mathscr{A}))$ *is non-degenerate if and only if* $\mathscr{S}(\mathscr{A})$ *is consistent.*

The proof, similar to that of 6.5, is left to the reader.

9.3. *Let* $\mathscr{S}(\mathscr{A})$ *be a consistent* **L**-*theory of zero order and let* $\mathfrak{A}(\mathscr{S}(\mathscr{A}))$ *be the algebra of* $\mathscr{S}(\mathscr{A})$. *Let* v^0 *be the canonical valuation of the formalized language* $\mathscr{L}$ *of* $\mathscr{S}(\mathscr{A})$, i.e.

(12) $$v^0(p) = ||p||_{\mathscr{A}} \quad \textit{for every propositional variable } p \textit{ of } \mathscr{L}.$$

Then v^0 *is an adequate* **L**-*model of* $\mathscr{S}(\mathscr{A})$, i.e. *for every formula* α *of* $\mathscr{L}$, α *is a theorem of* $\mathscr{S}(\mathscr{A})$ *if and only if* $\alpha_{\mathfrak{A}(\mathscr{S}(\mathscr{A}))}(v^0) = V$.

By 9.2 the algebra $\mathfrak{A}(\mathscr{S}(\mathscr{A}))$ is a non-degenerate $\mathscr{S}$-algebra. It follows from 3.4 that for every formula α of $\mathscr{L}$

(13) $$\alpha_{\mathfrak{A}(\mathscr{S}(\mathscr{A}))}(v^0) = ||\alpha||_{\mathscr{A}}.$$

Hence, by (9),

(14) $$\alpha_{\mathfrak{A}(\mathscr{S}(\mathscr{A}))}(v^0) = V \quad \text{if and only if} \quad \mathscr{A} \vdash \alpha \quad (\text{i.e. } \alpha \in C_{\mathscr{L}}(\mathscr{A})).$$

Since $\mathscr{A} \subset C_{\mathscr{L}}(\mathscr{A})$, we infer by (14) that v^0 is an **L**-model of $\mathscr{S}(\mathscr{A})$. Moreover, by (14) this **L**-model is adequate for $\mathscr{S}(\mathscr{A})$.

It follows from 9.3 that

[7] See the footnote (6), p. 183.

9.4. *For every consistent* **L**-*theory of zero order there exists an adequate* **L**-*model.*

As a corollary to 9.1 and 9.3 we get the following theorem.

9.5. *For any formula* α *of a consistent* **L**-*theory of zero order the following conditions are equivalent*:

(i) α *is a theorem of* $\mathscr{S}(\mathscr{A})$,

(ii) *every* **L**-*model of* $\mathscr{S}(\mathscr{A})$ *is an* **L**-*model of* α,

(iii) $\alpha_{\mathfrak{A}(\mathscr{S}(\mathscr{A}))}(v^0) = \vee$ *for the canonical valuation* v^0 *in the algebra* $\mathfrak{A}(\mathscr{S}(\mathscr{A}))$ *of the theory* $\mathscr{S}(\mathscr{A})$.

The above theorem completely characterizes theorems of any consistent **L**-theory $\mathscr{S}(\mathscr{A})$.

10. Standard systems of implicative extensional propositional calculi with semi-negation. Let $\mathscr{S} = (\mathscr{L}, C_{\mathscr{L}})$ be a system in **S**. If in the alphabet of $\mathscr{L}$ there appears a unary connective o such that for all formulas α, β of $\mathscr{L}$

(1) $$(o(\alpha \Rightarrow \alpha) \Rightarrow \beta) \in C_{\mathscr{L}}(0),$$

then the connective o is said to be a *semi-negation sign* and $\mathscr{S}$ a *standard system of implicative extensional propositional calculus with semi-negation* [(8)].

Let $\mathbf{S}_\nu$ be the class of all standard systems of implicative extensional propositional calculi with semi-negation.

10.1. *Let* $\mathscr{S} = (\mathscr{L}, C_{\mathscr{L}}) \in \mathbf{S}_\nu$ *and let* $\mathfrak{A}$ *be an* $\mathscr{S}$-*algebra. Then in the ordered set* $(A, \leqslant)$, *where* A *is the set of all elements of* $\mathfrak{A}$ *and where*

(2) $$a \leqslant b \quad \textit{if and only if} \quad a \Rightarrow b = \vee \quad \textit{for all } a, b \in A,$$

there exists a least element $\wedge = o(a \Rightarrow a)$, $a \in A$, i.e.

(3) $$o(a \Rightarrow a) \leqslant b \quad \textit{for all } a, b \in A.$$

Moreover,

(4) $$o\vee = \wedge.$$

By 6.2 and III 1 the relation $\leqslant$ defined by (2) is an ordering on A. By (1), for some propositional variables p, q of $\mathscr{L}$, the formula

(8) A connective satisfying (1) has been introduced by Rasiowa and Sikorski [3].

$(o(p \Rightarrow p) \Rightarrow q) \in C_{\mathscr{L}}(0)$. Hence, by 6.1, for every valuation v of $\mathscr{L}$ in $\mathfrak{A}$ we have

$$(5) \qquad V = (o(p \Rightarrow p) \Rightarrow q)_{\mathfrak{A}}(v) = o\big(v(p) \Rightarrow v(p)\big) \Rightarrow v(q).$$

Thus, for all $a, b \in A$, $o(a \Rightarrow a) \Rightarrow b = V$. Hence by (2) we get (3). Since $a \Rightarrow a = V$ for each $a \in A$ (see 6.2 and II 1.1), we get $oV \Rightarrow b = V$ for all $b \in A$, i.e. oV is the least element in $(A, \leqslant)$. Thus we get (4).

10.2. *Let $\mathscr{S} = (\mathscr{L}, C_{\mathscr{L}})$ be a consistent system in $\mathbf{S}_v$ and let $\mathbf{L}$ be the class of all systems logically equivalent to $\mathscr{S}$. Then for every $\mathbf{L}$-theory $\mathscr{S}(\mathscr{A}) = (\mathscr{L}, C_{\mathscr{L}}, \mathscr{A})$ of zero order the following conditions are equivalent:*

(i) *$\mathscr{S}(\mathscr{A})$ is consistent,*

(ii) *there exists an $\mathbf{L}$-model of $\mathscr{S}(\mathscr{A})$,*

(iii) *for any formula α of $\mathscr{L}$ either $\alpha \notin C_{\mathscr{L}}(\mathscr{A})$ or $o\alpha \notin C_{\mathscr{L}}(\mathscr{A})$, where o is a connective of semi-negation in $\mathscr{S}$.*

The condition (i) implies (ii) by 9.4. To prove that (ii) implies (iii) suppose that (ii) holds. Let v be an $\mathbf{L}$-*model of $\mathscr{S}(\mathscr{A})$ in a non-degenerate $\mathscr{S}$-algebra* $\mathfrak{A}$. By 10.1, $oV = \Lambda \neq V$, since $\mathfrak{A}$ is non-degenerate. If $\alpha \in C_{\mathscr{L}}(\mathscr{A})$ and $o\alpha \in C_{\mathscr{L}}(\mathscr{A})$, then by 6.1

$$o\alpha_{\mathfrak{A}}(v) = V \quad \text{and} \quad \alpha_{\mathfrak{A}}(v) = V.$$

Hence $oV = \Lambda = V$, which contradicts $\Lambda \neq V$. The condition (iii) clearly implies (i), since a theory is consistent if there exists a formula in the formalized language of that theory which is not a theorem.

11. Theorems on logically equivalent systems in S. Let $\mathscr{L} = (A^0, F)$ and $\mathscr{L}' = (A^{0\prime}, F')$ be similar formalized languages of zero order such that $\operatorname{card} V \leqslant \operatorname{card} V'$, where V and V' are the sets of all propositional variables in $\mathscr{L}$ and $\mathscr{L}'$, respectively. Let $\mathscr{S} = (\mathscr{L}, C_{\mathscr{L}})$ and $\mathscr{S}' = (\mathscr{L}', C_{\mathscr{L}'})$ be consistent systems in $\mathbf{S}$. For every $\mathscr{A} \subset F$ and every substitution $\mathfrak{s}$ from $\mathscr{L}$ into $\mathscr{L}'$ let $\mathfrak{s}\mathscr{A}$ be the set of all formulas $\mathfrak{s}\alpha$ such that $\alpha \in \mathscr{A}$. Under these hypotheses the following theorem holds.

11.1. *The following conditions are equivalent:*

(i) *the systems $\mathscr{S}$ and $\mathscr{S}'$ are logically equivalent,*

(ii) *for every one-one substitution* $\mathfrak{s}\colon V \to V'$ *and for every* $\mathscr{A} \subset F$

(1) $$\mathfrak{s}(C_{\mathscr{L}}(\mathscr{A})) = C_{\mathscr{L}'}(\mathfrak{s}\mathscr{A}) \cap \mathfrak{s}F,$$

(iii) *there exists a one-one substitution* $\mathfrak{s}\colon V \to V'$ *such that for every* $\mathscr{A} \subset F$ *equation* (1) *is satisfied.*

We write C and C' instead of $C_{\mathscr{L}}$ and $C_{\mathscr{L}'}$, respectively.

Proof that (i) *implies* (ii): Suppose that $\mathscr{S}$ and $\mathscr{S}'$ are logically equivalent. Let $\mathfrak{s}$ be a one-one substitution $\mathfrak{s}\colon V \to V'$ and let $\mathscr{A} \subset F$. Suppose that $\mathfrak{s}\beta \in \mathfrak{s}(C(\mathscr{A}))$ for some β in F. Consider the theory $\mathscr{S}'(\mathfrak{s}\mathscr{A}) = (\mathscr{L}', C', \mathfrak{s}\mathscr{A})$. If $\mathfrak{s}\beta \notin C'(\mathfrak{s}\mathscr{A})$, then by 9.5 there exists a valuation v of $\mathscr{L}'$ in a non-degenerate $\mathscr{S}$-algebra $\mathfrak{A}$ such that

(2) $$\mathfrak{s}\alpha_{\mathfrak{A}}(v) = \mathsf{V} \quad \text{for each } \alpha \in \mathscr{A} \text{ and } \mathfrak{s}\beta_{\mathfrak{A}}(v) \neq \mathsf{V}.$$

Let $\mathfrak{s}_{\mathfrak{A}}v$ be the valuation of $\mathscr{L}$ in $\mathfrak{A}$ defined as follows

(3) $$\mathfrak{s}_{\mathfrak{A}}v(p) = \mathfrak{s}p_{\mathfrak{A}}(v) \quad \text{for each } p \in V.$$

Then by (2), (3) and 3.3 we get

(4) $$\alpha_{\mathfrak{A}}(\mathfrak{s}_{\mathfrak{A}}v) = \mathsf{V} \quad \text{for each } \alpha \in \mathscr{A} \text{ and } \beta_{\mathfrak{A}}(\mathfrak{s}_{\mathfrak{A}}v) \neq \mathsf{V}.$$

Applying 9.5 to the theory $\mathscr{S}(\mathscr{A}) = (\mathscr{L}, C, \mathscr{A})$, we get, by (4), $\beta \notin C(\mathscr{A})$. Hence $\mathfrak{s}\beta \notin \mathfrak{s}(C(\mathscr{A}))$, which contradicts our hypothesis. Thus $\mathfrak{s}(C(\mathscr{A})) \subset C'(\mathfrak{s}\mathscr{A}) \cap \mathfrak{s}F$.

Conversely, suppose that a formula β' in F' belongs to $C'(\mathfrak{s}\mathscr{A}) \cap \mathfrak{s}F$. Then $\beta' = \mathfrak{s}\beta$ for a certain β in F and $\beta' \in C'(\mathfrak{s}\mathscr{A})$. If $\beta' \notin \mathfrak{s}(C(\mathscr{A}))$, then $\beta \notin C(\mathscr{A})$. Applying 9.5 to the theory $\mathscr{S}(\mathscr{A}) = (\mathscr{L}, C, \mathscr{A})$, we infer that there exists a valuation v of $\mathscr{L}$ in an $\mathscr{S}$-algebra $\mathfrak{A}$ such that

(5) $$\alpha_{\mathfrak{A}}(v) = \mathsf{V} \quad \text{for each } \alpha \in \mathscr{A} \text{ and } \beta_{\mathfrak{A}}(v) \neq \mathsf{V}.$$

Let v' be a valuation of $\mathscr{L}'$ in $\mathfrak{A}$ satisfying the condition

(6) $$v'(\mathfrak{s}p) = v(p) \quad \text{for every } p \in V.$$

Then, by 3.3,

(7) $$\mathfrak{s}\gamma_{\mathfrak{A}}(v') = \gamma_{\mathfrak{A}}(v) \quad \text{for every } \gamma \in F.$$

By (5) and (7)

$$\mathfrak{s}\alpha_{\mathfrak{A}}(v') = \mathsf{V} \quad \text{for each } \alpha \in \mathscr{A} \text{ and } \mathfrak{s}\beta_{\mathfrak{A}}(v') \neq \mathsf{V}.$$

Hence, applying 9.5 to the theory $\mathscr{S}'(\mathfrak{s}\mathscr{A}) = (\mathscr{L}', C', \mathfrak{s}\mathscr{A})$, we infer that $\beta' = \mathfrak{s}\beta \notin C'(\mathfrak{s}\mathscr{A})$, which contradicts our hypothesis. Thus $C'(\mathfrak{s}\mathscr{A}) \cap \cap \mathfrak{s}F \subset \mathfrak{s}(C(\mathscr{A}))$ and therefore (1) holds.

Clearly, (ii) implies (iii).

Proof that (iii) *implies* (i): Suppose that (iii) holds and let $\mathfrak{A}$ be an $\mathscr{S}$-algebra. We shall prove that $\mathfrak{A}$ is an $\mathscr{S}'$-algebra. By (1) we get

(8) $$\mathfrak{s}C(O) = C'(O) \cap \mathfrak{s}F.$$

If α' in F' is a logical axiom in $\mathscr{S}'$, then for a certain one-one substitution $\mathfrak{s}'\colon V' \to V'$ from V' onto V', $\mathfrak{s}'\alpha' \in \mathfrak{s}F$ and by 5 (s_1), $\mathfrak{s}'\alpha'$ is also a logical axiom of $\mathscr{S}'$. Consequently, $\mathfrak{s}'\alpha' \in C'(O)$ and $\mathfrak{s}'\alpha' = \mathfrak{s}\alpha$ for a certain α in F. By (8), $\mathfrak{s}\alpha \in \mathfrak{s}C(O)$ and hence $\alpha \in C(O)$. By 6.1

(9) $$\alpha_{\mathfrak{A}}(v) = V \quad \text{for every valuation } v \text{ of } \mathscr{L} \text{ in } \mathfrak{A}.$$

Now, let v be an arbitrary valuation of $\mathscr{L}'$ in $\mathfrak{A}$. Then by 3.3

(10) $$\mathfrak{s}\alpha_{\mathfrak{A}}(v) = \alpha_{\mathfrak{A}}(\mathfrak{s}_{\mathfrak{A}}v),$$

where $\mathfrak{s}_{\mathfrak{A}}v$ is the valuation of $\mathscr{L}$ in $\mathfrak{A}$ defined by (3). It follows from (9) and (10) that

(11) $$\mathfrak{s}'\alpha'_{\mathfrak{A}}(v) = \mathfrak{s}\alpha_{\mathfrak{A}}(v) = V \quad \text{for every valuation } v \text{ of } \mathscr{L}' \text{ in } \mathfrak{A}.$$

Given a valuation v of $\mathscr{L}'$ in $\mathfrak{A}$, let v' be the valuation of $\mathscr{L}'$ in $\mathfrak{A}$ defined as follows:

(12) $$v'(\mathfrak{s}'p') = v(p') \quad \text{for every } p' \in V'.$$

Then, by 3.3,

(13) $$\mathfrak{s}'\gamma'_{\mathfrak{A}}(v') = \gamma'_{\mathfrak{A}}(v) \quad \text{for every } \gamma' \text{ in } F'.$$

It follows from (11) and (13) that $\alpha'_{\mathfrak{A}}(v) = V$. Thus the condition 6 ($a_1$) is satisfied in $\mathfrak{A}$ for $\mathscr{S}'$.

Suppose that a rule of inference (r') of $\mathscr{S}'$ assigns to $\alpha'_1, \ldots, \alpha'_n$ in F' a formula β' in F'. Then for a certain one-one substitution $\mathfrak{s}'\colon V' \to V'$ from V' onto V', $\mathfrak{s}'\alpha'_1, \ldots, \mathfrak{s}'\alpha'_n, \mathfrak{s}'\beta' \in \mathfrak{s}F$. Thus $\mathfrak{s}'\alpha' = \mathfrak{s}\alpha_i$, $i = 1, \ldots, n$, for some α_i in F and $\mathfrak{s}'\beta' = \mathfrak{s}\beta$ for some β in F. By 5 (s_2) (r') associates with $\mathfrak{s}'\alpha'_1, \ldots, \mathfrak{s}'\alpha'_n$ the formula $\mathfrak{s}'\beta'$. Hence $\mathfrak{s}'\beta' \in C'(\{\mathfrak{s}'\alpha'_1, \ldots, \mathfrak{s}'\alpha'_n\})$. Thus, by (1), we get $\mathfrak{s}'\beta' = \mathfrak{s}\beta \in \mathfrak{s}C(\{\alpha_1, \ldots, \alpha_n\})$.

Consequently, $\beta \in C(\{\alpha_1, \ldots, \alpha_n\})$. Hence, by 9.1 applied to the theory $(\mathscr{L}, C, \{\alpha_1, \ldots, \alpha_n\})$, it follows that for each valuation v of $\mathscr{L}$ in $\mathfrak{A}$

(14) if $\alpha_{i\mathfrak{A}}(v) = V$ for each $i = 1, \ldots, n$, then $\beta_{\mathfrak{A}}(v) = V$.

Let v be an arbitrary valuation of $\mathscr{L}'$ in $\mathfrak{A}$ such that

(15) $$\alpha'_{i\mathfrak{A}}(v) = V \quad \text{for every } i = 1, \ldots, n.$$

Then, for the valuation v' of $\mathscr{L}'$ in $\mathfrak{A}$ defined by (12), we get by (13) and (15)

(16) $$\mathfrak{s}'\alpha'_{i\mathfrak{A}}(v') = \mathfrak{s}\alpha_{i\mathfrak{A}}(v') = V \quad \text{for every } i = 1, \ldots, n.$$

Hence, by 3.3, we get

(17) $$\alpha_{i\mathfrak{A}}(\mathfrak{s}_{\mathfrak{A}}v') = V \quad \text{for every } i = 1, \ldots, n,$$

where $\mathfrak{s}_{\mathfrak{A}}v'$ is the valuation of $\mathscr{L}$ in $\mathfrak{A}$ defined by the equation $\mathfrak{s}_{\mathfrak{A}}v'(p) = \mathfrak{s}p_{\mathfrak{A}}(v')$ for every $p \in V$. It follows from (14) and (17) that $\beta_{\mathfrak{A}}(\mathfrak{s}_{\mathfrak{A}}v') = V$. Hence, by 3.3, $\mathfrak{s}\beta_{\mathfrak{A}}(v') = \mathfrak{s}'\beta'_{\mathfrak{A}}(v') = V$. Applying (13), we get $\beta'_{\mathfrak{A}}(v) = V$. Thus (15) implies that $\beta'_{\mathfrak{A}}(v) = V$. This proves that the condition 6 (a_2) is satisfied in $\mathfrak{A}$ for $\mathscr{S}'$. Since the conditions 6 (a_3), 6 (a_4) are also satisfied, $\mathfrak{A}$ is an $\mathscr{S}'$-algebra.

Now let us suppose that $\mathfrak{A}$ is an $\mathscr{S}'$-algebra. We shall prove that $\mathfrak{A}$ is an $\mathscr{S}$-algebra. If α in F is a logical axiom of $\mathscr{S}$, then $\alpha \in C(O)$. Hence, by (8), $\mathfrak{s}\alpha \in C'(O) \cap \mathfrak{s}F$. Thus, by 6.1,

(18) $$\mathfrak{s}\alpha_{\mathfrak{A}}(v) = V \quad \text{for every valuation } v \text{ of } \mathscr{L}' \text{ in } \mathfrak{A}.$$

Let v be an arbitrary valuation of $\mathscr{L}$ in $\mathfrak{A}$. Let v' be a valuation of $\mathscr{L}'$ in $\mathfrak{A}$ satisfying the condition (6). Hence, by (7) and (18), $\alpha_{\mathfrak{A}}(v) = \mathfrak{s}\alpha_{\mathfrak{A}}(v') = V$. Thus, the condition 6 (a_1) is satisfied in $\mathfrak{A}$ for $\mathscr{S}$.

Suppose that a rule of inference (r) in $\mathscr{S}$ assigns to the formulas $\alpha_1, \ldots, \alpha_n$ in F a formula β in F. Then $\beta \in C(\{\alpha_1, \ldots, \alpha_n\})$. Hence, by (1), $\mathfrak{s}\beta \in C'(\{\mathfrak{s}\alpha_1, \ldots, \mathfrak{s}\alpha_n\}) \cap \mathfrak{s}F$. Consequently, by 9.5 applied to the theory $(\mathscr{L}', C', \{\mathfrak{s}\alpha_1, \ldots, \mathfrak{s}\alpha_n\})$, for every valuation v' of $\mathscr{L}'$ in $\mathfrak{A}$

(19) if $\mathfrak{s}\alpha_i(v') = V$ for each $i = 1, \ldots, n$, then $\mathfrak{s}\beta_{\mathfrak{A}}(v') = V$.

Let v be an arbitrary valuation of $\mathscr{L}$ in $\mathfrak{A}$ such that

(20) $$\alpha_{i\mathfrak{A}}(v) = V \quad \text{for every } i = 1, \ldots, n.$$

Then, for a valuation v' of $\mathscr{L}'$ in $\mathfrak{A}$ satisfying the condition (6), we get by (7) and (20)

(21) $$\mathfrak{s}\alpha_{i\mathfrak{A}}(v') = V \quad \text{for every } i = 1, \ldots, n.$$

By (19) and (21), $\mathfrak{s}\beta_{\mathfrak{A}}(v') = V$. Hence, by (7), $\beta_{\mathfrak{A}}(v) = V$. Thus the condition (20) implies that $\beta_{\mathfrak{A}}(v) = V$. This proves that the condition 6 (a_2) holds in $\mathfrak{A}$ for the system $\mathscr{S}$. Since the conditions 6 (a_3), 6 (a_4) are also satisfied in $\mathfrak{A}$, the algebra $\mathfrak{A}$ is an $\mathscr{S}$-algebra. This completes the proof of 11.1.

The following two theorems are corollaries to 11.1.

11.2. *If $\mathscr{L}$ and $\mathscr{L}'$ are similar languages of zero order and $\mathscr{L}'$ is an extension of $\mathscr{L}$, then for any consistent systems $\mathscr{S} = (\mathscr{L}, C_{\mathscr{L}})$, $\mathscr{S}' = (\mathscr{L}', C_{\mathscr{L}'})$ in* **S** *the following conditions are equivalent:*

(i) *$\mathscr{S}$ and $\mathscr{S}'$ are logically equivalent,*

(ii) *for every subset $\mathscr{A}$ of the set F of all formulas in $\mathscr{L}$*

(22) $$C_{\mathscr{L}}(\mathscr{A}) = C_{\mathscr{L}'}(\mathscr{A}) \cap F.$$

Indeed, if (i) holds, then by 11.1 (ii), for the identity substitution $\mathfrak{s}$ the equation (1) gives (22). On the other hand, if (22) holds, then for the identity substitution $\mathfrak{s}$ the equation (1) holds and therefore the condition (iii) of 11.1 is satisfied, which implies that $\mathscr{S}$ and $\mathscr{S}'$ are logically equivalent.

11.3. *For any consistent systems $\mathscr{S} = (\mathscr{L}, C_{\mathscr{L}})$, $\mathscr{S}' = (\mathscr{L}, C'_{\mathscr{L}})$ in* **S** *the following conditions are equivalent:*

(i) *the systems $\mathscr{S}$ and $\mathscr{S}'$ are logically equivalent,*

(ii) *for every subset $\mathscr{A}$ of the set of all formulas in $\mathscr{L}$*

(23) $$C_{\mathscr{L}}(\mathscr{A}) = C'_{\mathscr{L}}(\mathscr{A}).$$

Theorems 11.1, 11.2 and 11.3 explain the logical meaning of the definition, adopted in Section 8, of logically equivalent consistent systems in **S**.

11.4. *Let $\mathscr{S}' = (\mathscr{L}', C_{\mathscr{L}'})$ be a consistent system in* **S**. *Assume that the consequence operation $C_{\mathscr{L}'}$ is determined by a set $\mathscr{A}'_l$ of logical axioms and by a set $\{(r'_1), \ldots, (r'_k)\}$ of rules of inference. Let $\mathscr{L}$ be a sublanguage* *of $\mathscr{L}'$ similar to $\mathscr{L}'$ and let F be the set of all formulas of $\mathscr{L}$. Consider the consequence operation $C_{\mathscr{L}}$ in $\mathscr{L}$ determined by the set*

(24) $$\mathscr{A}_l = \mathscr{A}'_l \cap F$$

of logical axioms and by the set $\{(r_1), \ldots, (r_k)\}$ of rules of inference, where (r_i), $i = 1, \ldots, k$, is the restriction of (r'_i) to the language $\mathscr{L}$. Then $\mathscr{S} = (\mathscr{L}, C_{\mathscr{L}}) \in \mathbf{S}$ and $\mathscr{S}$ is logically equivalent to $\mathscr{S}'$.

To begin with, we shall prove that

(25) $$C_{\mathscr{L}}(\mathscr{A}) = C_{\mathscr{L}'}(\mathscr{A}) \cap F, \quad \text{for every } \mathscr{A} \subset F.$$

Clearly, $C_{\mathscr{L}}(\mathscr{A}) \subset C_{\mathscr{L}'}(\mathscr{A}) \cap F$. Suppose that $\alpha \in C_{\mathscr{L}'}(\mathscr{A}) \cap F$. Then there exists a formal proof $\alpha_1, \ldots, \alpha_n$ of α in the theory $\mathscr{S}'(\mathscr{A}) = (\mathscr{L}', C_{\mathscr{L}'}, \mathscr{A})$. Let $p'_1, \ldots, p'_m$ be all propositional variables, different from one another, in $\alpha_1, \ldots, \alpha_n$ which do not belong to the language $\mathscr{L}$ and let $p_1, \ldots, p_m$ be propositional variables in $\mathscr{L}$ not appearing in any α_i, $i = 1, \ldots, n$, and such that $p_i \neq p_j$, $i \neq j$ and $p_i \neq p'_j$ for all $i = 1, \ldots, m$, $j = 1, \ldots, m$. Let $\mathfrak{s}$ be the following substitution from $\mathscr{L}'$ into $\mathscr{L}'$:

(26) $\mathfrak{s}(p_i) = p'_i$, $\mathfrak{s}(p'_i) = p_i$, $i = 1, \ldots, m$, $\mathfrak{s}(p) = p$ for each propositional variable p of $\mathscr{L}'$ such that $p \neq p_i, p \neq p'_i, i = 1, \ldots, m$.

It follows from (26) that, for every γ in $\mathscr{L}'$,

(27) if neither p'_i nor p_i, $i = 1, \ldots, m$, appears in γ, then $\mathfrak{s}\gamma = \gamma$.

Moreover,

(28) $$\mathfrak{s}\alpha_i \in F \quad \text{for every } i = 1, \ldots, m,$$

and by (27)

(29) $$\text{if} \quad \alpha_i \in F, \quad \text{then} \quad \mathfrak{s}\alpha_i = \alpha_i, \quad i = 1, \ldots, m.$$

By (29), $\mathfrak{s}\alpha_n = \mathfrak{s}\alpha = \alpha$. The sequence $\mathfrak{s}\alpha_1, \ldots, \mathfrak{s}\alpha_n$ is a formal proof of α in the theory $\mathscr{S}(\mathscr{A}) = (\mathscr{L}, C_{\mathscr{L}}, \mathscr{A})$. Indeed, if $\alpha_i \in \mathscr{A}'_l$, then by 5 (s_1), (28) and (24), $\mathfrak{s}\alpha_i \in \mathscr{A}_l$. If $\alpha_i \in \mathscr{A}$ then by (29) $\mathfrak{s}\alpha_i \in \mathscr{A}$. If α_i, $i > 1$, is obtained from $\alpha_{\nu_1}, \ldots, \alpha_{\nu_r}$, $1 \leqslant \nu_1 < \ldots \nu_r < i$ by a rule of inference (r'_j), then $\mathfrak{s}\alpha_i$ is the conclusion from $\mathfrak{s}\alpha_{\nu_1}, \ldots, \mathfrak{s}\alpha_{\nu_r}$ by the same rule of inference by 5 (s_2) and consequently, by (28), $\mathfrak{s}\alpha_i$ is the conclusion from $\mathfrak{s}\alpha_{\nu_1}, \ldots, \mathfrak{s}\alpha_{\nu_r}$ by the rule (r_j) in $\mathscr{S}$. This completes the proof that $\alpha \in C_{\mathscr{L}}(\mathscr{A})$ and consequently that (25) holds. It follows from the hypothesis that $\mathscr{S}' \in \mathbf{S}$, from the definition of $C_{\mathscr{L}}$ and from (25) that the conditions 5 ((s_1)-(s_8)) are satisfied for the system $\mathscr{S}$, i.e. $\mathscr{S} \in \mathbf{S}$. By (25) and 11.2, the systems $\mathscr{S}$ and $\mathscr{S}'$ are logically equivalent.

12. Deductive filters. Let $\mathbf{L}$ be a class of consistent logically equivalent systems in $\mathbf{S}$ and let $\mathbf{K_L}$ be the class of all $\mathscr{S}$-algebras for systems $\mathscr{S}$ in $\mathbf{L}$.

Let us define the kernel $K(h)$ of an epimorphism h of an algebra $\mathfrak{A} = (A, V, \Rightarrow, o_1, \ldots, o_t, o^1, \ldots, o^s, e_0, \ldots, e_{m-1})$ in $\mathbf{K_L}$ onto an algebra $\mathfrak{B} = (B, V', \Rightarrow, o_1, \ldots, o_t, o^1, \ldots, o^s, e_0, \ldots, e_{m-1})$ in $\mathbf{K_L}$ as the set of all elements in A which are mapped by h onto V'. Thus, for every $a \in A$,

(1) $$a \in K(h) \quad \text{if and only if} \quad h(a) = V'.$$

12.1. *If $K(h)$ is the kernel of an epimorphism h of an algebra $\mathfrak{A}$ in $\mathbf{K_L}$ onto an algebra $\mathfrak{B}$ in $\mathbf{K_L}$, then the following conditions are satisfied:*

(2) $$V \in K(h),$$

(3) *if $a \in K(h)$ and $a \Rightarrow b \in K(h)$, then $b \in K(h)$, for all a, b in $\mathfrak{A}$,*

(4) *if $\mathscr{S} = (\mathscr{L}, C_{\mathscr{L}}) \in \mathbf{L}$ and $\beta \in C_{\mathscr{L}}(\mathscr{A})$, where $\mathscr{A}$ is a set of formulas in $\mathscr{L}$, then for every valuation v of $\mathscr{L}$ in $\mathfrak{A}$ the condition $\alpha_{\mathfrak{A}}(v) \in K(h)$ for all α in $\mathscr{A}$ implies that $\beta_{\mathfrak{A}}(v) \in K(h)$.*

Moreover, $h(a) = h(b)$ is equivalent to $a \Rightarrow b \in K(h)$ and $b \Rightarrow a \in K(h)$. Hence, by I 4.7, *the relation $\approx$ defined as*

(5) $$a \approx b \quad \text{if and only if} \quad a \Rightarrow b \in K(h) \text{ and } b \Rightarrow a \in K(h),$$

is a congruence in $\mathfrak{A}$. The quotient algebra $\mathfrak{A}/{\approx}$ is isomorphic to $\mathfrak{B}$.

Since $h(V) = V'$, by (1), the condition (2) is satisfied. Suppose that $a, a \Rightarrow b \in K(h)$. Then $h(a) = V'$ and $h(a) \Rightarrow h(b) = V'$. Hence, by 6.2 and II 1.2, $h(b) = V'$, i.e. $b \in K(h)$. Thus (3) holds. Suppose that the hypotheses of (4) are satisfied. If $\beta \in C_{\mathscr{L}}(\mathscr{A})$ and for a valuation v of $\mathscr{L}$ in $\mathfrak{A}$, $\alpha_{\mathfrak{A}}(v) \in K(h)$ for all α in $\mathscr{A}$, then by (1), $h\alpha_{\mathfrak{A}}(v) = V'$ for all α in $\mathfrak{A}$. Consequently, by 3.2,

(6) $$\alpha_{\mathfrak{B}}(hv) = V' \quad \text{for all } \alpha \text{ in } \mathscr{A}.$$

Hence, the valuation hv of $\mathscr{L}$ in $\mathfrak{B}$ is an $\mathbf{L}$-model of the $\mathbf{L}$-theory $\mathscr{S}(\mathscr{A}) = (\mathscr{L}, C_{\mathscr{L}}, \mathscr{A})$. By 9.1, $\beta_{\mathfrak{B}}(hv) = V'$. Hence, by 3.2, $h\beta_{\mathfrak{A}}(v) = V'$. Thus $\beta_{\mathfrak{A}}(v) \in K(h)$, which proves (4). If $h(a) = h(b)$, then $h(a \Rightarrow b) = h(a) \Rightarrow h(b) = V'$ and $h(b \Rightarrow a) = h(b) \Rightarrow h(a) = V'$ by 6.2 and II 1 (i_1). Thus $a \Rightarrow b$, $b \Rightarrow a \in K(h)$. Conversely, if $a \Rightarrow b$, $b \Rightarrow a \in K(h)$, then $h(a) \Rightarrow h(b) = h(a \Rightarrow b) = V'$ and $h(b) \Rightarrow h(a) = h(b \Rightarrow a) = V'$. Hence, by 6.2 and II 1 (i_3), $h(a) = h(b)$.

A subset ∇ of the set A of all elements of an algebra $\mathfrak{A}$ in $\mathbf{K}_{\mathbf{L}}$ will be said to be a *deductive filter* provided the following conditions are satisfied:

(f_1) $V \in \nabla$,

(f_2) if $a, a \Rightarrow b \in \nabla$, then $b \in \nabla$,

(f_3) if $\mathscr{S} = (\mathscr{L}, C_{\mathscr{L}}) \in \mathbf{L}$ and $\beta \in C_{\mathscr{L}}(\mathscr{A})$, where $\mathscr{A}$ is a set of formulas in $\mathscr{L}$, then for every valuation v of $\mathscr{L}$ in $\mathfrak{A}$ the condition $\alpha_{\mathfrak{A}}(v) \in \nabla$ for all α in $\mathscr{A}$ implies that $\beta_{\mathfrak{A}}(v) \in \nabla$.

It follows from the above definition and 12.1 that

12.2. *If $K(h)$ is the kernel of an epimorphism h of an algebra $\mathfrak{A}$ in $\mathbf{K}_{\mathbf{L}}$ onto an algebra $\mathfrak{B}$ in $\mathbf{K}_{\mathbf{L}}$, then $K(h)$ is a deductive filter.*

Observe that in any deductive filter ∇

(7) *if* $a \in \nabla$ *and* $a \leqslant b$, *then* $b \in \nabla$.

Indeed, this follows from (f_1), (f_2) and 6.2, II 1.1.

Let us note that the converse statement to 12.2 also holds (see the remark at the end of Sec. 13).

12.3. *Let $\mathscr{S} = (\mathscr{L}, C_{\mathscr{L}})$ be a system in $\mathbf{L}$ with an enumerable set V of propositional variables. Let $\mathfrak{A}$ be an algebra in $\mathbf{K}_{\mathbf{L}}$. Then a subset ∇ of the set of all elements in $\mathfrak{A}$ is a deductive filter if and only if the conditions* (f_1), (f_2) *and the following one are satisfied*:

(f_3^*) *if a rule of inference in $\mathscr{S}$ associates with formulas $\alpha_1, \ldots, \alpha_n$ a formula β, then, for every valuation v of $\mathscr{L}$ in $\mathfrak{A}$, if $\alpha_{1\mathfrak{A}}(v) \in \nabla, \ldots$ $\ldots, \alpha_{n\mathfrak{A}}(v) \in \nabla$, then $\beta_{\mathfrak{A}}(v) \in \nabla$.*

Suppose that ∇ is a deductive filter. If a rule of inference in $\mathscr{S}$ associates with $\alpha_1, \ldots, \alpha_n$ a formula β, then $\beta \in C_{\mathscr{L}}(\{\alpha_1, \ldots, \alpha_n\})$. Hence, by ($f_3$), the condition ($f_3^*$) is satisfied.

Now, let us assume that the conditions (f_1), (f_2), (f_3^*) are satisfied. We are going to prove that

(8) if $\beta \in C_{\mathscr{L}}(\mathscr{A})$, then for every valuation v of $\mathscr{L}$ in $\mathfrak{A}$ the condition $\alpha_{\mathfrak{A}}(v) \in \nabla$ for all α in $\mathscr{A}$ implies that $\beta_{\mathfrak{A}}(v) \in \nabla$.

Suppose that $\beta \in C_{\mathscr{L}}(\mathscr{A})$. Then there exists a formal proof $\beta_1, \ldots, \beta_m$ of β in the **L**-theory $\mathscr{S}(\mathscr{A})$. Suppose that for any $\alpha \in \mathscr{A}$, $\alpha_{\mathfrak{A}}(v) \in \nabla$, where v is a valuation of $\mathscr{L}$ in $\mathfrak{A}$. Then, by the definition of a formal proof (see Sec. 9), 6 (s_1), (f_1), (f_3^*) and the above condition, we get $\beta_{i\mathfrak{A}}(v) \in \nabla$ for all $i = 1, \ldots, m$. Hence $\beta_{\mathfrak{A}}(v) = \beta_{m\mathfrak{A}}(v) \in \nabla$. This completes the proof of (8).

Let $\mathscr{S}' = (\mathscr{L}', C_{\mathscr{L}'})$ be an arbitrary system in **L** with a set V' of propositional variables. Then $\operatorname{card} V \leqslant \operatorname{card} V'$ and systems $\mathscr{S}, \mathscr{S}'$ are logically equivalent. Suppose that $\beta' \in C_{\mathscr{L}'}(\mathscr{A}')$, where $\mathscr{A}'$ is a set of formulas in $\mathscr{L}'$ and let $\alpha'_{\mathfrak{A}}(v') \in \nabla$ for all α' in $\mathscr{A}'$, where v' is a valuation of $\mathscr{L}'$ in $\mathfrak{A}$. Since the consequence operation $C_{\mathscr{L}'}$ has a finite character (see Sec. 4), there exists a finite subset $\{\alpha'_1, \ldots, \alpha'_m\} \subset \mathscr{A}'$ such that $\beta' \in C_{\mathscr{L}'}(\{\alpha'_1, \ldots, \alpha'_m\})$. Let $\mathfrak{s}\colon V \to V'$ be a one-one substitution such that

(9) $\beta' = \mathfrak{s}\beta$ and $\alpha'_i = \mathfrak{s}\alpha_i$, $i = 1, \ldots, m$ for some β, α_i in $\mathscr{L}$.

By 11.1, $\mathfrak{s}C_{\mathscr{L}}(\{\alpha_1, \ldots, \alpha_m\}) = C_{\mathscr{L}'}(\{\mathfrak{s}\alpha_1, \ldots, \mathfrak{s}\alpha_m\}) \cap \mathfrak{s}F$, where F is the set of all formulas in $\mathscr{L}$. Hence $\mathfrak{s}\beta \in \mathfrak{s}C_{\mathscr{L}}(\{\alpha_1, \ldots, \alpha_m\})$, i.e.

(10) $$\beta \in C_{\mathscr{L}}(\{\alpha_1, \ldots, \alpha_m\}).$$

Let v be a valuation of $\mathscr{L}$ in $\mathfrak{A}$ defined as follows:

(11) $$v(p) = v'(\mathfrak{s}p) \quad \text{for all } p \in V.$$

Then, by 3.3, for every γ in F,

(12) $$\gamma_{\mathfrak{A}}(v) = \mathfrak{s}\gamma_{\mathfrak{A}}(v').$$

Hence,

(13) $$\beta_{\mathfrak{A}}(v) = \mathfrak{s}\beta_{\mathfrak{A}}(v') = \beta'_{\mathfrak{A}}(v')$$

and

(14) $$\alpha_{i\mathfrak{A}}(v) = \mathfrak{s}\alpha_{i\mathfrak{A}}(v') = \alpha'_{i\mathfrak{A}}(v') \in \nabla.$$

By (8), (10) and (14), $\beta_{\mathfrak{A}}(v) \in \nabla$. Thus, by (13), $\beta'_{\mathfrak{A}}(v') \in \nabla$. This completes the proof of the condition (f_3), and consequently, ∇ is a deductive filter.

It follows from 12.3 and 11.4 that if there exists in **L** a system $\mathscr{S}$ in which the only rule of inference is the following *rule of detachment* (modus ponens)

$$\text{(r)} \qquad \frac{\alpha, (\alpha \Rightarrow \beta)}{\beta} \quad \text{for all formulas } \alpha, \beta \text{ of } \mathscr{S},$$

then a subset ∇ of the set of all elements of an algebra in $\mathbf{K}_\mathbf{L}$ is a deductive filter if and only if the conditions (f_1) and (f_2) are satisfied. In that case the notion of a deductive filter coincides with the notion of an implicative filter (see II 1).

The simplest examples of deductive filters in any algebra $\mathfrak{A} \in \mathbf{K}_\mathbf{L}$ are the set $\{V\}$ and the set A of all elements in $\mathfrak{A}$. A deductive filter in $\mathfrak{A}$ is said to be *proper* if it is a proper subset of A.

Any non-empty class of deductive filters in an algebra $\mathfrak{A} \in \mathbf{K}_\mathbf{L}$ will be considered as an ordered set, the ordering relation being the set-theoretical inclusion. Thus by a *chain of deductive filters* we mean a non-empty class of deductive filters such that, for any ∇_1, ∇_2 in this class, either $\nabla_1 \subset \nabla_2$ or $\nabla_2 \subset \nabla_1$.

12.4. *The union of any chain of deductive filters in any algebra $\mathfrak{A} \in \mathbf{K}_\mathbf{L}$ is a deductive filter. The union of any chain of deductive filters which do not contain an element a_0 is a deductive filter which does not contain a_0.*

The proof, by an easy verification based on 12.3 and 11.4, is left to the reader.

The intersection of any non-empty class of deductive filters in an algebra $\mathfrak{A} \in \mathbf{K}_\mathbf{L}$ is a deductive filter. Hence, for any set A_0 of elements in $\mathfrak{A}$, there exists a least deductive filter containing A_0, viz. the intersection of all deductive filters containing A_0. This deductive filter is said to be the *deductive filter generated by* A_0, and A_0 is said to be the *set of generators* for this filter.

A proper deductive filter ∇ is said to be *irreducible* if for any two proper deductive filters ∇_1, ∇_2, the condition $\nabla = \nabla_1 \cap \nabla_2$ implies that either $\nabla = \nabla_1$ or $\nabla = \nabla_2$.

12.5. *If ∇_0 is a deductive filter in an algebra $\mathfrak{A} \in \mathbf{K}_\mathbf{L}$ such that $a_0 \notin \nabla_0$*

for some element a_0 *of* $\mathfrak{A}$, *then there exists an irreducible deductive filter* ∇ *such that* $\nabla_0 \subset \nabla$ *and* $a_0 \notin \nabla$.

The proof, based on 12.4 and I 3.1 and similar to that of II 1.4, is left to the reader.

A deductive filter in an algebra $\mathfrak{A} \in \mathbf{K_L}$ is said to be *maximal* provided it is proper and is not a proper subset of any proper deductive filter in $\mathfrak{A}$.

12.6. *Any maximal deductive filter is irreducible.*

This follows directly from the definions of maximal and irreducible deductive filters.

12.7. *If, in an algebra* $\mathfrak{A} \in \mathbf{K_L}$, *there exists an element* Λ *such that* $\Lambda \Rightarrow a = V$ *for every element* a *in* $\mathfrak{A}$, *then for every proper deductive filter* ∇ *there exists a maximal deductive filter* ∇^* *such that* $\nabla \subset \nabla^*$.

Let A be the set of all elements in $\mathfrak{A}$. By 6.2 and II 1.1, Λ is the least element in the ordered set $(A, \leqslant)$ where $\leqslant$ is defined by the equivalence $a \leqslant b$ if and only if $a \Rightarrow b = V$. Since ∇ is a proper deductive filter, by (7) it does not contain Λ. Consider the class of all proper deductive filters in $\mathfrak{A}$ containing ∇. No deductive filter in this class contains Λ. Thus the theorem follows from 12.4 and I 3.1.

13. The connection between L-theories and deductive filters [9]. Let $\mathbf{L}$ be a class of consistent, logically equivalent systems in $\mathbf{S}$ and let $\mathbf{K_L}$ be the class of all $\mathscr{S}$-algebras for the systems $\mathscr{S}$ in $\mathbf{L}$.

Consider an arbitrary consistent $\mathbf{L}$-theory $\mathscr{S}(\mathscr{A}) = (\mathscr{L}, C_{\mathscr{L}}, \mathscr{A})$ of zero order based on a system $\mathscr{S} = (\mathscr{L}, C_{\mathscr{L}})$ in $\mathbf{L}$. For any formula α of $\mathscr{L}$ let $||\alpha||_{\mathscr{A}}$ be the element of the algebra $\mathfrak{A}(\mathscr{S}(\mathscr{A}))$ of the theory $\mathscr{S}(\mathscr{A})$ determined by α (see Sec. 9). If $\mathscr{A} = O$, then instead of $||\alpha||_{\mathscr{A}}$ we shall write $||\alpha||$. For every set $\mathscr{A}'$ of formulas of $\mathscr{L}$ let $\Gamma_{\mathscr{A}'}$ be the set of all $||\alpha||_{\mathscr{A}}$ such that $\alpha \in \mathscr{A}'$ and let $\nabla_{\mathscr{A}'}$ be the set of all $||\alpha||_{\mathscr{A}}$ such that $\alpha \in C_{\mathscr{L}}(\mathscr{A} \cup \mathscr{A}')$. Thus we have

(1) $$||\alpha||_{\mathscr{A}} \in \nabla_{\mathscr{A}'} \quad \text{if and only if} \quad \alpha \in C_{\mathscr{L}}(\mathscr{A} \cup \mathscr{A}').$$

[9] The results of this section are a generalization of similar results by Tarski [5], [7] concerning filters in Boolean algebras and classical theories of zero order. Cf. [MM].

Under the above hypotheses and notation the following theorem holds:

13.1. *The set* $\nabla_{\mathscr{A}'}$ *is a deductive filter in* $\mathfrak{A}(\mathscr{S}(\mathscr{A}))$ *and* $\Gamma_{\mathscr{A}'}$ *is the set of generators for this deductive filter. The* **L**-*theory* $\mathscr{S}(\mathscr{A}\cup\mathscr{A}')$ *is consistent if and only if* $\nabla_{\mathscr{A}'}$ *is proper.*

It follows from 12.3 and 11.4 that in order to prove that $\nabla_{\mathscr{A}'}$ is a deductive filter it suffices to show that the conditions 12 $((f_1), (f_2))$ and the condition 12 (f_3^*) with respect to the rules of inference adopted in the system $\mathscr{S}$ in question are satisfied. If $||\alpha||_{\mathscr{A}} = V$, then by 9 (9), $\alpha \in C_{\mathscr{L}}(\mathscr{A}) \subset C_{\mathscr{L}}(\mathscr{A}\cup\mathscr{A}')$. Hence, by (1), $V \in \nabla_{\mathscr{A}'}$, i.e. the condition 12 (f_1) holds. Suppose that $||\alpha||_{\mathscr{A}} \in \nabla_{\mathscr{A}'}$ and $||\alpha||_{\mathscr{A}} \Rightarrow ||\beta||_{\mathscr{A}} = ||(\alpha\Rightarrow\beta)||_{\mathscr{A}} \in \nabla_{\mathscr{A}'}$. Hence, by (1), $\alpha \in C_{\mathscr{L}}(\mathscr{A}\cup\mathscr{A}')$ and $(\alpha\Rightarrow\beta) \in C_{\mathscr{L}}(\mathscr{A}\cup\mathscr{A}')$. Consequently, by 5 (s_4), $\beta \in C_{\mathscr{L}}(\mathscr{A}\cup\mathscr{A}')$, i.e. $||\beta||_{\mathscr{A}} \in \nabla_{\mathscr{A}'}$. Thus the condition 12 (f_2) is satisfied. Suppose that a rule of inference in $\mathscr{S}$ associates with the premises $\alpha_1, \ldots, \alpha_n$ the conclusion β and let v be a valuation in $\mathfrak{A}(\mathscr{S}(\mathscr{A}))$ such that

$$(2) \qquad \alpha_{i\mathfrak{A}(\mathscr{S}(\mathscr{A}))}(v) = V \quad \text{for} \quad i = 1, \ldots, n.$$

Let $\mathfrak{s}$ be a substitution in $\mathscr{L}$ such that $v(p) = ||\mathfrak{s}p||_{\mathscr{A}}$ for each propositional variable p of $\mathscr{L}$. It follows from (2) and 3.4 that $||\mathfrak{s}\alpha_i||_{\mathscr{A}} \in \nabla_{\mathscr{A}'}$ for every $i = 1, \ldots, n$. Hence, by (1), $\mathfrak{s}\alpha_i \in C_{\mathscr{L}}(\mathscr{A}\cup\mathscr{A}')$, $i = 1, \ldots, n$. In consequence, by 5 (s_2), $\mathfrak{s}\beta \in C_{\mathscr{L}}(\mathscr{A}\cup\mathscr{A}')$, i.e. $||\mathfrak{s}\beta||_{\mathscr{A}} \in \nabla_{\mathscr{A}'}$. Applying 3.4, we get $||\mathfrak{s}\beta||_{\mathscr{A}} = \beta_{\mathfrak{A}(\mathscr{S}(\mathscr{A}))}(v) \in \nabla_{\mathscr{A}'}$, which completes the proof that $\nabla_{\mathscr{A}'}$ is a deductive filter in the algebra $\mathfrak{A}(\mathscr{S}(\mathscr{A}))$.

The theory $\mathscr{S}(\mathscr{A}\cup\mathscr{A}')$ is consistent if and only if there exists a formula α of $\mathscr{L}$ such that $\alpha \notin C_{\mathscr{L}}(\mathscr{A}\cup\mathscr{A}')$, i.e. by (1) if and only if $\nabla_{\mathscr{A}'}$ is proper.

Clearly, $\Gamma_{\mathscr{A}'} \subset \nabla_{\mathscr{A}'}$. Suppose that ∇ is a deductive filter in $\mathfrak{A}(\mathscr{S}(\mathscr{A}))$ containing $\Gamma_{\mathscr{A}'}$. We shall prove that $\nabla_{\mathscr{A}'} \subset \nabla$. Suppose that $||\beta||_{\mathscr{A}} \in \nabla_{\mathscr{A}'}$. Hence $\beta \in C_{\mathscr{L}}(\mathscr{A}\cup\mathscr{A}')$. Let v^0 be the canonical valuation in $\mathfrak{A}(\mathscr{S}(\mathscr{A}))$. Then, by 3.4,

$$(3) \qquad ||\alpha||_{\mathscr{A}} = \alpha_{\mathfrak{A}(\mathscr{S}(\mathscr{A}))}(v^0) \quad \text{for every formula } \alpha \text{ of } \mathscr{L}.$$

If $\alpha \in \mathscr{A}$, then by 9.3, $\alpha_{\mathfrak{A}(\mathscr{S}(\mathscr{A}))}(v^0) = V \in \nabla$. If $\alpha \in \mathscr{A}'$, then $||\alpha||_{\mathscr{A}} \in \Gamma_{\mathscr{A}'}$. Hence, by (3), $\alpha_{\mathfrak{A}(\mathscr{S}(\mathscr{A}))}(v^0) \in \Gamma_{\mathscr{A}'} \subset \nabla$. Consequently, by 12 (f_3), $\beta_{\mathfrak{A}(\mathscr{S}(\mathscr{A}))}(v^0) \in \nabla$. Applying (3), we get $||\beta||_{\mathscr{A}} \in \nabla$. Thus $\nabla_{\mathscr{A}'} \subset \nabla$, which completes the proof that $\Gamma_{\mathscr{A}'}$ is the set of generators for the deductive filter $\nabla_{\mathscr{A}'}$.

The following theorem follows directly from 13.1.

13.2. *Let* $\mathscr{S}(\mathscr{A}') = (\mathscr{L}, C_{\mathscr{L}}, \mathscr{A}')$ *be an arbitrary* **L**-*theory. Then the set* $\nabla_{\mathscr{A}'}$ *of all elements* $||\alpha||$ *in the algebra* $\mathfrak{A}(\mathscr{S})$ *of the system* $\mathscr{S} = (\mathscr{L}, C_{\mathscr{L}})$ *such that* $\alpha \in C_{\mathscr{L}}(\mathscr{A}')$ *is a deductive filter in* $\mathfrak{A}(\mathscr{S})$ *generated by the set* $\Gamma_{\mathscr{A}'}$ *of all* $||\alpha||$ *such that* $\alpha \in \mathscr{A}'$. *The theory* $\mathscr{S}(\mathscr{A}')$ *is consistent if and only if* $\nabla_{\mathscr{A}'}$ *is proper.*

In fact, taking $\mathscr{A} = O$ in 13.1, we get 13.2.

13.3. *For every deductive filter* ∇ *in the algebra* $\mathfrak{A}(\mathscr{S}(\mathscr{A}))$ *of a consistent* **L**-*theory* $\mathscr{S}(\mathscr{A}) = (\mathscr{L}, C_{\mathscr{L}}, \mathscr{A})$ *there exists a set* $\mathscr{A}'$ *of formulas of* $\mathscr{L}$ *such that* $\nabla = \nabla_{\mathscr{A}'}$. *Moreover, for any set* A_0 *of generators of* ∇ *we can choose the set* $\mathscr{A}'$ *in such a way that* $\Gamma_{\mathscr{A}'} = A_0$.

For any element $a \in A_0$, let α_a be a formula of $\mathscr{L}$ such that $||\alpha_a|| = a$ and let $\mathscr{A}'$ be the set of all formulas α_a, $a \in A_0$. By definition, $\Gamma_{\mathscr{A}'} = A_0$. Consequently, $\nabla_{\mathscr{A}'} = \nabla$ by the first part of 13.1.

Taking $\mathscr{A} = O$ in 13.3, we get the following theorem:

13.4. *For any deductive filter* ∇ *in the algebra* $\mathfrak{A}(\mathscr{S})$ *of a system* $\mathscr{S} = (\mathscr{L}, C_{\mathscr{L}})$ *in* **L** *there exists an* **L**-*theory* $\mathscr{S}(\mathscr{A}') = (\mathscr{L}, C_{\mathscr{L}}, \mathscr{A}')$ *based on* $\mathscr{S}$ *with the set* $\mathscr{A}'$ *of specific axioms such that* $\nabla = \nabla_{\mathscr{A}'}$. *Moreover, for any set* A_0 *of generators for* ∇ *we can choose the set* $\mathscr{A}'$ *in such a way that* $\Gamma_{\mathscr{A}'} = A_0$.

An **L**-theory $\mathscr{S}(\mathscr{A}) = (\mathscr{L}, C_{\mathscr{L}}, \mathscr{A})$ will be said to be *irreducible* provided it is consistent and for any two consistent **L**-theories $\mathscr{S}(\mathscr{A}') = (\mathscr{L}, C_{\mathscr{L}}, \mathscr{A}')$, $\mathscr{S}(\mathscr{A}'') = (\mathscr{L}, C_{\mathscr{L}}, \mathscr{A}'')$ based on the same system $\mathscr{S} = (\mathscr{L}, C_{\mathscr{L}}) \in \mathbf{L}$ the condition $C_{\mathscr{L}}(\mathscr{A}) = C_{\mathscr{L}}(\mathscr{A}') \cap C_{\mathscr{L}}(\mathscr{A}'')$ implies that either $C_{\mathscr{L}}(\mathscr{A}) = C_{\mathscr{L}}(\mathscr{A}')$ or $C_{\mathscr{L}}(\mathscr{A}) = C_{\mathscr{L}}(\mathscr{A}'')$.

An **L**-theory $\mathscr{S}(\mathscr{A}) = (\mathscr{L}, C_{\mathscr{L}}, \mathscr{A})$ will be said to be *maximal* provided it is consistent and for each consistent **L**-theory $\mathscr{S}(\mathscr{A}') = (\mathscr{L}, C_{\mathscr{L}}, \mathscr{A}')$ based on the same system $\mathscr{S} = (\mathscr{L}, C_{\mathscr{L}}) \in \mathbf{L}$ the condition $C_{\mathscr{L}}(\mathscr{A}) \subset C_{\mathscr{L}}(\mathscr{A}')$ implies that $C_{\mathscr{L}}(\mathscr{A}) = C_{\mathscr{L}}(\mathscr{A}')$.

13.5. *An* **L**-*theory* $\mathscr{S}(\mathscr{A}) = (\mathscr{L}, C_{\mathscr{L}}, \mathscr{A})$ *is irreducible if and only if the deductive filter* $\nabla_{\mathscr{A}}$ *is irreducible.*

This theorem follows from the definition of irreducible theories and of irreducible deductive filters and from the natural one-one correspondence between deductive filters in $\mathfrak{A}(\mathscr{S})$ and **L**-theories based on the system $\mathscr{S}$ stated in 13.2, 13.4.

13.6. *An* **L**-*theory* $\mathscr{S}(\mathscr{A}) = (\mathscr{L}, C_{\mathscr{L}}, \mathscr{A})$ *is maximal if and only if the deductive filter* $\nabla_{\mathscr{A}}$ *in* $\mathfrak{A}(\mathscr{S})$ *is maximal.*

This theorem follows from the definition of maximal **L**-theories and of maximal deductive filters and from Theorems 13.2, 13.4.

13.7. *For every consistent* **L**-*theory* $\mathscr{S}(\mathscr{A}) = (\mathscr{L}, C_{\mathscr{L}}, \mathscr{A})$ *there exists an irreducible* **L**-*theory* $\mathscr{S}(\mathscr{A}') = (\mathscr{L}, C_{\mathscr{L}}, \mathscr{A}')$ *such that* $C_{\mathscr{L}}(\mathscr{A}) \subset C_{\mathscr{L}}(\mathscr{A}')$.

This theorem follows from 13.2, 12.5, 13.4, 13.5.

13.8. *If* **L** *is a class of consistent logically equivalent systems in* **S** *with semi-negation, then for every consistent* **L**-*theory* $\mathscr{S}(\mathscr{A}) = (\mathscr{L}, C_{\mathscr{L}}, \mathscr{A})$ *there exists a maximal* **L**-*theory* $\mathscr{S}(\mathscr{A}') = (\mathscr{L}, C_{\mathscr{L}}, \mathscr{A}')$ *such that* $C_{\mathscr{L}}(\mathscr{A}) \subset C_{\mathscr{L}}(\mathscr{A}')$.

This theorem follows from 13.2, 10.1, 12.7, 13.4, 13.6.

13.9. *Every maximal* **L**-*theory is irreducible.*

This follows from 13.6, 12.6, 13.5.

The next theorem states that every deductive filter in the algebra $\mathfrak{A}(\mathscr{S}(\mathscr{A}))$ of an **L**-theory is the kernel of an epimorphism.

13.10. *Let* ∇ *be a deductive filter in the algebra* $\mathfrak{A}(\mathscr{S}(\mathscr{A}))$ *of an* **L**-*theory* $\mathscr{S}(\mathscr{A}) = (\mathscr{L}, C_{\mathscr{L}}, \mathscr{A})$. *Then the relation* $\approx_{\nabla}$ *defined as*

$$(4) \qquad ||\alpha||_{\mathscr{A}} \approx_{\nabla} ||\beta||_{\mathscr{A}} \quad \textit{if and only if} \quad ||\alpha||_{\mathscr{A}} \Rightarrow ||\beta||_{\mathscr{A}} \in \nabla \textit{ and } ||\beta||_{\mathscr{A}} \Rightarrow ||\alpha||_{\mathscr{A}} \in \nabla$$

is a congruence in $\mathfrak{A}(\mathscr{S}(\mathscr{A}))$. *The quotient algebra which will be denoted by* $\mathfrak{A}(\mathscr{S}(\mathscr{A}))/\nabla$ *is an* $\mathscr{S}$-*algebra. The mapping defined by the equation*

$$(5) \qquad h||\alpha||_{\mathscr{A}} = |||\alpha||| \in \mathfrak{A}(\mathscr{S}(\mathscr{A}))/\nabla$$

is an epimorphism of $\mathfrak{A}(\mathscr{S}(\mathscr{A}))$ *onto* $\mathfrak{A}(\mathscr{S}(\mathscr{A}))/\nabla$ *and* ∇ *is the kernel of* h. *For each* $||\alpha||_{\mathscr{A}}$ *in* $\mathfrak{A}(\mathscr{S}(\mathscr{A}))$,

$$(6) \qquad ||\alpha||_{\mathscr{A}} \in \nabla \quad \textit{if and only if} \quad ||\alpha||_{\mathscr{A}} \approx_{\nabla} V.$$

$\mathfrak{A}(\mathscr{S}(\mathscr{A}))/\nabla$ *is degenerate if and only if* ∇ *is not proper.*

By 13.3 there exists a set $\mathscr{A}'$ of formulas of $\mathscr{L}$ such that $\nabla = \nabla_{\mathscr{A}'}$. Hence, by (1)

(7) $\quad ||\alpha||_{\mathscr{A}} \in \nabla_{\mathscr{A}'}$ if and only if $\alpha \in C_{\mathscr{L}}(\mathscr{A} \cup \mathscr{A}')$.

It follows from (7) and (4) that

(8) $\quad ||\alpha||_{\mathscr{A}} \approx_{\nabla} ||\beta||_{\mathscr{A}}$ if and only if

$$(\alpha \Rightarrow \beta), (\beta \Rightarrow \alpha) \in C_{\mathscr{L}}(\mathscr{A} \cup \mathscr{A}').$$

By 5 ((s_3), (s_5), (s_7), (s_8)) and (8), the relation $\approx_{\nabla}$ is a congruence in $\mathfrak{A}(\mathscr{S}(\mathscr{A}))$.

Let $\approx$ be the relation in the algebra $\mathfrak{F}$ of formulas of $\mathscr{L}$ defined as follows:

(9) $\quad \alpha \approx \beta$ if and only if $(\alpha \Rightarrow \beta), (\beta \Rightarrow \alpha) \in C_{\mathscr{L}}(\mathscr{A} \cup \mathscr{A}')$.

By (8) and (9) we get

(10) $\quad ||\alpha||_{\mathscr{A}} \approx_{\nabla} ||\beta||_{\mathscr{A}}$ if and only if $\alpha \approx \beta$.

Hence, by I 4.8 and Sec. 9 (p. 191), the algebra $\mathfrak{A}(\mathscr{S}(\mathscr{A} \cup \mathscr{A}'))$ of the L-theory $\mathscr{S}(\mathscr{A} \cup \mathscr{A}') = (\mathscr{L}, C_{\mathscr{L}}, \mathscr{A} \cup \mathscr{A}')$ is isomorphic to the quotient algebra $\mathfrak{A}(\mathscr{S}(\mathscr{A}))/\nabla$. Thus, by 9.2, $\mathfrak{A}(\mathscr{S}(\mathscr{A}))/\nabla$ is an $\mathscr{S}$-algebra. Clearly (see I 4 p. 12), the mapping h is an epimorphism of $\mathfrak{A}(\mathscr{S}(\mathscr{A}))$ onto $\mathfrak{A}(\mathscr{S}(\mathscr{A}))/\nabla$.

By 9 (9)

$$||\alpha||_{\mathscr{A}} \approx_{\nabla} V \text{ if and only if } ||\alpha||_{\mathscr{A}} \approx_{\nabla} ||\beta||_{\mathscr{A}}, \text{ where } \beta \in C_{\mathscr{L}}(\mathscr{A}).$$

Hence, by (8),

$$||\alpha||_{\mathscr{A}} \approx_{\nabla} V \text{ if and only ir } (\alpha \Rightarrow \beta), (\beta \Rightarrow \alpha) \in C_{\mathscr{L}}(\mathscr{A} \cup \mathscr{A}')$$

for a certain $\beta \in C_{\mathscr{L}}(\mathscr{A})$. Observe that by 5 ($(s_4)$, (s_6)) this condition is satisfied if and only if $\alpha \in C_{\mathscr{L}}(\mathscr{A} \cup \mathscr{A}')$. Hence, by (7), the condition (6) holds. Consequently, ∇ is the kernel of the epimorphism h. By (6), $\mathfrak{A}(\mathscr{S}(\mathscr{A}))/\nabla$ is degenerate if and only if ∇ is not proper.

13.11. *The algebra* $\mathfrak{A}(\mathscr{S}(\mathscr{A}'))$ *of an arbitrary* L-*theory* $\mathscr{S}(\mathscr{A}') = (\mathscr{L}, C_{\mathscr{L}}, \mathscr{A}')$ *is isomorphic to the quotient algebra* $\mathfrak{A}(\mathscr{S})/\nabla_{\mathscr{A}'}$.

In fact, it follows from the proof of 13.10 that the algebra $\mathfrak{A}(\mathscr{S}(\mathscr{A} \cup \mathscr{A}'))$ of an L-theory $\mathscr{S}(\mathscr{A} \cup \mathscr{A}') = (\mathscr{L}, C_{\mathscr{L}}, \mathscr{A} \cup \mathscr{A}')$ is isomorphic to the quotient algebra $\mathfrak{A}(\mathscr{S}(\mathscr{A}))/\nabla$, where $\nabla = \nabla_{\mathscr{A}'}$. Assuming $\mathscr{A} = O$, we get 13.11.

13.12. *For any algebra* $\mathfrak{A}$ *in* $\mathbf{K_L}$ *there exists an* **L**-*theory* $\mathscr{S}(\mathscr{A}) = (\mathscr{L}, C_{\mathscr{L}}, \mathscr{A})$ *based on a system* $\mathscr{S} = (\mathscr{L}, C_{\mathscr{L}})$ *in* **L** *such that* $\mathfrak{A}$ *is isomorphic to the algebra* $\mathfrak{A}(\mathscr{S}(\mathscr{A}))$ *of the theory* $\mathscr{S}(\mathscr{A})$.

Let A_0 be a set of generators for the algebra $\mathfrak{A}$ and let $\mathscr{S} = (\mathscr{L}, C_{\mathscr{L}})$ be a system in **L** such that the cardinal of the set V of all propositional variables of $\mathscr{L}$ is not less than the cardinal of A_0. The existence of such a system $\mathscr{S}$ in **L** follows from 8.2. Let f be a mapping from V onto A_0. By 6.6, there exists a homomorphism h from $\mathfrak{A}(\mathscr{S})$ in $\mathfrak{A}$ such that $h||p|| = f(p)$ for every p in V. Since A_0 generates the algebra $\mathfrak{A}$, h maps $\mathfrak{A}(\mathscr{S})$ onto $\mathfrak{A}$. By 12.1 and 12.2 the algebra $\mathfrak{A}$ is isomorphic to the quotient algebra $\mathfrak{A}(\mathscr{S})/\nabla$, where ∇ is the deductive filter in $\mathfrak{A}(\mathscr{S})$ of all $||\alpha||$ such that $h||\alpha|| = V_{\mathfrak{A}}$ (i.e. ∇ is the kernel of the epimorphism h). By 13.4 there exists an **L**-theory $\mathscr{S}(\mathscr{A}') = (\mathscr{L}, C_{\mathscr{L}}, \mathscr{A}')$ based on $\mathscr{S}$ such that $\nabla = \nabla_{\mathscr{A}'}$. Thus $\mathfrak{A}$ is isomorphic to $\mathfrak{A}(\mathscr{S})/\nabla_{\mathscr{A}'}$. Hence, by 13.11, $\mathfrak{A}$ is isomorphic to $\mathfrak{A}(\mathscr{S}(\mathscr{A}'))$. This completes the proof of 13.12.

Observe that by 13.10 every deductive filter in the algebra $\mathfrak{A}(\mathscr{S}(\mathscr{A}))$ of an **L**-theory $\mathscr{S}(\mathscr{A})$ is the kernel of the epimorphism determined by this deductive filter. Hence, by 13.12, every deductive filter in any algebra $\mathfrak{A} \in \mathbf{K_L}$ is also the kernel of an epimorphism and the theorem 13.10 can be generalized for arbitrary algebras in $\mathbf{K_L}$.

Exercises

1. Construct a propositional calculus $\mathscr{S} = (\mathscr{L}, C_{\mathscr{L}})$ belonging to the class **S** (see Sec. 5) and such that the class of all $\mathscr{S}$-algebras coincides with the class of all implicative algebras.

2. Prove that the notion of a deductive filter in implicative algebras coincides with the notion of a special filter as introduced in Ex. 1, Chap. II.

3. Let $\mathscr{S} = (\mathscr{L}, C_{\mathscr{L}})$ be a propositional calculus with implication as the only connective, characterized by the following logical axioms: $((\alpha \Rightarrow \beta) \Rightarrow ((\beta \Rightarrow \gamma) \Rightarrow (\alpha \Rightarrow \gamma)))$, $((\beta \Rightarrow \gamma) \Rightarrow ((\alpha \Rightarrow \beta) \Rightarrow (\alpha \Rightarrow \gamma)))$, $(\alpha \Rightarrow \alpha)$, for any formulas α, β, γ of $\mathscr{L}$. Assume that modus ponens and the rule of inference (r) $\dfrac{\alpha}{(\beta \Rightarrow \alpha)}$ are adopted in $\mathscr{S}$. Prove that $\mathscr{S}$ belongs to the class **S** (see Sec. 5).

4. Prove that the class of all $\mathscr{S}$-algebras for the propositional calculus $\mathscr{S}$ in Ex. 3, contains the class of all positive implication algebras as a proper subclass and is a proper subclass of the class of all implicative algebras.

5. Show that the class of all $\mathcal{S}$-algebras for any propositional calculus $\mathcal{S}$ in **S** is either equationally definable or definable by conditional equations, i.e. that the axioms have the following form: if E_1 and ... and E_n, $n = 0, 1, \ldots$, then E, where $E_1, \ldots, E_n$, E are equations.

6. Let **K** be the class of all algebras $(A, \vee, \Rightarrow, o_1, \ldots, o_t, o^1, \ldots, o^s, e_0, \ldots, e_{m-1})$, where t, s, m are any fixed positive integers, $(A, \vee, \Rightarrow)$ is an implicative algebra and a finite number of fixed conditional equations is satisfied. Is **K** the class of all $\mathcal{S}$-algebras for a propositional calculus $\mathcal{S}$ in the class **S** (see Sec. 5)?

CHAPTER IX

POSITIVE IMPLICATIVE LOGIC AND CLASSICAL IMPLICATIVE LOGIC

Introduction. This chapter deals with two examples of logics in the sense of the definition given in VIII 8. The propositional calculi of these logics are based on the simplest formalized languages, i.e. in their alphabets implication occurs as the only connective.

Positive implicative logic (Hilbert-Bernays [1]) — denoted here by $\mathbf{L}_{\pi\iota}$ — can be characterized by the following property (Horn [1]): exactly these formulas are derivable in the propositional calculi of $\mathbf{L}_{\pi\iota}$ which are derivable in propositional intuitionistic calculi. Thus, roughly speaking, positive implicative logic is the part of intuitionistic logic corresponding to formulas in which implication occurs as the only connective. However, it can be proved that the analogous statement for predicate calculi does not hold.

Positive implicative logic, more exactly a predicate calculus of that logic, was examined for the first time from the algebraic point of view by Henkin [2]. The class of all $\mathscr{S}$-algebras for $\mathscr{S}$ belonging to $\mathbf{L}_{\pi\iota}$ coincides with that of all positive implication algebras. Because of this the metalogical theorems of Chapter VIII can be formulated, in particular, for propositional calculi and formalized theories of positive implicative logic, in which the concepts of $\mathbf{L}_{\pi\iota}$-validity, $\mathbf{L}_{\pi\iota}$-models etc. are understood in a suitable way. Certain theorems of the theory of positive implication algebras give, in addition, other metalogical results. For instance, the representation theorem for those algebras enables us to formulate the completeness theorem for propositional calculi in $\mathbf{L}_{\pi\iota}$ and a theorem on $\mathbf{L}_{\pi\iota}$-models of $\mathbf{L}_{\pi\iota}$-theories in a stronger way (3.2, 4.2). Deductive filters coincide with implicative filters in positive implication algebras. Theorems II 3.4. and II 3.5

on the structure of implicative filters generated by an implicative filter and an element and generated by a subset of elements, respectively, are algebraic analogues of the deduction Theorems 5.5, 5.6 and are used for their proofs.

In order to bring out certain fine points about the various logics discussed in this book, the more important theorems will be formulated separately in each case.

Classical implicative logic, denoted here by $\mathbf{L}_{\varkappa\iota}$, constitutes the part of the classical logic corresponding to formulas in which implication occurs as the only connective [1]. More exactly, a formula in which no connective appears other than implication signs is derivable in a propositional calculus of $\mathbf{L}_{\varkappa\iota}$ if and only if it is derivable in classical propositional calculi (A. Tarski).

The class of all $\mathscr{S}$-algebras for $\mathscr{S}$ belonging to $\mathbf{L}_{\varkappa\iota}$ coincides with the class of all implication algebras. This enables us to formulate the metalogical theorems of Chapter VIII for propositional calculi in $\mathbf{L}_{\varkappa\iota}$ and for $\mathbf{L}_{\varkappa\iota}$-theories. Applying the representation theorem for implication algebras, we can prove a stronger form (8.4) of the completeness theorem, which states also that a formula is derivable in $\mathscr{S} \in \mathbf{L}_{\varkappa\iota}$ if and only if it is a propositional tautology. This leads to the relationship, mentioned above, between $\mathbf{L}_{\varkappa\iota}$ and classical logic. Among all $\mathbf{L}_{\varkappa\iota}$-models of $\mathbf{L}_{\varkappa\iota}$-theories we distinguish semantic models, i.e. models in the two-element implication algebra. A stronger form of the theorem on $\mathbf{L}_{\varkappa\iota}$-models of $\mathbf{L}_{\varkappa\iota}$-theories (9.2), which is also concerned with semantic models, is proved by making use of the representation theorem for implication algebras. The concept of a deductive filter coincides with that of an implicative filter in implication algebras. Several theorems on implicative filters in the algebras under consideration give additional metalogical results. For example, both the deduction theorems hold. The concept of an irreducible theory coincides with that of a maximal theory. Hence every consistent $\mathbf{L}_{\varkappa\iota}$-theory can be extended to a maximal theory. Moreover, we prove a characterization of maximal $\mathbf{L}_{\varkappa\iota}$-theories by means of semantic models (10.14).

[1] See Theorem 8.4. Propositional calculi of classical implicative logic have been studied by Tarski, Łukasiewicz and other authors.

Since disjunction is definable by means of implication, it is possible to introduce the concept of a prime $\mathbf{L}_{\varkappa\iota}$-theory (i.e. a consistent theory such that $(\alpha \cup \beta)$ is derivable if and only if at least one of the formulas α, β is derivable). This concept coincides with that of a maximal $\mathbf{L}_{\varkappa\iota}$-theory because of the equivalence of the corresponding notions of prime and maximal implicative filters in implication algebras.

1. Propositional calculus $\mathscr{S}_{\pi\iota}$ of positive implicative logic (2). Let $\mathscr{L} = (A^0, F)$ be a formalized language of zero order (see VIII 1) such that in the alphabet $A^0 = (V, L_0, L_1, L_2, U)$

(1) the set V is enumerable,

(2) the sets L_0, L_1 are empty,

(3) the set L_2 contains only one element denoted by $\Rightarrow$ and called the *implication sign*.

Let $\mathscr{A}_\iota$ be the set of all formulas in F of the form

(A$_1$) $$(\alpha \Rightarrow (\beta \Rightarrow \alpha)),$$

(A$_2$) $$((\alpha \Rightarrow (\beta \Rightarrow \gamma)) \Rightarrow ((\alpha \Rightarrow \beta) \Rightarrow (\alpha \Rightarrow \gamma))),$$

where α, β, γ are any formulas in F. Let (r) be the following rule of inference in $\mathscr{L}$, called *modus ponens*:

$$(r) \quad \frac{\alpha, (\alpha \Rightarrow \beta)}{\beta}, \qquad \text{for all formulas } \alpha, \beta \text{ in } F.$$

Consider the consequence operation $C_{\mathscr{L}}$ *determined by the set* $\mathscr{A}_\iota$ *of logical axioms and by the rule of inference* (r) (see VIII 4).

The system

(4) $$\mathscr{S}_{\pi\iota} = (\mathscr{L}, C_{\mathscr{L}})$$

will be said to be a *propositional calculus of positive implicative logic.*

In order to prove that the propositional calculus $\mathscr{S}_{\pi\iota}$ belongs to the class **S** of standard systems of implicative extensional propositional calculi (see VIII 5) we shall apply the following general theorem.

1.1. *Let* $\mathscr{S} = (\mathscr{L}', C_{\mathscr{L}'})$ *be an arbitrary propositional calculus, where the consequence operation* $C_{\mathscr{L}'}$ *is determined by a set* $\mathscr{A}'_\iota$ *of logical axioms*

(2) Hilbert and Bernays [1] vol. I. See also e.g. Henkin [2], Church [1].

and by a set $\{(r_1), \ldots, (r_k)\}$ *of rules of inference containing modus ponens. If for all formulas* α, β, γ *of* $\mathcal{L}'$ *the formulas*

(A$_1$) $(\alpha \Rightarrow (\beta \Rightarrow \alpha))$,

(A$_2$) $((\alpha \Rightarrow (\beta \Rightarrow \gamma)) \Rightarrow ((\alpha \Rightarrow \beta) \Rightarrow (\alpha \Rightarrow \gamma)))$

are in $\mathcal{A}'_l$, *then for any set* $\mathcal{A}$ *of formulas of* $\mathcal{L}'$ *and for all formulas* α, β, γ, δ *of* $\mathcal{L}'$ *the following conditions are satisfied:*

(i) $(\alpha \Rightarrow \alpha) \in C_{\mathcal{L}'}(O)$;

(ii) *if* $\alpha \in C_{\mathcal{L}'}(\mathcal{A})$, *then* $(\beta \Rightarrow \alpha) \in C_{\mathcal{L}'}(\mathcal{A})$;

(iii) *if* $(\alpha \Rightarrow \beta)$, $(\beta \Rightarrow \gamma) \in C_{\mathcal{L}'}(\mathcal{A})$, *then* $(\alpha \Rightarrow \gamma) \in C_{\mathcal{L}'}(\mathcal{A})$;

(iv) *if* $(\alpha \Rightarrow (\beta \Rightarrow \gamma)) \in C_{\mathcal{L}'}(\mathcal{A})$, *then* $(\beta \Rightarrow (\alpha \Rightarrow \gamma)) \in C_{\mathcal{L}'}(\mathcal{A})$;

(v) *if* $(\beta \Rightarrow \gamma) \in C_{\mathcal{L}'}(\mathcal{A})$, *then* $((\alpha \Rightarrow \beta) \Rightarrow (\alpha \Rightarrow \gamma)) \in C_{\mathcal{L}'}(\mathcal{A})$;

(vi) *if* $(\alpha \Rightarrow \beta) \in C_{\mathcal{L}'}(\mathcal{A})$, *then* $((\beta \Rightarrow \gamma) \Rightarrow (\alpha \Rightarrow \gamma)) \in C_{\mathcal{L}'}(\mathcal{A})$;

(vii) *if* $(\beta \Rightarrow \alpha)$, $(\gamma \Rightarrow \delta) \in C_{\mathcal{L}'}(\mathcal{A})$, *then*

$$((\alpha \Rightarrow \gamma) \Rightarrow (\beta \Rightarrow \delta)) \in C_{\mathcal{L}'}(\mathcal{A}).$$

The easy proof of 1.1 is left to the reader.

As a corollary of 1.1 the following theorem holds:

1.2. *If a propositional calculus* $\mathcal{S} = (\mathcal{L}', C_{\mathcal{L}'})$ *satisfies the hypotheses of* 1.1, *then for* $\mathcal{S}$ *the conditions* VIII 5 (s$_3$)-(s$_6$) *and the condition* VIII 5 (s$_8$) *with respect to the connective* $\Rightarrow$ *are satisfied.*

It is easy to verify by applying 1.2 that:

1.3. *The propositional calculus* $\mathcal{S}_{\pi\iota}$ *of the positive implicative logic belongs to the class* **S** *of standard systems of implicative extensional propositional calculi* (see VIII 5).

2. $\mathcal{S}_{\pi\iota}$-algebras. The aim of this section is to prove the following theorem.

2.1. *The class of all* $\mathcal{S}_{\pi\iota}$-*algebras coincides with the class of all positive implication algebras* (II 2).

Suppose that an abstract algebra $\mathfrak{A} = (A, V, \Rightarrow)$ is an $\mathcal{S}_{\pi\iota}$-algebra (see VIII 6). By 1 (A$_1$), 1 (A$_2$) and VIII 6 (a$_1$) for every valuation $v\colon V \to A$,

(1) $$(p \Rightarrow (q \Rightarrow p))_{\mathfrak{A}}(v) = V,$$

(2) $$((p \Rightarrow (q \Rightarrow r)) \Rightarrow ((p \Rightarrow q) \Rightarrow (p \Rightarrow r)))_{\mathfrak{A}}(v) = V,$$

where p, q, r denote any distinct propositional variables in V. Hence, by VIII 3 (3), for every valuation $v\colon V \to A$

$$v(p) \Rightarrow \big(v(q) \Rightarrow v(p)\big) = \vee, \tag{3}$$

$$\big(v(p) \Rightarrow \big(v(q) \Rightarrow v(r)\big)\big) \Rightarrow \big(\big(v(p) \Rightarrow v(q)\big) \Rightarrow \big(v(p) \Rightarrow v(r)\big)\big) = \vee. \tag{4}$$

Consequently, for all $a, b, c \in A$

$$a \Rightarrow (b \Rightarrow a) = \vee, \tag{5}$$

$$\big(a \Rightarrow (b \Rightarrow c)\big) \Rightarrow \big((a \Rightarrow b) \Rightarrow (a \Rightarrow c)\big) = \vee. \tag{6}$$

Thus the axioms II 2 (p_1), (p_2) hold in $\mathfrak{A}$. The axiom II 2 (p_3) holds too by VIII 6 (a_4). The axiom II 2 (p_4) is satisfied by 1.3, VIII 6.2 and II 1 (i_4). Consequently, $\mathfrak{A}$ is a positive implication algebra.

Conversely, assume that $\mathfrak{A} = (A, \vee, \Rightarrow)$ is a positive implication algebra. Then by II 2 $\big((p_1), (p_2)\big)$, for all a, b, c in A the equations (5), (6) are satisfied. Consequently, by VIII 3 (3), for all formulas α, β, γ of $\mathscr{S}_{\pi\iota}$ and for every valuation $v\colon V \to A$

$$(\alpha \Rightarrow (\beta \Rightarrow \alpha))_{\mathfrak{A}}(v) = \alpha_{\mathfrak{A}}(v) \Rightarrow \big(\beta_{\mathfrak{A}}(v) \Rightarrow \alpha_{\mathfrak{A}}(v)\big) = \vee,$$

$$\begin{aligned} ((\alpha \Rightarrow (\beta \Rightarrow \gamma)) \Rightarrow ((\alpha \Rightarrow \beta) \Rightarrow (\alpha \Rightarrow \gamma)))_{\mathfrak{A}}(v) \\ = \big(\alpha_{\mathfrak{A}}(v) \Rightarrow \big(\beta_{\mathfrak{A}}(v) \Rightarrow \gamma_{\mathfrak{A}}(v)\big)\big) \\ \Rightarrow \big(\big(\alpha_{\mathfrak{A}}(v) \Rightarrow \beta_{\mathfrak{A}}(v)\big) \Rightarrow \big(\alpha_{\mathfrak{A}}(v) \Rightarrow \gamma_{\mathfrak{A}}(v)\big)\big) = \vee. \end{aligned}$$

Thus the condition VIII 6 (a_1) holds.

Suppose that $\alpha_{\mathfrak{A}}(v) = \vee$ and $(\alpha \Rightarrow \beta)_{\mathfrak{A}}(v) = \alpha_{\mathfrak{A}}(v) \Rightarrow \beta_{\mathfrak{A}}(v) = \vee$ for some formulas α, β of $\mathscr{S}_{\pi\iota}$ and a valuation $v\colon V \to A$. Thus $\vee \Rightarrow \beta_{\mathfrak{A}}(v) = \vee$. Hence, by II 2 (1), $\beta_{\mathfrak{A}}(v) = \vee$. Thus the condition VIII 6 (a_2) with respect to the only rule of inference (r) of $\mathscr{S}_{\pi\iota}$ is satisfied. The condition VIII 6 (a_3) holds by II 2.1 and II 1 (i_2). Similarly, the condition VIII 6 (a_4) is satisfied by II 2 (p_3). Thus $\mathfrak{A}$ is an $\mathscr{S}_{\pi\iota}$-algebra.

The following theorem follows from 2.1 and VIII 6.1.

2.2. *If α is a formula derivable in $\mathscr{S}_{\pi\iota}$, then*

$$\alpha_{\mathfrak{A}}(v) = \vee \tag{7}$$

for every valuation $v\colon V \to A$ in every positive implication algebra $\mathfrak{A} = (A, \vee, \Rightarrow)$.

It follows from Theorem 2.2 that

2.3. *The propositional calculus* $\mathscr{S}_{\pi\iota}$ *is consistent.*

Indeed, for any propositional variable p of $\mathscr{L}$, p is not derivable in $\mathscr{S}_{\pi\iota}$.

3. Positive implicative logic $\mathbf{L}_{\pi\iota}$. By 1.3 and 2.3 the system $\mathscr{S}_{\pi\iota}$ described in Sec. 1 belongs to the class **S** of standard systems of implicative extensional propositional calculi (see VIII 5) and is consistent.

Let $\mathbf{L}_{\pi\iota}$ be the class of all consistent systems in **S** logically equivalent to $\mathscr{S}_{\pi\iota}$ (see VIII 8). The class $\mathbf{L}_{\pi\iota}$ will be called *positive implicative logic*.

By the definition of logically equivalent systems (VIII 8) and by 2.1, for every system $\mathscr{S}$ in $\mathbf{L}_{\pi\iota}$ the class of all $\mathscr{S}$-algebras coincides with the class of all positive implication algebras (II 2). Thus the class $\mathbf{K}_{\mathbf{L}_{\pi\iota}}$ of all $\mathscr{S}$-algebras for $\mathscr{S} \in \mathbf{L}_{\pi\iota}$ is the class of all positive implication algebras. By VIII 6.5, for any $\mathscr{S} \in \mathbf{L}_{\pi\iota}$, the algebra $\mathfrak{A}(\mathscr{S})$ is a non-degenerate positive implication algebra and by VIII 6.6 it is a free algebra in the class $\mathbf{K}_{\mathbf{L}_{\pi\iota}}$.

A formula α in any system $\mathscr{S} \in \mathbf{L}_{\pi\iota}$ is said to be $\mathbf{L}_{\pi\iota}$-*valid* (cf. VIII 8) provided that for every valuation v of the formalized language of $\mathscr{S}$ in any positive implication algebra $\mathfrak{A} = (A, \bigvee, \Rightarrow)$

(1) $$\alpha_{\mathfrak{A}}(v) = \bigvee .$$

The following theorem, called the *completeness theorem for systems in* $\mathbf{L}_{\pi\iota}$, follows directly from VIII 8.1.

3.1. *For any formula* α *in any system* $\mathscr{S} \in \mathbf{L}_{\pi\iota}$ *the following conditions are equivalent*:

(2) α *is derivable in* $\mathscr{S}$,

(3) α *is* $\mathbf{L}_{\pi\iota}$-*valid* [(3)].

Theorem 3.1 is included in:

3.2. *For any formula* α *in any system* $\mathscr{S} = (\mathscr{L}, C_{\mathscr{L}})$ *in* $\mathbf{L}_{\pi\iota}$ *the following conditions are equivalent*:

(4) α *is derivable in* $\mathscr{S}$,

(5) α *is* $\mathbf{L}_{\pi\iota}$-*valid*,

[(3)] For an analogous theorem concerning predicate calculi of positive implicative logic see Henkin [2].

(6) *α is valid in every positive implication algebra $\mathfrak{G}(X) = (G(X), X, \Rightarrow)$ of open subsets of any topological space X,*

(7) *$\alpha_{\mathfrak{A}(\mathscr{S})}(v^0) = V$ for the canonical valuation v^0 in the algebra $\mathfrak{A}(\mathscr{S})$ of the system $\mathscr{S}$.*

By 3.1 the conditions (4) and (5) are equivalent. Clearly, (5) implies (6). By the representation theorem II 4.1, for positive implication algebras and VIII 6.5, the condition (6) implies (7). By VIII 6 (8) and VIII 6 (15), the condition (7) is equivalent to (4). This completes the proof of 3.2.

3.3. *For every system $\mathscr{S} = (\mathscr{L}, C_{\mathscr{L}})$ of the positive implicative logic $\mathbf{L}_{\pi\iota}$ and for all formulas α, β, γ in $\mathscr{L}$ the following formulas are derivable in $\mathscr{S}$:*

(D_1) $(\alpha \Rightarrow (\beta \Rightarrow \alpha))$,

(D_2) $((\alpha \Rightarrow (\beta \Rightarrow \gamma)) \Rightarrow ((\alpha \Rightarrow \beta) \Rightarrow (\alpha \Rightarrow \gamma)))$,

(D_3) $(\alpha \Rightarrow \alpha)$,

(D_4) $(((\alpha \Rightarrow \beta) \Rightarrow (\alpha \Rightarrow \gamma)) \Rightarrow (\alpha \Rightarrow (\beta \Rightarrow \gamma)))$,

(D_5) $(\alpha \Rightarrow ((\alpha \Rightarrow \beta) \Rightarrow \beta))$,

(D_6) $((\alpha \Rightarrow (\alpha \Rightarrow \beta)) \Rightarrow (\alpha \Rightarrow \beta))$,

(D_7) $((\alpha \Rightarrow (\beta \Rightarrow \gamma)) \Rightarrow (\beta \Rightarrow (\alpha \Rightarrow \gamma)))$,

(D_8) $((\alpha \Rightarrow \beta) \Rightarrow ((\beta \Rightarrow \gamma) \Rightarrow (\alpha \Rightarrow \gamma)))$,

(D_9) $((\beta \Rightarrow \gamma) \Rightarrow ((\alpha \Rightarrow \beta) \Rightarrow (\alpha \Rightarrow \gamma)))$,

(D_{10}) $(((\alpha \Rightarrow \alpha) \Rightarrow \beta) \Rightarrow \beta)$,

(D_{11}) $(((\alpha \Rightarrow \beta) \Rightarrow ((\beta \Rightarrow \alpha) \Rightarrow \beta)) \Rightarrow ((\beta \Rightarrow \alpha) \Rightarrow ((\alpha \Rightarrow \beta) \Rightarrow \alpha)))$.

By 3.1 it suffices to prove that these formulas are $\mathbf{L}_{\pi\iota}$-valid. We shall indicate only the numbers of the statements in II from which their $\mathbf{L}_{\pi\iota}$-validity follows (the exact proof is left to the reader): (D_1) — II 2 (p_1); (D_2) — II 2 (p_2); (D_3) — II 2.1 (3); (D_4) — II 2.4 (18), II 2.1 (3); (D_5) — II 2.3 (9), II 2.2 (4); (D_6) — II 2.3 (13), II 2.1 (3); (D_7) — II 2.3 (14), II 2.1 (3); (D_8) — II 2.3 (15), II 2.2 (4); (D_9) — II 2.3 (16), II 2.2 (4); (D_{10}) — II 2.4 (17), II 2.1 (3); (D_{11}) — II 2.4 (19), II 2.1 (3).

3.4. *The following formula* (called the *Peirce law*) *is not derivable in $\mathscr{S} \in \mathbf{L}_{\pi\iota}$*, i.e. *is not $\mathbf{L}_{\pi\iota}$-valid:*

(8) $$(((p \Rightarrow q) \Rightarrow p) \Rightarrow p)$$

where p, q denote any distinct propositional variables of $\mathscr{S}$.

Let us take as a positive implication algebra the algebra $\mathfrak{G}(X) = \big(G(X), X, \Rightarrow\big)$ of all open subsets of the real line X. Let v be an arbitrary valuation of the language $\mathscr{L}$ of $\mathscr{S}$ in $\mathfrak{G}(X)$ such that

$$(9) \qquad v(p) = \{x \in X\colon\ x \neq 0\},$$

$$(10) \qquad v(q) = O.$$

Then by II 2 (6), $(p \Rightarrow q)_{\mathfrak{G}(X)}(v) = \mathrm{I}\big((X - v(p)) \cup v(q)\big) = O$. Hence, $((p \Rightarrow q) \Rightarrow p)_{\mathfrak{G}(X)}(v) = \mathrm{I}\big(X \cup v(p)\big) = X$. Consequently, $(((p \Rightarrow q) \Rightarrow p) \Rightarrow p)_{\mathfrak{G}(X)}(v) = \mathrm{I}v(p) = v(p) \neq X$. Thus the formula (8) is not $\mathbf{L}_{\pi\iota}$-valid, which completes the proof of 3.4.

4. $\mathbf{L}_{\pi\iota}$-theories of zero order. Let $\mathscr{S} = (\mathscr{L}, C_{\mathscr{L}})$ be a system in $\mathbf{L}_{\pi\iota}$. For every set $\mathscr{A}$ of formulas in $\mathscr{L}$ the system $\mathscr{S}(\mathscr{A}) = (\mathscr{L}, C_{\mathscr{L}}, \mathscr{A})$ is said to be an $\mathbf{L}_{\pi\iota}$*-theory of zero order based on* $\mathscr{S}$ (see VIII 9).

Following VIII 9 and Section 3 by an $\mathbf{L}_{\pi\iota}$*-model* of an $\mathbf{L}_{\pi\iota}$-theory $\mathscr{S}(\mathscr{A}) = (\mathscr{L}, C_{\mathscr{L}}, \mathscr{A})$ we mean any valuation v of $\mathscr{L}$ in any non-degenerate positive implication algebra $\mathfrak{A} = (A, V, \Rightarrow)$ such that for every α in $\mathscr{A}$

$$(1) \qquad \alpha_{\mathfrak{A}}(v) = V.$$

The following theorem follows directly from VIII 9.2 and VIII 9.3.

4.1. *For every consistent* $\mathbf{L}_{\pi\iota}$*-theory* $\mathscr{S}(\mathscr{A}) = (\mathscr{L}, C_{\mathscr{L}}, \mathscr{A})$ *of zero order the algebra* $\mathfrak{A}\big(\mathscr{S}(\mathscr{A})\big)$ *of the theory* $\mathscr{S}(\mathscr{A})$ (see VIII 9) *is a non-degenerate positive implication algebra. The canonical valuation* v^0 *of* $\mathscr{L}$ *in* $\mathfrak{A}\big(\mathscr{S}(\mathscr{A})\big)$, i.e. *defined as follows*

(2) $\quad v^0(p) = ||p||_{\mathscr{A}}$ *for every propositional variable p of* $\mathscr{L}$,

is an $\mathbf{L}_{\pi\iota}$*-model of* $\mathscr{S}(\mathscr{A})$. *Moreover, for any formula* α *in* $\mathscr{L}$,

(3) $\quad \alpha$ *is a theorem of* $\mathscr{S}(\mathscr{A})$ *if and only if* $\alpha_{\mathfrak{A}(\mathscr{S}(\mathscr{A}))}(v^0) = V$.

The next theorem characterizes all theorems of any consistent $\mathbf{L}_{\pi\iota}$-theory of zero order.

4.2. *For any formula* α *of any consistent* $\mathbf{L}_{\pi\iota}$*-theory* $\mathscr{S}(\mathscr{A}) = (\mathscr{L}, C_{\mathscr{L}}, \mathscr{A})$ *of zero order the following conditions are equivalent*:

(i) α *is a theorem of* $\mathscr{S}(\mathscr{A})$,

(ii) *every* $\mathbf{L}_{\pi\iota}$*-model* v *of* $\mathscr{S}(\mathscr{A})$ *in any non-degenerate positive implication algebra* $\mathfrak{A} = (A, V, \Rightarrow)$ *is an* $\mathbf{L}_{\pi\iota}$*-model of* α,

(iii) *every* $\mathbf{L}_{\pi\iota}$*-model* v *of* $\mathscr{S}(\mathscr{A})$ *in any non-degenerate positive implication algebra* $\mathfrak{G}(X) = (G(X), X, \Rightarrow)$ *of open subsets of any topological space* X, *is an* $\mathbf{L}_{\pi\iota}$*-model of* α,

(iv) $\alpha_{\mathfrak{A}(\mathscr{S}(\mathscr{A}))}(v^0) = V$ *for the canonical valuation* v^0 *in the algebra* $\mathfrak{A}(\mathscr{S}(\mathscr{A}))$ *of the theory* $\mathscr{S}(\mathscr{A})$.

By VIII 9.5 the conditions (i), (ii), (iv) are equivalent.

Clearly, (ii) implies (iii) and by the representation theorem II 4.1 the condition (iii) implies (ii), which completes the proof.

5. The connection between $\mathbf{L}_{\pi\iota}$-theories and implicative filters. Consider the propositional calculus $\mathscr{S}_{\pi\iota}$ of positive implicative logic described in Sec. 1. Let $\mathfrak{A} = (A, V, \Rightarrow)$ be an arbitrary positive implication algebra. Since the only rule of inference in $\mathscr{S}_{\pi\iota}$ is modus ponens, it follows from Sec. 3 and VIII 12.3 that a subset ∇ of A is a deductive filter in $\mathfrak{A}$ if and only if the following conditions are satisfied:

(f_1) $V \in \nabla$,

(f_2) if $a, a \Rightarrow b \in \nabla$, then $b \in \nabla$.

Hence, by II 1, p. 18, it follows that:

5.1. *The notion of a deductive filter in a positive implication algebra coincides with the notion of an implicative filter.*

By II 3.1, II 3.2, the kernels of epimorphisms from positive implication algebras onto similar algebras are implicative filters and, conversely, every implicative filter in a positive implication algebra is the kernel of the epimorphism determined by the filter. The above statements follow independently from Sec. 3, 5.1, VIII 12.2, VIII 13.10, VIII 13.12 and II 2.4, II 2.5.

Let $\mathscr{S} = (\mathscr{L}, C_{\mathscr{L}})$ be an arbitrary system in $\mathbf{L}_{\pi\iota}$. By Sec. 3 and VIII 6.5 the algebra $\mathfrak{A}(\mathscr{S})$ of the propositional calculus $\mathscr{S}$ is a non-degenerate positive implication algebra and by VIII 6.6, $\mathfrak{A}(\mathscr{S})$ is a free algebra in the class $\mathbf{K}_{\mathbf{L}_{\pi\iota}}$ of all positive implication algebras. For every set $\mathscr{A}$ of formulas in $\mathscr{L}$, let $\Gamma_{\mathscr{A}}$ be the set of all elements $||\gamma||$ in $\mathfrak{A}(\mathscr{S})$ such that $\gamma \in \mathscr{A}$ and let $\nabla_{\mathscr{A}}$ be the set of all $||\gamma||$ in $\mathfrak{A}(\mathscr{S})$ such that $\gamma \in C_{\mathscr{L}}(\mathscr{A})$. Thus we have

(1) $||\gamma|| \in \nabla_{\mathscr{A}}$ *if and only if* $\gamma \in C_{\mathscr{L}}(\mathscr{A})$, *for every* γ *in* $\mathscr{L}$.

It follows from 5.1 and VIII 13.2 that

5.2. *For every set* $\mathscr{A}$ *of formulas in the formalized language* $\mathscr{L}$ *of an arbitrary system* $\mathscr{S} = (\mathscr{L}, C_{\mathscr{L}})$ *in* $\mathbf{L}_{\pi\iota}$ *the set* $\nabla_{\mathscr{A}}$ *is an implicative filter in the algebra* $\mathfrak{A}(\mathscr{S})$ *of the system* $\mathscr{S}$ *and* $\Gamma_{\mathscr{A}}$ *is the set of generators of this implicative filter. The* $\mathbf{L}_{\pi\iota}$*-theory* $\mathscr{S}(\mathscr{A}) = (\mathscr{L}, C_{\mathscr{L}}, \mathscr{A})$ *is consistent if and only if the implicative filter* $\nabla_{\mathscr{A}}$ *is proper.*

The next theorem follows from 5.1 and VIII 13.4.

5.3. *For every implicative filter* ∇ *in the algebra* $\mathfrak{A}(\mathscr{S})$ *of a system* $\mathscr{S} \in \mathbf{L}_{\pi\iota}$ *there exists a set* $\mathscr{A}$ *of formulas of* $\mathscr{S}$ *such that* $\nabla = \nabla_{\mathscr{A}}$. *Moreover, for any set* A_0 *of generators of* ∇ *it is possible to choose the set* $\mathscr{A}$ *in such a way that* $\Gamma_{\mathscr{A}} = A_0$.

Theorems 5.2 and 5.3 state that there exists a natural one-one correspondence between implicative filters in the algebra $\mathfrak{A}(\mathscr{S})$ of a system $\mathscr{S} = (\mathscr{L}, C_{\mathscr{L}})$ in $\mathbf{L}_{\pi\iota}$ and formalized $\mathbf{L}_{\pi\iota}$-theories based on $\mathscr{S}$, two theories $\mathscr{S}(\mathscr{A}) = (\mathscr{L}, C_{\mathscr{L}}, \mathscr{A})$ and $\mathscr{S}(\mathscr{A}') = (\mathscr{L}, C_{\mathscr{L}}, \mathscr{A}')$ being identified if and only if $C_{\mathscr{L}}(\mathscr{A}) = C_{\mathscr{L}}(\mathscr{A}')$. These theorems explain also the algebraic interpretation of the set of axioms and the set of all theorems in an $\mathbf{L}_{\pi\iota}$-theory of zero order.

The following statement will be useful in the proof of the deduction theorem 5.5.

5.4. *For any formula* α *of the formalized language* $\mathscr{L}$ *of an arbitrary system* $\mathscr{S} = (\mathscr{L}, C_{\mathscr{L}})$ *in* $\mathbf{L}_{\pi\iota}$ *and for every set* $\mathscr{A}$ *of formulas in* $\mathscr{L}$, *the set* $\nabla_{\mathscr{A} \cup \{\alpha\}}$ *is the least implicative filter in the algebra* $\mathfrak{A}(\mathscr{S})$ *of the system* $\mathscr{S}$ *containing* $\nabla_{\mathscr{A}}$ *and* $||\alpha||$.

Let ∇ be the least implicative filter in $\mathfrak{A}(\mathscr{S})$ containing $\nabla_{\mathscr{A}}$ and $||\alpha||$. By 5.2, $\nabla_{\mathscr{A} \cup \{\alpha\}}$ is the implicative filter in $\mathfrak{A}(\mathscr{S})$ generated by $\Gamma_{\mathscr{A} \cup \{\alpha\}}$. Since $\nabla_{\mathscr{A}} \subset \nabla$, it follows from 5.2 that $\Gamma_{\mathscr{A}} \subset \nabla$. On the other hand, $||\alpha|| \in \nabla$. Hence, $\Gamma_{\mathscr{A} \cup \{\alpha\}} = \Gamma_{\mathscr{A}} \cup \{||\alpha||\} \subset \nabla$. Consequently, $\nabla_{\mathscr{A} \cup \{\alpha\}} \subset \nabla$. Since $\Gamma_{\mathscr{A}} \subset \Gamma_{\mathscr{A} \cup \{\alpha\}} \subset \nabla_{\mathscr{A} \cup \{\alpha\}}$, by 5.2 $\nabla_{\mathscr{A}} \subset \nabla_{\mathscr{A} \cup \{\alpha\}}$. Moreover, by (1), $||\alpha|| \in \nabla_{\mathscr{A} \cup \{\alpha\}}$. Hence $\nabla \subset \nabla_{\mathscr{A} \cup \{\alpha\}}$, which completes the proof of 5.4.

The following two theorems are said to be *deduction theorems.*

5.5. *For any formulas* α, β *in the formalized language* $\mathscr{L}$ *of an arbitrary system* $\mathscr{S} = (\mathscr{L}, C_{\mathscr{L}})$ *in* $\mathbf{L}_{\pi\iota}$ *and for every set* $\mathscr{A}$ *of formulas in* $\mathscr{L}$:

$$(2) \qquad \beta \in C_{\mathscr{L}}(\mathscr{A} \cup \{\alpha\}) \quad \textit{if and only if} \quad (\alpha \Rightarrow \beta) \in C_{\mathscr{L}}(\mathscr{A}).$$

If $(\alpha \Rightarrow \beta) \in C_{\mathscr{L}}(\mathscr{A})$, then $(\alpha \Rightarrow \beta) \in C_{\mathscr{L}}(\mathscr{A} \cup \{\alpha\})$. Hence, by modus ponens, $\beta \in C_{\mathscr{L}}(\mathscr{A} \cup \{\alpha\})$.

If $\beta \in C_{\mathscr{L}}(\mathscr{A} \cup \{\alpha\})$, then by (1), $||\beta|| \in \nabla_{\mathscr{A} \cup \{\alpha\}}$. Hence, by 5.4, $||\beta||$ belongs to the least implicative filter ∇ in $\mathfrak{A}(\mathscr{S})$ containing $\nabla_{\mathscr{A}}$ and $||\alpha||$. Consequently, by II 3.4, $||\alpha|| \Rightarrow ||\beta|| \in \nabla_{\mathscr{A}}$. Thus $||(\alpha \Rightarrow \beta)|| \in \nabla_{\mathscr{A}}$, which by (1) implies that $(\alpha \Rightarrow \beta) \in C_{\mathscr{L}}(\mathscr{A})$.

Observe that in the proof of 5.5 the essential fact used is Theorem II 3.4. Conversely, Theorem II 3.4 can be deduced from 5.5. Thus Theorem II 3.4 expresses the algebraic contents of the Deduction Theorem 5.5.

5.6. *For any formula* α *in the formalized language* $\mathscr{L}$ *of an arbitrary system* $\mathscr{S} = (\mathscr{L}, C_{\mathscr{L}})$ *in* $\mathbf{L}_{\pi\iota}$ *and for every non-empty set* $\mathscr{A}$ *of formulas in* $\mathscr{L}$, *the condition* $\alpha \in C_{\mathscr{L}}(\mathscr{A})$ *is equivalent to the condition that there exist formulas* $\alpha_1, \ldots, \alpha_n$ *in* $\mathscr{A}$ *such that*

$$(3) \qquad (\alpha_1 \Rightarrow (\ldots(\alpha_n \Rightarrow \alpha)\ldots)) \in C_{\mathscr{L}}(O).$$

Clearly, if (3) holds for some $\alpha_1, \ldots, \alpha_n \in \mathscr{A}$, then because $C_{\mathscr{L}}(O) \subset C_{\mathscr{L}}(\mathscr{A})$ and so, using modus ponens, $\alpha \in C_{\mathscr{L}}(\mathscr{A})$.

By (1), if $\alpha \in C_{\mathscr{L}}(\mathscr{A})$, then $||\alpha|| \in \nabla_{\mathscr{A}}$. Thus, by 5.2 and II 3.5, there exist $||\alpha_1||, \ldots, ||\alpha_n||$ in $\Gamma_{\mathscr{A}}$ such that

$$(4) \qquad ||\alpha_1|| \Rightarrow (\ldots(||\alpha_n|| \Rightarrow ||\alpha||)\ldots) = V.$$

Condition (4) implies (3) by VIII 6 (10) and VIII 6 (8).

In the proof of 5.6 an essential part is played by Theorem II 3.5. Conversely, Deduction Theorem 5.6 implies that II 3.5 holds. Thus II 3.5 can be treated as the algebraic analogue of 5.6.

By 5.1 the notion of an irreducible deductive filter (see VIII 12) in positive implication algebras coincides with the notion of an irreducible implicative filter (see II 3). Thus it follows from VIII 13.5 that:

5.7. *An* $\mathbf{L}_{\pi\iota}$-*theory* $\mathscr{S}(\mathscr{A}) = (\mathscr{L}, C_{\mathscr{L}}, \mathscr{A})$ *of zero order is irreducible if and only if the implicative filter* $\nabla_{\mathscr{A}}$ *in* $\mathfrak{A}(\mathscr{S})$ *is irreducible.*

Theorem 5.7 also follows independently from the definition of irreducible implicative filters in the positive implication algebras from the definition of irreducible $\mathbf{L}_{\pi\iota}$-theories and from Theorems 5.2, 5.3.

The next theorem is a particular case of VIII 13.7.

5.8. *For every consistent* $\mathbf{L}_{\pi\iota}$*-theory* $\mathscr{S}(\mathscr{A}) = (\mathscr{L}, C_{\mathscr{L}}, \mathscr{A})$ *of zero order there exists an irreducible* $\mathbf{L}_{\pi\iota}$*-theory* $\mathscr{S}(\mathscr{A}') = (\mathscr{L}, C_{\mathscr{L}}, \mathscr{A}')$ *such that* $C_{\mathscr{L}}(\mathscr{A}) \subset C_{\mathscr{L}}(\mathscr{A}')$.

Theorem 5.8 follows independently from 5.2, II 3.6, 5.3 and 5.7. In addition we shall quote the following theorem, which is a particular case of VIII 13.11.

5.9. *The algebra* $\mathfrak{A}(\mathscr{S}(\mathscr{A}))$ *of an arbitrary* $\mathbf{L}_{\pi\iota}$*-theory* $\mathscr{S}(\mathscr{A}) = (\mathscr{L}, C_{\mathscr{L}}, \mathscr{A})$ *is isomorphic to the quotient algebra* $\mathfrak{A}(\mathscr{S})/\nabla_{\mathscr{A}}$.

6. Propositional calculus $\mathscr{S}_{\varkappa\iota}$ of classical implicative logic. Consider the propositional calculus $\mathscr{S}_{\pi\iota}$ of the positive implicative logic described in Sec. 1. Let $\mathscr{L} = (A^0, F)$ be the formalized language of $\mathscr{S}_{\pi\iota}$ and let $C_{\mathscr{L}\pi\iota}$ denote in this section the consequence operation in the propositional calculus $\mathscr{S}_{\pi\iota}$. Thus $\mathscr{S}_{\pi\iota} = (\mathscr{L}, C_{\mathscr{L}\pi\iota})$.

We shall introduce a new consequence operation $C_{\mathscr{L}}$ in $\mathscr{L}$, taking as the set $\mathscr{A}_l$ of logical axioms the set of all formulas in F of the form [4]

(A$_1$) $(\alpha \Rightarrow (\beta \Rightarrow \alpha))$,

(A$_2$) $((\alpha \Rightarrow (\beta \Rightarrow \gamma)) \Rightarrow ((\alpha \Rightarrow \beta) \Rightarrow (\alpha \Rightarrow \gamma)))$,

(A$_3$) $(((\alpha \Rightarrow \beta) \Rightarrow \alpha) \Rightarrow \alpha)$,

where α, β, γ are any formulas in F, and adopting modus ponens as the only rule of inference.

The system

(1) $$\mathscr{S}_{\varkappa\iota} = (\mathscr{L}, C_{\mathscr{L}})$$

will be said to be a *propositional calculus of classical implicative logic.*

Since the set $\mathscr{A}_l$ of logical axioms of $\mathscr{S}_{\varkappa\iota}$ contains the set of logical axioms of $\mathscr{S}_{\pi\iota}$ (see Sec. 1) and in both systems the only rule of inference is modus ponens, it is easy to see that

(2) $$C_{\mathscr{L}\pi\iota}(\mathscr{A}) \subset C_{\mathscr{L}}(\mathscr{A}), \quad \text{for every set } \mathscr{A} \subset F.$$

[4] Tarski took instead of (A$_2$) the scheme $((\alpha \Rightarrow \beta) \Rightarrow ((\beta \Rightarrow \gamma) \Rightarrow (\alpha \Rightarrow \gamma)))$ and instead of (A$_3$) the scheme (A$_3'$) $(((\alpha \Rightarrow \beta) \Rightarrow \gamma) \Rightarrow ((\alpha \Rightarrow \gamma) \Rightarrow \gamma))$. Bernays simplified Tarski's system of axioms replacing (A$_3'$) by (A$_3$). See Łukasiewicz [3].

In particular,

(3) $$C_{\mathscr{L}_{\pi\iota}}(O) \subset C_{\mathscr{L}}(O),$$

i.e. each formula derivable in $\mathscr{S}_{\pi\iota}$ is also derivable in $\mathscr{S}_{\varkappa\iota}$.

By 3.4 and (A_3), $C_{\mathscr{L}_{\pi\iota}}(O) \neq C_{\mathscr{L}}(O)$.

Let us note that by the definition of the set $\mathscr{A}_\iota$ of logical axioms of $\mathscr{S}_{\varkappa\iota}$ and by VIII 2.4 (9):

6.1. *The set* $\mathscr{A}_\iota$ *of logical axioms of* $\mathscr{S}_{\varkappa\iota}$ *is closed under substitutions,* i.e. *condition* VIII 5 (s_1) *is satisfied.*

Similarly, by the definition of modus ponens and VIII 2.4 (9):

6.2. *The only rule of inference of* $\mathscr{S}_{\varkappa\iota}$ *is invariant with respect to substitutions,* i.e. *condition* VIII 5 (s_2) *is fulfilled.*

The main theorem of this section is the following one.

6.3. *The system* $\mathscr{S}_{\varkappa\iota}$ *belongs to the class* **S** (see VIII 5) *of standard systems of implicative extensional propositional calculi.*

Indeed, by 6.1, 6.2 and 1.2, conditions VIII 5 (s_1)–(s_8) for $\mathscr{S}_{\varkappa\iota}$ are satisfied.

7. $\mathscr{S}_{\varkappa\iota}$-algebras. The aim of this section is to prove the following theorem.

7.1. *The class of all* $\mathscr{S}_{\varkappa\iota}$*-algebras coincides with the class of all implication algebras.*

Suppose that $\mathfrak{A} = (\mathscr{A}, V, \Rightarrow)$ is an $\mathscr{S}_{\varkappa\iota}$-algebra (see VIII 6). Analogously as in the proof of 2.1, it follows from 6 (A_1), 6 (A_2), 6 (A_3) and VIII 6 (a_1) that

(1) $$a \Rightarrow (b \Rightarrow a) = V,$$

(2) $$(a \Rightarrow (b \Rightarrow c)) \Rightarrow ((a \Rightarrow b) \Rightarrow (a \Rightarrow c)) = V,$$

(3) $$((a \Rightarrow b) \Rightarrow a) \Rightarrow a = V,$$

for all a, b, c in A.

By VIII 6 (a_4),

(4) for all $a, b \in A$, if $a \Rightarrow b = V$ and $b \Rightarrow a = V$, then $a = b$.

By 6.3, VIII 6.2 and II 1 (i_4),

(5) $$a \Rightarrow V = V \quad \text{for each } a \in A.$$

By (1), (2), (4), (5) and II 2, $\mathfrak{A}$ is a positive implication algebra.

It follows from (1) that $a \Rightarrow ((a \Rightarrow b) \Rightarrow a) = V$ for all $a, b \in A$. Hence, by (3) and (4), we get

(6) $$(a \Rightarrow b) \Rightarrow a = a, \quad \text{for all } a, b \in A.$$

Since $\mathfrak{A}$ is a positive implication algebra in which (6) holds, it is an implication algebra (see II 5).

Conversely, let us suppose that $\mathfrak{A} = (A, V, \Rightarrow)$ is an implication algebra. Thus for all a, b, c in A the conditions (1), (2), (4), (5), (6) are satisfied, i.e. $\mathfrak{A}$ is a positive implication algebra in which the condition (6) holds. Hence, by (6) and II 2 (3), the condition (3) is also satisfied. It follows from (1), (2), (3) that, for every valuation v of $\mathcal{L}$ (i.e. of the formalized language of $\mathcal{S}_{\varkappa\iota}$) in $\mathfrak{A}$ for every logical axiom α of $\mathcal{S}_{\varkappa\iota}$, we have $\alpha_{\mathfrak{A}}(v) = V$. Thus condition VIII 6 (a_1) holds. Condition VIII 6 (a_2) is satisfied by II 2 (1). Condition VIII 6 (a_3) follows by II 2.1 and II 1 (i_2). Condition VIII 6 (a_4) is satisfied by (4). Thus $\mathfrak{A}$ is an $\mathcal{S}_{\varkappa\iota}$-algebra.

The following theorem follows from 7.1 and VIII 6.1.

7.2. *If α is a formula derivable in $\mathcal{S}_{\varkappa\iota}$, then*

(7) $$\alpha_{\mathfrak{A}}(v) = V$$

for every valuation v of the formalized language of $\mathcal{S}_{\varkappa\iota}$ in any implication algebra $\mathfrak{A} = (\mathcal{A}, V, \Rightarrow)$.

It follows from the above theorem that:

7.3. *The propositional calculus $\mathcal{S}_{\varkappa\iota}$ is consistent.*

In fact, no propositional variable is derivable in $\mathcal{S}_{\varkappa\iota}$.

8. Classical implicative logic $\mathbf{L}_{\varkappa\iota}$. By 6.3 and 7.3 the system $\mathcal{S}_{\varkappa\iota}$ described in Sec. 6 belongs to the class **S** of standard systems of implicative extensional propositional calculi (see VIII 5) and is consistent.

Let $\mathbf{L}_{\varkappa\iota}$ be the class of all consistent systems in **S** logically equivalent to $\mathcal{S}_{\varkappa\iota}$ (see VIII 8). The class $\mathbf{L}_{\varkappa\iota}$ will be said to be *classical implicative logic*.

By the definition of logically equivalent systems (VIII 8) and by 7.1, for every system $\mathcal{S}$ in $\mathcal{L}_{\varkappa\iota}$ the class of all $\mathcal{S}$-algebras coincides with the class of all implication algebras (II 5). Thus the class $\mathbf{K}_{\mathbf{L}_{\varkappa\iota}}$ of all $\mathscr{S}$-algebras for $\mathscr{S} \in \mathbf{L}_{\varkappa\iota}$ is the class of all implication algebras.

Hence, by VIII 6.5 and VIII 6.6 respectively, we obtain the following two theorems:

8.1. *For every $\mathscr{S} \in \mathbf{L}_{\varkappa\iota}$, the algebra $\mathfrak{A}(\mathscr{S})$ of the propositional calculus $\mathscr{S}$ is a non-degenerate implication algebra.*

8.2. *For every $\mathscr{S} \in \mathbf{L}_{\varkappa\iota}$, the algebra $\mathfrak{A}(\mathscr{S})$ of the propositional calculus $\mathscr{S}$ is a free algebra in the class $\mathbf{K}_{\mathbf{L}_{\varkappa\iota}}$ of all implication algebras.*

As in VIII 8 a formula α in any system $\mathscr{S} \in \mathbf{L}_{\varkappa\iota}$ is said to be $\mathbf{L}_{\varkappa\iota}$-*valid* if for every valuation v of the formalized language of $\mathscr{S}$ in any implication algebra $\mathfrak{A} = (A, \vee, \Rightarrow)$

(1) $$\alpha_{\mathfrak{A}}(v) = \vee.$$

A formula α of any system $\mathscr{S} \in \mathbf{L}_{\varkappa\iota}$ is said to be a *propositional tautology* if for every valuation v of the formalized language of $\mathscr{S}$ in the two-element implication algebra $\mathfrak{A}_0 = (\{\wedge, \vee\}, \vee, \Rightarrow)$ (see II 5, p. 30) condition (1) holds.

The following *completeness theorem* follows directly from VIII 8.1.

8.3. *For any formula α in any system $\mathscr{S} \in \mathbf{L}_{\varkappa\iota}$ the following conditions are equivalent:*

(2) *α is derivable in $\mathscr{S}$,*

(3) *α is $\mathbf{L}_{\varkappa\iota}$-valid.*

The above theorem is included in:

8.4 [5]. *For any formula α in any system $\mathscr{S} = (\mathscr{L}, C_{\mathscr{L}})$ in $\mathbf{L}_{\varkappa\iota}$ the following conditions are equivalent:*

(4) *α is derivable in $\mathscr{S}$,*

(5) *α is $\mathbf{L}_{\varkappa\iota}$-valid,*

(6) *α is valid in every implication algebra $\mathfrak{A}(X) = (A(X), X, \Rightarrow)$ of subsets of any space $X \neq O$,*

[5] The equivalence of conditions (4) and (8) has been proved by Tarski (see Łukasiewicz [3]) for the propositional calculus based on his system of axioms (see footnote (4)).

(7) $\alpha_{\mathfrak{A}(\mathscr{S})}(v^0) = \vee$ *for the canonical valuation* v^0 *in the algebra* $\mathfrak{A}(\mathscr{S})$ *of the system* $\mathscr{S}$,

(8) α *is a propositional tautology.*

By 8.3, (4) implies (5). Clearly, (5) implies (6). By the Representation Theorem II 7.1, condition (6) implies (7). By VIII 7.2, condition (7) implies (4). Thus (4), (5), (6), (7) are equivalent. Clearly, (5) implies (8). We shall prove that (8) implies (4). Suppose that α is not derivable in $\mathscr{S} \in \mathbf{L}_{\varkappa\iota}$. Hence, by VIII 6 (8), in the algebra $\mathfrak{A}(\mathscr{S})$ of the system $\mathscr{S}$

(9) $$||\alpha|| \neq \vee.$$

By II 7.1 there exists a monomorphism h from $\mathfrak{A}(\mathscr{S})$ into an implication algebra $\mathfrak{A}(X) = (A(X), X, \Rightarrow)$ of all subsets of a space $X \neq O$. Thus $h(\vee) = X$ and for any $||\beta||$, $||\gamma||$ in $\mathfrak{A}(\mathscr{S})$:

(10) $$h(||\beta|| \Rightarrow ||\gamma||) = h(||\beta||) \Rightarrow h(||\gamma||) = \big(X - h(||\beta||)\big) \cup h(||\gamma||).$$

It follows from (9), since h is a monomorphism, that

(11) $$h(||\alpha||) \neq X.$$

Thus there exists an element $x \in X$ such that

(12) $$x \notin h(||\alpha||).$$

Let us put for any $||\beta||$ in $\mathfrak{A}(\mathscr{S})$

(13) $$h_0(||\beta||) = h(||\beta||) \cap \{x\}.$$

The mapping h_0 defined by (13) is a homomorphism from $\mathfrak{A}(\mathscr{S})$ onto the two-element implication algebra $\mathfrak{A}^* = (\{O, \{x\}\}, \{x\}, \Rightarrow)$ of all subsets of the set $\{x\}$. In fact, we have $h_0(\vee) = h(\vee) \cap \{x\} = X \cap \{x\} = \{x\}$ and by (10), (13), II 5,

$$h_0(||\beta|| \Rightarrow ||\gamma||) = \big(\{x\} - h_0(||\beta||)\big) \cup h_0(||\gamma||)$$
$$= h_0(||\beta||) \Rightarrow h_0(||\gamma||).$$

It is easy to verify that the implication algebra $\mathfrak{A}^*$ is isomorphic to the two-element implication algebra $\mathfrak{A}_0 = (\{\wedge, \vee\}, \vee, \Rightarrow)$ (see II 5), the required isomorphism g being defined as follows:

(14) $$g(\{x\}) = \vee, \quad g(O) = \wedge.$$

Let v^0 be the canonical valuation of $\mathscr{L}$ in the algebra $\mathfrak{A}(\mathscr{S})$, i.e

(15) $v^0(p) = ||p||$ for every propositional variable p of $\mathscr{L}$.

Let v be the valuation of $\mathscr{L}$ in $\mathfrak{A}_0 = (\{\Lambda, V\}, V, \Rightarrow)$ defined as follows:

(16)
$$v(p) = g\big(h_0(v^0(p))\big).$$

Thus (see VIII 3, p. 175)

(17)
$$v = gh_0 v^0.$$

By (17), VIII 3.2 and VIII 6 (15) we get for every formula β in $\mathscr{L}$

(18) $\beta_{\mathfrak{A}_0}(v) = \beta_{\mathfrak{A}_0}(gh_0 v^0) = g\beta_{\mathscr{A}^*}(h_0 v^0) = gh_0 \beta_{\mathfrak{A}(\mathscr{S})}(v^0) = gh_0(||\beta||)$.

It follows from (18), (13), (12) and (14) that

(19) $\alpha_{\mathfrak{A}_0}(v) = gh_0(||\alpha||) = g\big(\{x\} \cap h(||\alpha||)\big) = g(O) = \Lambda$.

By (19), α is not a propositional tautology. This completes the proof that (8) implies (4) and hence the proof of 8.3.

From the equivalence of the conditions (4), (8) it follows that

8.5. *Every propositional calculus $\mathscr{S}$ of the classical implicative logic is decidable*, i.e. *there exists a method which, for every formula α of $\mathscr{S}$, enables us to verify in a finite number of steps whether it is derivable in $\mathscr{S}$ or is not derivable.*

Let us note that

8.6. *For every system $\mathscr{S} = (\mathscr{L}, C_{\mathscr{L}})$ of the classical implicative logic* $\mathbf{L}_{\varkappa\iota}$ *and for all formulas α, β, γ in $\mathscr{L}$ the formulas of the form* (D_1)-(D_{11}) *in Sec.* 3 *and, moreover, the following ones are derivable in $\mathscr{S}$*:

(D_{12}) $(((\alpha \Rightarrow \beta) \Rightarrow \alpha) \Rightarrow \alpha)$,

(D_{13}) $(((\alpha \Rightarrow \beta) \Rightarrow \beta) \Rightarrow ((\beta \Rightarrow \alpha) \Rightarrow \alpha))$,

(D_{14}) $((\alpha \Rightarrow \gamma) \Rightarrow ((\beta \Rightarrow \gamma) \Rightarrow (((\alpha \Rightarrow \beta) \Rightarrow \beta) \Rightarrow \gamma)))$,

(D_{15}) $(((\alpha \Rightarrow \beta) \Rightarrow (\beta \Rightarrow \alpha)) \Rightarrow (\beta \Rightarrow \alpha))$.

9. $\mathbf{L}_{\varkappa\iota}$-theories of zero order. Let $\mathscr{S} = (\mathscr{L}, C_{\mathscr{L}})$ be a propositional calculus of classical implicative logic $\mathbf{L}_{\varkappa\iota}$. For every set $\mathscr{A}$ of formulas in $\mathscr{L}$ the system $\mathscr{S}(\mathscr{A}) = (\mathscr{L}, C_{\mathscr{L}}, \mathscr{A})$ is said to be an $\mathbf{L}_{\varkappa\iota}$-*theory of zero order based on* $\mathscr{S}$ (see VIII 9). The set $C_{\mathscr{L}}(\mathscr{A})$, i.e. the set of all formulas in $\mathscr{L}$ for which there exists a formal proof in $\mathscr{S}$ from $\mathscr{A}$, is said to be the *set of theorems of* $\mathscr{S}(\mathscr{A})$.

By an *$\mathbf{L}_{\varkappa\iota}$-model of an $\mathbf{L}_{\varkappa\iota}$-theory* $\mathscr{S}(\mathscr{A}) = (\mathscr{L}, C_{\mathscr{L}}, \mathscr{A})$ of zero order we mean any valuation v of $\mathscr{L}$ in any non-degenerate implication algebra $\mathfrak{A} = (A, V, \Rightarrow)$ such that

(1) $$\alpha_{\mathfrak{A}}(v) = V \quad \text{for every } \alpha \text{ in } \mathscr{A}.$$

Any $\mathbf{L}_{\varkappa\iota}$-model of an $\mathbf{L}_{\varkappa\iota}$-theory in the two-element implication algebra $\mathfrak{A}_0 = (\{\Lambda, V\}, V, \Rightarrow)$ is said to be a *semantic model* of that theory.

The following theorem follows directly from VIII 9.2, VIII 9.3 and from Sec. 8.

9.1. *For every consistent $\mathbf{L}_{\varkappa\iota}$-theory $\mathscr{S}(\mathscr{A}) = (\mathscr{L}, C_{\mathscr{L}}, \mathscr{A})$ of zero order the algebra $\mathfrak{A}(\mathscr{S}(\mathscr{A}))$ of $\mathscr{S}(\mathscr{A})$* (see VIII 9) *is a non-degenerate implication algebra. The canonical valuation v^0 of $\mathscr{L}$ in $\mathfrak{A}(\mathscr{S}(\mathscr{A}))$ defined as follows:*

(2) $v^0(p) = ||p||_{\mathscr{A}}$, *for every propositional variable p of $\mathscr{L}$,*

is an $\mathbf{L}_{\varkappa\iota}$-model of $\mathscr{S}(\mathscr{A})$. Moreover, for any formula α in $\mathscr{L}$

(3) *α is a theorem of $\mathscr{S}(\mathscr{A})$ if and only if $\alpha_{\mathfrak{A}(\mathscr{S}(\mathscr{A}))}(v^0) = V$.*

The next theorem characterizes all theorems of any consistent $\mathbf{L}_{\varkappa}$-theory of zero order.

9.2. *For any formula α of any consistent $\mathbf{L}_{\varkappa\iota}$-theory of zero order the following conditions are equivalent:*

(i) *α is a theorem of $\mathscr{S}(\mathscr{A})$,*

(ii) *every $\mathbf{L}_{\varkappa\iota}$-model v of $\mathscr{S}(\mathscr{A})$ is an $\mathbf{L}_{\varkappa\iota}$-model of α,*

(iii) *every $\mathbf{L}_{\varkappa\iota}$-model v of $\mathscr{S}(\mathscr{A})$ in any non-degenerate implication algebra $\mathfrak{A}(X) = (A(X), X, \Rightarrow)$ of subsets of any set $X \neq O$ is an $\mathbf{L}_{\varkappa\iota}$-model of α,*

(iv) *$\alpha_{\mathfrak{A}(\mathscr{S}(\mathscr{A}))}(v^0) = V$ for the canonical valuation v^0 in the algebra $\mathfrak{A}(\mathscr{S}(\mathscr{A}))$ of $\mathscr{S}(\mathscr{A})$,*

(v) *every semantic model v of $\mathscr{S}(\mathscr{A})$ is a semantic model of α.*

By VIII 9.5 conditions (i), (ii), (iv) are equivalent. Clearly, (ii) implies (iii) and by the Representation Theorem II 7.1, (iii) implies (ii). Thus conditions (i), (ii), (iii) and (iv) are equivalent. Obviously, (ii) implies (v). To complete the proof of 9.2 it suffices to show that (v) implies (i). The proof can be carried out analogously to the proof of the statement that condition 8 (8) implies condition 8 (4) (see the proof of 8.4).

10. The connection between $\mathbf{L}_{\varkappa\iota}$-theories of zero order and implicative filters. Consider the propositional calculus $\mathscr{S}_{\varkappa\iota}$ of classical implicative logic $\mathbf{L}_{\varkappa\iota}$, described in Section 6. Let $\mathfrak{A} = (A, V, \Rightarrow)$ be an arbitrary implication algebra. Since the only rule of inference in $\mathscr{S}_{\varkappa\iota}$ is modus ponens, it follows from Section 8 and VIII 12.3 that a subset ∇ of A is a deductive filter in $\mathfrak{A}$ if and only if the following conditions are satisfied:

$$(f_1) \qquad V \in \nabla,$$

$$(f_2) \qquad \text{if} \quad a, a \Rightarrow b \in \nabla, \quad \text{then} \quad b \in \nabla.$$

Hence, by the definition of implicative filters (see II 1), it follows that:

10.1. *The notion of a deductive filter in implication algebras coincides with the notion of an implicative filter.*

By II 6, II 3.1 and II 6.3, the kernels of epimorphisms from implication algebras onto similar algebras are implicative filters and, conversely, every implicative filter in any implication algebra is the kernel of the epimorphism determined by that implicative filter. The above statements follow independently from Section 8, 10.1, VIII 12.2, VIII 13.10, VIII 13.12 and II 5.1.

Let $\mathscr{S} = (\mathscr{L}, C_{\mathscr{L}})$ be an arbitrary system in $\mathbf{L}_{\varkappa\iota}$. By 8.1 and 8.2 the algebra $\mathfrak{A}(\mathscr{S})$ is a non-degenerate implication algebra, free in the class $\mathbf{K}_{\mathbf{L}_{\varkappa\iota}}$ of all implication algebras. For every set $\mathscr{A}$ of formulas in $\mathscr{L}$, let $\Gamma_{\mathscr{A}}$ be the set of all elements $||\gamma||$ in $\mathfrak{A}(\mathscr{S})$ such that $\gamma \in \mathscr{A}$ and let $\nabla_{\mathscr{A}}$ be the set of all $||\gamma||$ in $\mathfrak{A}(\mathscr{S})$ such that $\gamma \in C_{\mathscr{L}}(\mathscr{A})$. Thus we have for every formula γ in $\mathscr{L}$

$$(1) \qquad ||\gamma|| \in \nabla_{\mathscr{A}} \quad \text{if and only if} \quad \gamma \in C_{\mathscr{L}}(\mathscr{A}).$$

It follows from 10.1 and VIII 3.2 that:

10.2. *For every set $\mathscr{A}$ of formulas in the formalized language $\mathscr{L}$ of an arbitrary system $\mathscr{S} = (\mathscr{L}, C_{\mathscr{L}})$ in $\mathbf{L}_{\varkappa\iota}$ the set $\nabla_{\mathscr{A}}$ is an implicative filter in the algebra $\mathfrak{A}(\mathscr{S})$ of the system $\mathscr{S}$ and $\Gamma_{\mathscr{A}}$ is the set of generators* *of this implicative filter. The* $\mathbf{L}_{\varkappa\iota}$*-theory* $\mathscr{S}(\mathscr{A}) = (\mathscr{L}, C_{\mathscr{L}}, \mathscr{A})$ *is consistent if and only if the implicative filter* $\nabla_{\mathscr{A}}$ *is proper.*

The next theorem follows from 10.1 and VIII 13.4.

10.3. *For every implicative filter* ∇ *in the algebra* $\mathfrak{A}(\mathscr{S})$ *of a system* $\mathscr{S} \in \mathbf{L}_{\varkappa\iota}$ *there exists a set* $\mathscr{A}$ *of formulas of* $\mathscr{S}$ *such that* $\nabla = \nabla_{\mathscr{A}}$. *Moreover, for any set* A_0 *of generators of* ∇ *it is possible to choose the set* $\mathscr{A}$ *in such a way that* $\Gamma_{\mathscr{A}} = A_0$.

Theorems 10.2 and 10.3 state that there exists a natural one-one correspondence between implicative filters in the algebra $\mathfrak{A}(\mathscr{S})$ of a system $\mathscr{S} = (\mathscr{L}, C_{\mathscr{L}})$ in $\mathbf{L}_{\varkappa\iota}$ and formalized $\mathbf{L}_{\varkappa\iota}$-theories of zero order based on $\mathscr{S}$, two theories $\mathscr{S}(\mathscr{A}) = (\mathscr{L}, C_{\mathscr{L}}, \mathscr{A})$ and $\mathscr{S}(\mathscr{A}') = (\mathscr{L}, C_{\mathscr{L}}, \mathscr{A}')$ being identified if and only if $C_{\mathscr{L}}(\mathscr{A}) = C_{\mathscr{L}}(\mathscr{A}')$. These theorems explain also the algebraic interpretation of the set of axioms and the set of theorems in any $\mathbf{L}_{\varkappa\iota}$-theory of zero order.

The following statement will be applied in proof of the Deduction Theorem 10.5.

10.4. *For any formula* α *of the formalized language* $\mathscr{L}$ *of an arbitrary system* $\mathscr{S} = (\mathscr{L}, C_{\mathscr{L}})$ *in* $\mathbf{L}_{\varkappa\iota}$ *and for every set* $\mathscr{A}$ *of formulas in* $\mathscr{L}$, *the set* $\nabla_{\mathscr{A} \cup \{\alpha\}}$ *is the least implicative filter in the algebra* $\mathfrak{A}(\mathscr{S})$ *of the system* $\mathscr{S}$ *containing* $\nabla_{\mathscr{A}}$ *and* $||\alpha||$.

The proof, analogous to that of Theorem 5.4, is left to the reader.

The following two theorems are said to be *deduction theorems.*

10.5. *For any formulas* α, β *in the formalized language* $\mathscr{L}$ *of an arbitrary system* $\mathscr{S} = (\mathscr{L}, C_{\mathscr{L}})$ *in* $\mathbf{L}_{\varkappa\iota}$ *and for every set* $\mathscr{A}$ *of formulas in* $\mathscr{L}$:

(2) $\quad \beta \in C_{\mathscr{L}}(\mathscr{A} \cup \{\alpha\})$ *if and only if* $(\alpha \Rightarrow \beta) \in C_{\mathscr{L}}(\mathscr{A})$.

The proof, analogous to that of 5.5 and based on (1), 10.4, the definition of implication algebras (II 5) and II 3.4, is left to the reader.

10.6. *For any formula* α *in the formalized language* $\mathscr{L}$ *of an arbitrary system* $\mathscr{S} = (\mathscr{L}, C_{\mathscr{L}})$ *in* $\mathbf{L}_{\varkappa\iota}$ *and for every non-empty set* $\mathscr{A}$ *of formulas in* $\mathscr{L}$, *the condition* $\alpha \in C_{\mathscr{L}}(\mathscr{A})$ *is equivalent to the condition that there exist formulas* $\alpha_1, \ldots, \alpha_n$ *in* $\mathscr{A}$ *such that*

(3) $$(\alpha_1 \Rightarrow (\ldots(\alpha_n \Rightarrow \alpha)\ldots)) \in C_{\mathscr{L}}(O).$$

The proof, analogous to that of 5.6 and based on (1), 10.2, the definition of implication algebras (II 5), II 3.5, VIII 6 (10) and VIII 6 (8), is left to the reader.

Let us note (cf. the remarks in Section 5 following the proofs of 5.5 and 5.6) that Theorems II 3.4 and II 3.5 concerning the implication algebras are algebraic analogues of Theorems 10.5 and 10.6, respectively.

Let us adopt the notions of *maximal* and *irreducible* $\mathbf{L}_{\varkappa\iota}$*-theories* as given in VIII 13. By 10.1, the notion of a maximal (an irreducible) deductive filter coincides with the notion of a maximal (an irreducible) implicative filter. Thus the following two statements follow from VIII 13.5 and VIII 13.6.

10.7. *An* $\mathbf{L}_{\varkappa\iota}$*-theory* $\mathscr{S}(\mathscr{A}) = (\mathscr{L}, C_{\mathscr{L}}, \mathscr{A})$ *of zero order is irreducible if and only if the implicative filter* $\nabla_{\mathscr{A}}$ *in* $\mathfrak{A}(\mathscr{S})$ *is irreducible.*

Theorem 10.7 can also be deduced from the definitions of irreducible implicative filters in implication algebras and of irreducible $\mathbf{L}_{\varkappa\iota}$-theories and from Theorems 10.2, 10.3.

10.8. *An* $\mathbf{L}_{\varkappa\iota}$*-theory* $\mathscr{S}(\mathscr{A}) = (\mathscr{L}, C_{\mathscr{L}}, \mathscr{A})$ *is maximal if and only if the implicative filter* $\nabla_{\mathscr{A}}$ *in* $\mathfrak{A}(\mathscr{S})$ *is maximal.*

This theorem can also be deduced from the definitions of maximal implicative filters and maximal $\mathbf{L}_{\varkappa\iota}$-theories and from Theorems 10.2, 10.3.

The next theorem is a particular case of VIII 13.7.

10.9. *For every consistent* $\mathbf{L}_{\varkappa\iota}$*-theory* $\mathscr{S}(\mathscr{A}) = (\mathscr{L}, C_{\mathscr{L}}, \mathscr{A})$ *there exists an irreducible* $\mathbf{L}_{\varkappa\iota}$*-theory* $\mathscr{S}(\mathscr{A}') = (\mathscr{L}, C_{\mathscr{L}}, \mathscr{A}')$ *such that* $C_{\mathscr{L}}(\mathscr{A}) \subset C_{\mathscr{L}}(\mathscr{A}')$.

Theorem 10.9 also follows from 10.2, II 3.6, the definition of implication algebras, 10.3 and 10.7.

The next theorem follows from 10.7, 10.8 and II 6.4.

10.10. *An* $\mathbf{L}_{\varkappa\iota}$*-theory is maximal if and only if it is irreducible.*

Hence, by 10.9,

10.11. *For every consistent* $\mathbf{L}_{\varkappa\iota}$*-theory* $\mathscr{S}(\mathscr{A}) = (\mathscr{L}, C_{\mathscr{L}}, \mathscr{A})$ *there exists a maximal* $\mathbf{L}_{\varkappa\iota}$*-theory* $\mathscr{S}(\mathscr{A}') = (\mathscr{L}, C_{\mathscr{L}}, \mathscr{A}')$ *such that* $C_{\mathscr{L}}(\mathscr{A}) \subset C_{\mathscr{L}}(\mathscr{A}')$.

For any $\mathscr{S} = (\mathscr{L}, C_{\mathscr{L}})$ in $\mathbf{L}_{\varkappa\iota}$ and, for any formulas α, β in $\mathscr{L}$, we shall write

(4) $\quad (\alpha \cup \beta)$ instead of $((\alpha \Rightarrow \beta) \Rightarrow \beta)$.

10.12. *For every* $\mathbf{L}_{\chi\iota}$-*theory* $\mathscr{S}(\mathscr{A}) = (\mathscr{L}, C_{\mathscr{L}}, \mathscr{A})$ *of zero order the following conditions are equivalent*:

(5) $\mathscr{S}(\mathscr{A})$ *is maximal*,

(6) $\mathscr{S}(\mathscr{A})$ *is consistent and, for any two formulas* α, β *in* $\mathscr{L}$, *if* $(\alpha \cup \beta) \in C_{\mathscr{L}}(\mathscr{A})$, *then either* $\alpha \in C_{\mathscr{L}}(\mathscr{A})$ *or* $\beta \in C_{\mathscr{L}}(\mathscr{A})$.

If (5) holds, then $\mathscr{S}(\mathscr{A})$ is consistent and by 10.8 the implicative filter $\nabla_{\mathscr{A}}$ in $\mathfrak{A}(\mathscr{S})$ is maximal. Hence, by II 6.4, $\nabla_{\mathscr{A}}$ is a prime implicative filter in $\mathfrak{A}(\mathscr{S})$. Consequently, by (4), and the definition of prime implicative filters in implication algebras (see II 6) and VIII 6 (10), for any formulas α, β in $\mathscr{L}$

(7) $\quad ||(\alpha \cup \beta)|| = (||\alpha|| \Rightarrow ||\beta||) \Rightarrow ||\beta|| \in \nabla_{\mathscr{A}}$ implies that either $||\alpha|| \in \nabla_{\mathscr{A}}$ or $||\beta|| \in \nabla_{\mathscr{A}}$.

Hence, by (1), if $(\alpha \cup \beta) \in C_{\mathscr{L}}(\mathscr{A})$, then either $\alpha \in C_{\mathscr{L}}(\mathscr{A})$ or $\beta \in C_{\mathscr{L}}(\mathscr{A})$. Thus (5) implies (6). Suppose that (6) holds. Then, by 10.2, the implicative filter $\nabla_{\mathscr{A}}$ in $\mathfrak{A}(\mathscr{S})$ is proper and by (1), (4), VIII 6 (10), condition (7) is satisfied. Thus $\nabla_{\mathscr{A}}$ is a prime implicative filter in $\mathfrak{A}(\mathscr{S})$. Consequently, by II 6.4, $\nabla_{\mathscr{A}}$ is maximal. Hence, by 10.8, $\mathscr{S}(\mathscr{A})$ is maximal.

We quote the following theorem, which is a particular case of VIII 13.11.

10.13. *The algebra* $\mathfrak{A}(\mathscr{S}(\mathscr{A}))$ *of an arbitrary* $\mathbf{L}_{\chi\iota}$-*theory* $\mathscr{S}(\mathscr{A}) = (\mathscr{L}, C_{\mathscr{L}}, \mathscr{A})$ *is isomorphic to the quotient algebra* $\mathfrak{A}(\mathscr{S})/\nabla_{\mathscr{A}}$.

This theorem will be applied in the proof of the next theorem which gives another characterization of maximal $\mathbf{L}_{\chi\iota}$-theories.

10.14. *For any* $\mathbf{L}_{\chi\iota}$-*theory* $\mathscr{S}(\mathscr{A}) = (\mathscr{L}, C_{\mathscr{L}}, \mathscr{A})$ *of zero order the following conditions are equivalent*:

(8) $\mathscr{S}(\mathscr{A})$ *is maximal*,

(9) $\mathscr{S}(\mathscr{A})$ *is consistent and there exists a semantic adequate model* v *of* $\mathscr{S}(\mathscr{A})$, i.e. *a model in the two-element implication algebra* $\mathfrak{A}_0 = (\{\wedge, \vee\}, \vee, \Rightarrow)$ *such that for every formula* α *in* $\mathscr{L}$, $\alpha_{\mathfrak{A}}(v) = \vee$ *if* $\alpha \in C_{\mathscr{L}}(\mathscr{A})$ *and* $\alpha_{\mathfrak{A}}(v) = \wedge$ *if* $\alpha \notin C_{\mathscr{L}}(\mathscr{A})$.

Suppose that (8) holds. Then by 10.8, the implicative filter $\nabla_{\mathscr{A}}$ in $\mathfrak{A}(\mathscr{S})$ is maximal. Hence, by 10.13 and II 6.4, the algebra $\mathfrak{A}(\mathscr{S}(\mathscr{A}))$

is isomorphic to the two-element implication algebra $\mathfrak{A}_0$. Let h be the isomorphism from $\mathfrak{A}(\mathscr{S}(\mathscr{A}))$ onto $\mathfrak{A}_0$. Then by 9.1 and VIII 3.2, the valuation hv^0 (where v^0 is the canonical valuation in $\mathfrak{A}(\mathscr{S}(\mathscr{A}))$) is the required semantic adequate model of $\mathscr{S}(\mathscr{A})$. Thus condition (8) implies (9).

Conversely, suppose that (9) holds. We shall prove that condition (6) is satisfied. Indeed, if $(\alpha \cup \beta) \in C_{\mathscr{L}}(\mathscr{A})$, then by (4) and (9), $((\alpha \Rightarrow \beta) \Rightarrow \beta)_{\mathfrak{A}_0}(v) = V$ in a semantic adequate model v of $\mathscr{S}(\mathscr{A})$. Hence, by the definition of $\mathfrak{A}_0$, either $\beta_{\mathfrak{A}_0}(v) = V$ or $(\alpha \Rightarrow \beta)_{\mathfrak{A}_0}(v) = \Lambda$. Hence, either $\beta_{\mathfrak{A}_0}(v) = V$ or $\alpha_{\mathfrak{A}_0}(v) = V$. Since v is an adequate semantic model, it follows from the above conditions that either $\beta \in C_{\mathscr{L}}(\mathscr{A})$ or $\alpha \in C_{\mathscr{L}}(\mathscr{A})$. Thus (9) implies (6). By 10.12, condition (6) implies (8). This completes the proof of 10.14.

Exercises

1. Adjoining a unary connective $\sim_c$ to the alphabet of the formalized language of $\mathscr{S}_{\pi\iota}$ (see Sec. 1), taking as logical axioms all formulas of the form $(\alpha \Rightarrow (\beta \Rightarrow \alpha))$, $((\alpha \Rightarrow (\beta \Rightarrow \gamma)) \Rightarrow ((\alpha \Rightarrow \beta) \Rightarrow (\alpha \Rightarrow \gamma)))$, $((\alpha \Rightarrow \sim_c \beta) \Rightarrow (\beta \Rightarrow \sim_c \alpha))$, where α, β, γ are any formulas in the extended language, and taking modus ponens as the only rule of inference, we obtain the propositional calculus $\mathscr{S}$ which is an extension of $\mathscr{S}_{\pi\iota}$. Prove that $\mathscr{S}$ belongs to the class **S** (see VIII 5).

2. Prove that the class of all $\mathscr{S}$-algebras for $\mathscr{S}$ as defined in Ex. 1 coincides with the class of all contrapositionally complemented positive implication algebras (see II Ex. 5). Show that $(((p \Rightarrow q) \Rightarrow p) \Rightarrow p)$ is not derivable in $\mathscr{S}$.

3. Prove an analogue of 3.2 for propositional calculi of the logic **L** determined by $\mathscr{S}$ from Ex. 1 and analogues of the theorems in Sections 4, 5.

4. Prove that for each formula α in the formalized language of $\mathscr{S}_{\pi\iota}$, α is derivable in $\mathscr{S}_{\pi\iota}$ if and only if α is derivable in the propositional calculus $\mathscr{S}$ defined in Ex. 1.

5. Adjoin a unary connective $\dot{-}$ to the alphabet of the formalized language of $\mathscr{S}_{\pi\iota}$, take as logical axioms all formulas $(\alpha \Rightarrow (\beta \Rightarrow \alpha))$, $((\alpha \Rightarrow (\beta \Rightarrow \gamma)) \Rightarrow ((\alpha \Rightarrow \beta) \Rightarrow (\alpha \Rightarrow \gamma)))$, $(\dot{-}(\alpha \Rightarrow \alpha) \Rightarrow \beta)$, where α, β, γ are any formulas of the extended language, and take modus ponens and the rule (*r*) $\dfrac{(\alpha \Rightarrow \beta), (\beta \Rightarrow \alpha)}{(\dot{-}\alpha \Rightarrow \dot{-}\beta)}$ as rules of inference. Prove that the propositional calculus $\mathscr{S} = (\mathscr{L}, C_{\mathscr{L}})$ obtained in this way belongs to the class **S** (see VIII 5).

6. Prove that the class of all $\mathscr{S}$-algebras for $\mathscr{S}$ as described in Ex. 5 coincides with the class of all semi-complemented positive implication algebras (see II Ex. 11). Show that $\mathscr{S}$ is consistent.

7. Prove the analogue of 3.2 for propositional calculi of the logic **L** determined by the propositional calculus $\mathcal{S}$ of Ex. 5. Prove also the analogues of the theorems in Sections 4, 5.

8. Prove that for each formula α of the formalized language of $\mathcal{S}_{\pi\iota}$, α is derivable in $\mathcal{S}_{\pi\iota}$ if and only if α is derivable in the propositional calculus $\mathcal{S}$ defined in Ex. 5.

9. Adjoin a unary connective $\neg$ to the alphabet of the formalized language of $\mathcal{S}_{\pi\iota}$, take as logical axioms all formulas of the form mentioned in Ex. 1 and Ex. 5, where the signs $\sim_c$, $\dot{-}$ are both replaced by $\neg$, and take modus ponens as the only rule of inference. Prove that: 1) the propositional calculus $\mathcal{S} = (\mathcal{L}, C_{\mathcal{L}})$ obtained in this way belongs to **S**; 2) the class of all $\mathcal{S}$-algebras coincides with the class of all pseudo-complemented positive implication algebras (see II Ex. 9) and $\mathcal{S}$ is consistent; 3) the analogue of 3.2 holds for propositional calculi of the logic **L** determined by $\mathcal{S}$; 4) any **L**-theory is consistent if and only if it has an **L**-model; 5) any consistent **L**-theory can be extended to a maximal **L**-theory; 6) for each formula α without $\neg$, α is derivable in $\mathcal{S}_{\pi\iota}$ if and only if α is derivable is $\mathcal{S}$ (6).

10. Examine the propositional calculi obtained from $\mathcal{S}_{\kappa\iota}$ in the same way as the propositional calculus $\mathcal{S}$ was obtained in Ex. 1 and Ex. 5 from $\mathcal{S}_{\pi\iota}$ (i.e. take additionally as logical axioms all formulas of the form $(((\alpha \Rightarrow \beta) \Rightarrow \alpha) \Rightarrow \alpha)$. See II Ex. 12.

(6) For this equivalence see Horn [1].

CHAPTER X

POSITIVE LOGIC

Introduction. This chapter deals with the positive logic of Hilbert and Bernays [1], which can be formulated according to the definition of a logic adopted in VIII 8, and will be denoted here by $\mathbf{L}_\pi$. The formalized languages of the propositional calculi of $\mathbf{L}_\pi$ are richer than those of the propositional calculi of $\mathbf{L}_{\pi\iota}$ and $\mathbf{L}_{\varkappa\iota}$ (see IX). Their alphabets contain besides the implication sign two more connectives: the disjunction sign and the conjunction sign. Positive logic constitutes also a part of intuitionistic logic, namely that part which corresponds to formulas in which the negation sign does not occur. This means that a formula of an arbitrary propositional calculus $\mathscr{S}$ of $\mathbf{L}_\pi$ is derivable in $\mathscr{S}$ if and only if it is derivable in intuitionistic propositional calculi (1).

Positive logic has been discussed from the algebraic point of view by the present author and Sikorski ([3], [MM]). For the sake of continuity the algebraic treatment of $\mathbf{L}_\pi$ will also be expounded here. The class of all $\mathscr{S}$-algebras for propositional calculi $\mathscr{S} \in \mathbf{L}_\pi$ coincides with that of all relatively pseudo-complemented lattices. This statement enables us to formulate the metalogical theorems of Chapter VIII especially for the propositional calculi of $\mathbf{L}_\pi$ and for $\mathbf{L}_\pi$-theories. By the application of the representation theorem (IV 3.1) for relatively pseudo-complemented lattices and of Lemma VI 11.2 (on imbedding for these algebras) the completeness theorem can be formulated in a much stronger way (3.4). The decidability of the propositional calculi of $\mathbf{L}_\pi$ is a corollary of the above theorem. Similarly, the representation theorem mentioned above leads to a stronger form of the theorem on $\mathbf{L}_\pi$-models of $\mathbf{L}_\pi$-theories (5.2). By IV 1.4 every relatively pseudo-complemented lattice is a positive implication algebra. On the other

(1) See Rasiowa and Sikorski [3], [MM].

hand, it easily follows from the representation theorems for positive implication algebras and for relatively pseudo-complemented lattices that each positive implication algebra can be extended to a relatively pseudo-complemented lattice. These two theorems lead to the metalogical Theorem 3.7, which states that $\mathbf{L}_{\pi\iota}$ is the part of $\mathbf{L}_{\pi}$ corresponding to formulas in which there occur no connective except implication signs. An analogous relationship holds between $\mathbf{L}_{\pi\iota}$-theories and certain $\mathbf{L}_{\pi}$-theories (5.3). Applying the representation theorem IV 3.1 and the theorem on one-point strong compactification of topological spaces, we can obtain a metalogical result on disjunctions derivable in the propositional calculi of $\mathbf{L}_{\pi}$.

Deductive filters in the algebras under consideration coincide with implicative filters. The theorems on the structure of an implicative filter generated by a filter and an element and generated by a subset of elements, respectively, are algebraic analogues of the deduction theorems and can be used for their proofs. Since the notion of an irreducible implicative filter coincides with that of an implicative prime filter, the concept of an irreducible $\mathbf{L}_{\pi}$-theory and of a prime $\mathbf{L}_{\pi}$-theory also coincide. Consequently, each consistent $\mathbf{L}_{\pi}$-theory can be extended to a prime $\mathbf{L}_{\pi}$-theory (but not to a maximal one).

1. Propositional calculus $\mathscr{S}_{\pi}$ of positive logic. Let $A^0 = (V, L_0, L_1, L_2, U)$ be an alphabet of zero order (see VIII 1) such that:

(1) the set V is enumerable,

(2) the sets L_0, L_1 are empty,

(3) the set L_2 contains three elements, denoted by $\Rightarrow$, $\cup$, $\cap$ and called the *implication sign*, the *disjunction sign* and the *conjunction sign*, respectively.

It will be convenient to assume that A^0 is an extension of the alphabet of the propositional calculus $\mathscr{S}_{\pi\iota}$ (see IX 1) and that both alphabets have the same set V of propositional variables. Let F be the set of all formulas over A^0 and let F_0 denote, in this section, the set of all formulas of $\mathscr{S}_{\pi\iota}$. Thus we have

(4) $$F_0 \subset F.$$

Let $\mathscr{A}_l$ be the set of all formulas in F of the form

(A$_1$) $(\alpha \Rightarrow (\beta \Rightarrow \alpha))$,

(A$_2$) $((\alpha \Rightarrow (\beta \Rightarrow \gamma)) \Rightarrow ((\alpha \Rightarrow \beta) \Rightarrow (\alpha \Rightarrow \gamma)))$,

(A$_3$) $(\alpha \Rightarrow (\alpha \cup \beta))$,

(A$_4$) $(\beta \Rightarrow (\alpha \cup \beta))$,

(A$_5$) $((\alpha \Rightarrow \gamma) \Rightarrow ((\beta \Rightarrow \gamma) \Rightarrow ((\alpha \cup \beta) \Rightarrow \gamma)))$,

(A$_6$) $((\alpha \cap \beta) \Rightarrow \alpha)$,

(A$_7$) $((\alpha \cap \beta) \Rightarrow \beta)$,

(A$_8$) $((\alpha \Rightarrow \beta) \Rightarrow ((\alpha \Rightarrow \gamma) \Rightarrow (\alpha \Rightarrow (\beta \cap \gamma))))$,

where α, β, γ are any formulas in F.

Let us introduce the consequence operation $C_{\mathscr{L}}$ in $\mathscr{L} = (A^0, F)$ determined by the set $\mathscr{A}_l$ of logical axioms and by *modus ponens* as the only rule of inference.

The system

(5) $$\mathscr{S}_\pi = (\mathscr{L}, C_{\mathscr{L}})$$

will be said to be a *propositional calculus of positive logic*.

In order to prove that the propositional calculus $\mathscr{S}_\pi$ belongs to the class **S** (see VIII 5) of standard systems of implicative extensional propositional calculi we shall apply the following general theorem.

1.1. *Let* $\mathscr{S} = (\mathscr{L}', C_{\mathscr{L}'})$ *be an arbitrary propositional calculus, where the consequence operation* $C_{\mathscr{L}'}$ *is determined by a set* $\mathscr{A}'_l$ *of logical axioms and by a set* $\{(r_1), \ldots, (r_k)\}$ *of rules of inference containing modus ponens. If there appear in the alphabet of* $\mathscr{L}'$ (besides implication) *two binary connectives* $\cup$, $\cap$ *and, for any formulas* α, β, γ *in* $\mathscr{L}'$, *all formulas of the form*:

(A$_1$) $(\alpha \Rightarrow (\beta \Rightarrow \alpha))$,

(A$_2$) $((\alpha \Rightarrow (\beta \Rightarrow \gamma)) \Rightarrow ((\alpha \Rightarrow \beta) \Rightarrow (\alpha \Rightarrow \gamma)))$,

(A$_3$) $(\alpha \Rightarrow (\alpha \cup \beta))$,

(A$_4$) $(\beta \Rightarrow (\alpha \cup \beta))$,

(A$_5$) $((\alpha \Rightarrow \gamma) \Rightarrow ((\beta \Rightarrow \gamma) \Rightarrow ((\alpha \cup \beta) \Rightarrow \gamma)))$,

(A$_6$) $((\alpha \cap \beta) \Rightarrow \alpha)$,

(A_7) $((\alpha \cap \beta) \Rightarrow \beta)$,

(A_8) $((\alpha \Rightarrow \beta) \Rightarrow ((\alpha \Rightarrow \gamma) \Rightarrow (\alpha \Rightarrow (\beta \cap \gamma))))$,

are in $\mathcal{A}'$, *then, for any formulas* $\alpha, \beta, \gamma, \delta$ *of* $\mathcal{L}'$ *and for any set* $\mathcal{A}$ *of formulas of* $\mathcal{L}'$, *the condition* $(\alpha \Rightarrow \beta), (\gamma \Rightarrow \delta) \in C_{\mathcal{L}'}(\mathcal{A})$ *implies that* $((\alpha \cup \gamma) \Rightarrow (\beta \cup \delta)) \in C_{\mathcal{L}'}(\mathcal{A})$ *and* $((\alpha \cap \gamma) \Rightarrow (\beta \cap \delta)) \in C_{\mathcal{L}'}(\mathcal{A})$.

An easy proof is left to the reader.

The following corollary follows from 1.1, IX 1.2 and IX 1.1.

1.2. *If a propositional calculus* $\mathcal{S} = (\mathcal{L}', C_{\mathcal{L}'})$ *satisfies the hypotheses of* 1.1, *then conditions* VIII 5 $((s_3)\text{-}(s_6))$ *and condition* VIII 5 (s_8) *with respect to the connectives* $\Rightarrow, \cup, \cap$, *are satisfied.*

1.3. *The propositional calculus* $\mathcal{S}_\pi$ *belongs to the class* **S** (see VIII 5) *of standard systems of implicative extensional propositional calculi.*

By the definition of the set $\mathcal{A}_l$ of logical axioms and by VIII 2.4 (9), the set $\mathcal{A}_l$ is closed under substitutions. By the definition of modus ponens and VIII 2.4 (9), the only rule of inference of $\mathcal{S}_\pi$ is invariant with respect to substitutions. Thus conditions VIII 5 $((s_1), (s_2))$ are satisfied. Since $\mathcal{S}_\pi$ satisfies the hypotheses of 1.1, it follows from 1.2 that conditions VIII 5 $((s_3)\text{–}(s_6))$ and VIII 5 (s_8) hold. On the other hand, in the alphabet of $\mathcal{S}_\pi$ there appears no unary connective. Thus condition VIII 5 (s_7) does not concern $\mathcal{S}_\pi$. This completes the proof of 1.3.

2. $\mathcal{S}_\pi$**-algebras.** The aim of this section is to prove the following theorem.

2.1. *The class of all* $\mathcal{S}_\pi$*-algebras coincides with the class of all relatively pseudo-complemented lattices.*

Suppose that $\mathfrak{A} = (A, V, \Rightarrow, \cup, \cap)$ is an $\mathcal{S}_\pi$-algebra. By an analogous argument to that used in the proof of IX 2.1, it follows from 1 $((A_1)\text{–}(A_8))$ and VIII 6 (a_1) that for all a, b, c in A:

(1) $a \Rightarrow (b \Rightarrow a) = V$,

(2) $(a \Rightarrow (b \Rightarrow c)) \Rightarrow ((a \Rightarrow b) \Rightarrow (a \Rightarrow c)) = V$,

(3) $a \Rightarrow (a \cup b) = V$,

(4) $b \Rightarrow (a \cup b) = V$,

(5) $(a \Rightarrow c) \Rightarrow ((b \Rightarrow c) \Rightarrow ((a \cup b) \Rightarrow c)) = V$,

(6) $(a \cap b) \Rightarrow a = V$,

(7) $(a \cap b) \Rightarrow b = V$,

(8) $(a \Rightarrow b) \Rightarrow \big((a \Rightarrow c) \Rightarrow (a \Rightarrow (b \cap c))\big) = V$.

By VIII 6 (a_4),

(9) for all $a, b \in A$, if $a \Rightarrow b = V$ and $b \Rightarrow a = V$, then $a = b$.

Moreover, on account of 1.3, VIII 6.2 and II 1 (i_4),

(10) $$a \Rightarrow V = V \quad \text{for each } a \in A.$$

It follows from (1), (2), (9), (10) and II 2 ((p_1)-(p_4)) that $(A, V, \Rightarrow)$ is a positive implication algebra. Hence, by (3)-(8) and IV 1.4, $\mathfrak{A}$ is a relatively pseudo-complemented lattice.

Conversely, if $\mathfrak{A} = (A, V, \Rightarrow, \cup, \cap)$ is a relatively pseudo-complemented lattice, then by IV 1.4, $(A, V, \Rightarrow)$ is a positive implication algebra and the equations (3)-(8) hold. Consequently, by II 2 ((p_1), (p_2)) the equations (1), (2) also hold. Thus condition VIII 6 (a_1) is satisfied, i.e., for each logical axiom α of $\mathscr{S}_\pi$ and for every valuation v of the formalized language $\mathscr{L}$ of $\mathscr{S}_\pi$ in $\mathfrak{A}$, $\alpha_{\mathfrak{A}}(v) = V$. Condition VIII 6 ($a_2$) is satisfied for the only rule of inference in $\mathscr{S}_\pi$, namely modus ponens, by II 2 (1). Condition VIII 6 (a_3) holds by II 2.1 and II 1 (i_2). Condition VIII 6 (a_4) is satisfied by II 2 (p_3). Thus $\mathfrak{A}$ is an $\mathscr{S}_\pi$-algebra.

The following theorem follows directly from 2.1 and VIII 6.1.

2.2. *If α is a formula derivable in $\mathscr{S}_\pi$, then*

(11) $$\alpha_{\mathfrak{A}}(v) = V$$

for every valuation $v\colon V \to A$ in any relatively pseudo-complemented lattice $\mathfrak{A} = (A, V, \Rightarrow, \cup, \cap)$.

It follows from 2.2 that no propositional variable is derivable in $\mathscr{S}_\pi$, i.e. that

2.3. *The propositional calculus $\mathscr{S}_\pi$ is consistent.*

3. Positive logic $\mathbf{L}_\pi$. By 1.3 and 2.3 the propositional calculus $\mathscr{S}_\pi$ belongs to the class **S** (see VIII 5) of standard systems of implicative extensional propositional calculi and is consistent.

Let $\mathbf{L}_\pi$ be the class of all consistent systems in **S** logically equivalent (see VIII 8) to $\mathscr{S}_\pi$. The class $\mathbf{L}_\pi$ will be called *positive logic*.

By the definition of logically equivalent systems (VIII 8) and by 2.1, for every system $\mathcal{S}$ in $\mathbf{L}_\pi$ the class of all $\mathcal{S}$-algebras coincides with the class of all relatively pseudo-complemented lattices (IV 1). Thus the class $\mathbf{K}_{\mathbf{L}_\pi}$ of all $\mathcal{S}$-algebras for $\mathcal{S} \in \mathbf{L}_\pi$ is the class of all relatively pseudo-complemented lattices. Hence, by VIII 6.5:

3.1. *For every $\mathcal{S} \in \mathbf{L}_\pi$, the algebra $\mathfrak{A}(\mathcal{S})$ of the propositional calculus $\mathcal{S}$ is a non-degenerate relatively pseudo-complemented lattice.*

Moreover, by VIII 6.6:

3.2. *For every $\mathcal{S} \in \mathbf{L}_\pi$, the algebra $\mathfrak{A}(\mathcal{S})$ is a free algebra in the class $\mathbf{K}_{\mathbf{L}_\pi}$ of all relatively pseudo-complemented lattices.*

A formula α in any system $\mathcal{S} \in \mathbf{L}_\pi$ is said to be $\mathbf{L}_\pi$-*valid* provided that for every valuation v of the formalized language of $\mathcal{S}$ in every relatively pseudo-complemented lattice $\mathfrak{A} = (A, V, \Rightarrow, \cup, \cap)$

(1) $$\alpha_{\mathfrak{A}}(v) = V.$$

The following *completeness theorem* follows immediately from VIII 8.1.

3.3. *For any formula α in any system $\mathcal{S} \in \mathbf{L}_\pi$ the following conditions are equivalent:*

(2) *α is derivable in $\mathcal{S}$,*

(3) *α is $\mathbf{L}_\pi$-valid.*

The above theorem is included in:

3.4. *For any formula α of any propositional calculus $\mathcal{S} = (\mathcal{L}, C_{\mathcal{L}})$ of the positive logic $\mathbf{L}_\pi$ the following conditions are equivalent:*

(4) *α is derivable in $\mathcal{S}$,*

(5) *α is $\mathbf{L}_\pi$-valid,*

(6) *α is valid in every relatively pseudo-complemented lattice $\mathfrak{G}(X) = (G(X), X, \Rightarrow, \cup, \cap)$ of open subsets of any topological space X,*

(7) *$\alpha_{\mathfrak{A}(\mathcal{S})}(v^0) = V$ for the canonical valuation v^0 in the algebra $\mathfrak{A}(\mathcal{S})$ of the system $\mathcal{S}$,*

(8) *α is valid in every relatively pseudo-complemented lattice $\mathfrak{A}$ with at most 2^{2^r} elements, where r is the number of all subformulas of α* [2].

[2] See Rasiowa and Sikorski [3], [MM].

Conditions (4), (5) are equivalent by 3.3. Clearly, (5) implies (6) and, by the representation theorem IV 3.1, condition (6) implies (7). By VIII 7.2, condition (7) implies (4). Thus (4), (5), (6), (7) are equivalent. Clearly, (5) implies (8). To complete the proof of 3.4 it suffices to show that (8) implies (4). Suppose that α is not derivable in $\mathscr{S}$. Hence, by VIII 6 (8), in the algebra $\mathfrak{A}(\mathscr{S})$ of the system $\mathscr{S}$,

(9) $$||\alpha|| \neq V.$$

Suppose that α contains r subformulas. Let A_0 be the set consisting of all $||\beta||$ in $\mathfrak{A}(\mathscr{S})$ such that β is a subformula of α. It follows from VI 11.2 that there exists a finite relatively pseudo-complemented lattice $\mathfrak{A}$ consisting of at most 2^{2^r} elements such that $\mathfrak{A}$ contains $A_0 \cup \{V\}$ and for all $||\beta||$, $||\gamma||$, $||\delta||$ in $A_0 \cup \{V\}$:

(10) if $||\beta||$ is the join of $||\gamma||$, $||\delta||$ in $\mathfrak{A}(\mathscr{S})$, then $||\beta||$ is the join of $||\gamma||$, $||\delta||$ in $\mathfrak{A}$,

(11) if $||\beta||$ is the meet of $||\gamma||$, $||\delta||$ in $\mathfrak{A}(\mathscr{S})$, then $||\beta||$ is the meet of $||\gamma||$, $||\delta||$ in $\mathfrak{A}$,

(12) if $||\beta||$ is the pseudo-complement of $||\gamma||$ relative to $||\delta||$ in $\mathfrak{A}(\mathscr{S})$, then $||\beta||$ is the pseudo-complement of $||\gamma||$ relative to $||\delta||$ in $\mathfrak{A}$.

Let v be the valuation of $\mathscr{L}$ in $\mathfrak{A}$ defined as follows:

(13) $$v(p) = \begin{cases} ||p|| & \text{for each propositional variable } p \text{ in } \alpha, \\ V & \text{for each propositional variable } p \text{ which does not occur in } \alpha. \end{cases}$$

Then, for every subformula β of α,

(14) $$\beta_{\mathfrak{A}}(v) = ||\beta|| \in A_0 \cup \{V\}.$$

The easy proof by induction on the length of β, based on (13), (10), (11), (12), VIII 3 (3), VIII 6 (10), is left to the reader. In particular, it follows from (14) and (9) that

$$\alpha_{\mathfrak{A}}(v) = ||\alpha|| \neq V,$$

i.e. condition (8) does not hold for α. Thus (8) implies (4).

It follows from the equivalence of the conditions (4) and (8) that:

3.5. *Every propositional calculus $\mathscr{S}$ of the positive logic $\mathbf{L}_\pi$ is decidable* [(3)], i.e. *there exists a method which, for every formula α of $\mathscr{S}$, enables us to determine in a finite number of steps whether or not it is derivable in $\mathscr{S}$.*

To give examples of formulas derivable in any propositional calculus of positive logic we shall prove the following theorem.

3.6. *For every system $\mathscr{S} = (\mathscr{L}, C_{\mathscr{L}})$ of the positive logic $\mathbf{L}_\pi$ and for all formulas α, β, γ in $\mathscr{L}$ the formulas of the form* (D_3)–(D_{11}) *in* IX 3, (A_1)-(A_8) *in Sec.* 1 *and the following ones are derivable in $\mathscr{S}$:*

(D_{16}) $((\alpha \cap (\alpha \Rightarrow \beta)) \Rightarrow \beta)$,

(D_{17}) $(((\alpha \Rightarrow \gamma) \cap (\beta \Rightarrow \gamma)) \Rightarrow ((\alpha \cup \beta) \Rightarrow \gamma))$,

(D_{18}) $(((\alpha \Rightarrow \beta) \cap (\alpha \Rightarrow \gamma)) \Rightarrow (\alpha \Rightarrow (\beta \cap \gamma)))$,

(D_{19}) $(\alpha \Rightarrow (\beta \Rightarrow (\alpha \cap \beta)))$,

(D_{20}) $((\alpha \Rightarrow (\beta \Rightarrow \gamma)) \Rightarrow ((\alpha \cap \beta) \Rightarrow \gamma))$,

(D_{21}) $(((\alpha \cap \beta) \Rightarrow \gamma) \Rightarrow (\alpha \Rightarrow (\beta \Rightarrow \gamma)))$,

(D_{22}) $(((\alpha \Rightarrow \beta) \cap (\beta \Rightarrow \gamma)) \Rightarrow (\alpha \Rightarrow \gamma))$,

(D_{23}) $((\alpha \cup \beta) \Rightarrow (\beta \cup \alpha))$,

(D_{24}) $(((\alpha \cup \beta) \cup \gamma) \Rightarrow (\alpha \cup (\beta \cup \gamma)))$,

(D_{25}) $((\alpha \cup (\beta \cup \gamma)) \Rightarrow ((\alpha \cup \beta) \cup \gamma))$,

(D_{26}) $((\alpha \cap \beta) \Rightarrow (\beta \cap \alpha))$,

(D_{27}) $(((\alpha \cap \beta) \cap \gamma) \Rightarrow (\alpha \cap (\beta \cap \gamma)))$,

(D_{28}) $((\alpha \cap (\beta \cap \gamma)) \Rightarrow ((\alpha \cap \beta) \cap \gamma))$,

(D_{29}) $((\alpha \cap (\beta \cup \gamma)) \Rightarrow ((\alpha \cap \beta) \cup (\alpha \cap \gamma)))$,

(D_{30}) $(((\alpha \cap \beta) \cup (\alpha \cap \gamma)) \Rightarrow (\alpha \cap (\beta \cup \gamma)))$,

(D_{31}) $((\alpha \cup (\beta \cap \gamma)) \Rightarrow ((\alpha \cup \beta) \cap (\alpha \cup \gamma)))$,

(D_{32}) $(((\alpha \cup \beta) \cap (\alpha \cup \gamma)) \Rightarrow (\alpha \cup (\beta \cap \gamma)))$.

By 3.3 it suffices to prove that the above formulas are $\mathbf{L}_\pi$-valid. We shall indicate only the numbers of statements from which their $\mathbf{L}_\pi$-validity follows (the exact proof is left to the reader): (D_3)–(D_{11}) in IX 3 — the proof is the same as that of IX 3.3 by IV 1.4; (A_1) — IV 1.3 (2), (4); (A_2) — IV 1.3 (2), (6); (A_3)–(A_8) — IV 1.4; (D_{16}) —

(3) See Hilbert and Bernays [1].

IV 1.3 (2), (5); (D_{17}) — IV 1.3 (2), (8); (D_{18}) — IV 1.3 (2), (9); (D_{19}) — IV 1.5 (21), IV 1.3 (2); (D_{20}) — IV 1.5 (18), IV 1 (1); (D_{21}) — IV 1.5 (18), IV 1 (1); (D_{22}) — IV 1.5 (20), IV 1.3 (2); (D_{23})–(D_{28}) — III 1 (l_1), (l_2), IV 1 (1) and the definition of relatively pseudo-complemented lattices (IV 1); (D_{29})–(D_{32}) — IV 1.2, III 2 (d), IV 1 (1).

We shall complete this section by proving the following theorem, which states a relationship between the positive implicative logic $\mathbf{L}_{\pi\iota}$ and the positive logic $\mathbf{L}_\pi$.

3.7. *Let $\mathscr{S} = (\mathscr{L}, C_{\mathscr{L}})$ be a propositional calculus of the positive logic $\mathbf{L}_\pi$ and let $\mathscr{S}_0 = (\mathscr{L}_0, C_{\mathscr{L}_0})$ be a propositional calculus of the positive implicative logic $\mathbf{L}_{\pi\iota}$ such that $\mathscr{L}$ is an extension of $\mathscr{L}_0$. Then for any formula α in $\mathscr{L}_0$ the following conditions are equivalent*:

(15) *α is derivable in $\mathscr{S}$*,

(16) *α is derivable in $\mathscr{S}_0$* [4].

If a formula α in $\mathscr{L}_0$ is not derivable in $\mathscr{S}$, then by 3.3 there exist a relatively pseudo-complemented lattice $\mathfrak{A} = (A, V, \Rightarrow, \cup, \cap)$ and a valuation v of $\mathscr{L}$ in $\mathfrak{A}$ such that $\alpha_{\mathfrak{A}}(v) \neq V$. By IV 1.4, $\mathfrak{A}_0 = (A, V, \Rightarrow)$ is a positive implication algebra. Let $v_0(p) = v(p)$ for any propositional variable p in $\mathscr{L}_0$. Then v_0 is a valuation of $\mathscr{L}_0$ in $\mathfrak{A}_0$ and, for any formula β in $\mathscr{L}_0$, $\beta_{\mathfrak{A}_0}(v_0) = \beta_{\mathfrak{A}}(v)$. In particular, $\alpha_{\mathfrak{A}_0}(v_0) = \alpha_{\mathfrak{A}}(v) \neq V$. Hence, by IX 3.1, α is not derivable in $\mathscr{S}_0$. Conversely, suppose that a formula α in $\mathscr{L}_0$ is not derivable in $\mathscr{S}_0$. Then, by IX 3.2, there exist a positive implication algebra $\mathfrak{G}_0(X) = (G_0(X), X, \Rightarrow)$ of some open subsets of a topological space X and a valuation v_0 of $\mathscr{L}_0$ in $\mathfrak{G}_0(X)$ such that

(17) $$\alpha_{\mathfrak{G}_0(X)}(v_0) \neq X.$$

Let $\mathfrak{G}(X) = (G(X), X, \Rightarrow, \cup, \cap)$ be the relatively pseudo-complemented lattice of all open subsets of X. It follows from IV 1 (12) and II 2 (6) that $\mathfrak{G}_0(X)$ is a subalgebra of $(G(X), X, \Rightarrow)$. Consequently, for the valuation v of $\mathscr{L}$ in $\mathfrak{G}(X)$, defined by

(18) $$v(p) = \begin{cases} X & \text{for any propositional variable } p \text{ in } \mathscr{L} \text{ which does not occur in } \mathscr{L}_0, \\ v_0(p) & \text{for any propositional variable } p \text{ in } \mathscr{L}_0, \end{cases}$$

[4] See Horn [1].

and for any formula β in $\mathscr{L}_0$

(19) $$\beta_{\mathfrak{G}(X)}(v) = \beta_{\mathfrak{G}_0(X)}(v_0).$$

The easy proof by induction on the length of β is left to the reader. In particular, $\alpha_{\mathfrak{G}(X)}(v) = \alpha_{\mathfrak{G}_0(X)}(v_0) \neq X$. Hence, by 3.3, α is not derivable in $\mathscr{S}$.

It follows from 3.7 and 3.5 that *every propositional calculus of the positive implicative logic is decidable* [5].

4. On disjunctions derivable in the propositional calculi of positive logic. The aim of this section is to prove the following theorem.

4.1. *Let* $\mathscr{S} = (\mathscr{L}, C_{\mathscr{L}})$ *be a propositional calculus of the positive logic. Then for any formulas* α, β *in* $\mathscr{L}$,

(1) $(\alpha \cup \beta) \in C_{\mathscr{L}}(O)$ *if and only if either* $\alpha \in C_{\mathscr{L}}(O)$ *or* $\beta \in C_{\mathscr{L}}(O)$ [6].

The sufficiency follows from Theorem 3.6, which states that for any α, β the formulas $(\alpha \Rightarrow (\alpha \cup \beta))$, $(\beta \Rightarrow (\alpha \cup \beta))$ are derivable in $\mathscr{S}$.

By 3.1 and IV 3.1 there exists a monomorphism h from $\mathfrak{A}(\mathscr{S})$ into a relatively pseudo-complemented set lattice $\mathfrak{G}(X) = (G(X), X, \Rightarrow, \cup, \cap)$ of all open subsets of a topological space X. By VIII 3.2, for the canonical valuation v^0 of $\mathscr{L}$ in $\mathfrak{A}(\mathscr{S})$ and for any formula γ in $\mathscr{L}$,

(2) $$h\gamma_{\mathfrak{A}(\mathscr{S})}(v^0) = \gamma_{\mathfrak{G}(X)}(hv^0).$$

Let X_0 be the one-point strong compactification of X (see VI 8) and let $\mathfrak{G}_0(X_0) = (G_0(X_0), X_0, \Rightarrow, \cup, \cap)$ be the relatively pseudo-complemented set lattice of all open subsets of X_0. Then $G_0(X_0)$ consists of all open subsets of X and of the whole space X_0 (see VI 8.1). Let us put

(3) $$g(Y) = Y \cap X \quad \text{for any} \quad Y \in G_0(X_0).$$

It can be proved by an easy verification that $g \colon G_0(X_0) \to G(X)$ is an epimorphism from $\mathfrak{G}_0(X_0)$ onto $\mathfrak{G}(X)$. Let v be a valuation in $\mathfrak{G}_0(X_0)$ satisfying the following condition:

(4) $g(v(p)) = h(v^0(p)) = h(\|p\|)$ for each propositional variable p in $\mathscr{L}$. Thus $gv = hv^0$.

[5] See Hilbert and Bernays [1].

[6] See Rasiowa and Sikorski [5], [MM].

Suppose that $(\alpha \cup \beta) \in C_{\mathscr{L}}(O)$. Then by 3.3

(5) $$(\alpha \cup \beta)_{\mathfrak{G}_0(X_0)}(v) = \alpha_{\mathfrak{G}_0(X_0)}(v) \cup \beta_{\mathfrak{G}_0(X_0)}(v) = X_0 .$$

Since X_0 is strongly compact, it follows from (5) that at least one of the summands is equal to X_0. Assume that $\alpha_{\mathfrak{G}_0(X_0)}(v) = X_0$. Hence, by (3), (4), VIII 3.2, (2), we get

(6) $$X = g(X_0) = g\alpha_{\mathfrak{G}_0(X_0)}(v) = \alpha_{\mathfrak{G}(X)}(gv) = \alpha_{\mathfrak{G}(X)}(hv^0) = h\alpha_{\mathfrak{A}(\mathscr{S})}(v^0).$$

Since h is a monomorphism, it follows from (6) that $\alpha_{\mathfrak{A}(\mathscr{S})}(v^0) = V$ and consequently, by 3.3, α is derivable in $\mathscr{S}$. Thus the necessity is proved.

5. $\mathbf{L}_\pi$-theories of zero order. Let $\mathscr{S} = (\mathscr{L}, C_{\mathscr{L}})$ be a system in $\mathbf{L}_\pi$. For every set $\mathscr{A}$ of formulas in $\mathscr{L}$ the system $\mathscr{S}(\mathscr{A}) = (\mathscr{L}, C_{\mathscr{L}}, \mathscr{A})$ is said to be an $\mathbf{L}_\pi$-*theory of zero order based on* $\mathscr{S}$ (see VIII 9).

By an $\mathbf{L}_\pi$-*model of an* $\mathbf{L}_\pi$-*theory* $\mathscr{S}(\mathscr{A}) = (\mathscr{L}, C_{\mathscr{L}}, \mathscr{A})$ we mean any valuation v of $\mathscr{L}$ in any non-degenerate relatively pseudo-complemented lattice $\mathfrak{A} = (A, V, \Rightarrow, \cup, \cap)$ such that for every α in $\mathscr{A}$

(1) $$\alpha_{\mathfrak{A}}(v) = V .$$

The following theorem follows directly from VIII 9.2, VIII 9.3 and 2.1.

5.1. *For every consistent* $\mathbf{L}_\pi$-*theory* $\mathscr{S}(\mathscr{A}) = (\mathscr{L}, C_{\mathscr{L}}, \mathscr{A})$ *of zero order the algebra* $\mathfrak{A}(\mathscr{S}(\mathscr{A}))$ (see VIII 9) *is a non-degenerate relatively pseudo-complemented lattice. The canonical valuation* v^0 *of* $\mathscr{L}$ *in* $\mathfrak{A}(\mathscr{S}(\mathscr{A}))$ i.e. *the valuation defined by*

(2) $$v^0(p) = ||p||_{\mathscr{A}} \quad \text{for every propositional variable } p \text{ in } \mathscr{L},$$

is an $\mathbf{L}_\pi$-*model of* $\mathscr{S}(\mathscr{A})$. *Moreover, for any formula* α *in* $\mathscr{L}$

(3) α *is a theorem of* $\mathscr{S}(\mathscr{A})$ *if and only if* $\alpha_{\mathfrak{A}(\mathscr{S}(\mathscr{A}))}(v^0) = V$.

The next theorem characterizes all theorems of any consistent $\mathbf{L}_\pi$-theory of zero order.

5.2. *For any formula* α *of every consistent* $\mathbf{L}_\pi$-*theory* $\mathscr{S}(\mathscr{A}) = (\mathscr{L}, C_{\mathscr{L}}, \mathscr{A})$ *the following conditions are equivalent*:

(i) α *is a theorem of* $\mathscr{S}(\mathscr{A})$,

(ii) *every* $\mathbf{L}_\pi$-*model* v *of* $\mathscr{S}(\mathscr{A})$ *is an* $\mathbf{L}_\pi$-*model of* α,

(iii) *every* $\mathbf{L}_\pi$-*model* v *of* $\mathscr{S}(\mathscr{A})$ *in any non-degenerate relatively pseudo-complemented lattice* $\mathfrak{G}(X) = (G(X), X, \Rightarrow, \cup, \cap)$ *of open subsets of any topological space* X *is an* $\mathbf{L}_\pi$-*model of* α,

(iv) $\alpha_{(\mathfrak{A}\mathcal{S}(\mathcal{A}))}(v^0) = V$ *for the canonical valuation* v^0 *in the algebra* $\mathfrak{A}(\mathcal{S}(\mathcal{A}))$ *of the theory* $\mathcal{S}(\mathcal{A})$ (7).

By VIII 9.5 conditions (i), (ii), (iv) are equivalent. Clearly, (ii) implies (iii) and by the representation theorem IV 3.1 condition (iii) implies (ii), which completes the proof.

The next theorem states the connection between $\mathbf{L}_\pi$-theories and $\mathbf{L}_{\pi\iota}$-theories of zero order.

5.3. *Let* $\mathcal{S}(\mathcal{A}) = (\mathcal{L}, C_{\mathcal{L}}, \mathcal{A})$ *be an* $\mathbf{L}_\pi$-*theory and let* $\mathcal{S}_0 = (\mathcal{L}_0, C_{\mathcal{L}_0})$ *be a propositional calculus of the positive implicative logic* $\mathbf{L}_{\pi\iota}$ *such that* $\mathcal{L}$ *is an extension of* $\mathcal{L}_0$. *If each* α *in* $\mathcal{A}$ *is a formula in* $\mathcal{L}_0$, *then for every* β *in* $\mathcal{L}_0$ *the following conditions are equivalent*:

(i) β *is a theorem of* $\mathcal{S}(\mathcal{A})$,

(ii) β *is a theorem of the* $\mathbf{L}_{\pi\iota}$-*theory* $\mathcal{S}_0(\mathcal{A}) = (\mathcal{L}_0, C_{\mathcal{L}_0}, \mathcal{A})$.

The proof, analogous to that of 3.7, is left to the reader. Instead of Theorems 3.3 and IX 3.2 applied in the proof of 3.7, we apply Theorems 5.2 and IX 4.2 in the proof of 5.3.

6. The connection between $\mathbf{L}_\pi$-theories and filters. Consider the propositional calculus $\mathcal{S}_\pi$ of the positive logic described in Section 1. Let $\mathfrak{A} = (A, V, \Rightarrow, \cup, \cap)$ be an arbitrary relatively pseudo-complemented lattice. Since the only rule of inference in $\mathcal{S}_\pi$ is modus ponens, it follows from 3 and VIII 12.3 that a subset ∇ of A is a deductive filter in $\mathfrak{A}$ if and only if the following conditions are satisfied:

(f_1) $V \in \nabla$,

(f_2) if $a, a \Rightarrow b \in \nabla$, then $b \in \nabla$.

Hence, by IV 1.4, II 1 p. 18, and IV 2.1,

6.1. *For any set* ∇ *of elements in a relatively pseudo-complemented lattice* $\mathfrak{A} = (A, V, \Rightarrow, \cup, \cap)$ *the following conditions are equivalent*:

(i) ∇ *is a deductive filter*,

(ii) ∇ *is an implicative filter*,

(iii) ∇ *is a filter*.

(7) See [MM].

By IV 2.3 and IV 2.4 the kernels of epimorphisms from a relatively pseudo-complemented lattice onto similar algebras are filters, and conversely, every filter is the kernel of the epimorphism determined by itself. The above statements follow independently from Section 3, 6.1, VIII 12.2, VIII 13.10, VIII 13.12 and the equational definability of the class of all relatively pseudo-complemented lattices (see IV 1.4 and II 2.4, II 2.5).

Let $\mathscr{S} = (\mathscr{L}, C_{\mathscr{L}})$ be an arbitrary system in $\mathbf{L}_\pi$. By Section 3, VIII 6.5 and VIII 6.6 the non-degenerate relatively pseudo-complemented lattice $\mathfrak{A}(\mathscr{S})$ is a free algebra in the class $\mathbf{K}_{\mathbf{L}_\pi}$ of all relatively pseudo-complemented lattices. For any set $\mathscr{A}$ of formulas in $\mathscr{L}$, let $\Gamma_{\mathscr{A}}$ be the set of all elements $||\gamma||$ in $\mathfrak{A}(\mathscr{S})$ such that $\gamma \in \mathscr{A}$ and let $\nabla_{\mathscr{A}}$ be the set o all $||\gamma||$ in $\mathfrak{A}(\mathscr{S})$ such that $\gamma \in C_{\mathscr{L}}(\mathscr{A})$.

It follows from 6.1 and VIII 13.2 that:

6.2. *For each set $\mathscr{A}$ of formulas in the formalized language $\mathscr{L}$ of any system $\mathscr{S} = (\mathscr{L}, C_{\mathscr{L}})$ in $\mathbf{L}_\pi$ the set $\nabla_{\mathscr{A}}$ is a filter in $\mathfrak{A}(\mathscr{S})$ and $\Gamma_{\mathscr{A}}$ is the set of generators of this filter. The $\mathbf{L}_\pi$-theory $\mathscr{S}(\mathscr{A}) = (\mathscr{L}, C_{\mathscr{L}}, \mathscr{A})$ is consistent if and only if the filter $\nabla_{\mathscr{A}}$ is proper.*

The next theorem follows from 6.1 and VIII 13.4.

6.3. *For every filter ∇ in the algebra $\mathfrak{A}(\mathscr{S})$ of a system $\mathscr{S} \in \mathbf{L}_\pi$, there exists a set $\mathscr{A}$ of formulas of $\mathscr{S}$ such that $\nabla = \nabla_{\mathscr{A}}$. Moreover, for any set A_0 of generators of ∇ it is possible to choose the set $\mathscr{A}$ in such a way that $\Gamma_{\mathscr{A}} = A_0$.*

Theorems 6.2 and 6.3 state that there exists a natural one-one correspondence between filters in the algebra $\mathfrak{A}(\mathscr{S})$ of a system $\mathscr{S}$ in $\mathbf{L}_\pi$ and formalized $\mathbf{L}_\pi$-theories based on $\mathscr{S}$, two theories $\mathscr{S}(\mathscr{A}) = (\mathscr{L}, C_{\mathscr{L}}, \mathscr{A})$ and $\mathscr{S}(\mathscr{A}') = (\mathscr{L}, C_{\mathscr{L}}, \mathscr{A}')$ being identified if and only if $C_{\mathscr{L}}(\mathscr{A}) = C_{\mathscr{L}}(\mathscr{A}')$. The above theorems explain also the algebraic sense of the set of axioms and the set of all theorems in $\mathbf{L}_\pi$-theories of zero order.

Let us note that for any system $\mathscr{S}$ of positive logic there holds a theorem corresponding to IX 5.4. Moreover, deduction theorems analogous to IX 5.5 and IX 5.6 also hold. The similar proofs are left to the reader. By 6.1 we also get the following form of the deduction theorem corresponding to IX 5.6.

6.4. *For every formula* α *in the formalized language* $\mathscr{L}$ *of any system* $\mathscr{S} = (\mathscr{L}, C_{\mathscr{L}})$ *in* $\mathbf{L}_\pi$ *and for each non-empty set* $\mathscr{A}$ *of formulas in* $\mathscr{L}$, *the condition* $\alpha \in C_{\mathscr{L}}(\mathscr{A})$ *is equivalent to the condition that there exist formulas* $\alpha_1, \ldots, \alpha_n$ *in* $\mathscr{A}$ *such that*

(1) $$((\alpha_1 \cap \ldots \cap (\alpha_{n-1} \cap \alpha_n) \ldots) \Rightarrow \alpha) \in C_{\mathscr{L}}(O).$$

Clearly, if (1) holds, then by 3.6 (D_{19}) and modus ponens, $\alpha \in C_{\mathscr{L}}(\mathscr{A})$. Conversely, if $\alpha \in C_{\mathscr{L}}(\mathscr{A})$, then $||\alpha|| \in \nabla_{\mathscr{A}}$. Thus, by 6.2 and III 1.6, there exist $||\alpha_1||, \ldots, ||\alpha_n||$ in $\Gamma_{\mathscr{A}}$ such that $||\alpha_1|| \cap \ldots \cap ||\alpha_n|| \leqslant ||\alpha||$, which is equivalent (see IV 1.3 and VIII 9 (9)) to (1).

By 6.1 the notion of an irreducible deductive filter (see VIII 12) coincides with the notion of an irreducible filter in relatively pseudo-complemented lattices. Thus the following two theorems follow from VIII 13.5 and VIII 13.7.

6.5. *An* $\mathbf{L}_\pi$-*theory* $\mathscr{S}(\mathscr{A}) = (\mathscr{L}, C_{\mathscr{L}}, \mathscr{A})$ *is irreducible if and only if the filter* $\nabla_{\mathscr{A}}$ *in* $\mathfrak{A}(\mathscr{S})$ *is irreducible.*

6.6. *For every consistent* $\mathbf{L}_\pi$-*theory* $\mathscr{S}(\mathscr{A}) = (\mathscr{L}, C_{\mathscr{L}}, \mathscr{A})$ *there exists an irreducible* $\mathbf{L}_\pi$-*theory* $\mathscr{S}(\mathscr{A}') = (\mathscr{L}, C_{\mathscr{L}}, \mathscr{A}')$ *such that* $C_{\mathscr{L}}(\mathscr{A}) \subset C_{\mathscr{L}}(\mathscr{A}')$.

An $\mathbf{L}_\pi$-theory $\mathscr{S}(\mathscr{A}) = (\mathscr{L}, C_{\mathscr{L}}, \mathscr{A})$ is said to be *prime* provided for any formulas α, β in $\mathscr{L}$

(2) if $(\alpha \cup \beta) \in C_{\mathscr{L}}(\mathscr{A})$, then either $\alpha \in C_{\mathscr{L}}(\mathscr{A})$ or $\beta \in C_{\mathscr{L}}(\mathscr{A})$.

It follows from the above definition, 6.2 and the definition of a prime filter that

6.7. *An* $\mathbf{L}_\pi$-*theory* $\mathscr{S}(\mathscr{A}) = (\mathscr{L}, C_{\mathscr{L}}, \mathscr{A})$ *is prime if and only if the filter* $\nabla_{\mathscr{A}}$ *in* $\mathfrak{A}(\mathscr{S})$ *is prime.*

As a corollary of 6.5, 6.7 and IV 2.2, we get the following statement.

6.8. *An* $\mathbf{L}_\pi$-*theory of zero order is irreducible if and only if it is prime.*

Hence, by 6.6, it follows that

6.9. *For every consistent* $\mathbf{L}_\pi$-*theory* $\mathscr{S}(\mathscr{A}) = (\mathscr{L}, C_{\mathscr{L}}, \mathscr{A})$ *there exists a prime* $\mathbf{L}_\pi$-*theory* $\mathscr{S}(\mathscr{A}') = (\mathscr{L}, C_{\mathscr{L}}, \mathscr{A}')$ *such that* $C_{\mathscr{L}}(\mathscr{A}) \subset C_{\mathscr{L}}(\mathscr{A}')$.

Prime $\mathbf{L}_\pi$-theories will be characterized in the next chapter.

Exercises

1. Let $\mathscr{L}_d$ be the formalized language obtained from the formalized language of $\mathscr{S}_\pi$ (see Sec. 1) by eliminating the conjunction sign. Take as logical axioms all formulas in $\mathscr{L}_d$ of the form (A_1)–(A_5) in Section 1 and modus ponens as the only rule of inference. Prove that the propositional calculus $\mathscr{S}_d = (\mathscr{L}_d, C_{\mathscr{L}_d})$ obtained in this way belongs to the class **S** (see VIII 5).

2. Prove that the class of all $\mathscr{S}_d$-algebras for the propositional calculus $\mathscr{S}_d$ of Ex. 1 coincides with the class of all algebras $(A, V, \Rightarrow, \cup)$ such that $(A, V, \Rightarrow)$ is a positive implication algebra and $a \Rightarrow (a \cup b) = V$, $b \Rightarrow (a \cup b) = V$, $(a \Rightarrow c) \Rightarrow ((b \Rightarrow c) \Rightarrow ((a \cup b) \Rightarrow c)) = V$ for all a, b, c in A. Show that $\mathscr{S}_d$ is consistent.

3. Prove that each irreducible implicative filter ∇ in any $\mathscr{S}_d$-algebra is prime i.e. if $a \cup b \in \nabla$, then either $a \in \nabla$ or $b \in \nabla$. Use this statement for the proof of the following representation theorem: every $\mathscr{S}_d$-algebra is isomorphic with an $\mathscr{S}_d$-algebra $(G_0(X), X, \Rightarrow, \cup)$ of some open subsets of a topological space X (the operation $\Rightarrow$ is defined as in positive implication algebras of sets (II 2 (6)) and $\cup$ is the set-theoretical union). Apply a similar argument to that used in II 4.1.

4. Prove an analogue of 3.4 for propositional calculi of the logic $\mathbf{L}_d$ determined by $\mathscr{S}_d$. Prove analogues of theorems in Sections 5, 6.

5. Prove that for any formula α of $\mathscr{S}_d$, α is derivable in $\mathscr{S}_d$ if and only if α is derivable in $\mathscr{S}_\pi$ [8].

6. Prove that for any formula α of $\mathscr{S}_{\pi\iota}$ (see IX 1), α is derivable in $\mathscr{S}_{\pi\iota}$ if and only if α is derivable in $\mathscr{S}_d$.

7. Let $\mathscr{L}_c$ be the formalized language obtained from the formalized language of $\mathscr{S}_\pi$ by eliminating the disjunction sign. Take as logical axioms all formulas of $\mathscr{L}_c$ of the form (A_1), (A_2), (A_6), (A_7), (A_8) in Sec. 1 and modus ponens as the only rule of inference. Prove that the propositional calculus $\mathscr{S}_c = (\mathscr{L}_c, C_{\mathscr{L}_c})$ obtained in this way belongs to the class **S** (see VIII 5).

8. Prove that the class of all $\mathscr{S}_c$-algebras for the propositional calculus $\mathscr{S}_c$ of Ex. 7 coincides with the class of all algebras $(A, V, \Rightarrow, \cap)$ such that $(A, V, \Rightarrow)$ is a positive implication algebra and, moreover, $(a \cap b) \Rightarrow a = V$, $(a \cap b) \Rightarrow b = V$, $(a \Rightarrow b) \Rightarrow ((a \Rightarrow c) \Rightarrow (a \Rightarrow (b \cap c))) = V$ for all a, b, c in A. Show that $\mathscr{S}_c$ is consistent.

9. Prove that in any $\mathscr{S}_c$-algebra $a \Rightarrow (b \Rightarrow (a \cap b)) = V$ holds identically. Use this statement for the proof that, for any implicative filter ∇ in each $\mathscr{S}_c$-algebra, $a \cap b \in \nabla$ if and only if $a \in \nabla$ and $b \in \nabla$. Applying this result and an argument similar to that of II 4.1 prove the following representation theorem: every $\mathscr{S}_c$-algebra is isomorphic to an $\mathscr{S}_c$-algebra $(G_0(X), X, \Rightarrow, \cap)$ of some open subsets of a top-

[8] Cf. Horn [1].

ological space X (the operation $\Rightarrow$ is defined as in positive implication algebras of sets (II 2 (6)) and $\cap$ is the set-theoretical intersection).

10. Prove an analogue of 3.4 for propositional calculi of the logic $\mathbf{L}_c$ determined by $\mathcal{S}_c$ of Ex. 7. Prove analogues of theorems in Sections 5 and 6.

11. Prove that for any formula α of $\mathcal{S}_c$ (see Ex. 7), α is derivable in $\mathcal{S}_c$ if and only if α is derivable in $\mathcal{S}_\pi$[9].

12. Prove that, for any formula α of $\mathcal{S}_{\pi\iota}$ (see IX 1), α is derivable in $\mathcal{S}_{\pi\iota}$ if and only if α is derivable in $\mathcal{S}_c$ (see Ex. 7).

(9) Cf. Horn [1].

CHAPTER XI

MINIMAL LOGIC, POSITIVE LOGIC WITH SEMI-NEGATION AND INTUITIONISTIC LOGIC

Introduction. Intuitionistic logic is closely connected with certain philosophical views on the foundations of mathematics, known as intuitionism (1). This trend in the foundations of mathematics, anticipated by Kant and later by such mathematicians as Kronecker and Poincaré, and developed in a systematic manner by Brouwer and his school, is a radical form of constructivism. Brouwer's objections to the application of the principle of the excluded third (2) in mathematical reasoning have caused its elimination in intuitionistic logic and hence the elimination of certain other laws of classical logic. Thus intuitionistic logic is weaker than classical logic. The formalization of intuitionistic logic is due to Heyting [1], [2]. The intuitionistic propositional and predicate calculi have been investigated from various points of view by many authors (3).

Stone [4] and Tarski [8] established connections between intuitionistic propositional calculi and the algebra of open (closed) subsets of topological spaces; these results were greatly developed and presented in an algebraic form by McKinsey and Tarski [2], [3]. Mostowski [2]

(1) For an exposition of the principles of intuitionism cf. Heyting [3], Beth [4], Kleene and Vesley [1], Kleene [2].

(2) See 7.5.

(3) Cf. Beth [3], Chandrasekharan [1], Dyson and Kreisel [1], Gentzen [1], Glivenko [1], Gödel [2], [3], [4], Henkin [2], Jaśkowski [1], Kleene [1], Kleene and Vesley [1], Kreisel and Putnam [1], Markov [1], McKinsey and Tarski [2], [3], Mostowski [2], Nelson [1], [2], Rasiowa [2]–[5], Rasiowa and Sikorski [3]–[6], Rieger [1], Stone [4], Tarski [8]; certain of above-mentioned authors in other papers and many other authors. For propositional calculi of *intermediate logics*, i.e. logics between intuitionistic and classical, see a survey by Hosoi and Ono [1].

indicated means of extending the algebraic methods of McKinsey and Tarski to intuitionistic predicate calculi [4]. His idea has been taken up by Henkin [2] and the present author [2]. The algebraic treatment of intuitionistic predicate calculi and elementary intuitionistic theories has been developed in several papers by the present author and Sikorski [5]. Since a comprehensive exposition of research on the algebraic approach to intuitionistic logic is given in [MM], it is presented here in a cursory manner and as far as this book is concerned is restricted to intuitionistic propositional calculi and theories of zero order.

Minimal logic and positive logic with semi-negation are still weaker than intuitionistic logic but stronger than positive logic. The first, formulated by Johansson [1], has rather peculiar metalogical properties. Note that the contrapositional negation occurring in the propositional calculi of this logic is characterized by the contraposition law (1 (A_9)). The second logic was constructed by Rasiowa and Sikorski [3]. It is based on the idea that the weakest property that can be required for negation is the following one: the negation of a true statement is a false statement. This property, formulated by means of the scheme of logical axioms (A_{10}) in Section 4, is characteristic for semi-negation, and together with the above-mentioned contraposition law characterizes intuitionistic negation in terms of positive implicative logic. Both minimal logic and positive logic with semi-negation have been examined from the algebraic point of view by Rasiowa and Sikorski [3], [5].

Minimal logic, which will be denoted by $\mathbf{L}_\mu$, is algebraically characterized by the class of all contrapositionally complemented lattices, i.e. the class of all $\mathscr{S}$-algebras for $\mathscr{S} \in \mathbf{L}_\mu$ coincides with that of all contrapositionally complemented lattices. Positive logic with semi-negation, denoted by $\mathbf{L}_\nu$, is characterized from the algebraic point of view by the class of all semi-complemented lattices. Pseudo-Boolean algebras play an analogous part for intuitionistic logic, to be denoted by $\mathbf{L}_\chi$.

The representation theorems (IV 4.4, IV 5.1, IV 6.7) and the lemmas on imbeddings (VI 11.3, VI 11.4, VI 11.1) for these three kinds of

[4] Cf. also Chandrasekharan [1].

[5] See [MM].

algebras enable us to obtain stronger forms of the completeness theorems (2.2, 5.2, 8.2) for the propositional calculi of $\mathbf{L}_\mu$, $\mathbf{L}_\nu$ and $\mathbf{L}_\chi$, respectively. The decidability of the corresponding propositional calculi follows from these completeness theorems. The representation theorems lead also to stronger forms of the theorems characterizing the formulas which are theorems of the formalized theories based on these logics (3.2, 6.3, 9.3). The consistency of $\mathbf{L}_\nu$-theories and of $\mathbf{L}_\chi$-theories is equivalent to the existence, for these theories, of $\mathbf{L}_\nu$-models and of $\mathbf{L}_\chi$-models, respectively. An analogous statement for $\mathbf{L}_\mu$-theories does not hold. In the propositional calculi of all the three logics a disjunction is derivable if and only if at least one of its parts is derivable.

The examination of deductive filters in these three kinds of algebras leads to the following metalogical results. The deduction theorems hold for formalized theories based on $\mathbf{L}_\mu$ and $\mathbf{L}_\chi$ but do not hold for $\mathbf{L}_\nu$-theories. All consistent theories can be extended to prime theories, but only those based on $\mathbf{L}_\nu$ and $\mathbf{L}_\chi$ can be extended to maximal ones.

A characterization of maximal intuitionistic theories is formulated in Theorem 9.7. Section 10 is concerned with a characterization of prime intuitionistic theories which can also be applied to prime theories based on positive logic.

1. Propositional calculus $\mathscr{S}_\mu$ of minimal logic[6]**.** Let $\mathscr{S}_\mu = (\mathscr{L}, C_{\mathscr{L}})$ be the propositional calculus described briefly as follows. The alphabet $A^0 = (V, L_0, L_1, L_2, U)$ differs from the alphabet of the formalized language of $\mathscr{S}_\pi$ (see X 1) only in admitting in L_1 one element denoted by $\sim_c$ and called the *contrapositional negation sign*. Let F be the set of all formulas over A^0. Assume that the set $\mathscr{A}_l$ of logical axioms consists of all formulas of the form (A_1)–(A_8) in X 1 (where α, β, γ, are any formulas in F and, moreover, of all formulas

$$(A_9) \qquad ((\alpha \Rightarrow \sim_c \beta) \Rightarrow (\beta \Rightarrow \sim_c \alpha)) \quad \text{for any } \alpha, \beta \in F.$$

The *consequence operation* $C_{\mathscr{L}}$ in the language described above $\mathscr{L} = (A^0, F)$ is determined by the set $\mathscr{A}_l$ of logical axioms and by modus ponens as the only rule of inference. The system $\mathscr{S}_\mu = (\mathscr{L}, C_{\mathscr{L}})$ is said to be a *propositional calculus of minimal logic*.

[6] Johansson [1].

Let us note that

1.1. *For any formulas* $\alpha, \beta \in F$

(i) $(\alpha \Rightarrow \alpha) \in C_{\mathscr{L}}(O)$,

(ii) $(\beta \Rightarrow \sim_c \sim_c \beta) \in C_{\mathscr{L}}(O)$,

(iii) $((\alpha \Rightarrow \beta) \Rightarrow (\sim_c \beta \Rightarrow \sim_c \alpha)) \in C_{\mathscr{L}}(O)$.

(i) follows from IX 1.1. By (A_9), (i) and modus ponens we get (ii). By (ii) and IX 1.1 (v)

(1) $$((\alpha \Rightarrow \beta) \Rightarrow (\alpha \Rightarrow \sim_c \sim_c \beta)) \in C_{\mathscr{L}}(O).$$

On the other hand, by (A_9)

(2) $$((\alpha \Rightarrow \sim_c \sim_c \beta) \Rightarrow (\sim_c \beta \Rightarrow \sim_c \alpha)) \in C_{\mathscr{L}}(O).$$

Thus (iii) follows from (1), (2) and IX 1.1 (iii).

1.2. *The system* $\mathscr{S}_\mu = (\mathscr{L}, C_{\mathscr{L}})$ *belongs to the class* **S** *of standard systems of implicative extensional propositional calculi.*

Clearly, the conditions VIII 5 (s_1), (s_2) are satisfied. By X 1.1 and X 1.2 the conditions VIII 5 (s_3)–(s_6), (s_8) hold. The condition VIII 5 (s_7) follows from 1.1 (iii). This completes the proof.

The next step is to characterize the class of all $\mathscr{S}_\mu$-algebras.

1.3. *The class of all* $\mathscr{S}_\mu$-*algebras coincides with the class of all contrapositionally complemented lattices* (see IV 4) [7].

The proof, by an easy verification similar to that of X 2.1, is left to the reader.

The next theorem follows from 1.3 and VIII 6.1.

1.4. *If* α *is a formula derivable in* $\mathscr{S}_\mu$, *then*

(3) $$\alpha_{\mathfrak{A}}(v) = V$$

for every valuation $v\colon V \to A$ *in every contrapositionally complemented lattice* $\mathfrak{A} = (A, V, \Rightarrow, \cup, \cap, \sim_c)$.

It follows from 1.4 that

1.5. *The propositional calculus* $\mathscr{S}_\mu$ *is consistent.*

2. Minimal logic $\mathbf{L}_\mu$. By the *minimal logic* $\mathbf{L}_\mu$ we mean the class of all consistent standard systems of implicative extensional prop-

[7] Rasiowa and Sikorski [3].

ositional calculi logically equivalent (see VIII 8) to $\mathcal{S}_\mu$. Thus, because of 1.3, the class $\mathbf{K}_{\mathbf{L}_\mu}$ of all $\mathcal{S}$-algebras for $\mathcal{S} \in \mathbf{L}_\mu$ is the class of all contrapositionally complemented lattices. Hence, by VIII 6.5 and VIII 6.6,

2.1. *For every $\mathcal{S} \in \mathbf{L}_\mu$, the algebra $\mathfrak{A}(\mathcal{S})$ is a non-degenerate contrapositionally complemented lattice. Moreover, $\mathfrak{A}(\mathcal{S})$ is a free algebra in the class of all contrapositionally complemented lattices.*

A formula α in $\mathcal{S} \in \mathbf{L}_\mu$ is said to be $\mathbf{L}_\mu$-*valid* provided that for any valuation v of the formalized language of $\mathcal{S}$ in any contrapositionally complemented lattice $\mathfrak{A} = (A, V, \Rightarrow, \cup, \cap, \sim_c)$,

$$\alpha_{\mathfrak{A}}(v) = V.$$

The following theorem characterizes all formulas derivable in any system $\mathcal{S} \in \mathbf{L}_\mu$.

2.2. *For any formula α of any system $\mathcal{S} \in \mathbf{L}_\mu$, the following conditions are equivalent*:

(i) *α is derivable in $\mathcal{S}$*,

(ii) *α is $\mathbf{L}_\mu$-valid*,

(iii) *α is valid in every contrapositionally complemented lattice $\mathfrak{G}(X) = (G(X), X, \Rightarrow, \cup, \cap, \sim_c)$ of open subsets of any topological space X*,

(iv) *$\alpha_{\mathfrak{A}(\mathcal{S})}(v^0) = V$ for the canonical valuation v^0 in the algebra $\mathfrak{A}(\mathcal{S})$ of the system $\mathcal{S}$*,

(v) *α is valid in every contrapositionally complemented lattice $\mathfrak{A}$ with at most $2^{2^{r+1}}$ elements, where r is the number of all subformulas of α* [(8)].

The equivalence of the conditions (i) and (ii) follows from VIII 8.1 and is called the *completeness theorem for propositional calculi of minimal logic*. Obviously, (ii) implies (iii) and, by the representation theorem IV 4.4, (iii) implies (iv).

By VIII 7.2, (i) follows from (iv). Clearly, (ii) implies (v). The easy proof that (v) implies (i), based on VI 11.3 and similar to the proof of X 3.4, is left to the reader.

The following statement follows from the equivalence of (i) and (v).

(8) See Rasiowa and Sikorski [3].

2.3. *Every propositional calculus* $\mathscr{S} \in \mathbf{L}_\mu$ *is decidable.*

The next theorem follows from 2.2, X 3.4, X 3.6 and 1.1.

2.4. *For any system* $\mathscr{S}$ *of the minimal logic, the formulas of the form* (A_1)–(A_8) *in* X 1, (A_9) *in Sec.* 1, (D_3)–(D_{11}) *in* IX 3, (D_{16})–(D_{32}) *in* X 3 *and, moreover,*

(D_{33}) $$(\alpha \Rightarrow \sim_c \sim_c \alpha),$$

(D_{34}) $$((\alpha \Rightarrow \beta) \Rightarrow (\sim_c \beta \Rightarrow \sim_c \alpha)),$$

are derivable in $\mathscr{S}$.

2.5. *Let* $\mathscr{S} = (\mathscr{L}, C_{\mathscr{L}})$ *be a propositional calculus of the minimal logic* $\mathbf{L}_\mu$ *and let* $\mathscr{S}_0 = (\mathscr{L}_0, C_{\mathscr{L}_0})$ *be a propositional calculus of the positive logic* $\mathbf{L}_\pi$ *such that* $\mathscr{L}$ *is an extension of* $\mathscr{L}_0$. *Then for every formula* α *in* $\mathscr{L}_0$ *the following conditions are equivalent*:

(i) α *is derivable in* $\mathscr{S}$,

(ii) α *is derivable in* $\mathscr{S}_0$.

The proof, similar to that of X 3.7 and based on 2.2, the definition of the contrapositionally complemented lattice (IV 4), X 3.4 and IV 4.1, is left to the reader.

Let us note that the following theorem, analogous to X 4.1, holds.

2.6. *Let* $\mathscr{S} = (\mathscr{L}, C_{\mathscr{L}})$ *be a propositional calculus of minimal logic. Then for any formulas* α, β *in* $\mathscr{L}$,

$(\alpha \cup \beta) \in C_{\mathscr{L}}(O)$ *if and only if either* $\alpha \in C_{\mathscr{L}}(O)$ *or* $\beta \in C_{\mathscr{L}}(O)$ [9].

The proof, analogous that of X 4.1 and based on 2.1, IV 4.4, VIII 3.2, 2.2, VI 8.1, is left to the reader. The following remark will be helpful. Let $\sim_c X = Z$ in $\mathfrak{G}(X) = (G(X), X, \Rightarrow, \cup, \cap, \sim_c)$. Then we define the operation $\sim_c$ in $\mathfrak{G}_0(X_0)$ as follows: $\sim_c Y = Y \Rightarrow Z$ for any $Y \in G_0(X_0)$, where $\Rightarrow$ on the right side is taken in $\mathfrak{G}_0(X_0)$ (not in $\mathfrak{G}(X)$).

3. $\mathbf{L}_\mu$-theories of zero order and their connection with filters. For any system $\mathscr{S} = (\mathscr{L}, C_{\mathscr{L}})$ in $\mathbf{L}_\mu$ and for any set $\mathscr{A}$ of formulas in $\mathscr{L}$ the system $\mathscr{S}(\mathscr{A}) = (\mathscr{L}, C_{\mathscr{L}}, \mathscr{A})$ is said to be an $\mathbf{L}_\mu$-*theory of zero order based on* $\mathscr{S}$ (cf. VIII 9). Every valuation v of $\mathscr{L}$ in any non-degenerate

[9] See Rasiowa and Sikorski [5].

contrapositionally complemented lattice $\mathfrak{A} = (A, V, \Rightarrow, \cup, \cap, \sim_c)$ such that, for all α in $\mathscr{A}$,

(1) $$\alpha_{\mathfrak{A}}(v) = V$$

is said to be an $\mathbf{L}_\mu$*-model of* $\mathscr{S}(\mathscr{A})$.

The following theorem follows from VIII 9.2, VIII 9.3 and 1.3.

3.1. *For every consistent* $\mathbf{L}_\mu$*-theory* $\mathscr{S}(\mathscr{A}) = (\mathscr{L}, C_{\mathscr{L}}, \mathscr{A})$ *of zero order, the algebra* $\mathfrak{A}(\mathscr{S}(\mathscr{A}))$ (see VIII 9) *is a non-degenerate contrapositionally complemented lattice. The canonical valuation* v^0 *of* $\mathscr{L}$ *in* $\mathfrak{A}(\mathscr{S}(\mathscr{A}))$, i.e. *a valuation defined by*

(1) $v^0(p) = ||p||_{\mathscr{A}}$ *for each propositional variable* p *in* $\mathscr{L}$,

is an $\mathbf{L}_\mu$*-model of* $\mathscr{S}(\mathscr{A})$. *Moreover, for any formula* α *in* $\mathscr{L}$

(2) $\alpha \in C_{\mathscr{L}}(\mathscr{A})$ *if and only if* $\alpha_{\mathfrak{A}(\mathscr{S}(\mathscr{A}))}(v^0) = V$.

The next theorem characterizes all theorems of any consistent $\mathbf{L}_\mu$-theory of zero order.

3.2. *For any formula* α *of every consistent* $\mathbf{L}_\mu$*-theory* $\mathscr{S}(\mathscr{A}) = (\mathscr{L}, C_{\mathscr{L}}, \mathscr{A})$ *of zero order the following conditions are equivalent*:

(i) $\alpha \in C_{\mathscr{L}}(\mathscr{A})$,

(ii) *every* $\mathbf{L}_\mu$*-model of* $\mathscr{S}(\mathscr{A})$ *is an* $\mathbf{L}_\mu$*-model of* α,

(iii) *every* $\mathbf{L}_\mu$*-model of* $\mathscr{S}(\mathscr{A})$ *in any non-degenerate contrapositionally complemented lattice* $\mathfrak{G}(X) = (G(X), X, \Rightarrow, \cup, \cap, \sim_c)$ *of open subsets of any topological space* X *is an* $\mathbf{L}_\mu$*-model of* α,

(iv) $\alpha_{\mathfrak{A}(\mathscr{S}(\mathscr{A}))}(v^0) = V$ *for the canonical valuation* v^0 *in the algebra* $\mathfrak{A}(\mathscr{S}(\mathscr{A}))$ *of the theory* $\mathscr{S}(\mathscr{A})$.

The conditions (i), (ii), (iv) are equivalent by VIII 9.5. Clearly, (ii) implies (iii) and, by the representation theorem IV 4.4, (iii) implies (ii), which completes the proof.

The next theorem states the connection between $\mathbf{L}_\mu$-theories and $\mathbf{L}_\pi$-theories of zero order.

3.3. *Let* $\mathscr{S}(\mathscr{A}) = (\mathscr{L}, C_{\mathscr{L}}, \mathscr{A})$ *be an* $\mathbf{L}_\mu$*-theory and let* $\mathscr{S}_0 = (\mathscr{L}_0, C_{\mathscr{L}_0})$ *be a propositional calculus of the positive logic* $\mathbf{L}_\pi$ *such that*

$\mathscr{L}$ is an extension of $\mathscr{L}_0$. If each formula α in $\mathscr{A}$ is a formula in $\mathscr{L}_0$, then for every formula β in $\mathscr{L}_0$ the following conditions are equivalent:

(i) *β is a theorem of $\mathscr{S}(\mathscr{A})$,*
(ii) *β is a theorem of the $\mathbf{L}_\pi$-theory $\mathscr{S}_0(\mathscr{A}) = (\mathscr{L}_0, C_{\mathscr{L}_0}, \mathscr{A})$.*

If β in $\mathscr{L}_0$ is not a theorem of $\mathscr{S}(\mathscr{A})$, then by 3.2 there exists an $\mathbf{L}_\mu$-model v of $\mathscr{S}(\mathscr{A})$ in a contrapositionally complemented lattice $\mathfrak{A} = (A, V, \Rightarrow, \cup, \cap, \sim_c)$ such that $\beta_{\mathfrak{A}}(v) \neq V$. The valuation v, restricted to $\mathscr{L}_0$ and treated as a valuation in the relatively pseudo-complemented lattice $\mathfrak{A}' = (A, V, \Rightarrow, \cup, \cap)$, is an $\mathbf{L}_\pi$-model of $\mathscr{S}_0(\mathscr{A})$ and $\beta_{\mathfrak{A}'}(v) = \beta_{\mathfrak{A}}(V) \neq V$. Thus, by X 5.2, β is not a theorem of $\mathscr{S}_0(\mathscr{A})$. Conversely, if $\beta \notin C_{\mathscr{L}_0}(\mathscr{A})$, then by X 5.2 there exists an $\mathbf{L}_\pi$-model v' of $\mathscr{S}_0(\mathscr{A})$ in a relatively pseudo-complemented lattice $\mathfrak{A}' = (A, V, \Rightarrow, \cup, \cap)$ such that $\beta_{\mathfrak{A}'}(v') \neq V$. By IV 4.1, $\mathfrak{A}'$ can be extended to a contrapositionally complemented lattice $\mathfrak{A} = (A, V, \Rightarrow, \cup, \cap, \sim_c)$. Let v be a valuation of $\mathscr{L}$ in $\mathfrak{A}$ such that $v(p) = v'(p)$ for every propositional variable p in $\mathscr{L}_0$. Then v is an $\mathbf{L}_\mu$-model of $\mathscr{S}(\mathscr{A})$ and $\beta_{\mathfrak{A}}(v) = \beta_{\mathfrak{A}'}(v') \neq V$, which by 3.2 implies that $\beta \notin C_{\mathscr{L}}(\mathscr{A})$.

Since the only rule of inference of the propositional calculus $\mathscr{S}_\mu$ is modus ponens, it follows from VIII 12.3, IV 2.1 and the definition of a contrapositionally complemented lattice that

3.4. *For any set ∇ of elements in a contrapositionally complemented lattice the following conditions are equivalent*:

(i) *∇ is a deductive filter,*
(ii) *∇ is an implicative filter,*
(iii) *∇ is a filter.*

It follows from 3.4, VIII 12.2, VIII 13.10, VIII 13.12 and the equational definability of the class of all contrapositionally complemented lattices (see IV 4) that the kernels of epimorphisms from contrapositionally complemented lattices onto similar algebras are filters and, conversely, every filter is the kernel of the epimorphism determined by itself.

By 3.4, VIII 13.2 and VIII 13.4 we get two theorems analogous to X 6.2 and X 6.3, which state that there exists a natural one-one correspondence between filters in the algebra $\mathfrak{A}(\mathscr{S})$ of a system $\mathscr{S}$

in $\mathbf{L}_\mu$ and formalized $\mathbf{L}_\mu$-theories based on $\mathscr{S}$, two theories $\mathscr{S}(\mathscr{A})$ and $\mathscr{S}(\mathscr{A}')$ being identified if and only if they have the same set of theorems.

Let us note that for any system $\mathscr{S}$ of minimal logic there holds a theorem corresponding to IX 5.4. Moreover, deduction theorems analogous to IX 5.5, IX 5.6, X 6.4 also hold. They can be proved in a similar way. Clearly, theorems analogous to X 6.5–X 6.9 can also be deduced by a similar argument. The formulation and proofs are left to the reader.

4. Propositional calculus $\mathscr{S}_\nu$ of positive logic with semi-negation [10]**.** Let $\mathscr{S}_\nu = (\mathscr{L}, C_{\mathscr{L}})$ be the propositional calculus described briefly as follows. The alphabet $A^0 = (V, L_0, L_1, L_2, U)$ of $\mathscr{L}$ differs from the alphabet of the formalized language of $\mathscr{S}_\pi$ (see X 1) only in admitting in L_1 one element denoted by $\dot{-}$ and called the *semi-negation sign*. Let F be the set of all formulas over A^0.

By definition $\mathscr{L} = (A^0, F)$. Assume that the set $\mathscr{A}_l$ of logical axioms consists of all formulas of the form (A_1)–(A_8) in X 1 (where α, β, γ are any formulas in F) and, moreover, of all formulas

$$(\mathrm{A}_{10}) \qquad (\dot{-}(\alpha \Rightarrow \alpha) \Rightarrow \beta) \qquad \text{for any } \alpha,\ \beta \text{ in } F.$$

The consequence operation $C_{\mathscr{L}}$ in $\mathscr{L}$ is determined by the set $\mathscr{A}_l$ of logical axioms and by the following rules of inference: modus ponens and

$$(r_\nu)\ \frac{(\alpha \Rightarrow \beta),\ (\beta \Rightarrow \alpha)}{(\dot{-}\alpha \Rightarrow \dot{-}\beta)} \qquad \text{for any } \alpha,\ \beta \text{ in } F.$$

The system $\mathscr{S}_\nu = (\mathscr{L}, C_{\mathscr{L}})$ is said to be a *propositional calculus of positive logic with semi-negation*.

Note that

4.1. *The system $\mathscr{S}_\nu = (\mathscr{L}, C_{\mathscr{L}})$ belongs to the class* $\mathbf{S}$ *of standard systems of implicative extensional propositional calculi.*

Clearly, conditions VIII 5 $((\mathrm{s}_1), (\mathrm{s}_2))$ are satisfied. By X 1.1 and X 1.2 conditions VIII 5 $((\mathrm{s}_3)$–$(\mathrm{s}_6), (\mathrm{s}_8))$ hold. Condition VIII 5 (s_7) follows from (r_ν). This completes the proof.

The next theorem characterizes the class of all $\mathscr{S}_\nu$-algebras.

[10] Rasiowa and Sikorski [3].

4.2. *The class of all $\mathscr{S}_\nu$-algebras coincides with the class of all semi-complemented lattices* (see IV 5) [11].

The proof, by an easy verification similar to that of X 2.1, is left to the reader.

The next theorem follows from 4.2 and VIII 6.1.

4.3. *If α is a formula derivable in $\mathscr{S}_\nu$, then $\alpha_{\mathfrak{A}}(v) = \vee$ for every valuation v of $\mathscr{L}$ in any semi-complemented lattice $\mathfrak{A} = (A, \vee, \Rightarrow, \cup, \cap, \dot{-})$.*

It follows from 4.3 that

4.4. *The propositional calculus $\mathscr{S}_\nu$ is consistent.*

5. Positive logic with semi-negation $\mathbf{L}_\nu$. The class of all consistent standard systems of implicative extensional propositional calculi logically equivalent (see VIII 8) to $\mathscr{S}_\nu$ will be said to be the *positive logic with semi-negation* and will be denoted by $\mathbf{L}_\nu$. By 4.2 the class $\mathbf{K}_{\mathbf{L}_\nu}$ of all $\mathscr{S}$-algebras for $\mathscr{S} \in \mathbf{L}_\nu$ is the class of all semi-complemented lattices. Hence, by VIII 6.5 and VIII 6.6, the following theorem holds.

5.1. *For every $\mathscr{S} \in \mathbf{L}_\nu$ the algebra $\mathfrak{A}(\mathscr{S})$ is a non-degenerate semi-complemented lattice. Moreover, $\mathfrak{A}(\mathscr{S})$ is a free algebra in the class of all semi-complemented lattices.*

A formula α of $\mathscr{S} \in \mathbf{L}_\nu$ is said to be $\mathbf{L}_\nu$-*valid* provided that for any valuation v of the formalized language of $\mathscr{S}$ in any semi-complemented lattice $\mathfrak{A} = (A, \vee, \Rightarrow, \cup, \cap, \dot{-})$

$$\alpha_{\mathfrak{A}}(v) = \vee .$$

The following theorem characterizes all formulas derivable in any system $\mathscr{S} \in \mathbf{L}_\nu$.

5.2. *For any formula α of any system $\mathscr{S} \in \mathbf{L}_\nu$, the following conditions are equivalent*:

(i) *α is derivable in $\mathscr{S}$,*

(ii) *α is $\mathbf{L}_\nu$-valid,*

(iii) *α is valid in every semi-complemented lattice $\mathfrak{G}(X) = (G(X), X, \Rightarrow, \cup, \cap, \dot{-})$ of open subsets of any topological space X,*

(iv) *$\alpha_{\mathfrak{A}(\mathscr{S})}(v^0) = \vee$ for the canonical valuation v^0 in the algebra $\mathfrak{A}(\mathscr{S})$ of the system $\mathscr{S}$,*

[11] Rasiowa and Sikorski [3].

(v) *α is valid in every semi-complemented lattice $\mathfrak{A}$ with at most 2^{2^r} elements, where r is the number of all subformulas of α* [12].

The equivalence of (i) and (ii) follows from VIII 8.1 and is called the *completeness theorem for propositional calculi of positive logic with semi-negation.* Clearly, (ii) implies (iii) and by the representation theorem IV 5.1, (iii) implies (iv). By VIII 7.2, (i) follows from (iv). Obviously, (ii) implies (v). The easy proof that (v) implies (i), based on VI 11.4 and similar to the proof of X 3.4, is left to the reader.

As a corollary of the equivalence of (i) and (v) we get the following result.

5.3. *Every propositional calculus $\mathcal{S} \in \mathbf{L}_\nu$ is decidable.*

The next theorem follows from 5.2, X 3.4, X 3.6 and IV 5 (s).

5.4. *For any system $\mathcal{S} \in \mathbf{L}_\nu$ the formulas of the form* (A_1)-(A_8) *in* X 1, (A_{10}) *in Sec.* 4, (D_3)–(D_{11}) *in* IX 3, (D_{16})–(D_{32}) *in* X 3 *are derivable in $\mathcal{S}$.*

5.5. *Let $\mathcal{S} = (\mathcal{L}, C_{\mathcal{L}})$ be a propositional calculus in $\mathbf{L}_\nu$ and let $\mathcal{S}_0 = (\mathcal{L}_0, C_{\mathcal{L}_0})$ be a propositional calculus of the positive logic $\mathbf{L}_\pi$ such that $\mathcal{L}$ is an extension of $\mathcal{L}_0$. Then for every formula α in $\mathcal{L}_0$ the following conditions are equivalent*:

(i) *α is derivable in $\mathcal{S}$*,

(ii) *α is derivable in $\mathcal{S}_0$*.

The proof, similar to that of X 3.7 and based on 5.2 and the definition of semi-complemented lattices (IV 5), X 3.4, is left to the reader.

Let us note the validity of the following theorem, analogous to X 4.1.

5.6. *Let $\mathcal{S} = (\mathcal{L}, C_{\mathcal{L}})$ be a propositional calculus in $\mathbf{L}_\nu$. Then for any formulas α, β in $\mathcal{L}$,*

$(\alpha \cup \beta) \in C_{\mathcal{L}}(O)$ if and only if either $\alpha \in C_{\mathcal{L}}(O)$ or $\beta \in C_{\mathcal{L}}(O)$ [13].

The proof, analogous to that of X 4.1 and based on 5.1, IV 5.1, VIII 3.2, VI 8.1, 5.2, is left to the reader. The following remark will be helpful. If $\mathfrak{G}(X) = (G(X), X, \Rightarrow, \cup, \cap, \dot{-})$ is a semi-complemented lattice of all open subsets of a topological space X and X_0 is the one-

[12] Cf. Rasiowa and Sikorski [3].

[13] Rasiowa and Sikorski [5].

point strong compactification of X, then we define the operation $\dot{-}$ in $\mathfrak{G}_0(X_0) = (G_0(X_0), X_0, \Rightarrow, \cup, \cap, \dot{-})$ as follows: if $Y \in G_0(X_0)$, then $\dot{-}Y = \dot{-}(Y \cap X)$, where $\dot{-}$ on the right side is taken in $\mathfrak{G}(X)$.

6. $\mathbf{L}_\nu$-theories of zero order and their connection with filters. For any system $\mathscr{S} = (\mathscr{L}, C_{\mathscr{L}}) \in \mathbf{L}_\nu$ and for any set $\mathscr{A}$ of formulas in $\mathscr{L}$, the system $\mathscr{S}(\mathscr{A}) = (\mathscr{L}, C_{\mathscr{L}}, \mathscr{A})$ is said to be an $\mathbf{L}_\nu$-*theory of zero order based on* $\mathscr{S}$. Every valuation v of $\mathscr{L}$ in any non-degenerate semi-complemented lattice $\mathfrak{A} = (A, V, \Rightarrow, \cup, \cap, \dot{-})$ such that for all $\alpha \in \mathscr{A}$

(1) $$\alpha_{\mathfrak{A}}(v) = V$$

is said to be an $\mathbf{L}_\nu$-*model of* $\mathscr{S}(\mathscr{A})$.

The following theorem follows from VIII 9.2, VIII 9.3 and 4.2.

6.1. *For every consistent* $\mathbf{L}_\nu$-*theory* $\mathscr{S}(\mathscr{A}) = (\mathscr{L}, C_{\mathscr{L}}, \mathscr{A})$ *the algebra* $\mathfrak{A}(\mathscr{S}(\mathscr{A}))$ (see VIII 9) *is a non-degenerate semi-complemented lattice. The canonical valuation* (cf. 3 (1)) v_0 *of* $\mathscr{L}$ *in* $\mathfrak{A}(\mathscr{S}(\mathscr{A}))$ *is an* $\mathbf{L}_\nu$-*model of* $\mathscr{S}(\mathscr{A})$. *Moreover, for any formula* α *in* $\mathscr{L}$

(2) $$\alpha \in C_{\mathscr{L}}(\mathscr{A}) \quad \textit{if and only if} \quad \alpha_{\mathfrak{A}}(v^0) = V.$$

By 5.4 for any $\mathscr{S} \in \mathbf{L}_\nu$ all formulas of the form $(\dot{-}(\alpha \Rightarrow \alpha) \Rightarrow \beta)$ (where α, β are any formulas of $\mathscr{S}$) are derivable in $\mathscr{S}$. Hence, every system $\mathscr{S}$ in $\mathbf{L}_\nu$ belongs to the class $\mathbf{S}_\nu$ of standard systems of implicative extensional propositional calculi with semi-negation (see VIII 10). Consequently, by VIII 10.2 and IV 5.1, the following theorem holds.

6.2. *For any* $\mathbf{L}_\nu$-*theory* $\mathscr{S}(\mathscr{A}) = (\mathscr{L}, C_{\mathscr{L}}, \mathscr{A})$ *of zero order the following conditions are equivalent*:

(i) $\mathscr{S}(\mathscr{A})$ *is consistent,*

(ii) *there exists an* $\mathbf{L}_\nu$-*model of* $\mathscr{S}(\mathscr{A})$,

(iii) *there exists an* $\mathbf{L}_\nu$-*model of* $\mathscr{S}(\mathscr{A})$ *in a semi-complemented lattice of open subsets of a topological space,*

(iv) *for any formula* α *of* $\mathscr{S}(\mathscr{A})$ *either* $\alpha \notin C_{\mathscr{L}}(\mathscr{A})$ *or* $\dot{-}\alpha \notin C_{\mathscr{L}}(\mathscr{A})$.

The next theorem characterizes all theorems of any consistent $\mathbf{L}_\nu$-theory of zero order.

6.3. *For any formula* α *of any consistent* $\mathbf{L}_\nu$-*theory of zero order the following conditions are equivalent*:

(i) $\alpha \in C_{\mathscr{L}}(\mathscr{A})$,

(ii) *every* $\mathbf{L}_\nu$-*model of* $\mathscr{S}(\mathscr{A})$ *is an* $\mathbf{L}_\nu$-*model of* α,

(iii) *every* $\mathbf{L}_\nu$-*model of* $\mathscr{S}(\mathscr{A})$ *in any semi-complemented lattice* $\mathfrak{G}(X) = (G(X), X, \Rightarrow, \cup, \cap, \dot{-})$ *of open subsets of any topological space* X *is an* $\mathbf{L}_\nu$-*model of* α,

(iv) $\alpha_{\mathfrak{A}(\mathscr{S}(\mathscr{A}))}(v^0) = V$ *for the canonical valuation* v^0 *in the algebra* $\mathfrak{A}(\mathscr{S}(\mathscr{A}))$ *of the theory* $\mathscr{S}(\mathscr{A})$.

Conditions (i), (ii), (iv) are equivalent by VIII 9.5. Clearly, (ii) implies (iii) and by the representation theorem IV 5.1 condition (iii) implies (ii), which completes the proof.

In the next theorem the connection between $\mathbf{L}_\nu$-theories and $\mathbf{L}_\pi$-theories of zero order is given.

6.4. *Let* $\mathscr{S}(\mathscr{A}) = (\mathscr{L}, C_{\mathscr{L}}, \mathscr{A})$ *be an* $\mathbf{L}_\nu$-*theory of zero order and let* $\mathscr{S}_0 = (\mathscr{L}_0, C_{\mathscr{L}_0})$ *be a propositional calculus of the positive logic* $\mathbf{L}_\pi$ *such that* $\mathscr{L}$ *is an extension of* $\mathscr{L}_0$. *If each formula* α *in* $\mathscr{A}$ *is a formula in* $\mathscr{L}_0$, *then for every formula* β *in* $\mathscr{L}_0$ *the following conditions are equivalent*:

(i) β *is a theorem of* $\mathscr{S}(\mathscr{A})$,

(ii) β *is a theorem of the* $\mathbf{L}_\pi$-*theory* $\mathscr{S}_0(\mathscr{A}) = (\mathscr{L}_0, C_{\mathscr{L}_0}, \mathscr{A})$.

The proof, similar to the proof of 3.3 and based on 6.3, X 5.2, and the definition of semi-complemented lattices, if left to the reader.

A filter ∇ in a semi-complemented lattice $\mathfrak{A} = (A, V, \Rightarrow, \cup, \cap, \dot{-})$ will be said to be a ν-*filter*, provided the following condition is satisfied:

$$(\mathrm{f}_\nu) \qquad \text{if} \quad a \Rightarrow b,\ b \Rightarrow a \in \nabla, \quad \text{then} \quad \dot{-}a \Rightarrow \dot{-}b \in \nabla.$$

6.5. *A subset* ∇ *of the set of elements in a semi-complemented lattice* $\mathfrak{A}$ *is a deductive filter if and only if* ∇ *is a* ν-*filter*.

By VIII 12.3, Sec. 4 and IV 2.1, ∇ is a deductive filter if and only if ∇ is a filter such that for any formulas α, β of $\mathscr{S}_\nu$ and for every valuation v of the formalized language of $\mathscr{S}_\nu$ in $\mathfrak{A}$

$$(3) \qquad \alpha_{\mathfrak{A}}(v) \Rightarrow \beta_{\mathfrak{A}}(v),\ \beta_{\mathfrak{A}}(v) \Rightarrow \alpha_{\mathfrak{A}}(v) \in \nabla \quad \text{implies that} \quad \dot{-}\alpha_{\mathfrak{A}}(v) \Rightarrow \dot{-}\beta_{\mathfrak{A}}(v) \in \nabla.$$

Clearly, conditions (f_ν) and (3) are equivalent.

It follows from 6.5, VIII 12.2, VIII 13.10, VIII 13.12 and the equational definability of the class of all semi-complemented lattices (see IV

5) that the kernels of epimorphisms from semi-complemented lattices onto similar algebras are ν-filters and, conversely, every ν-filter is the kernel of the epimorphism determined by itself.

By 6.1, VIII 13.2 and VIII 13.4 we get two theorems analogous to X 6.2 and X 6.3, which state that there exists a natural one-one correspondence between ν-filters in the algebra $\mathfrak{A}(\mathcal{S})$ of a system $\mathcal{S}$ in $\mathbf{L}_\nu$ and formalized $\mathbf{L}_\nu$-theories based on $\mathcal{S}$, two theories being identified if and only if they have the same set of theorems.

Note that a theorem analogous to IX 5.5 (the deduction theorem) does not hold. Indeed, let $\mathcal{S} = (\mathcal{L}, C_{\mathcal{L}})$ be a system in $\mathbf{L}_\nu$ and let p, q be any distinct propositional variables of $\mathcal{L}$. It is easy to see that $(\dot{-}p \Rightarrow \dot{-}q) \in C_{\mathcal{L}}(\{((p \Rightarrow q) \cap (q \Rightarrow p))\})$, but $(((p \Rightarrow q) \cap (q \Rightarrow p)) \Rightarrow (\dot{-}p \Rightarrow \dot{-}q)) \notin C_{\mathcal{L}}(O)$. Let $\mathfrak{G}(X) = (G(X), X, \Rightarrow, \cup, \cap, \dot{-})$ be the semi-complemented lattice of all open subsets of the real line X such that $\dot{-}X = O$ and $\dot{-}Y = X$ for every $Y \neq X$, $Y \in G(X)$. Then by a valuation v of $\mathcal{L}$ in $G(X)$ such that $v(p) = \{x \in X\colon 0 < x\}$ and $v(q) = X$ the above formula does not admit the value X, which by 5.2 proves our statement. Consequently, theorems corresponding to IX 5.6 and X 6.4 do not hold, either.

Theorems which are special cases of VIII 12.5, VIII 12.6 and VIII 12.7 for ν-filters follow from the above-mentioned theorems and 6.5.

Since every system $\mathcal{S}$ in $\mathbf{L}_\nu$ belongs to $\mathbf{S}_\nu$ (see VIII 10), by VIII 13.8 the following theorem holds.

6.6. *For every consistent* $\mathbf{L}_\nu$-*theory* $\mathcal{S}(\mathcal{A}) = (\mathcal{L}, C_{\mathcal{L}}, \mathcal{A})$ *there exists a maximal* $\mathbf{L}_\nu$-*theory* $\mathcal{S}(\mathcal{A}') = (\mathcal{L}, C_{\mathcal{L}}, \mathcal{A}')$ *such that* $C_{\mathcal{L}}(\mathcal{A}) \subset C_{\mathcal{L}}(\mathcal{A}')$.

Clearly, Theorems VIII 13.7 and VIII 13.9 hold also for $\mathbf{L}_\nu$-theories.

7. Propositional calculus $\mathcal{S}_\chi$ of intuitionistic logic[14]**.** Let $\mathcal{S}_\chi = (\mathcal{L}, C_{\mathcal{L}})$ be a propositional calculus described briefly as follows. The alphabet $A^0 = (V, L_0, L_1, L_2, U)$ of $\mathcal{L}$ differs from the alphabet of the formalized language of $\mathcal{S}_\pi$ (see X 1) only in admitting in L_1 one element denoted by $\urcorner$ and called the *intuitionistic negation sign*. Let F be the set of all formulas over A^0. By definition $\mathcal{L} = (A^0, F)$. Assume

[14] See Heyting [1].

that the set $\mathscr{A}_l$ of logical axioms consists of all formulas of the form (A_1)–(A_8) in X 1 (where α, β, γ are any formulas in F) and, moreover, of all formulas

(A_9) $((\alpha \Rightarrow \neg\beta) \Rightarrow (\beta \Rightarrow \neg\alpha))$,

(A_{10}) $(\neg(\alpha \Rightarrow \alpha) \Rightarrow \beta)$ (15)

for any α, β in F.

The consequence operation $C_{\mathscr{L}}$ in $\mathscr{L}$ is determined by the set $\mathscr{A}_l$ of logical axioms and by modus ponens as the only rule of inference.

The system $\mathscr{S}_\chi = (\mathscr{L}, C_{\mathscr{L}})$ is said to be a *propositional calculus of intuitionistic logic*.

By the same argument as that used in the proof of 1.1 we get

7.1. *For any formulas* α, β *in* F

(1) $$((\alpha \Rightarrow \beta) \Rightarrow (\neg\beta \Rightarrow \neg\alpha)) \in C_{\mathscr{L}}(O).$$

7.2. *The system* $\mathscr{S}_\chi = (\mathscr{L}, C_{\mathscr{L}})$ *belongs to the class* **S** *of standard systems of implicative extensional propositional calculi.*

Clearly, conditions VIII 5 $((s_1), (s_2))$ are satisfied. By X 1.1 and X 1.2 conditions VIII 5 $((s_3)$–$(s_6), (s_8))$ hold. Condition VIII 5 (s_7) is satisfied by 7.1. This completes the proof.

The next theorem characterizes the class of all $\mathscr{S}_\chi$-algebras.

7.3. *The class of all* $\mathscr{S}_\chi$*-algebras coincides with the class of all pseudo-Boolean algebras* (see IV 6) (16).

The proof, by an easy verification using X 2.1, is left to the reader.

The next theorem follows from 7.3 and VIII 6.1.

7.4. *If a formula* α *is derivable in* $\mathscr{S}_\chi$, *then*

(2) $$\alpha_{\mathfrak{A}}(v) = V$$

for every valuation $v\colon V \to A$ *in any pseudo-Boolean algebra* $\mathfrak{A} = (A, V, \Rightarrow, \cup, \cap, \neg)$.

It easily follows from 7.4 that

7.5. *The propositional calculus* $\mathscr{S}_\chi$ *is consistent.*

Note that the formula $(p \cup \neg p)$, where p is a propositional variable of $\mathscr{L}$, is not derivable in $\mathscr{S}_\chi$. Let $\mathfrak{G}(X) = (G(X), X, \Rightarrow, \cup, \cap, \neg)$

(15) A. Mostowski observed that by adjoining to the system of axioms (A_1)-(A_9) the scheme (A_{10}) we obtain a system of axioms for intuitionistic propositional calculi.

(16) Stone [4], Tarski [8], McKinsey and Tarski [3].

be the pseudo-Boolean algebra of all open subsets of the real line X. Then for the valuation v such that $v(p) = \{x \in X\colon 0 < x\}$, $v(q) = X$ for any propositional variable $q \neq p$, we get $(p \cup \neg p)_{\mathfrak{G}(X)}(v) = X - \{O\} \neq X$, which by 7.4 proves our statement.

8. Intuitionistic logic $\mathbf{L}_\chi$. By the *intuitionistic logic* $\mathbf{L}_\chi$ we mean the class of all consistent standard systems of implicative extensional propositional calculi logically equivalent (see VIII 8) to $\mathscr{S}_\chi$. Thus by 7.3 the class $\mathbf{K}_{\mathbf{L}_\chi}$ of all $\mathscr{S}$-algebras for $\mathscr{S} \in \mathbf{L}_\chi$ is the class of all pseudo-Boolean algebras. Hence, by VIII 6.5 and VIII 6.6,

8.1. *For every $\mathscr{S} \in \mathbf{L}_\chi$, the algebra $\mathfrak{A}(\mathscr{S})$ is a non-degenerate pseudo-Boolean algebra. Moreover, $\mathfrak{A}(\mathscr{S})$ is a free algebra in the class of all pseudo-Boolean algebras.*

A formula α in $\mathscr{S} \in \mathbf{L}_\chi$ is said to be $\mathbf{L}_\chi$-*valid* (*intuitionistically valid*) provided that for any valuation v of the formalized language of $\mathscr{S}$ in any pseudo-Boolean algebra $\mathfrak{A} = (A, V, \Rightarrow, \cup, \cap, \neg)$

$$\alpha_{\mathfrak{A}}(v) = V.$$

The next theorem characterizes all formulas derivable in any system $\mathscr{S}$ of intuitionistic logic $\mathbf{L}_\chi$.

8.2. *For any formula α of any system $\mathscr{S} \in \mathbf{L}_\chi$ the following conditions are equivalent:*

(i) *α is derivable in $\mathscr{S}$,*

(ii) *α is intuitionistically valid,*

(iii) *α is valid in every pseudo-Boolean algebra $\mathfrak{G}(X) = (G(X), X, \Rightarrow, \cup, \cap, \neg)$ of open subsets of any topological space X,*

(iv) *$\alpha_{\mathfrak{A}(\mathscr{S})}(v^0) = V$ for the canonical valuation v^0 in the algebra $\mathfrak{A}(\mathscr{S})$ of the system $\mathscr{S}$,*

(v) *α is valid in every pseudo-Boolean algebra $\mathfrak{A}$ with at most 2^{2^r} elements, where r is the number of all subformulas of α* [17].

The equivalence of (i) and (ii) follows from VIII 8.1 and is called the *completeness theorem for propositional calculi of intuitionistic logic*. Obviously, (ii) implies (iii) and by the representation theorem IV 6.7,

[17] Tarski [8], McKinsey and Tarski [2], [3]. See also Rieger [1], for some partial results, and Stone [4].

condition (iii) implies (iv). By VIII 7.2, (i) follows from (iv). Clearly, (ii) implies (v). The easy proof that (v) implies (i), based on VI 11.1 and similar to the proof of X 3.4, is left to the reader.

Let us note that each of the conditions (i)–(v) is equivalent to the following one:

(vi) *α is valid in the pseudo-Boolean algebra $\mathfrak{G}(X)$ of all open subsets of a dense-in-itself metric space $X \neq O$* (in particular, of a *Euclidean space X*)[18].

The following theorem follows from the equivalence of conditions (i) and (v).

8.3. *Every propositional calculus $\mathscr{S}$ of the intuitionistic logic $\mathbf{L}_\chi$ is decidable*[19].

The next theorem gives examples of intuitionistically valid formulas.

8.4. *For any system $\mathscr{S}$ of the intuitionistic logic $\mathbf{L}_\chi$ the following formulas are derivable in $\mathscr{S}$: formulas of the form* (A_1)–(A_8) *in* X 1, (A_9), (A_{10}) *in Section* 7, *formulas of the form* (D_3)–(D_{11}) *in* IX 3, (D_{16})–(D_{32}) *in* X 3 *and the following formulas*:

(D_{33}) $(\alpha \Rightarrow \neg\neg\alpha)$,

(D_{34}) $((\alpha \Rightarrow \beta) \Rightarrow (\neg\beta \Rightarrow \neg\alpha))$,

(D_{35}) $\neg(\alpha \cap \neg\alpha)$,

(D_{36}) $((\neg\alpha \cup \beta) \Rightarrow (\alpha \Rightarrow \beta))$,

(D_{37}) $(\neg(\alpha \cup \beta) \Rightarrow (\neg\alpha \cap \neg\beta))$,

(D_{38}) $((\neg\alpha \cap \neg\beta) \Rightarrow \neg(\alpha \cup \beta))$,

(D_{39}) $((\neg\alpha \cup \neg\beta) \Rightarrow \neg(\alpha \cap \beta))$,

(D_{40}) $(\neg\neg\neg\alpha \Rightarrow \neg\alpha)$,

where α, β, γ are any formulas of $\mathscr{S}$.

The proof for formulas of the form (A_1)–(A_8), (D_3)–(D_{11}), (D_{16})-(D_{32}) is analogous to that of X 3.6, making use of Theorem 8.2 and the definition of pseudo-Boolean algebras. Formulas of the form (D_{33})–(D_{40}) are intuitionistically valid by IV 1.3 (2), IV 6.1 and IV 6.2; formulas

(18) McKinsey and Tarski [2], [3].

(19) Jaśkowski [1]. This also follows from the results of Gentzen [1]. See also Wajsberg [2], McKinsey and Tarski [2], Rieger [1].

of the form (A_9), (A_{10}) are intuitionistically valid by IV 6 (c) and IV 6 (s). Thus by 8.2 they are derivable in $\mathscr{S}$.

8.5. *Let* $\mathscr{S} = (\mathscr{L}, C_{\mathscr{L}})$ *be a propositional calculus of the intuitionistic logic and let* $\mathscr{S}_0 = (\mathscr{L}_0, C_{\mathscr{L}_0})$ *be a propositional calculus of the positive logic* $\mathbf{L}_\pi$ *such that* $\mathscr{L}$ *is an extension of* $\mathscr{L}_0$. *Then for every formula* α *in* $\mathscr{L}_0$ *the following conditions are equivalent*:

(i) α *is derivable in* $\mathscr{S}$,

(ii) α *is derivable in* $\mathscr{S}_0$ [20].

The proof, analogous to that of X 3.7 and based on 8.2, the definition of pseudo-Boolean algebras (see IV 6), X 3.4, IV 1.6 and IV 6.6, is left to the reader.

Let us note the validity of the following theorem on intuitionistically derivable disjuntions, similar to X 4.1.

8.6. *Let* $\mathscr{S} = (\mathscr{L}, C_{\mathscr{L}})$ *be a propositional calculus of the intuitionistic logic. Then for any formulas* α, β *in* $\mathscr{L}$

$(\alpha \cup \beta) \in C_{\mathscr{L}}(O)$ *if and only if either* $\alpha \in C_{\mathscr{L}}(O)$ *or* $\beta \in C_{\mathscr{L}}(O)$ [21].

The proof, analogous to that of X 4.1 and based on 8.1, IV 6.7, VIII 3.2, 8.2, VIII 3.2, is left to the reader.

9. $\mathbf{L}_\chi$-theories of zero order and their connection with filters [22]. For any standard system $\mathscr{S} = (\mathscr{L}, C_{\mathscr{L}}) \in \mathbf{L}_\chi$ and for any set $\mathscr{A}$ of formulas in $\mathscr{L}$ the system $\mathscr{S}(\mathscr{A}) = (\mathscr{L}, C_{\mathscr{L}}, \mathscr{A})$ is said to be an $\mathbf{L}_\chi$-*theory* (an *intuitionistic theory*) *of zero order based on* $\mathscr{S}$. Any valuation v of $\mathscr{L}$ in any non-degenerate pseudo-Boolean algebra $\mathfrak{A} = (A, \vee, \Rightarrow, \cup, \cap, \neg)$ such that for all $\alpha \in \mathscr{A}$

(1) $$\alpha_{\mathfrak{A}}(v) = \vee$$

is said to be an $\mathbf{L}_\chi$-*model* (an *intuitionistic model*) of $\mathscr{S}(\mathscr{A})$. An $\mathbf{L}_\chi$-model of $\mathscr{S}(\mathscr{A})$ in the two-element pseudo-Boolean algebra (i.e. in the two-element Boolean algebra) will be said to be a *semantic model.*

(20) See Rasiowa and Sikorski [3], [MM].

(21) This theorem was announced without proof by Gödel [4]. It follows from Gentzen [1]. The proof given by Rasiowa and Sikorski [5] is a modification of a proof by McKinsey and Tarski [2], [3]. See also Rieger [1].

(22) The exposition in Sec. 9 is a slight modification of that in [MM].

The following theorem follows from VIII 9.2, VIII 9.3 and the fact that the class of all $\mathcal{S}$-algebras for $\mathcal{S} \in \mathbf{L}_\chi$ coincides with the class of all pseudo-Boolean algebras.

9.1. *For every consistent intuitionistic theory* $\mathcal{S}(\mathcal{A}) = (\mathcal{L}, C_{\mathcal{L}}, \mathcal{A})$ *of zero order the algebra* $\mathfrak{A}(\mathcal{S}(\mathcal{A}))$ (see VIII 9) *is a non-degenerate pseudo-Boolean algebra. The canonical valuation* v^0 *of* $\mathcal{L}$ *in* $\mathfrak{A}(\mathcal{S}(\mathcal{A}))$ (cf. VIII 9 (14)) *is an* $\mathbf{L}_\chi$*-model of* $\mathcal{S}(\mathcal{A})$. *This* $\mathbf{L}_\chi$*-model is adequate*, i.e. *for any formula* α *in* $\mathcal{L}$

(2) $\alpha \in C_{\mathcal{L}}(\mathcal{A})$ *if and only if* $\alpha_{\mathfrak{A}(\mathcal{S}(\mathcal{A}))}(v^0) = V$.

By 8.4 for any $\mathcal{S} \in \mathbf{L}_\chi$ all formulas of the form $(\neg(\alpha \Rightarrow \alpha) \Rightarrow \beta)$, where α, β are any formulas in $\mathcal{L}$, are derivable in $\mathcal{S}$. Consequently, each system $\mathcal{S}$ in $\mathbf{L}_\chi$ belongs to the class $\mathbf{S}_\nu$ of standard systems of implicative extensional propositional calculi with semi-negation (see VIII 10). Thus

(3) $$\mathbf{L}_\chi \subset \mathbf{S}_\nu.$$

Let us note the following theorem:

9.2. *For any intuitionistic theory* $\mathcal{S}(\mathcal{A}) = (\mathcal{L}, C_{\mathcal{L}}, \mathcal{A})$ *of zero order the following conditions are equivalent*:

(i) $\mathcal{S}(\mathcal{A})$ *is consistent*,

(ii) *for any formula* α *in* $\mathcal{L}$, *either* $\alpha \notin C_{\mathcal{L}}(\mathcal{A})$ *or* $\neg\alpha \notin C_{\mathcal{L}}(\mathcal{A})$,

(iii) *there exists an* $\mathbf{L}_\chi$*-model of* $\mathcal{S}(\mathcal{A})$,

(iv) *there exists an* $\mathbf{L}_\chi$*-model of* $\mathcal{S}(\mathcal{A})$ *in a pseudo-Boolean algebra of open subsets of a topological space*,

(v) *there exists a semantic model of* $\mathcal{S}(\mathcal{A})$.

The equivalence of conditions (i), (ii), (iii), (iv) follows from (3), VIII 10.2 and the representation theorem IV 6.7. It is sufficient to show that (iii) implies (v), since clearly (v) implies (iii). Suppose that a valuation v is an $\mathbf{L}_\chi$-model of $\mathcal{S}(\mathcal{A})$ in a non-degenerate pseudo-Boolean algebra $\mathfrak{A}$. By III 1.11 there exists a maximal filter ∇ in $\mathfrak{A}$. By IV 6.5 the quotient algebra $\mathfrak{A}/\nabla$ is the two-element pseudo-Boolean algebra. Let h be the epimorphism from $\mathfrak{A}$ onto $\mathfrak{A}/\nabla$. It follows from VIII 3.2 that the valuation hv of $\mathcal{L}$ in $\mathfrak{A}/\nabla$ is a semantic model of $\mathcal{S}(\mathcal{A})$.

The next theorem characterizes all theorems of any consistent intuitionistic theory of zero order.

9.3. *For any formula* α *of any consistent intuitionistic theory* $\mathscr{S}(\mathscr{A}) = (\mathscr{L}, C_{\mathscr{L}}, \mathscr{A})$ *of zero order the following conditions are equivalent*:

(i) $\alpha \in C_{\mathscr{L}}(\mathscr{A})$,

(ii) *every intuitionistic model of* $\mathscr{S}(\mathscr{A})$ *is an intuitionistic model of* α,

(iii) *every intuitionistic model of* $\mathscr{S}(\mathscr{A})$ *in any non-degenerate pseudo-Boolean algebra* $\mathfrak{G}(X)$ *of open subsets of any topological space* X *is an intuitionistic model of* α,

(iv) $\alpha_{\mathfrak{A}(\mathscr{S}(\mathscr{A}))}(v^0) = V$ *for the canonical valuation* v^0 *in the algebra* $\mathfrak{A}(\mathscr{S}(\mathscr{A}))$ *of the theory* $\mathscr{S}(\mathscr{A})$.

The conditions (i), (ii), (iv) are equivalent by VIII 9.5. Clearly, (ii) implies (iii) and by the representation theorem IV 6.7 condition (iii) implies (ii), which completes the proof.

The connection between $\mathbf{L}_\chi$*-theories and* $\mathbf{L}_\pi$*-theories of zero order is stated in the following theorem.*

9.4. *Let* $\mathscr{S}(\mathscr{A}) = (\mathscr{L}, C_{\mathscr{L}}, \mathscr{A})$ *be an* $\mathbf{L}_\chi$*-theory and let* $\mathscr{S}_0 = (\mathscr{L}_0, C_{\mathscr{L}_0})$ *be a propositional calculus of the positive logic* $\mathbf{L}_\pi$ *such that* $\mathscr{L}$ *is an extension of* $\mathscr{L}_0$. *If each formula* α *in* $\mathscr{A}$ *is a formula in* $\mathscr{L}_0$, *then for every formula* β *in* $\mathscr{L}_0$ *the following conditions are equivalent*:

(i) β *is a theorem of* $\mathscr{S}(\mathscr{A})$,

(ii) β *is a theorem of the* $\mathbf{L}_\pi$*-theory* $\mathscr{S}_0(\mathscr{A}) = (\mathscr{L}_0, C_{\mathscr{L}_0}, \mathscr{A})$.

The proof, similar to that of 3.3 and based on 9.3, X 5.2, IV 1.6 and IV 6.6, is left to the reader.

Since the only rule of inference of the propositional calculus $\mathscr{S}_\chi$ (see Sec. 7) is modus ponens, it follows from VIII 12.3, IV 2.1 and the definition of pseudo-Boolean algebras (cf. IV 6) that

9.5. *For any set* ∇ *of elements in a pseudo-Boolean algebra the following conditions are equivalent*:

(i) ∇ *is a deductive filter*,

(ii) ∇ *is an implicative filter*,

(iii) ∇ *is a filter*.

By the definition of pseudo-Boolean algebras, IV 2.3 and IV 6.3, the kernels of epimorphisms from pseudo-Boolean algebras onto similar algebras are filters and, conversely, every filter is the kernel of the epimorphism determined by itself. The above statements follow independently from 9.5, VIII 12.2, VIII 13.10, VIII 13.12 and from the equational definability of pseudo-Boolean algebras.

By 9.5, VIII 13.2 and VIII 13.4 we get two theorems analogous to X 6.2 and X 6.3, which state that there exists a natural one-one correspondence between filters in the pseudo-Boolean algebra $\mathfrak{A}(\mathscr{S})$ of a system $\mathscr{S}$ in $\mathbf{L}_\chi$ and formalized $\mathbf{L}_\chi$-theories based on $\mathscr{S}$, two theories $\mathscr{S}(\mathscr{A})$ and $\mathscr{S}(\mathscr{A}')$ being identified if and only if they have the same set of theorems.

Let us note that for any system $\mathscr{S}$ of the intuitionistic logic $\mathbf{L}_\chi$ there holds a theorem corresponding to IX 5.4. Moreover, the deduction theorems analogous to IX 5.5, IX 5.6, X 6.4 also hold.

They can be proved by the same argument. Clearly, theorems corresponding to X 6.5-X 6.9 can be deduced analogously. The formulation and the proofs are left to the reader.

The following theorem follows from (3) and VIII 13.8.

9.6. *For every consistent intuitionistic theory* $\mathscr{S}(\mathscr{A}) = (\mathscr{L}, C_{\mathscr{L}}, \mathscr{A})$ *of zero order there exists a maximal intuitionistic theory* $\mathscr{S}(\mathscr{A}') = (\mathscr{L}, C_{\mathscr{L}}, \mathscr{A}')$ *such that* $C_{\mathscr{L}}(\mathscr{A}) \subset C_{\mathscr{L}}(\mathscr{A}')$.

The next theorem characterizes maximal intuitionistic theories of zero order.

9.7. *The following conditions are equivalent for any* $\mathbf{L}_\chi$*-theory* $\mathscr{S}(\mathscr{A}) = (\mathscr{L}, C_{\mathscr{L}}, \mathscr{A})$ *of zero order*:

(i) $\mathscr{S}(\mathscr{A})$ *is maximal*,

(ii) *for each formula* α *in* $\mathscr{L}$ *exactly one of the formulas* α *and* $\neg\alpha$ *is a theorem of* $\mathscr{S}(\mathscr{A})$,

(iii) $\mathfrak{A}(\mathscr{S}(\mathscr{A}))$ *is the two-element Boolean algebra*,

(iv) $\mathscr{S}(\mathscr{A})$ *has an adequate semantic model*,

(v) $\mathscr{S}(\mathscr{A})$ *is consistent and every semantic model of* $\mathscr{S}(\mathscr{A})$ *is adequate.*

Suppose that $\mathscr{S}(\mathscr{A})$ is maximal. Let $\nabla_{\mathscr{A}}$ be the set of all elements $||\alpha||$ in the algebra $\mathfrak{A}(\mathscr{S})$ such that $\alpha \in C_{\mathscr{L}}(\mathscr{A})$. By VIII 13.1, VIII 13.6 and 9.5, $\nabla_{\mathscr{A}}$ is a maximal filter in the pseudo-Boolean algebra $\mathfrak{A}(\mathscr{S})$. Hence, by IV 6.5 for any formula α of $\mathscr{L}$ exactly one of the elements $||\alpha||$, $\neg||\alpha||$ in $\mathfrak{A}(\mathscr{S})$ belongs to $\nabla_{\mathscr{A}}$, which implies (ii) by the definition of $\nabla_{\mathscr{A}}$. Condition (ii) implies that for any $||\alpha||$ in $\mathfrak{A}(\mathscr{S})$ exactly one of the elements $||\alpha||$ and $\neg||\alpha||$ is in the filter $\nabla_{\mathscr{A}}$. Thus by IV 6.5, the filter $\nabla_{\mathscr{A}}$ is maximal and consequently $\mathfrak{A}(\mathscr{S})/\nabla_{\mathscr{A}}$ is the two-element pseudo-Boolean algebra (i.e. the two-element Boolean algebra). Hence, by VIII 13.11, $\mathfrak{A}(\mathscr{S}(\mathscr{A}))$ is the two-element Boolean algebra, i.e. (iii) holds. Clearly, (iii) implies (iv) by 9.1. If (iv) holds, then by 9.2, $\mathscr{S}(\mathscr{A})$ is consistent. Since in every semantic model v of $\mathscr{S}(\mathscr{A})$ in $\mathfrak{B}_0$ for every formula α exactly one of the following conditions holds:

$$\alpha_{\mathfrak{B}_0}(v) = V, \qquad \neg\alpha_{\mathfrak{B}_0}(v) = V,$$

the same applies to an adequate semantic model v_0 of $\mathscr{S}(\mathscr{A})$. Hence for every formula α exactly one of the formulas α and $\neg\alpha$ is a theorem of $\mathscr{S}(\mathscr{A})$. If $\alpha \in C_{\mathscr{L}}(\mathscr{A})$, then $\alpha_{\mathfrak{B}_0}(v) = V$ by 9.3. If $\alpha \notin C_{\mathscr{L}}(\mathscr{A})$, then $\neg\alpha \in C_{\mathscr{L}}(\mathscr{A})$ and consequently $\neg\alpha_{\mathfrak{B}_0}(v) = V$, which implies that $\alpha_{\mathfrak{B}_0}(v) = \Lambda$. Thus each semantic model v of $\mathscr{S}(\mathscr{A})$ is adequate. It remains to prove that (v) implies (i). Suppose that (v) is satisfied. Since $\mathscr{S}(\mathscr{A})$ is consistent, by 9.2 there exists a semantic model v of $\mathscr{S}(\mathscr{A})$ in $\mathfrak{B}_0$. By (v) it is adequate. If $\alpha \notin C_{\mathscr{L}}(\mathscr{A})$, then $\alpha_{\mathfrak{B}_0}(v) = \Lambda$ and hence $\neg\alpha_{\mathfrak{B}_0}(v) = V$. Thus $\neg\alpha \in C_{\mathscr{L}}(\mathscr{A})$. Consequently, $\mathscr{S}(\mathscr{A}) \cup \{\alpha\})$ is not consistent by 9.2. This implies that $\mathscr{S}(\mathscr{A})$ is maximal.

10. Prime $\mathbf{L}_\chi$-theories [23]. An $\mathbf{L}_\chi$-theory of zero order $\mathscr{S}(\mathscr{A}) = (\mathscr{L}, C_{\mathscr{L}}, \mathscr{A})$ is said to be *prime* (cf. X 6) provided it is consistent and for any two formulas α, β in $\mathscr{L}$ the following condition is satisfied:

(p) if $(\alpha \cup \beta) \in C_{\mathscr{L}}(\mathscr{A})$, then either $\alpha \in C_{\mathscr{L}}(\mathscr{A})$ or $\beta \in C_{\mathscr{L}}(\mathscr{A})$.

We mentioned at the end of Sec. 9 that a theorem analogous to X 6.7 holds for $\mathbf{L}_\chi$-theories. It states that an $\mathbf{L}_\chi$-theory $\mathscr{S}(\mathscr{A}) = (\mathscr{L}, C_{\mathscr{L}}, \mathscr{A})$ is prime if and only if the filter $\nabla_{\mathscr{A}}$ (composed of all $||\alpha||$ in $\mathfrak{A}(\mathscr{S})$ such that $\alpha \in C_{\mathscr{L}}(\mathscr{A})$) in the pseudo-Boolean algebra $\mathfrak{A}(\mathscr{S})$ is prime. Hence by VIII 13.6, 9.5, IV 2.2 we get the following result.

10.1. *Every maximal* $\mathbf{L}_\chi$*-theory of zero order is prime.*

Note that the converse statement does not hold. The aim of this

[23] Theorems 10.2, 10.3, 10.4, 10.5, 10.6, 10.7 are analogous to the theorems presented by Rasiowa [5] (see also Rasiowa [4]) for elementary intuitionistic theories. The exposition in this section is analogous to that in [MM].

section is to characterize prime $\mathbf{L}_\chi$-theories. To begin with, we shall define the characteristic valuation for consistent $\mathbf{L}_\chi$-theories.

Let $\mathcal{S}(\mathcal{A}) = (\mathcal{L}, C_{\mathcal{L}}, \mathcal{A})$ be a consistent intuitionistic theory. By 9.1 and the representation theorem IV 6.7 there exists a monomorphism h from the pseudo-Boolean algebra $\mathfrak{A}(\mathcal{S}(\mathcal{A}))$ into the pseudo-field $\mathfrak{G}(X)$ of all open subsets of a topological space X. It follows from VIII 9 (15) and VIII 3.2 that for the canonical valuation v^0 of $\mathcal{L}$ in $\mathfrak{A}(\mathcal{S}(\mathcal{A}))$ and for each formula α in $\mathcal{L}$

$$(1) \qquad h\|\alpha\|_{\mathcal{A}} = h\alpha_{\mathfrak{A}(\mathcal{S}(\mathcal{A}))}(v^0) = \alpha_{\mathfrak{G}(X)}(hv^0).$$

Let X_0 be the one-point strong compactification of the space X (see VI 8). Putting

$$(2) \qquad g(Y) = X \cap Y \quad \text{for any open subset } Y \text{ of } X_0,$$

we define an epimorphism g from the pseudo-field $\mathfrak{G}(X_0)$ of all open subsets of X_0 onto $\mathfrak{G}(X)$. The valuation v of $\mathcal{L}$ in $\mathfrak{G}(X_0)$ defined by

$$(3) \qquad v(p) = \begin{cases} X_0 \text{ if } v^0(p) = \|p\|_{\mathcal{A}} = V, \\ h(v^0(p)) \text{ if } v^0(p) = \|p\|_{\mathcal{A}} \neq V \end{cases}$$

for any propositional variable p of $\mathcal{L}$, will be called the *characteristic valuation for* $\mathcal{S}(\mathcal{A})$.

By (2) and (3) we get

$$(4) \qquad gv = hv^0.$$

10.2. *For the characteristic valuation v of a consistent $\mathbf{L}_\chi$-theory $\mathcal{S}(\mathcal{A}) = (\mathcal{L}, C_{\mathcal{L}}, \mathcal{A})$ and for every formula α in $\mathcal{L}$*

$$(5) \qquad \textit{either} \quad \alpha_{\mathfrak{G}(X_0)}(v) = X_0 \quad \textit{or} \quad \alpha_{\mathfrak{G}(X_0)}(v) = h\|\alpha\|_{\mathcal{A}}.$$

Moreover,

$$(6) \quad \textit{if} \quad \alpha_{\mathfrak{G}(X_0)}(v) = X_0, \quad \textit{then} \quad h\|\alpha\|_{\mathcal{A}} = X, \text{ i.e. } \|\alpha\|_{\mathcal{A}} = V.$$

By VIII 3.2, (4), (1) and VIII 9 (15)

$$g(\alpha_{\mathfrak{G}(X_0)}(v)) = \alpha_{\mathfrak{G}(X)}(gv) = \alpha_{\mathfrak{G}(X)}(hv^0) = h(\alpha_{\mathfrak{A}(\mathcal{S}(\mathcal{A}))}(v^0)) = h\|\alpha\|_{\mathcal{A}}.$$

Hence, by (2),

$$(7) \qquad \alpha_{\mathfrak{G}(X_0)}(v) \cap X = h\|\alpha\|_{\mathcal{A}}.$$

Since the class of all open subsets of X_0 consists of X_0 and of all open subsets of X and, moreover, $\alpha_{\mathfrak{G}(X_0)}(v)$ is an open subset of X_0, equation (7) implies (5) and (6).

10.3. *If the characteristic valuation* v *for a consistent intuitionistic theory* $\mathcal{S}(\mathcal{A}) = (\mathcal{L}, C_{\mathcal{L}}, \mathcal{A})$ *is an intuitionistic model of* $\mathcal{S}(\mathcal{A})$, *then the theory* $\mathcal{S}(\mathcal{A})$ *is prime.*

Indeed, if $(\alpha \cup \beta) \in C_{\mathcal{L}}(\mathcal{A})$, then by 9.3

(8) $$\alpha_{\mathfrak{S}(X_0)}(v) \cup \beta_{\mathfrak{S}(X_0)}(v) = (\alpha \cup \beta)_{\mathfrak{S}(X_0)}(v) = X_0.$$

Since both summands on the right side of equation (8) are open subsets of X_0 and X_0 is a strongly compact topological space, it follows from (8) that either $\alpha_{\mathfrak{S}(X_0)}(v) = X_0$ or $\beta_{\mathfrak{S}(X_0)}(v) = X_0$. Hence by (6) either $||\alpha||_{\mathcal{A}} = V$ or $||\beta||_{\mathcal{A}} = V$, which implies by VIII 9 (9) that either $\alpha \in C_{\mathcal{L}}(\mathcal{A})$ or $\beta \in C_{\mathcal{L}}(\mathcal{A})$.

For any consistent $\mathbf{L}_\chi$-theory $\mathcal{S}(\mathcal{A}) = (\mathcal{L}, C_{\mathcal{L}}, \mathcal{A})$, let $Z_{\mathcal{S}(\mathcal{A})}$ be the set of all formulas α in $\mathcal{L}$ which have the following property:

(9) if $\alpha \in C_{\mathcal{L}}(\mathcal{A})$, then $\alpha_{\mathfrak{S}(X_0)}(v) = X_0$ for the characteristic valuation v for $\mathcal{S}(\mathcal{A})$.

10.4. *The set* $Z_{\mathcal{S}(\mathcal{A})}$ *satisfies the following conditions:*

(i) *each propositional variable is in* $Z_{\mathcal{S}(\mathcal{A})}$,

(ii) *if* $\beta, \gamma \in Z_{\mathcal{S}(\mathcal{A})}$, *then* $(\alpha \cap \beta) \in Z_{\mathcal{S}(\mathcal{A})}$,

(iii) *for every formula* β *of* $\mathcal{S}(\mathcal{A})$, $\neg\beta \in Z_{\mathcal{S}(\mathcal{A})}$,

(iv) *if* $\gamma \in Z_{\mathcal{S}(\mathcal{A})}$, *then for each formula* β *of* $\mathcal{S}(\mathcal{A})$, $(\beta \Rightarrow \gamma) \in Z_{\mathcal{S}(\mathcal{A})}$.

The proof, by verification, is left to the reader. It uses 10.2, IV 6.2 (6), IV 1.3 (7), VIII 9 (9) and VIII 9 (7).

10.5. *If a consistent* $\mathbf{L}_\chi$*-theory* $\mathcal{S}(\mathcal{A})$ *is prime, then the set* F *of all formulas of* $\mathcal{S}(\mathcal{A})$ *is contained in* $Z_{\mathcal{S}(\mathcal{A})}$.

It is easy to verify that if $\mathcal{S}(\mathcal{A})$ is prime, then for any formulas $\beta, \gamma \in F$:

(10) $$\text{if} \quad \beta, \gamma \in Z_{\mathcal{S}(\mathcal{A})}, \quad \text{then} \quad (\beta \cup \gamma) \in Z_{\mathcal{S}(\mathcal{A})}.$$

Since $Z_{\mathcal{S}(\mathcal{A})}$ satisfies the conditions (i), (ii), (iii), (iv) of 10.4 and the condition (10), it follows that $F \subset Z_{\mathcal{S}(\mathcal{A})}$.

The following main theorem follows from 10.3 and 10.5.

10.6. *A consistent intuitionistic theory* $\mathcal{S}(\mathcal{A})$ *of zero order is prime if and only if the characteristic valuation for* $\mathcal{S}(\mathcal{A})$ *is an intuitionistic model of* $\mathcal{S}(\mathcal{A})$.

By 9.4, Theorem 10.6 can also be applied to $\mathbf{L}_\pi$-theories of zero order.

Given an intuitionistic theory $\mathscr{S}(\mathscr{A})$, let $Z^*_{\mathscr{S}(\mathscr{A})}$ be the least set of formulas of $\mathscr{L}$ satisfying the conditions (i), (ii), (iii), (iv) of 10.4.

It follows from 10.3 and 10.4 that

10.7. *For any consistent intuitionistic theory* $\mathscr{S}(\mathscr{A})$, *if* $\mathscr{A} \subset Z^*_{\mathscr{S}(\mathscr{A})}$, *then the theory* $\mathscr{S}(\mathscr{A})$ *is prime. In particular, if no* α *in* $\mathscr{A}$ *contains the disjunction sign, then* $\mathscr{S}(\mathscr{A})$ *is prime.*

The above theorem can also be applied to $\mathbf{L}_\pi$-theories of zero order, since 9.4 holds.

Exercises

1. Prove that the following formulas are not derivable in any intuitionistic propositional calculus: $(p \cup \neg p)$, $(\neg\neg p \Rightarrow p)$, $((p \Rightarrow q) \Rightarrow (\neg p \cup q))$, $(\neg(p \cap q) \Rightarrow (\neg p \cup \neg q))$, $((\neg p \Rightarrow q) \Rightarrow (\neg q \Rightarrow p))$, $((\neg p \Rightarrow \neg q) \Rightarrow (q \Rightarrow p))$, $(((p \Rightarrow q) \Rightarrow p) \Rightarrow p)$, where p, q are any distinct propositional variables. (Cf. [MM]).

2. Prove that there exists no finite pseudo-Boolean algebra $\mathfrak{A}$ with the following property: for any formula α in an intuitionistic propositional calculus $\mathscr{S} = (\mathscr{L}, C_{\mathscr{L}})$, α is derivable in $\mathscr{S}$ if and only if $\alpha_{\mathfrak{A}}(v) = V$ for each valuation v of $\mathscr{L}$ in $\mathfrak{A}$ (24).

3. Prove that the following conditions are equivalent for each formula α of an intuitionistic propositional calculus $\mathscr{S} = (\mathscr{L}, C_{\mathscr{L}})$: (i) α is semantically valid (see XIII 2); (ii) for every pseudo-Boolean algebra $\mathfrak{A}$ and every valuation v of $\mathscr{L}$ in $\mathfrak{A}$, $\alpha_{\mathfrak{A}}(v)$ is a dense element in $\mathfrak{A}$ (see IV Ex. 3); (iii) in a non-degenerate pseudo-Boolean algebra $\mathfrak{A}$, the element $\alpha_{\mathfrak{A}}(v)$ is dense for every valuation v of $\mathscr{L}$ in $\mathfrak{A}$ (25). Apply IV Ex. 5, IV Ex. 6 and VIII 3.2.

4. Prove that a formula α of an intuitionistic propositional calculus $\mathscr{S} = (\mathscr{L}, C_{\mathscr{L}})$ is semantically valid if and only if the element $||\alpha||$ of $\mathfrak{A}(\mathscr{S})$ is dense (see [MM] IX 5.2). Apply Ex. 3, VIII 6 (15), IV Ex. 3 (i), (ii), 8.2.

5. Prove that a formula α of an intuitionistic propositional calculus $\mathscr{S} = (\mathscr{L}, C_{\mathscr{L}})$ is semantically valid if and only if $\neg\neg\alpha$ is derivable in $\mathscr{S}$ (26). Apply Ex. 4, IV Ex. 3 (i), (ii), VIII 6 (15), 8.2.

6. Prove that for each formula α of an intuitionistic propositional calculus $\mathscr{S} = (\mathscr{L}, C_{\mathscr{L}})$, $\neg\alpha$ is semantically valid if and only if $\neg\alpha$ is derivable in $\mathscr{S}$.

(24) This result is due to Gödel [2].
(25) Tarski [8].
(26) Glivenko [1], Tarski [8].

7. Prove that for each formula α of an intuitionistic propositional calculus $\mathcal{S} = (\mathcal{L}, C_{\mathcal{L}})$ which contains no connectives except $\cap$ and $\neg$, α is semantically valid if and only if α is derivable in $\mathcal{S}$ [27].

8. For any formula α of an intuitionistic propositional calculus $\mathcal{S} = (\mathcal{L}, C_{\mathcal{L}})$, let α' be the formula defined by induction as follows: 1) if α is a propositional variable, then α' is α; 2) if α is a disjunction $(\beta \cup \gamma)$, then α' is the formula $\neg(\neg\beta' \cap \neg\gamma')$; 3) if α is a conjunction $(\beta \cap \gamma)$, then α' is the formula $(\beta' \cap \gamma')$; if α is an implication $(\beta \Rightarrow \gamma)$, then α' is the formula $\neg(\beta' \cap \neg\gamma')$; 5) if α is a negation $\neg\beta$, then α' is the formula $\neg\beta'$. Prove that α is semantically valid if and only if α' is derivable in $\mathcal{S}$ [28].

9. For every formula α of an intuitionistic propositional calculus $\mathcal{S} = (\mathcal{L}, C_{\mathcal{L}})$, let α^* be the formula defined by induction as follows: 1) if α is a propositional variable, then α^* is the formula $\neg\neg\alpha$; 2) if α is a disjunction $(\beta \cup \gamma)$, then α^* is the formula $\neg(\neg\beta^* \cap \neg\gamma^*)$; 3) if α is a conjunction $(\beta \cap \gamma)$, then α^* is the formula $(\beta^* \cap \gamma^*)$; 4) if α is and implication $(\beta \Rightarrow \gamma)$, then α^* is the formula $(\beta^* \Rightarrow \gamma^*)$; 5) if α is a negation $\neg\beta$, then α^* is the formula $\neg\beta^*$. Prove that α is semantically valid if and only if α^* is derivable in $\mathcal{S}$ [29].

10. Let $\mathcal{S} = (\mathcal{L}, C_{\mathcal{L}})$ be the propositional calculus obtained from $\mathcal{S}_\chi$ (see Sec. 7) by adjoining to the logical axioms of $\mathcal{S}_\chi$ all formulas of the form $((\alpha \Rightarrow \beta) \cup (\beta \Rightarrow \alpha))$. Show that the class of all $\mathcal{S}$-algebras coincides with the class of all linear pseudo-Boolean algebras (see IV Ex. 9). Prove that for each formula α, α is derivable in $\mathcal{S}$ if and only if $\alpha_{\mathfrak{A}}(v) = V$ for every valuation v of $\mathcal{L}$ in every linearly ordered pseudo-Boolean algebra $\mathfrak{A}$ [30].

11. Let $\mathcal{S} = (\mathcal{L}, C_{\mathcal{L}})$ be an intuitionistic propositional calculus. Let $\mathcal{A}$ be the set of all formulas of the form $((\neg\alpha \Rightarrow (\beta \cup \gamma)) \Rightarrow ((\neg\alpha \Rightarrow \beta) \cup (\neg\alpha \Rightarrow \gamma)))$. Prove that $\mathcal{S}(\mathcal{A}) = (\mathcal{L}, C_{\mathcal{L}}, \mathcal{A})$ is a prime intuitionistic theory of zero order [31].

12. Prove that the intuitionistic theory $\mathcal{S}(\mathcal{A}) = (\mathcal{L}, C_{\mathcal{L}}, \mathcal{A})$ of zero order, where $\mathcal{A}$ is the set of all formulas of $\mathcal{L}$ of the form $(((\neg\neg\alpha \Rightarrow \alpha) \Rightarrow (\alpha \cup \neg\alpha)) \Rightarrow (\neg\neg\alpha \cup \neg\alpha))$, is prime [32].

13. Prove that a formula in $\mathcal{S}_\chi$ is a theorem of $\mathcal{S}_\chi$ if and only if it can be proved by using only logical axioms for implication and those for the logical connectives actually occuring in this formula (Horn [1]). Apply the method as indicated in 8.5 and in X 3.7. Use IX Ex. 9, X, Ex. 4, X Ex. 10.

(27) Gödel [3].
(28) Gödel [3].
(29) See Kleene [2], p. 495.
(30) Dummet [1], Horn [2].
(31) Kreisel and Putnam [1].
(32) This result is due to D. Scott (see Kreisel and Putnam [1]).

CHAPTER XII

CONSTRUCTIVE LOGIC WITH STRONG NEGATION

Introduction. Constructive logic with strong negation was formulated by Nelson [2] and independently by Markov [1] as a result of certain philosophical objections to intuitionistic negation. The criticism of intuitionistic negation concerns its disadvantageous non-constructive property, namely, that the derivability of the formula $\neg(\alpha \cap \beta)$ in an intuitionistic propositional calculus (or predicate calculus) does not imply that at least one of the formulas $\neg\alpha$, $\neg\beta$ is derivable (similarly the derivability of the formula $\neg\bigcap\xi\alpha(\xi)$ in an intuitionistic predicate calculus does not imply the derivability of the formula $\neg\alpha(\tau)$ for some term τ). In this sense falsity is non-constructive in intuitionistic logic. In order to obtain a constructive logic with the property of constructive falsity, the strong negation sign has been introduced as well as the connectives of the formalized languages of the intuitionistic propositional calculi. Falsity expressed by means of strong negation should be constructive, i.e. the derivability of the formula $\sim(\alpha \cap \beta)$ should be equivalent to the derivability of at least one of the formulas $\sim\alpha$, $\sim\beta$ (similarly the derivability of $\sim\bigcap\xi\alpha(\xi)$ in predicate calculi should be equivalent to the derivability of the formula $\sim\alpha(\tau)$ for some term τ). Moreover, the derivability of any formula without $\sim$ should be equivalent to its derivability in intuitionistic propositional calculi. Constructive propositional calculi with strong negation as created by Nelson and Markov have these properties [1], but unfortunately they do not belong to the class **S** (VIII 5). This is because intuitionistic implication in these systems (to be denoted later by $\rightarrow$) does not satisfy condition VIII 5 (a_7) with respect to strong negation. On the other hand,

[1] Białynicki-Birula and Rasiowa [2].

it is possible to define in these systems implication $\Rightarrow$ (2), in terms of $\rightarrow$, $\cap$ and $\sim$, in such a way that the conditions VIII 5 ((s_1)–(s_8)) are satisfied, i.e. the systems obtained belong to the class **S**. For the sake of uniformity in treating propositional calculi in the present book, the implication sign $\Rightarrow$ is admitted in formalized languages of propositional calculi of constructive logic with strong negation together with the intuitionistic (weak) implication sign $\rightarrow$. The logical axiom (A 17) replaces the definition of $\Rightarrow$ mentioned above.

The logic under consideration, to be denoted by $\mathbf{L}_{\mathcal{N}}$, has been studied by Vorobiev [1], [2], Thomason [1] and from the algebraic point of view by the present author in [6], [7] and jointly with Białynicki-Birula in [2]. The present chapter gives a comprehensive exposition of an algebraic approach to this logic.

Constructive logic with strong negation is characterized algebraically by the class of all quasi-pseudo-Boolean algebras. Besides the general metalogical theorems of Chapter VIII, which can be formulated for propositional calculi and formalized theories based on $\mathbf{L}_{\mathcal{N}}$, certain other theorems are obtained by applying the theory of quasi-pseudo-Boolean algebras, in particular the representation theorem V 5.5 and some theorems concerning special filters of the first kind.

The completeness theorem for propositional calculi of $\mathbf{L}_{\mathcal{N}}$ (3.2, 5.3) is formulated in a much stronger form. In particular, it gives the following topological characterization of formulas derivable in $\mathcal{S} \in \mathbf{L}_{\mathcal{N}}$: exactly those formulas are derivable which are valid in a quasi-pseudo-Boolean algebra $\mathfrak{B}(X)$ connected with the quasi-field $\mathfrak{G}(E)$ of all open subsets of the real line E (where $X = E \cup g(E)$, $E \cap g(E) = O$ and $\mathfrak{B}(X)$ consists of all subsets of X satisfying the conditions (b_1), (b_2), (b_3) of V 3). Instead of the real line we can take n-dimensional Euclidean space or the Cantor discontinuum (3). This is a generalization of Mc Kinsey's and Tarski's [2], [3] result concerning formulas derivable in intuitionistic propositional calculi.

Theorem 4.3 states that the derivability in $\mathcal{S} \in \mathbf{L}_{\mathcal{N}}$ of formulas in

(2) See Vorobiev [1]. In the propositional calculi discussed by Nelson [2] and Markov [1] the implication sign $\Rightarrow$ does not occur; $\rightarrow$, $\cup$, $\cap$, $\neg$, $\sim$ are all primitive connectives in those systems.

(3) Białynicki-Birula and Rasiowa [2].

which $\sim$ and $\Rightarrow$ do not appear is equivalent to their derivability in intuitionistic propositional calculi. Another connection between the derivability of formulas in the propositional calculi of $\mathbf{L}_{\mathcal{N}}$ and their derivability in intuitionistic propositional calculi is formulated in 4.5. The decidability of the propositional calculi of constructive logic with strong negation [4] follows from this theorem.

Disjunctions derivable in the propositional calculi of $\mathbf{L}_{\mathcal{N}}$ have the same property as disjunctions derivable in the intuitionistic propositional calculi, i.e. $(\alpha \cup \beta)$ is derivable if and only if at least one of the formulas α, β is derivable. By the De Morgan laws and the double negation law which hold for strong negation, this implies the property of constructive falsity for strong negation (4.8).

The consistency of any $\mathbf{L}_{\mathcal{N}}$-theory is equivalent to the existence for that theory of an $\mathbf{L}_{\mathcal{N}}$-model (of an $\mathbf{L}_{\mathcal{N}}$-model in a quasi-pseudo-Boolean algebra of sets), in particular to the existence of an $\mathbf{L}_{\mathcal{N}}$-model in a non-degenerate quasi-pseudo-Boolean algebra which has at most three elements. Theorem 6.3, on a characterization of formulas which are theorems of any consistent $\mathbf{L}_{\mathcal{N}}$-theory, is also formulated in a stronger form than the general theorem VIII 9.5.

Deduction filters in quasi-pseudo-Boolean algebras coincide with special filters of the first kind. Two deduction theorems (7.2, 7.3) follow from the corresponding theorems on the structure of those filters. Every consistent $\mathbf{L}_{\mathcal{N}}$-theory can be extended to a maximal one. A characterization of maximal $\mathbf{L}_{\mathcal{N}}$-theories is formulated in 7.7. The concept of prime $\mathbf{L}_{\mathcal{N}}$-theories coincides with that of irreducible $\mathbf{L}_{\mathcal{N}}$-theories. Note that a consistent $\mathbf{L}_{\mathcal{N}}$-theory is prime if and only if, for any formulas α, β, $\sim(\alpha \cap \beta)$ is a theorem if and only if either $\sim \alpha$ is a theorem or $\sim \beta$ is a theorem. Section 8 is concerned with a characterization of prime $\mathbf{L}_{\mathcal{N}}$-theories by means of characteristic valuations. In particular, the following corollary is obtained: if no axiom of a consistent $\mathbf{L}_{\mathcal{N}}$-theory contains $\cup$, $\sim$, $\Rightarrow$, then it is prime. The characterization of prime $\mathbf{L}_{\mathcal{N}}$-theories expounded in Section 8 generalizes analogous results concerning prime intuitionistic theories (see XI 10).

[4] Vorobiev [2]. The algebraic proof presented here is due to Białynicki-Birula and Rasiowa [2].

An interesting relationship between constructive logic with strong negation and three-valued logic of Łukasiewicz [2], observed by Monteiro [2], is mentioned in Exercises 4–7 at the end of this chapter.

1. Propositional calculus $\mathcal{S}_{\mathcal{N}}$ of constructive logic with strong negation. Let $\mathcal{S}_{\mathcal{N}} = (\mathcal{L}, C_{\mathcal{L}})$ be a propositional calculus described briefly as follows. The alphabet $A^0 = (V, L_0, L_1, L_2, U)$ of $\mathcal{L}$ differs from the alphabet of $\mathcal{S}_\chi$ (see XI 7) only in admitting in L_1 a second element, denoted by $\sim$ and called the *strong negation sign*, and by admitting in L_2 a fourth element, denoted by $\rightarrow$ and called the *weak implication sign*. The element in L_1 denoted by $\urcorner$ will be called in the above language $\mathcal{L}$ the *weak negation sign*. Let F be the set of all formulas in $\mathcal{L}$. By definition, $\mathcal{L} = (A^0, F)$.

For any α, β in F, instead of $((\alpha \Rightarrow \beta) \cap (\beta \Rightarrow \alpha))$ we shall write for brevity $(\alpha \equiv \beta)$.

Assume that the set $\mathcal{A}_l$ of logical axioms of $\mathcal{S}_{\mathcal{N}}$ consists of all formulas of the following form, where α, β, γ are any formulas in F:

(A1) $(\alpha \rightarrow (\beta \rightarrow \alpha))$,

(A2) $((\alpha \rightarrow (\beta \rightarrow \gamma)) \rightarrow ((\alpha \rightarrow \beta) \rightarrow (\alpha \rightarrow \gamma)))$,

(A3) $(\alpha \rightarrow (\alpha \cup \beta))$,

(A4) $(\beta \rightarrow (\alpha \cup \beta))$,

(A5) $((\alpha \rightarrow \gamma) \rightarrow ((\beta \rightarrow \gamma) \rightarrow ((\alpha \cup \beta) \rightarrow \gamma)))$,

(A6) $((\alpha \cap \beta) \rightarrow \alpha)$,

(A7) $((\alpha \cap \beta) \rightarrow \beta)$,

(A8) $((\alpha \rightarrow \beta) \rightarrow ((\alpha \rightarrow \gamma) \rightarrow (\alpha \rightarrow (\beta \cap \gamma))))$,

(A9) $((\alpha \rightarrow \urcorner\beta) \rightarrow (\beta \rightarrow \urcorner\alpha))$,

(A10) $(\urcorner(\alpha \rightarrow \alpha) \rightarrow \beta)$,

(A11) $(\sim \alpha \rightarrow (\alpha \rightarrow \beta))$,

(A12) $((\sim(\alpha \rightarrow \beta) \rightarrow (\alpha \cap \sim \beta)) \cap ((\alpha \cap \sim \beta) \rightarrow \sim (\alpha \rightarrow \beta)))$,

(A13) $(\sim(\alpha \cap \beta) \equiv (\sim \alpha \cup \sim \beta))$,

(A14) $(\sim(\alpha \cup \beta) \equiv (\sim \alpha \cap \sim \beta))$,

(A15) $((\sim \urcorner\alpha \rightarrow \alpha) \cap (\alpha \rightarrow \sim \urcorner\alpha))$,

(A16) $(\sim \sim \alpha \equiv \alpha)$,

(A17) $((\alpha \Rightarrow \beta) \equiv ((\alpha \to \beta) \cap (\sim \beta \to \sim \alpha)))$,

(A18) $((\alpha \Rightarrow \beta) \to (\alpha \to \beta))$,

(A19) $(((\alpha \to \alpha) \to \beta) \equiv \beta)$,

(A20) $(((\alpha \Rightarrow \beta) \to ((\beta \Rightarrow \alpha) \to \beta)) \equiv ((\beta \Rightarrow \alpha) \to ((\alpha \Rightarrow \beta) \to \alpha)))$.

Let $(r_{\mathcal{N}})$ be the following rule of inference

$$(r_{\mathcal{N}}) \qquad \frac{\alpha, (\alpha \to \beta)}{\beta} \qquad \text{for any } \alpha, \beta \text{ in } F.$$

The consequence operation $C_{\mathcal{L}}$ is determined by the set $\mathcal{A}_l$ of logical axioms defined above and by $(r_{\mathcal{N}})$ as the only rule of inference. The system $\mathcal{S}_{\mathcal{N}} = (\mathcal{L}, C_{\mathcal{L}})$ is said to be a *propositional calculus of constructive logic with strong negation.*

Replacing in IX 1.1 the implication sign $\Rightarrow$ by the weak implication sign $\to$ and treating $(r_{\mathcal{N}})$ as modus ponens with respect to weak implication, we get the following theorem.

1.1. *For any formulas* $\alpha, \beta, \gamma, \delta$ *in* F *and for any* $\mathcal{A} \subset F$:

(i) $(\alpha \to \alpha) \in C_{\mathcal{L}}(\mathcal{A})$,

(ii) *if* $\alpha \in C_{\mathcal{L}}(\mathcal{A})$, *then* $(\beta \to \alpha) \in C_{\mathcal{L}}(\mathcal{A})$,

(iii) *if* $(\alpha \to \beta), (\beta \to \gamma) \in C_{\mathcal{L}}(\mathcal{A})$, *then* $(\alpha \to \gamma) \in C_{\mathcal{L}}(\mathcal{A})$,

(iv) *if* $(\alpha \to (\beta \to \gamma)) \in C_{\mathcal{L}}(\mathcal{A})$, *then* $(\beta \to (\alpha \to \gamma)) \in C_{\mathcal{L}}(\mathcal{A})$,

(v) *if* $(\beta \to \gamma) \in C_{\mathcal{L}}(\mathcal{A})$, *then* $((\alpha \to \beta) \to (\alpha \to \gamma)) \in C_{\mathcal{L}}(\mathcal{A})$,

(vi) *if* $(\alpha \to \beta) \in C_{\mathcal{L}}(\mathcal{A})$, *then* $((\beta \to \gamma) \to (\alpha \to \gamma)) \in C_{\mathcal{L}}(\mathcal{A})$,

(vii) *if* $(\beta \to \alpha), (\gamma \to \delta) \in C_{\mathcal{L}}(\mathcal{A})$, *then* $((\alpha \to \gamma) \to (\beta \to \delta)) \in C_{\mathcal{L}}(\mathcal{A})$.

Similarly, replacing in X 1.1 the implication sign $\Rightarrow$ by the weak implication sign $\to$ and treating $(r_{\mathcal{N}})$ as modus ponens with respect to weak implication, we get the following result.

1.2. *For all formulas* $\alpha, \beta, \gamma, \delta$ *in* F *and for every set* $\mathcal{A} \subset F$: *if* $(\alpha \to \beta)$, $(\gamma \to \delta) \in C_{\mathcal{L}}(\mathcal{A})$, *then* $((\alpha \cup \gamma) \to (\beta \cup \delta)) \in C_{\mathcal{L}}(\mathcal{A})$ *and* $((\alpha \cap \gamma) \to (\beta \cap \delta)) \in C_{\mathcal{L}}(\mathcal{A})$.

The following theorem will be useful for proving that $\mathcal{S}_{\mathcal{N}}$ belongs to the class **S** (cf. VIII 5).

1.3. *For all formulas* $\alpha, \beta, \gamma, \delta$ *in* F *and for each set* $\mathcal{A} \subset F$:

(i) $\beta, \gamma \in C_{\mathcal{L}}(\mathcal{A})$ *if and only if* $(\beta \cap \gamma) \in C_{\mathcal{L}}(\mathcal{A})$,

(ii) $(\alpha \Rightarrow \beta) \in C_{\mathcal{L}}(\mathcal{A})$ *if and only if* $(\alpha \to \beta), (\sim \beta \to \sim \alpha) \in C_{\mathcal{L}}(\mathcal{A})$,

(iii) $(\alpha \Rightarrow \alpha) \in C_{\mathcal{L}}(\mathcal{A})$,

(iv) *if* $\alpha, (\alpha \Rightarrow \beta) \in C_{\mathcal{L}}(\mathcal{A})$, *then* $\beta \in C_{\mathcal{L}}(\mathcal{A})$,

(v) *if* $(\alpha \Rightarrow \beta), (\beta \Rightarrow \gamma) \in C_{\mathcal{L}}(\mathcal{A})$, *then* $(\alpha \Rightarrow \gamma) \in C_{\mathcal{L}}(\mathcal{A})$,

(vi) *if* $\alpha \in C_{\mathcal{L}}(\mathcal{A})$, *then* $(\beta \Rightarrow \alpha) \in C_{\mathcal{L}}(\mathcal{A})$,

(vii) $((\alpha \to \beta) \to (\neg\beta \to \neg\alpha)) \in C_{\mathcal{L}}(\mathcal{A})$,

(viii) *if* $(\alpha \Rightarrow \beta) \in C_{\mathcal{L}}(\mathcal{A})$, *then* $(\neg\beta \Rightarrow \neg\alpha) \in C_{\mathcal{L}}(\mathcal{A})$,

(ix) *if* $(\alpha \Rightarrow \beta) \in C_{\mathcal{L}}(\mathcal{A})$, *then* $(\sim \beta \Rightarrow \sim \alpha) \in C_{\mathcal{L}}(\mathcal{A})$,

(x) *if* $(\sim \beta \to \sim \alpha), (\sim \delta \to \sim \gamma) \in C_{\mathcal{L}}(\mathcal{A})$, *then* $(\sim(\beta \cup \delta) \to \sim(\alpha \cup \gamma)) \in C_{\mathcal{L}}(\mathcal{A})$,

(xi) *if* $(\sim \beta \to \sim \alpha), (\sim \delta \to \sim \gamma) \in C\ (\mathcal{A})$, *then* $(\sim(\beta \cap \delta) \to \sim(\alpha \cap \gamma)) \in C_{\mathcal{L}}(\mathcal{A})$,

(xii) *if* $(\alpha \Rightarrow \beta), (\gamma \Rightarrow \delta) \in C_{\mathcal{L}}(\mathcal{A})$, *then* $((\alpha \cup \gamma) \Rightarrow (\beta \cup \delta) \in C_{\mathcal{L}}(\mathcal{A})$,

(xiii) *if* $(\alpha \Rightarrow \beta), (\gamma \Rightarrow \delta) \in C_{\mathcal{L}}(\mathcal{A})$, *then* $((\alpha \cap \gamma) \Rightarrow (\beta \cap \delta)) \in C_{\mathcal{L}}(\mathcal{A})$,

(xiv) *if* $(\beta \Rightarrow \alpha), (\gamma \Rightarrow \delta) \in C_{\mathcal{L}}(\mathcal{A})$, *then* $(\sim(\beta \to \delta) \to \sim (\alpha \to \gamma)) \in C_{\mathcal{L}}(\mathcal{A})$,

(xv) *if* $(\beta \Rightarrow \alpha), (\gamma \Rightarrow \delta) \in C_{\mathcal{L}}(\mathcal{A})$, *then* $((\alpha \to \gamma) \Rightarrow (\beta \to \delta)) \in C_{\mathcal{L}}(\mathcal{A})$,

(xvi) *if* $(\beta \Rightarrow \alpha), (\gamma \Rightarrow \delta) \in C_{\mathcal{L}}(\mathcal{A})$, *then* $((\alpha \Rightarrow \gamma) \Rightarrow (\beta \Rightarrow \delta)) \in C_{\mathcal{L}}(\mathcal{A})$.

PROOF OF (i): If $\beta, \gamma \in C_{\mathcal{L}}(\mathcal{A})$, then by 1.1 (ii), $(\alpha \to \beta), (\alpha \to \gamma) \in C_{\mathcal{L}}(\mathcal{A})$ for a formula $\alpha \in \mathcal{A}_l$. Hence, by (A8), ($r_{\mathcal{N}}$), we get $(\alpha \cap \beta) \in C_{\mathcal{L}}(\mathcal{A})$. The converse statement follows by (A6), (A7), ($r_{\mathcal{N}}$).

(ii) follows from (A17), (i), (A18), ($r_{\mathcal{N}}$).

(iii) follows from (ii), 1.1 (i).

(iv) follows from (ii) and ($r_{\mathcal{N}}$).

(v) follows from (ii) and 1.1 (iii).

(vi) follows from (A11), 1.1 (iv), ($r_{\mathcal{N}}$), 1.1 (ii), (ii).

PROOF OF (vii): By (A9), 1.1 (i), ($r_{\mathcal{N}}$), $(\beta \to \neg\neg\beta) \in C_{\mathcal{L}}(\mathcal{A})$. Hence, by 1.1 (v), $((\alpha \to \beta) \to (\alpha \to \neg\neg\beta)) \in C_{\mathcal{L}}(\mathcal{A})$. On the other hand, by (A9), $((\alpha \to \neg\neg\beta) \to (\neg\beta \to \neg\alpha)) \in C_{\mathcal{L}}(\mathcal{A})$. Applying 1.1 (iii), we get (vii).

(viii) follows from (ii), (vii), ($r_{\mathcal{N}}$), (A15), (i), 1.1 (iii) and (ii).

(ix) follows by (ii), (A16), (i), (ii), 1.1 (iii) and (ii).

(x) follows by application of 1.2, (A14), (i), (ii), 1.1 (iii).

(xi) follows by application of 1.2, (A13), (i), (ii), 1.1 (iii).

(xii) follows by application of (ii), 1.2, (x), (ii).

(xiii) follows by application of (ii), 1.2, (xi), (ii).

(xiv) follows by application of (ii), 1.2, (A12), (i), 1.1 (iii).

(xv) follows from (ii), 1.1 (vii), (xiv).

PROOF OF (xvi): By (ix), $(\sim \delta \Rightarrow \sim \gamma), (\sim \alpha \Rightarrow \sim \beta) \in C_{\mathcal{L}}(\mathcal{A})$; hence, by (xv), $((\sim \gamma \to \sim \alpha) \Rightarrow (\sim \delta \to \sim \beta)) \in C_{\mathcal{L}}(\mathcal{A})$. On the other hand, by (xv), $((\alpha \to \gamma) \Rightarrow (\beta \to \delta)) \in C_{\mathcal{L}}(\mathcal{A})$. Applying (xiii), (A17), (i), (v) we get $((\alpha \Rightarrow \gamma) \Rightarrow (\beta \Rightarrow \delta)) \in C_{\mathcal{L}}(\mathcal{A})$.

1.4. *The propositional calculus $\mathcal{S}_{\mathcal{N}}$ belongs to the class* S *of standard systems of implicative extensional propositional calculi.*

Clearly, conditions VIII 5 ((s_1), (s_2)) are satisfied. Conditions VIII 5 ((s_3)-(s_8)) follow from 1.3 (iii), (iv), (v), (vi), (viii), (ix), (xii), (xiii), (xv), (xvi).

2. $\mathcal{S}_{\mathcal{N}}$-algebras. The main purpose of this section is to prove the following theorem.

2.1. *The class of all $\mathcal{S}_{\mathcal{N}}$-algebras coincides with the class of all quasi-pseudo-Boolean algebras* [5].

Suppose that an abstract algebra $\mathfrak{A} = (A, V, \Rightarrow, \cup, \cap, \to, \neg, \sim)$ is an $\mathcal{S}_{\mathcal{N}}$-algebra. It follows from VIII 6 (a_1) and axioms (A1)–(A12), (A15) in Section 1 that the equations V 2 ((qpB$_1^*$), (qpB$_2^*$), (qpB$_6^*$)–(qpB$_{15}^*$), (qpB$_{19}^*$)) hold in $\mathfrak{A}$.

By VIII 6 (a_2) and ($r_{\mathcal{N}}$) we get

(1) if $a = V$ and $(a \to b) = V$, then $b = V$, for all a, b in A.

By (qpB$_9^*$) and (qpB$_{10}^*$) and (1), for any $a, b \in A$:

(2) if $a \cap b = V$, then $a = V$ and $b = V$.

(5) Rasiowa [6], [7].

Moreover, it follows from 1.1 (i) and VIII 6.1 that

(3) $$a \to a = \bigvee, \quad \text{for every } a \in A.$$

By VIII 6 (a_1), axioms (A13), (A14), (A16), (A17), (A19), (A20) in Section 1 and (2), VIII 6 (a_4), (3) we infer that also equations V 2 ((qpB^*_{16})–(qpB^*_{18}), (qpB^*_4), (qpB^*_3), (qpB^*_5)) hold in $\mathfrak{A}$. Hence, by V 2.1, $\mathfrak{A}$ is a quasi-pseudo-Boolean algebra.

Conversely, if $\mathfrak{A}$ is a quasi-pseudo-Boolean algebra, then by V 2.1 equations V 2 ((qpB^*_1)–(qpB^*_{19})) hold in $\mathfrak{A}$. Moreover, by V 1.2 (5) and V 1.3 (9), $a \Rightarrow a = \bigvee$ and $a \to a = \bigvee$ for each element a of $\mathfrak{A}$. Hence, condition VIII 6 (a_1) is satisfied for axioms (A1)–(A17), (A19), (A20). This condition holds also for axiom (A18) by (qpB^*_3) and (qpB^*_4). Conditions VIII 6 ((a_2), (a_3), (a_4)) are satisfied in $\mathfrak{A}$ by V 1.3 (23), V 1.2. Thus $\mathfrak{A}$ is an $\mathcal{S}_{\mathcal{N}}$-algebra.

It follows from 2.1 and VIII 6.1 that

2.2. *If α is a formula derivable in $\mathcal{S}_{\mathcal{N}}$, then*

(4) $$\alpha_{\mathfrak{A}}(v) = \bigvee$$

for every valuation v of $\mathcal{L}$ in any quasi-pseudo-Boolean algebra $\mathfrak{A}$.

2.3. *By* 2.2 *the propositional calculus $\mathcal{S}_{\mathcal{N}}$ is consistent.*

Observe that the formula $((p \to \sim q) \to (q \to \sim p))$, where p, q are any distinct propositional variables of $\mathcal{S}_{\mathcal{N}}$, is not derivable in $\mathcal{S}_{\mathcal{N}}$. Indeed, let v be a valuation of $\mathcal{L}$ in the quasi-pseudo-Boolean algebra $\mathfrak{C}_0$ (see V 3) such that $v(p) = a$ and $v(q) = \bigvee$. Then $((p \to \sim q) \to (q \to \sim p))_{\mathfrak{C}_0}(v) = (a \to \bigwedge) \to (\bigvee \to a) = \bigvee \to a = a \neq \bigvee$, which by 2.2 completes the proof.

3. Constructive logic with strong negation $\mathbf{L}_{\mathcal{N}}$. By *constructive logic with strong negation*, denoted by $\mathbf{L}_{\mathcal{N}}$, we shall understand the class of all consistent standard systems of implicative extensional propositional calculi logically equivalent (see VIII 8) to $\mathcal{S}_{\mathcal{N}}$. Thus, by 2.1, the class $\mathbf{K}_{\mathbf{L}_{\mathcal{N}}}$ of all $\mathcal{S}$-algebras for $\mathcal{S} \in \mathbf{L}_{\mathcal{N}}$ coincides with the class of all quasi-pseudo-Boolean algebras. Hence, by VIII 6.5 and VIII 6.6,

3.1. *For every $\mathcal{S} \in \mathbf{L}_{\mathcal{N}}$, the algebra $\mathfrak{A}(\mathcal{S})$ is a non-degenerate quasi-pseudo-Boolean algebra. Moreover, $\mathfrak{A}(\mathcal{S})$ is a free algebra in the class of all quasi-pseudo-Boolean algebras.*

A formula α in $\mathscr{S} \in \mathbf{L}_{\mathscr{N}}$ is said to be $\mathbf{L}_{\mathscr{N}}$-*valid* provided that for any valuation v of the formalized language of $\mathscr{S}$ in any quasi-pseudo-Boolean algebra $\mathfrak{A} = (A, V, \Rightarrow, \cup, \cap, \rightarrow, \neg, \sim)$

$$\alpha_{\mathfrak{A}}(v) = V.$$

The following theorem characterizes all formulas derivable in any propositional calculus $\mathscr{S} \in \mathbf{L}_{\mathscr{N}}$.

3.2. *For any formula* α *of any system* $\mathscr{S} = (\mathscr{L}, C_{\mathscr{L}})$ *in* $\mathbf{L}_{\mathscr{N}}$ *the following conditions are equivalent*:

(i) α *is derivable in* $\mathscr{S}$,

(ii) α *is* $\mathbf{L}_{\mathscr{N}}$-*valid,*

(iii) α *is valid in every quasi-pseudo-Boolean algebra* $\mathfrak{B}(X)$ *of sets,* i.e. *connected with any pseudo-field* $\mathfrak{G}(X_1)$ *of open subsets of any topological space* X_1,

(iv) $\alpha_{\mathfrak{A}(\mathscr{S})}(v^0) = V$ *for the canonical valuation* v^0 *of* $\mathscr{L}$ *in the algebra* $\mathfrak{A}(\mathscr{S})$ *of the system* $\mathscr{S}$.

The equivalence of the conditions (i) and (ii) follows from VIII 8.1 and is called the *completeness theorem for propositional calculi of the constructive logic with strong negation.*

Clearly, (ii) implies (iii) and, by the representation theorem V 5.6, condition (iii) implies (iv). By VIII 7.2, (iv) implies (i), which completes the proof.

We shall prove in Section 5 that each of conditions (i)-(iv) of 3.2 is equivalent to the following one:

(v) $\alpha_{\mathfrak{B}(X)}(v) = V$ *for every valuation* v *of* $\mathscr{L}$ *in a quasi-pseudo-Boolean algebra* $\mathfrak{B}(X)$ *connected with the pseudo-field* $\mathfrak{G}(X_1)$ *of all open subsets of the real line* X_1 (*where* $\mathfrak{B}(X)$ *consists of all subsets of* X *satisfying the conditions* (b_1), (b_2), (b_3) *in* V 3 *and* $X = X_1 \cup g(X_1)$, $X_1 \cap g(X_1) = O$)[6].

The next theorem presents some examples of formulas derivable in any $\mathscr{S} \in \mathbf{L}_{\mathscr{N}}$.

[6] Białynicki-Birula and Rasiowa [2].

3.3. *For any system $\mathscr{S} \in \mathbf{L}_{\mathscr{N}}$ and for all formulas α, β, γ of $\mathscr{S}$ the formulas of the form* (A1)–(A20) *and, moreover, the following ones are derivable in $\mathscr{S}$*:

(D$_1$) $(\neg\alpha \equiv (\alpha \to \sim(\beta \to \beta)))$,

(D$_2$) $(\alpha \to \alpha)$,

(D$_3$) $(\alpha \Rightarrow \alpha)$,

(D$_4$) $((\alpha \cap (\alpha \to \beta)) \to \beta)$,

(D$_5$) $((\alpha \to (\beta \to \gamma)) \equiv (\beta \to (\alpha \to \gamma)))$,

(D$_6$) $((\alpha \to (\beta \to \gamma)) \equiv ((\alpha \cap \beta) \to \gamma))$,

(D$_7$) $(\alpha \to ((\alpha \to \beta) \to \beta))$,

(D$_8$) $((\alpha \cap \sim \alpha) \Rightarrow (\beta \cup \sim \beta))$,

(D$_9$) $(\alpha \Rightarrow (\beta \Rightarrow (\alpha \cap \beta)))$,

(D$_{10}$) $(\alpha \to (\beta \to (\alpha \cap \beta)))$,

(D$_{11}$) $((\alpha \to \neg\beta) \equiv (\beta \to \neg\alpha))$,

(D$_{12}$) $((\alpha \to \beta) \to (\neg\beta \to \neg\alpha))$.

Any formulas of the form (A1)–(A20) are $\mathbf{L}_{\mathscr{N}}$-valid (see the proof of 2.1) and by 3.2 derivable in any $\mathscr{S} \in \mathbf{L}_{\mathscr{N}}$. The formulas of the form (D$_1$)–(D$_{12}$) are $\mathbf{L}_{\mathscr{N}}$-valid by: V 1 (qpB$_{13}$), V 1.3 (9), V 1.2, V 1.3 (13), V 1.3 (14), V 1.3 (15), V 1.3 (16), V 1.3 (18) and V 1.1, V 1.3 (18), V 1 (46), V 1 (19), 1.3 (vii), 2.2. Thus, by 3.2, they are derivable in any $\mathscr{S} \in \mathbf{L}_{\mathscr{N}}$.

3.4. (The rule of replacement). *Let α, β, γ be any formulas of $\mathscr{S} \in \mathbf{L}_{\mathscr{N}}$ such that β is a subformula of α and $(\beta \equiv \gamma) \in C_{\mathscr{L}}(O)$. Then for the formula α^* obtained from α by the replacement of β by γ the following condition is satisfied*:

(1) $$(\alpha \equiv \alpha^*) \in C_{\mathscr{L}}(O).$$

If $(\beta \equiv \gamma) \in C_{\mathscr{L}}(O)$, then by 3.2, for every valuation v in every quasi-pseudo-Boolean algebra $\mathfrak{A}$, $\beta_{\mathfrak{A}}(v) \Rightarrow \gamma_{\mathfrak{A}}(v) = V$ and $\gamma_{\mathfrak{A}}(v) \Rightarrow \beta_{\mathfrak{A}}(v) = V$. Hence, by V 1.2, $\beta_{\mathfrak{A}}(v) = \gamma_{\mathfrak{A}}(v)$ and, consequently, $\alpha_{\mathfrak{A}}(v) = \alpha^*_{\mathfrak{A}}(v)$. Thus $(\alpha \equiv \alpha^*)$ is $\mathbf{L}_{\mathscr{N}}$-valid and by 3.2 derivable in $\mathscr{S}$.

Let F be the set of all formulas of a system $\mathscr{S} = (\mathscr{L}, C_{\mathscr{L}})$ in $\mathbf{L}_{\mathscr{N}}$ and let $F_0 \subset F$ be the least set of formulas in F satisfying the following conditions:

(n_1) if p is a propositional variable in $\mathscr{L}$, then $p, \sim p \in F_0$,

(n_2) if $\alpha, \beta \in F_0$, then $(\alpha \cup \beta), (\alpha \cap \beta), (\alpha \rightarrow \beta), \neg\alpha \in F_0$.

A formula in F will be said to be of *normal form* provided it belongs to F_0.

3.5. *For every formula α of any system $\mathscr{S} \in \mathbf{L}_{\mathscr{N}}$ there exists a formula α^0 of normal form having the same propositional variables, such that the formulas $(\alpha \rightarrow \alpha^0)$, $(\alpha^0 \rightarrow \alpha)$ are both derivable in $\mathscr{S}$.*

By 3.4 and 3.3, 1 (A 17) it is sufficient to prove 3.5 for formulas without the implication sign $\Rightarrow$. It is easy to see that by 3.2, Theorems 1.1, 1.2, 1.3 with $\mathscr{A} = 0$ hold for any system $\mathscr{S} \in \mathbf{L}_{\mathscr{N}}$. The easy proof of 3.5, making use of these theorems and of 3.3, 3.4, is left to the reader.

The following corollary follows from 3.5.

3.6. *Let α be a formula of a system $\mathscr{S} \in \mathbf{L}_{\mathscr{N}}$ such that $p_1, \ldots, p_n$ are the distinct propositional variables appearing in α. Then there exists a formula α^* of $\mathscr{S}$ satisfying the following conditions:*

(i) *the signs $\sim$, $\Rightarrow$ do not appear in α^*,*

(ii) *α^* contains at most the propositional variables $p_1, \ldots, p_n, q_1, \ldots, q_n$, all different from each other, and at least one of the variables p_k, q_k for every $k = 1, \ldots, n$.*

(iii) *for any substitution $\mathfrak{s}$ such that $\mathfrak{s}q_k = \sim p_k$ and $\mathfrak{s}p_k = p_k, k = 1, \ldots, n$, the formulas $(\alpha \rightarrow \mathfrak{s}\alpha^*)$, $(\mathfrak{s}\alpha^* \rightarrow \alpha)$ are derivable in $\mathscr{S}$.*

4. Connections between constructive logic with strong negation and intuitionistic logic [7]. Let us start with some lemmas which will be useful for the investigations of this section.

4.1. *Let $\mathfrak{G}(X_1) = (G(X_1), X_1, \Rightarrow_1, \cup, \cap, \neg_1)$ be a pseudo-Boolean algebra of open subsets of a topological space X_1 and let $\mathfrak{B}(X) = (B(X), X, \Rightarrow, \cup, \cap, \rightarrow, \neg, \sim)$ be a quasi-pseudo-Boolean algebra of subsets of X connected with $\mathfrak{G}(X_1)$* (see V 3). *Let*

(1) $$h(Y) = Y \cap X_1 \quad \textit{for any } Y \in B(X).$$

[7] For all the results in this section except 4.3 and 4.6 see Białynicki-Birula and Rasiowa [2]. Theorems 4.3 and 4.6 were first stated and proved by Vorobiev [2].

Then h is a mapping from $B(X)$ into $G(X_1)$ satisfying the following conditions:

(i) $h(X) = X_1$, $h(O) = O$,

(ii) *if* $Y \in B(X)$ *and* $Y \neq X$, *then* $h(Y) \neq X_1$,

(iii) $h(Y \cup Z) = h(Y) \cup h(Z)$,

(iv) $h(Y \cap Z) = h(Y) \cap h(Z)$,

(v) $h(Y \to Z) = h(Y) \Rightarrow_1 h(Z)$,

(vi) $h(\neg Y) = \neg_1 h(Y)$.

By V 3 (b_1), h maps $B(X)$ into $G(X_1)$. (i), (iii) and (iv) are obvious. (ii) follows from V 3 $\big((b_3), (6), (7)\big)$. (v) follows from V 3 (14). (vi) follows by (v) and V 3 (22).

4.2. *Suppose that the hypotheses of* 4.1 *hold and, moreover,* $B(X)$ *is composed of all subsets of* $X = X_1 \cup g(X_1)$, *where* $X_1 \cap g(X_1) = O$, *which satisfy the conditions* V 3 $\big((b_1), (b_2), (b_3)\big)$. *Then*

(i) *if* $Y \in G(X_1)$, *then* $Y \cup g(X_1) \in B(X)$,

(ii) $\big(Y \cup g(X_1)\big) \cup \big(Z \cup g(X_1)\big) = (Y \cup Z) \cup g(X_1)$ *for any* $Y, Z \in G(X_1)$,

(iii) $\big(Y \cup g(X_1)\big) \cap \big(Z \cup g(X_1)\big) = (Y \cap Z) \cup g(X_1)$ *for any* $Y, Z \in G(X_1)$,

(iv) $\big(Y \cup g(X_1)\big) \to \big(Z \cup g(X_1)\big) = (Y \Rightarrow_1 Z) \cup g(X_1)$ *for any* $Y, Z \in G(X_1)$,

(v) $\neg\big(Y \cup g(X_1)\big) = \neg_1 Y \cup g(X_1)$, *for any* $Y \in G(X_1)$,

(vi) *if* $Y \in G(X_1)$ *and* $Y \neq X_1$, *then* $Y \cup g(X_1) \neq X$.

If $Y \in G(X_1)$, then $Y \cup g(X_1)$ satisfies conditions V 3 $\big((b_1), (b_2), (b_3)\big)$; thus (i) holds. (ii) and (iii) are obvious. (iv) follows from V 3 (10). (v) follows from (iv), V 3 (22) and V 3 (2). (vi) is obvious.

Let $\mathcal{S} = (\mathcal{L}, C_{\mathcal{L}})$ be in this section a propositional calculus of constructive logic with strong negation. Let F^* be the set of all formulas of $\mathcal{L}$ without the implication sign $\Rightarrow$ and without the strong negation sign $\sim$. Treating $\to$ in formulas in F^* as the intuitionistic implication sign, we can think of F^* as the set of all formulas of the corresponding formalized language $\mathcal{L}^*$ of an intuitionistic propositional calculus $\mathcal{S}^* = (\mathcal{L}^*, C_{\mathcal{L}^*})$.

With these hypotheses and notation the following theorem holds.

4.3. *For any formula* α *in* F^* *the following conditions are equivalent:*

(i) α *is derivable in the propositional calculus* $\mathcal{S} \in \mathbf{L}_{\mathcal{N}}$,

(ii) α *is derivable in the propositional calculus* $\mathcal{S}^* \in \mathbf{L}_{\chi}$ (8).

Suppose that α is not derivable in $\mathcal{S}_{\mathcal{N}}$. Then, by 3.2, there exists a quasi-pseudo-Boolean algebra $\mathfrak{B}(X)$ of subsets of X connected with a pseudo-Boolean algebra $\mathfrak{G}(X_1)$ of open subsets of a topological space X_1 and a valuation v of $\mathcal{L}$ in $\mathfrak{B}(X)$ such that

(2) $$\alpha_{\mathfrak{B}(X)}(v) \neq X.$$

Let us put $v_1(p) = v(p) \cap X_1 = h(v(p))$ for every propositional variable p of $\mathcal{L}$. By 4.1, v_1 is a valuation of $\mathcal{L}^*$ in $\mathfrak{G}(X_1)$ and for every formula β of F^*, $\beta_{\mathfrak{B}(X)}(v) \cap X_1 = h(\beta_{\mathfrak{B}(X)}(v)) = \beta_{\mathfrak{G}(X_1)}(hv) = \beta_{\mathfrak{G}(X_1)}(v_1)$. Hence, by (2) and 4.1 (ii), $\alpha_{\mathfrak{G}(X_1)}(v_1) \neq X_1$. Thus, by XI 8.2, α is not derivable in $\mathcal{S}^*$.

Suppose that α is not derivable in $\mathcal{S}^*$. Then, by XI 8.2, there exists a pseudo-Boolean algebra $\mathfrak{G}(X_1)$ of open subsets of a topological space X_1 and a valuation v_1 of $\mathcal{L}^*$ such that

(3) $$\alpha_{\mathfrak{G}(X_1)}(v_1) \neq X_1.$$

Let $\mathfrak{B}(X)$ be a quasi-pseudo-Boolean algebra of subsets of X connected with $\mathfrak{G}(X_1)$ and such that all subsets of X satisfying conditions V 3 ((b$_1$), (b$_2$), (b$_3$)) belong to $\mathfrak{B}(X)$ and $X_1 \cap g(X_1) = O$. Let $v(p) = v_1(p) \cup \cup g(X_1)$, for every propositional variable p of $\mathcal{L}^*$. By 4.2, v is a valuation of $\mathcal{L}$ in $\mathfrak{B}(X)$ and, moreover, for any formula β in F^*, $\beta_{\mathfrak{B}(X)}(v) = \beta_{\mathfrak{G}(X_1)}(v_1) \cup g(X_1)$. Hence, by (3) and 4.2 (vi), $\alpha_{\mathfrak{B}(X)}(v) \neq X$, which by 3.2 proves that α is not derivable in $\mathcal{S}_{\mathcal{N}}$.

For any formula α in $\mathcal{L}$ such that $p_1, \ldots, p_n$ are the distinct propositional variables appearing in α, let α^* be a formula in F^* satisfying conditions (i), (ii) and (iii) of Theorem 3.6. With these hypotheses and notation the following theorem holds.

4.4. *A formula* α *of* $\mathcal{L}$ *if derivable in* $\mathcal{S} = (\mathcal{L}, C_{\mathcal{L}}) \in \mathbf{L}_{\mathcal{N}}$ *if and only if* α^* *of* $\mathcal{L}^*$ *is a theorem of the intuitionistic theory* $\mathcal{S}^*(\mathcal{A}^*) = (\mathcal{L}^*, C_{\mathcal{L}^*}, \mathcal{A}^*)$, *where* $\mathcal{A}^*$ *is the set of all formulas* $\neg(p_i \cap q_i)$, $i = 1, \ldots, n$,

(8) Vorobiev [2]. For the proof given here see Białynicki-Birula and Rasiowa [2].

and α, α^*, $p_1, \ldots, p_n$, $q_1, \ldots, q_n$ *satisfy conditions* (i), (ii) *and* (iii) *of Theorem* 3.6 [(9)].

PROOF OF SUFFICIENCY: Suppose that $\alpha \notin C_{\mathscr{L}}(O)$. Then, by 3.6 (iii), for a substitution $\mathfrak{s}$ from $\mathscr{L}$ into $\mathscr{L}$ such that

$$(4) \qquad \mathfrak{s}p_i = p_i \quad \text{and} \quad \mathfrak{s}q_i = \sim p_i, \quad i = 1, \ldots, n,$$

$$(5) \qquad \mathfrak{s}\alpha^* \notin C_{\mathscr{L}}(O).$$

Hence, by 3.2, there exists a quasi-pseudo-Boolean algebra $\mathfrak{B}(X)$ of subsets of X connected with a pseudo-Boolean algebra $\mathfrak{G}(X_1)$ of open subsets of a topological space X_1 and a valuation v of $\mathscr{L}$ in $\mathfrak{B}(X)$ such that

$$(6) \qquad \mathfrak{s}\alpha^*_{\mathfrak{B}(X)}(v) \neq X.$$

Let $\mathfrak{s}_{\mathfrak{B}(X)}v$ be the valuation of $\mathscr{L}$ in $\mathfrak{B}(X)$ defined as follows:

$$(7) \qquad \mathfrak{s}_{\mathfrak{B}(X)}v(p) = \mathfrak{s}p_{\mathfrak{B}(X)}(v) \quad \text{for every propositional variable } p.$$

In particular, by (4),

$$(8) \qquad \mathfrak{s}_{\mathfrak{B}(X)}v(p_i) = v(p_i), \quad \mathfrak{s}_{\mathfrak{B}(X)}v(q_i) = \sim v(p_i), \quad i = 1, \ldots, n.$$

By VIII 3.3 and (6),

$$(9) \qquad \alpha^*_{\mathfrak{B}(X)}(\mathfrak{s}_{\mathfrak{B}(X)}v) = \mathfrak{s}\alpha^*_{\mathfrak{B}(X)}(v) \neq X.$$

For any $Y \in B(X)$, let

$$(10) \qquad h(Y) = Y \cap X_1.$$

Moreover, let

$$(11) \qquad v_1(p) = h\big(\mathfrak{s}_{\mathfrak{B}(X)}v(p)\big) \quad \text{for each propositional variable } p.$$

By 4.1, v_1 is a valuation of $\mathscr{L}^*$ in $\mathfrak{G}(X_1)$ and, since $\alpha^* \in F^*$, we have by (11), 4.1 and (9),

$$(12) \qquad \alpha^*_{\mathfrak{G}(X_1)}(v_1) = \alpha^*_{\mathfrak{G}(X_1)}(h\mathfrak{s}_{\mathfrak{B}(X)}v) = h\big(\alpha^*_{\mathfrak{B}(X)}(\mathfrak{s}_{\mathfrak{B}(X)}v)\big) \neq X_1.$$

On the other hand, by V 3 (23), for each $Y \in B(X)$,

$$(13) \qquad (Y \cap \sim Y) \cap X_1 = O.$$

Moreover, by (11) and (8),

$$(14) \qquad v_1(p_i) = h\big(v(p_i)\big), \quad v_1(q_i) = h\big(\sim v(p_i)\big), \quad i = 1, \ldots, n.$$

(9) Białynicki-Birula and Rasiowa [2].

It follows from (14), 4.1, (10), (13), V 3 (2) that for each $i = 1, \ldots, n$

$$(15) \quad \neg(p_i \cap q_i)_{\mathfrak{G}(X_1)}(v_1) = \neg_1\big(v_1(p_i) \cap v_1(q_i)\big) = \neg_1 h\big(v(p_i) \cap \sim v(p_i)\big) = \neg_1 O = X_1 .$$

By (15), v_1 is an intuitionistic model of $\mathscr{S}^*(\mathscr{A}^*)$ in $\mathfrak{G}(X_1)$. Hence, by (12) and XI 9.3, $\alpha^* \notin C_{\mathscr{L}^*}(\mathscr{A}^*)$.

Proof of necessity: Suppose that $\alpha^* \notin C_{\mathscr{L}^*}(\mathscr{A}^*)$. Then, by XI 9.3, there exists a pseudo-Boolean algebra $\mathfrak{G}(X_1) = \big(G(X_1), X_1, \Rightarrow_1, \cup, \cap, \neg_1\big)$ of open subsets of a topological space X_1 and a valuation v_1 of $\mathscr{L}^*$ in $\mathfrak{G}(X_1)$ such that the following conditions are satisfied:

$$(16) \quad \neg(p_i \cap q_i)_{\mathfrak{G}(X_1)}(v_1) = \neg_1\big(v_1(p_i) \cap v_1(q_i)\big) = X_1, \quad i = 1, \ldots, n,$$

$$(17) \quad \alpha^*_{\mathfrak{G}(X_1)}(v_1) \neq X_1 .$$

By (16) we have

$$(18) \quad v_1(p_i) \cap v_1(q_i) = O, \quad i = 1, \ldots, n.$$

Let $\mathfrak{B}(X)$ be a quasi-pseudo-Boolean algebra of subsets of X connected with $\mathfrak{G}(X_1)$, such that $X_1 \cap g(X_1) = O$ (see V 3) and consisting of all subsets of X satisfying conditions (b_1), (b_2) and (b_3) of V 3. By these hypotheses and (18), the subsets of X

$$(19) \quad Y_i = v_1(p_i) \cup g\big(X_1 - v_1(q_i)\big), \quad i = 1, \ldots, n,$$

satisfy conditions (b_1), (b_2) and (b_3) of V 3 and thus belong to $B(X)$. Let v be the valuation of $\mathscr{L}$ in $\mathfrak{B}(X)$ defined as follows:

$$(20) \quad \begin{aligned} & v(p_i) = Y_i, \quad v(q_i) = \sim Y_i, \quad i = 1, \ldots, n, \\ & v(p) = v_1(p) \cup g(X_1), \quad \text{for any propositional variable } p \\ & \qquad \text{different from } p_i,\ q_i,\ i = 1, \ldots, n. \end{aligned}$$

Let $h(Y) = Y \cap X_1$ for every Y in $B(X)$. It is easy to verify that by (20), (19), the hypotheses concerning $\mathfrak{B}(X)$, and V 3 $\big((7), (8), (9)\big)$ we have

$$(21) \quad h\big(v(p_i)\big) = v_1(p_i), \quad h\big(v(q_i)\big) = v_1(q_i), \quad i = 1, \ldots, n,$$

and $h\big(v(p)\big) = v_1(p)$ for any propositional variable p different from p_i, q_i, $i = 1, \ldots, n$. Hence, by 4.1 and (17),

$$(22) \quad h\big(\alpha^*_{\mathfrak{B}(X)}(v)\big) = \alpha^*_{\mathfrak{G}(X_1)}(hv) = \alpha^*_{\mathfrak{G}(X_1)}(v_1) \neq X_1 .$$

Thus, by the definition of h, $X_1 \cap \alpha^*_{\mathfrak{B}(X)}(v) \neq X_1$. Since $X = X_1 \cup g(X_1)$ and $X_1 \cap g(X_1) = O$, we get

$$(23) \qquad \alpha^*_{\mathfrak{B}(X)}(v) \neq X.$$

Let $\mathfrak{s}$ be a substitution from $\mathscr{L}$ into $\mathscr{L}$ satisfying the condition (4) and let $\mathfrak{s}_{\mathfrak{B}(X)}v$ be the valuation of $\mathscr{L}$ in $\mathfrak{B}(X)$ defined by the equation (7). Then the conditions (8) are satisfied. Moreover, by (8), (20) we get

$$(24) \qquad \mathfrak{s}_{\mathfrak{B}(X)}v(p_i) = v(p_i), \quad \mathfrak{s}_{\mathfrak{B}(X)}v(q_i) = v(q_i), \quad i = 1, \dots, n.$$

By VIII 3.3, (24) and (23) we get

$$(25) \qquad \mathfrak{s}\alpha^*_{\mathfrak{B}(X)}(v) = \alpha^*_{\mathfrak{B}(X)}(\mathfrak{s}_{\mathfrak{B}(X)}v) = \alpha^*_{\mathfrak{B}(X)}(v) \neq X.$$

Consequently, by 3.2, the formula $\mathfrak{s}\alpha^*$ is not derivable in $\mathscr{S} = (\mathscr{L}, C_{\mathscr{L}}) \in \mathbf{L}_{\mathcal{N}}$. Hence, by 3.6 (iii), α is not derivable in $\mathscr{S}$, which completes the proof.

It follows from 4.4 and the deduction theorem analogous to X 6.4, which holds for the systems in $\mathbf{L}_{\chi}$ (see XI 9) that

4.5. *A formula α of $\mathscr{L}$ is derivable in $\mathscr{S} = (\mathscr{L}, C_{\mathscr{L}}) \in \mathbf{L}_{\mathcal{N}}$ if and only if the formula*

$$((\neg(p_1 \cap q_1) \cap \dots \cap (\neg(p_{n-1} \cap q_{n-1}) \cap \neg(p_n \cap q_n)) \dots) \to \alpha^*),$$

(*where α, α^*, $p_1, \dots, p_n$, $q_1, \dots, q_n$ satisfy conditions* (i), (ii), (iii) *of* 3.6) *is derivable in the intuitionistic propositional calculus $\mathscr{S}^* = (\mathscr{L}^*, C_{\mathscr{L}^*})$* [10].

Since every intuitionistic propositional calculus is decidable (see XI 8.3), we get the following theorem.

4.6. *Every propositional calculus $\mathscr{S} = (\mathscr{L}, C_{\mathscr{L}})$ of constructive logic with strong negation is decidable* [11].

We shall apply Theorem 4.4 to prove the following theorem on disjunctions derivable in the propositional calculi of constructive logic with strong negation.

[10] Białynicki-Birula and Rasiowa [2].

[11] Vorobiev [2].

4.7. *For any formulas* α, β *of a propositional calculus* $\mathscr{S} = (\mathscr{L}, C_{\mathscr{L}}) \in \mathbf{L}_{\mathscr{N}}$

(26) $(\alpha \cup \beta) \in C_{\mathscr{L}}(O)$ *if and only if either* $\alpha \in C_{\mathscr{L}}(O)$ *or* $\beta \in C_{\mathscr{L}}(O)$[12].

The sufficiency follows from Theorem 3.3, which states that for any formulas α, β the formulas $(\alpha \to (\alpha \cup \beta))$, $(\beta \to (\alpha \cup \beta))$ are derivable in $\mathscr{S}$, from 3.2 and V 1.3 (23).

Suppose that $p_1, \ldots, p_k, p_{k+1}, \ldots, p_n$ are the distinct propositional variables appearing in α and $p_{k+1}, \ldots, p_n, p_{n+1}, \ldots, p_m$ are the distinct propositional variables appearing in β. Let α^*, $p_1, \ldots, p_n$, $q_1, \ldots, q_n$ satisfy conditions (i), (ii), (iii) of 3.6 with respect to the formula α and let β^*, $p_{k+1}, \ldots, p_m$, $q_{k+1}, \ldots, q_m$ satisfy these conditions with respect to the formula β. Then the formula $(\alpha^* \cup \beta^*)$ satisfies the same conditions for the formula $(\alpha \cup \beta)$, since 1.2 with $\mathscr{A} = O$ holds for any $\mathscr{S} \in \mathbf{L}_{\mathscr{N}}$. Thus

(27)
$$((\alpha \cup \beta) \to \mathfrak{s}(\alpha^* \cup \beta^*)) \in C_{\mathscr{L}}(O) \quad \text{and} \quad (\mathfrak{s}(\alpha^* \cup \beta^*) \to (\alpha \cup \beta)) \in C_{\mathscr{L}}(O).$$

Suppose that $(\alpha \cup \beta) \in C_{\mathscr{L}}(0)$. Then, by 4.4,

(28)
$$(\alpha^* \cup \beta^*) \in C_{\mathscr{L}^*}(\mathscr{A}^*),$$

where $\mathscr{A}^*$ is the set of all formulas $\neg(p_i \cap q_i)$, $i = 1, \ldots, m$. By XI 10.7 the intuitionistic theory $\mathscr{S}^*(\mathscr{A}^*) = (\mathscr{L}^*, C_{\mathscr{L}^*}, \mathscr{A}^*)$ is prime. Hence, (28) implies that

(29)
$$\text{either} \quad \alpha^* \in C_{\mathscr{L}^*}(\mathscr{A}^*) \quad \text{or} \quad \beta^* \in C_{\mathscr{L}^*}(\mathscr{A}^*).$$

Let us suppose that $\alpha^* \in C_{\mathscr{L}^*}(\mathscr{A}^*)$. It is easy to see that $\alpha^* \in C_{\mathscr{L}^*}(\mathscr{A}_1^*)$, where $\mathscr{A}_1^*$ is the set of all formulas $\neg(p_i \cap q_i)$, $i = 1, \ldots, n$. For, if not, there exists (by XI 9.2) a pseudo-Boolean algebra $\mathfrak{G}(X_1)$ of open subsets of a topological space X_1 and a valuation v of $\mathscr{L}^*$ in $\mathfrak{G}(X_1)$ such that

(30)
$$\neg(p_i \cap q_i)_{\mathfrak{G}(X_1)}(v) = X_1, \quad i = 1, \ldots, n,$$

(31)
$$\alpha^*_{\mathfrak{G}(X_1)}(v) \neq X_1.$$

[12] Białynicki-Birula and Rasiowa [2].

Let us put

(32) $v_1(p_i) = v(p_i)$, $v_1(q_i) = v(q_i)$, $i = 1, \ldots, n$, $v_1(p) = O$ for each propositional variable p different from p_i, q_i, $i = 1, \ldots, n$.

Then $\neg(p_i \cap q_i)_{\mathfrak{G}(X_1)}(v_1) = X_1$ for $i = 1, \ldots, m$ and $\alpha^*_{\mathfrak{G}(X_1)}(v_1) \neq X_1$. Thus, by XI 9.2, $\alpha^* \notin C_{\mathcal{S}^*}(\mathcal{A}^*)$, which contradicts our hypothesis. Applying 4.4, we infer that $\alpha \in C_{\mathcal{S}}(O)$, which completes the proof.

By 1 (A13), 3.3, 3.2, for any formulas α, β of $\mathcal{S} \in \mathbf{L}_{\mathcal{N}}$ the formulas $(\sim(\alpha \cap \beta) \Rightarrow (\sim \alpha \cup \sim \beta))$, $((\sim \alpha \cup \sim \beta) \Rightarrow \sim(\alpha \cap \beta))$ are derivable in $\mathcal{S}$. Hence, by VIII 5 (s_4), $\sim(\alpha \cap \beta)$ is derivable in $\mathcal{S}$ if and only if $(\sim \alpha \cup \sim \beta)$ is derivable in $\mathcal{S}$. Applying 4.7, we get the following theorem *on constructible falsity in constructive logic with strong negation*:

4.8. *For any formulas α, β of a system $\mathcal{S} \in \mathbf{L}_{\mathcal{N}}$, $\sim(\alpha \cap \beta)$ is derivable in $\mathcal{S}$ if and only if at least one of the formulas $\sim \alpha$, $\sim \beta$ is derivable in $\mathcal{S}$* [13].

5. A topological characterization of formulas derivable in propositional calculi of constructive logic with strong negation [14]. Let $\mathcal{S} = (\mathcal{L}, C_{\mathcal{L}})$ be a propositional calculus in $\mathbf{L}_{\mathcal{N}}$ and let $\mathcal{S}^* = (\mathcal{L}^*, C_{\mathcal{L}^*})$ be a propositional calculus of intuitionistic logic $\mathbf{L}_{\chi}$ such that the set F^* of all formulas in $\mathcal{L}^*$ consists of all formulas in $\mathcal{L}$ without $\Rightarrow$, $\sim$ (see Sec. 4).

A pseudo-Boolean algebra $\mathfrak{G}(X_1) = (G(X_1), X_1, \Rightarrow_1, \cup, \cap, \neg_1)$ of open subsets of a topological space X_1 will be said to satisfy the *condition Adq* provided that for any formula α^* of $\mathcal{L}^*$, if $p_1, \ldots, p_n$, $q_1, \ldots, q_n$ are distinct propositional variables which satisfy condition (ii) of 3.6 and $\alpha^* \notin C_{\mathcal{L}^*}(\mathcal{A}^*)$, where $\mathcal{A}^*$ is the set of all formulas $\neg(p_i \cap q_i)$, $i = 1, \ldots, n$, then there exists a valuation v_1 of $\mathcal{L}^*$ in $\mathfrak{G}(X_1)$ such that

(1) $$\neg(p_i \cap q_i)_{\mathfrak{G}(X_1)}(v_1) = X_1, \quad i = 1, \ldots, n,$$

and

(2) $$\alpha^*_{\mathfrak{G}(X_1)}(v_1) \neq X_1.$$

(13) Białynicki-Birula and Rasiowa [2].

(14) For all the results in this section see Białynicki-Birula and Rasiowa [2].

5.1. *Let* $\mathfrak{B}(X) = (B(X), X, \Rightarrow, \cup, \cap, \rightarrow, \neg, \sim)$ *be a quasi-pseudo-Boolean algebra of subsets of* X, *connected with a pseudo-Boolean algebra* $\mathfrak{G}(X_1)$ *satisfying the condition Adq and such that* $X = X_1 \cup g(X_1)$ *where* $X_1 \cap g(X_1) = O$ *and* $B(X)$ *consists of all subsets of* X *satisfying the conditions* (b_1), (b_2), (b_3) *of* V 3. *Then, for every formula* α *of* $\mathcal{S} \in \mathbf{L}_{\mathcal{N}}$,

(3) α *is derivable in* $\mathcal{S}$ *if and only if* $\alpha_{\mathfrak{B}(X)}(v) = V$ *for every valuation* v *in* $\mathfrak{B}(X)$.

The necessity follows from 3.2. If α is not derivable in $\mathcal{S}$, then by 4.4 the corresponding formula α^* in F^* is not a theorem of the intuitionistic theory $\mathcal{S}^*(\mathcal{A}^*)$, where $\mathcal{A}^*$ consists of all formulas $\neg(p_i \cap q_i)$, $i = 1, \ldots, n$. Since $\mathfrak{G}(X_1)$ satisfies the condition *Adq*, there exists a valuation v_1 of $\mathcal{L}^*$ in $\mathfrak{G}(X_1)$ such that (1) and (2) hold. By the proof of 4.4 (the proof of necessity), there exists a valuation v of $\mathcal{L}$ in $\mathfrak{B}(X)$ such that $\alpha_{\mathfrak{B}(X)}(v) \neq X$. Thus, by 3.2, α is not derivable in $\mathcal{S}$.

5.2. *The pseudo-Boolean algebra* $\mathfrak{G}(E)$ *of all open subsets of the real line* E *satisfies the condition Adq.*

Suppose that a formula α^* in F^* is not a theorem of $\mathcal{S}^*(\mathcal{A}^*)$, where $\mathcal{A}^*$ consists of all formulas $\neg(p_i \cap q_i)$, $i = 1, \ldots, n$, and $p_1, \ldots, p_n$, $q_1, \ldots, q_n$ are propositional variables satisfying the condition (ii) of 3.6 for the formula α^*. By the deduction theorem (analogous to X 6.4) for the intuitionistic propositional calculi, the following formula β is not derivable in $\mathcal{S}^*$

$$((\neg(p_1 \cap q_1) \cap \ldots \cap (\neg(p_{n-1} \cap q_{n-1}) \cap \neg(p_n \cap q_n)) \ldots) \rightarrow \alpha^*).$$

By XI 8.2 (vi), there exists a valuation v of $\mathcal{L}^*$ in $\mathfrak{G}(E)$ such that $\beta_{\mathfrak{G}(E)}(v) \neq E$. Consequently,

$$(\neg(p_1 \cap q_1) \cap \ldots \cap (\neg(p_{n-1} \cap q_{n-1}) \cap \neg(p_n \cap q_n)) \ldots)_{\mathfrak{G}(E)}(v) \not\subset \alpha^*_{\mathfrak{G}(E)}(v).$$

Thus there exists a point $x \in E$ such that

$$x \in (\neg(p_1 \cap q_1) \cap \ldots \cap (\neg(p_{n-1} \cap q_{n-1}) \cap \neg(p_n \cap q_n)) \ldots)_{\mathfrak{G}(E)}(v) - \alpha^*_{\mathfrak{G}(E)}(v).$$

Hence, there exists an open interval $X_1 \subset E$ such that $x \in X_1$ and

$$X_1 \cap \bigcup_{i=1}^{n} (v(p_i) \cap v(q_i)) = O. \tag{4}$$

Let $\mathfrak{G}(X_1)$ be the pseudo-Boolean algebra of all open subsets of X_1 and let us put $h(Y) = Y \cap X_1$ for each open subset Y of E. The mapping h is an epimorphism from $\mathfrak{G}(E)$ onto $\mathfrak{G}(X_1)$. Moreover,

$$(5) \qquad x \in X_1 - \alpha^*_{\mathfrak{G}(E)}(v).$$

Let $v_1 = hv$. Then by (5)

$$(6) \quad \alpha^*_{\mathfrak{G}(X_1)}(v_1) = \alpha^*_{\mathfrak{G}(X_1)}(hv) = h(\alpha^*_{\mathfrak{G}(E)}(v)) = X_1 \cap \alpha^*_{\mathfrak{G}(E)}(v) \neq X_1.$$

On the other hand, by (4), for each $i = 1, \ldots, n$ we have

$$\neg(p_i \cap q_i)_{\mathfrak{G}(X_1)}(v_1) = \mathbf{I}(X_1 - (v_1(p_i) \cap v_1(q_i))) = \mathbf{I}X_1 = X_1.$$

The pseudo-Boolean algebras $\mathfrak{G}(E)$ and $\mathfrak{G}(X_1)$ being isomorphic, there exists a valuation v_0 of $\mathscr{L}^*$ in $\mathfrak{G}(E)$ such that

$$\neg(p_i \cap q_i)_{\mathfrak{G}(E)}(v_0) = E \quad \text{for each } i = 1, \ldots, n,$$

and

$$\alpha^*_{\mathfrak{G}(E)}(v_0) \neq E.$$

The following theorem follows directly from 5.1 and 5.2.

5.3. *Let $\mathfrak{G}(E)$ be the pseudo-Boolean algebra of all open subsets of the real line E and let $\mathfrak{B}(X)$ be a quasi-pseudo-Boolean algebra of subsets of X, connected with $\mathfrak{G}(E)$ and satisfying the conditions that $E \cap g(E) = O$, $X = E \cup g(E)$, and $\mathfrak{B}(X)$ consists of all subsets of X satisfying the conditions* (b_1), (b_2), (b_3) *in* V 3. *Then for every formula α of a propositional calculus $\mathscr{S} = (\mathscr{L}, C_{\mathscr{L}})$ of constructive logic with strong negation,*

(6) $\alpha \in C_{\mathscr{L}}(O)$ *if and only if* $\alpha_{\mathfrak{B}(X)}(v) = X$ *for every valuation v of $\mathscr{L}$ in* $\mathfrak{B}(X)$[15].

6. $\mathbf{L}_{\mathscr{N}}$-theories of zero order. Let $\mathscr{S} = (\mathscr{L}, C_{\mathscr{L}})$ be a system in $\mathbf{L}_{\mathscr{N}}$. For every set $\mathscr{A}$ of formulas of $\mathscr{L}$ the system $\mathscr{S}(\mathscr{A}) = (\mathscr{L}, C_{\mathscr{L}}, \mathscr{A})$ will be called an $\mathbf{L}_{\mathscr{N}}$-*theory of zero order based on* $\mathscr{S}$ (see VIII 9).

By an $\mathbf{L}_{\mathscr{N}}$-*model* of an $\mathbf{L}_{\mathscr{N}}$-theory $\mathscr{S}(\mathscr{A}) = (\mathscr{L}, C_{\mathscr{L}}, \mathscr{A})$ we mean any valuation v of $\mathscr{L}$ in any non-degenerate quasi-pseudo-Boolean algebra $\mathfrak{A} = (A, V, \Rightarrow, \cup, \cap, \rightarrow, \neg, \sim)$ such that for each α in $\mathscr{A}$

$$(1) \qquad \alpha_{\mathfrak{A}}(v) = V.$$

The following theorem follows from VIII 9.2, VIII 9.3 and 2.1.

(15) Białynicki-Birula and Rasiowa [2]. As E we can take n-dimensional Euclidean space or the Cantor discontinuum.

6.1. *For every consistent* $\mathbf{L}_{\mathcal{N}}$*-theory* $\mathcal{S}(\mathcal{A}) = (\mathcal{L}, C_{\mathcal{L}}, \mathcal{A})$ *of zero order the algebra* $\mathfrak{A}(\mathcal{S}(\mathcal{A}))$ (see VIII 9) *is a non-degenerate quasi-pseudo-Boolean algebra. The canonical valuation* v^0 *of* $\mathcal{L}$ *in* $\mathfrak{A}(\mathcal{S}(\mathcal{A}))$ (cf. VIII 9 (14)) *is an* $\mathbf{L}_{\mathcal{N}}$*-model of* $\mathcal{S}(\mathcal{A})$. *This* $\mathbf{L}_{\mathcal{N}}$*-model is adequate,* i.e. *for any formula* α *of* $\mathcal{L}$

(2) $$\alpha \in C_{\mathcal{L}}(\mathcal{A}) \quad \textit{if and only if} \quad \alpha_{\mathfrak{A}(\mathcal{S}(\mathcal{A}))}(v^0) = \bigvee.$$

For any formulas α, β of $\mathcal{S} \in \mathbf{L}_{\mathcal{N}}$ the formula $(\neg(\alpha \Rightarrow \alpha) \Rightarrow \beta)$ is derivable in $\mathcal{S}$. Indeed, by V 1.2, V 1 (qpB_{13}), V 1 (qpB_1), III 3 (2), V 1.3 (22) and V 1.1, for every valuation v in every quasi-pseudo-Boolean algebra $\mathfrak{A}$,

$$(\neg(\alpha \Rightarrow \alpha) \Rightarrow \beta)_{\mathfrak{A}}(v) = \neg \bigvee \Rightarrow \beta_{\mathfrak{A}}(v)$$
$$= (\bigvee \to \sim \bigvee) \Rightarrow \beta_{\mathfrak{A}}(v) = (\bigvee \to \bigwedge) \Rightarrow \beta_{\mathfrak{A}}(v) = \bigwedge \Rightarrow \beta_{\mathfrak{A}}(v) = \bigvee.$$

Thus by 3.2 the above formula is derivable in $\mathcal{S}$. Consequently, each system $\mathcal{S}$ in $\mathbf{L}_{\mathcal{N}}$ belongs to the class $\mathbf{S}_{\nu}$ of standard systems of implicative extensional propositional calculi with a semi-negation (see VIII 10). Thus

(3) $$\mathbf{L}_{\mathcal{N}} \subset \mathbf{S}_{\nu}.$$

Observe that for any formulas α, β of $\mathcal{S} \in \mathbf{L}_{\mathcal{N}}$ the formula $(\sim(\alpha \Rightarrow \alpha) \Rightarrow \beta)$ is also derivable in $\mathcal{S}$. In fact, by V 1.2, V 1 (qpB_1), III 3 (2) and V 1.1, $(\sim(\alpha \Rightarrow \alpha) \Rightarrow \beta)_{\mathfrak{A}}(v) = \sim \bigvee \Rightarrow \beta_{\mathfrak{A}}(v) = \bigwedge \Rightarrow \beta_{\mathfrak{A}}(v) = \bigvee$ for every valuation v in any quasi-pseudo-Boolean algebra $\mathfrak{A}$. Thus by 3.2 this formula is derivable in $\mathcal{S}$. Consequently, strong negation has also the property required for semi-negation (see VIII 10).

The following theorem characterizes consistent $\mathbf{L}_{\mathcal{N}}$-theories.

6.2. *For any* $\mathbf{L}_{\mathcal{N}}$*-theory* $\mathcal{S}(\mathcal{A}) = (\mathcal{L}, C_{\mathcal{L}}, \mathcal{A})$ *of zero order the following conditions are equivalent*:

(i) $\mathcal{S}(\mathcal{A})$ *is consistent,*

(ii) *for any formula* α *of* $\mathcal{L}$, *either* $\alpha \notin C_{\mathcal{L}}(\mathcal{A})$ *or* $\neg\alpha \notin C_{\mathcal{L}}(\mathcal{A})$,

(iii) *for any formula* α *of* $\mathcal{L}$, *either* $\alpha \notin C_{\mathcal{L}}(\mathcal{A})$ *or* $\sim \alpha \notin C_{\mathcal{L}}(\mathcal{A})$,

(iv) *there exists an* $\mathbf{L}_{\mathcal{N}}$*-model of* $\mathcal{S}(\mathcal{A})$,

(v) *there exists an* $\mathbf{L}_{\mathcal{N}}$*-model of* $\mathcal{S}(\mathcal{A})$ *in a quasi-pseudo-Boolean algebra of sets,*

(vi) *there exists an* $\mathbf{L}_{\mathcal{N}}$*-model of* $\mathscr{S}(\mathscr{A})$ *in a quasi-pseudo-Boolean algebra which has at most three elements*, i.e. *either in* $\mathfrak{C}_0$ *or in* $\mathfrak{B}_0$ (see V 3).

The equivalence of the conditions (i), (ii), (iii), (iv) follows by VIII 10.2, since $\sim$, $\neg$ are both semi-negations. (iv) implies (v) by the representation theorem V 5.6 and (v) clearly implies (iv). It is sufficient to prove that (iv) implies (vi), since obviously (vi) implies (iv). Suppose that a valuation v of $\mathscr{L}$ in a non-degenerate quasi-pseudo-Boolean algebra $\mathfrak{A}$ is an $\mathbf{L}_{\mathcal{N}}$-model of $\mathscr{S}(\mathscr{A})$. By V 4.15 there exists a maximal s.f.f.k. ∇ in $\mathfrak{A}$. By V 4.22, the quotient algebra $\mathfrak{A}/\nabla$ is non-degenerate and has at most three elements. Let h be an epimorphism from $\mathfrak{A}$ onto $\mathfrak{A}/\nabla$. It follows from VIII 3.2 that hv is a model of $\mathscr{S}(\mathscr{A})$ in $\mathfrak{A}/\nabla$.

The next theorem characterizes all theorems of any consistent $\mathbf{L}_{\mathcal{N}}$-theory of zero order.

6.3. *For any formula* α *of any consistent* $\mathbf{L}_{\mathcal{N}}$*-theory* $\mathscr{S}(\mathscr{A}) = (\mathscr{L}, C_{\mathscr{L}}, \mathscr{A})$ *of zero order the following conditions are equivalent*:

(i) $\alpha \in C_{\mathscr{L}}(\mathscr{A})$,

(ii) $\alpha_{\mathfrak{A}}(v) = \vee$ *for every* $\mathbf{L}_{\mathcal{N}}$*-model* v *of* $\mathscr{S}(\mathscr{A})$ *in any non-degenerate quasi-pseudo-Boolean algebra* $\mathfrak{A}$,

(iii) $\alpha_{\mathfrak{B}(X)}(v) = X$ *for every* $\mathbf{L}_{\mathcal{N}}$*-model* v *of* $\mathscr{S}(\mathscr{A})$ *in any non-degenerate quasi-pseudo-Boolean algebra* $\mathfrak{B}(X)$ *of sets* (i.e. *an algebra connected with any pseudo-field* $\mathfrak{G}(X_1)$ *of open subsets of any topological space* X_1),

(iv) $\alpha_{\mathfrak{A}(\mathscr{S}(\mathscr{A}))}(v^0) = \vee$ *for the canonical valuation* v^0 *in the algebra* $\mathfrak{A}(\mathscr{S}(\mathscr{A}))$ *of the theory* $\mathscr{S}(\mathscr{A})$.

Conditions (i), (ii), (iv) are equivalent by VIII 9.5. Obviously, (ii) implies (iii) and, by the representation theorem V 5.6, (iii) implies (ii), which completes the proof.

Let us note the following simple result.

6.4. *For any* $\mathbf{L}_{\mathcal{N}}$*-theory* $\mathscr{S}(\mathscr{A}) = (\mathscr{L}, C_{\mathscr{L}}, \mathscr{A})$ *and for all formulas* α, β *of* $\mathscr{L}$:

(i) *if* α, $(\alpha \to \beta) \in C_{\mathscr{L}}(\mathscr{A})$, *then* $\beta \in C_{\mathscr{L}}(\mathscr{A})$,

(ii) *if* $(\alpha \equiv \beta) \in C_{\mathscr{L}}(\mathscr{A})$, *then* $(\alpha \to \beta)$, $(\beta \to \alpha) \in C_{\mathscr{L}}(\mathscr{A})$.

This follows easily by 6.3 and V 1.3 (23), V 1 (qpB_1), V 1 (1), V 1.1.

The connection between $\mathbf{L}_{\mathcal{N}}$-theories and intuitionistic theories of zero order will now be considered.

6.5. *Let* $\mathcal{S}(\mathcal{A}) = (\mathcal{L}, C_{\mathcal{L}}, \mathcal{A})$ *be an* $\mathbf{L}_{\mathcal{N}}$*-theory and let* $\mathcal{S}^* = (\mathcal{L}^*, C_{\mathcal{L}^*})$ *be a propositional calculus of intuitionistic logic such that the set* F^* *of all formulas in* $\mathcal{L}^*$ *coincides with the set of all formulas in* $\mathcal{L}$ *without* $\Rightarrow$ *and* $\sim$ (see Sec. 4). *If each formula* α *in* $\mathcal{A}$ *belongs to* F^*, *then for every formula* β *in* F^* *the following conditions are equivalent*:

(i) β *is a theorem of* $\mathcal{S}(\mathcal{A})$,

(ii) β *is a theorem of the intuitionistic theory* $\mathcal{S}^*(\mathcal{A}) = (\mathcal{L}^*, C_{\mathcal{L}^*}, \mathcal{A})$.

The proof is analogous to the proof of 4.3. Instead of 3.2 and XI 8.2, we should apply 6.3 and XI 9.3, respectively.

7. The connection between $\mathbf{L}_{\mathcal{N}}$-theories of zero order and special filters of the first kind. Consider the propositional calculus $\mathcal{S}_{\mathcal{N}}$ of constructive logic with strong negation described in Section 1. Let $\mathfrak{A} = (A, V, \Rightarrow, \cup, \cap, \rightarrow, \neg, \sim)$ be a quasi-pseudo-Boolean algebra. Since the only rule of inference in $\mathcal{S}_{\mathcal{N}}$ is $(r_{\mathcal{N}})$, it follows from Section 3 and VIII 12.3, that a subset ∇ of A is a deductive filter if and only if the following conditions hold:

(f_1) $V \in \nabla$,

(f_2) if a, $a \Rightarrow b \in \nabla$, then $b \in \nabla$,

(f_3) if a, $a \rightarrow b \in \nabla$, then $b \in \nabla$.

Observe that, by V 1 (qpB_4) and V 1.3 (29), conditions (f_1) and (f_3) imply (f_2). Hence, by V 4.2, we infer that:

7.1. *For any set* ∇ *of elements of a quasi-pseudo-Boolean algebra* $\mathfrak{A}$ *the following conditions are equivalent*:

(i) ∇ *is a deductive filter*,

(ii) ∇ *is a special filter of the first kind* (s.f.f.k.).

By V 4.4 and V 4.5 the kernels of epimorphisms from quasi-pseudo-Boolean algebras onto similar algebras are special filters of the first kind and, conversely, every s.f.f.k. is the kernel of the epimorphism determined by itself. This follows also from Section 3, 7.1, VIII 12.2, VIII

13.10 and the equational definability of the class of all quasi-pseudo-Boolean algebras (see V 2.1).

By 7.1, VIII 13.2 and VIII 13.4 we get two theorems analogous to X 6.2 and X 6.3, which state that there exists a natural one-one correspondence between special filters of the first kind in the quasi-pseudo-Boolean algebra $\mathfrak{A}(\mathcal{S})$ of a system $\mathcal{S}$ in $\mathbf{L}_{\mathcal{N}}$ and formalized $\mathbf{L}_{\mathcal{N}}$-theories based on $\mathcal{S}$, two theories $\mathcal{S}(\mathcal{A})$ and $\mathcal{S}(\mathcal{A}')$ being identified if and only if they have the same set of theorems.

Let us note that for any system $\mathcal{S}$ in $\mathbf{L}_{\mathcal{N}}$ there holds a theorem corresponding to IX 5.4. Moreover, it is easy to see by V 4.9, V 1 (qpB_3) and V 1.3 (9) that the s.f.f.k. generated by an element a and a s.f.f.k. ∇ in a quasi-pseudo-Boolean algebra $\mathfrak{A}$ is the set of all x in $\mathfrak{A}$ such that $a \to x \in \nabla$. Applying 6.4 and the theorems mentioned above, we prove, similarly to IX 5.5, the following *deduction theorem.*

7.2. *For any formulas* α, β *of each system* $\mathcal{S} = (\mathcal{L}, C_{\mathcal{L}})$ *in* $\mathbf{L}_{\mathcal{N}}$ *and for each set* $\mathcal{A}$ *of formulas of* $\mathcal{L}$

$$\beta \in C_{\mathcal{L}}(\mathcal{A} \cup \{\alpha\}) \quad \textit{if and only if} \quad (\alpha \to \beta) \in C_{\mathcal{L}}(\mathcal{A}).$$

The second *deduction theorem* can be formulated as follows.

7.3. *For every formula* α *in the formalized language* $\mathcal{L}$ *of any system* $\mathcal{S} = (\mathcal{L}, C_{\mathcal{L}})$ *in* $\mathbf{L}_{\mathcal{N}}$ *and for every non-empty set* $\mathcal{A}$ *of formulas of* $\mathcal{L}$, *the condition* $\alpha \in C_{\mathcal{L}}(\mathcal{A})$ *is equivalent to the condition that there exist formulas* $\alpha_1, \ldots, \alpha_n$ *in* $\mathcal{A}$ *such that*

$$((\alpha_1 \cap \ldots \cap (\alpha_{n-1} \cap \alpha_n) \ldots) \to \alpha) \in C_{\mathcal{L}}(O).$$

The proof, analogous to that of X 6.4 and based on 3.3 (D_{10}), 6.4, a theorem analogous to X 6.2, V 4.8, V 1 (1), and VIII 6 (8), is left to the reader.

By 7.1 the notions of an irreducible deductive filter and of a maximal deductive filter coincide with the notion of an irreducible s.f.f.k. and of a maximal s.f.f.k., respectively.

The following theorems follow from 7.1, VIII 13.5, VIII 13.6, VIII 13.7.

7.4. *An* $\mathbf{L}_{\mathcal{N}}$-*theory* $\mathcal{S}(\mathcal{A}) = (\mathcal{L}, C_{\mathcal{L}}, \mathcal{A})$ *is irreducible* (*maximal*) *if and only if the* s.f.f.k. $\nabla_{\mathcal{A}}$ *in* $\mathfrak{A}(\mathcal{S})$, *consisting of all* $||\gamma||$ *in* $\mathfrak{A}(\mathcal{S})$ *such that* $\gamma \in C_{\mathcal{L}}(\mathcal{A})$, *is irreducible* (*maximal*).

7.5. *For every consistent* $\mathbf{L}_{\mathscr{N}}$*-theory* $\mathscr{S}(\mathscr{A}) = (\mathscr{L}, C_{\mathscr{L}}, \mathscr{A})$ *there exists an irreducible* $\mathbf{L}_{\mathscr{N}}$*-theory* $\mathscr{S}(\mathscr{A}') = (\mathscr{L}, C_{\mathscr{L}}, \mathscr{A}')$ *such that* $C_{\mathscr{L}}(\mathscr{A}) \subset C_{\mathscr{L}}(\mathscr{A}')$.

The following theorem follows from (3) and VIII 13.8.

7.6. *For every consistent* $\mathbf{L}_{\mathscr{N}}$*-theory* $\mathscr{S}(\mathscr{A}) = (\mathscr{L}, C_{\mathscr{L}}, \mathscr{A})$ *of zero order there exists a maximal* $\mathbf{L}_{\mathscr{N}}$*-theory* $\mathscr{S}(\mathscr{A}') = (\mathscr{L}, C_{\mathscr{L}}, \mathscr{A}')$ *such that* $C_{\mathscr{L}}(\mathscr{A}) \subset C_{\mathscr{L}}(\mathscr{A}')$.

The next theorem characterizes maximal $\mathbf{L}_{\mathscr{N}}$-theories.

7.7. *The following conditions are equivalent for each* $\mathbf{L}_{\mathscr{N}}$*-theory* $\mathscr{S}(\mathscr{A}) = (\mathscr{L}, C_{\mathscr{L}}, \mathscr{A})$:

(i) $\mathscr{S}(\mathscr{A})$ *is maximal,*

(ii) *for each formula* α *of* $\mathscr{L}$ *exactly one of the formulas* $\alpha, \urcorner\alpha$ *is a theorem of* $\mathscr{S}(\mathscr{A})$,

(iii) *for each formula* α *of* $\mathscr{L}$ *exactly one of the following conditions is satisfied*:

(a) $\alpha \in C_{\mathscr{L}}(\mathscr{A}),\quad \urcorner\alpha \notin C_{\mathscr{L}}(\mathscr{A}),\quad \sim\alpha \notin C_{\mathscr{L}}(\mathscr{A}),\quad \urcorner\sim\alpha \in C_{\mathscr{L}}(\mathscr{A})$,

(b) $\alpha \notin C_{\mathscr{L}}(\mathscr{A}),\quad \urcorner\alpha \in C_{\mathscr{L}}(\mathscr{A}),\quad \sim\alpha \in C_{\mathscr{L}}(\mathscr{A}),\quad \urcorner\sim\alpha \notin C_{\mathscr{L}}(\mathscr{A})$,

(c) $\alpha \notin C_{\mathscr{L}}(\mathscr{A}),\quad \urcorner\alpha \in C_{\mathscr{L}}(\mathscr{A}),\quad \sim\alpha \notin C_{\mathscr{L}}(\mathscr{A}),\quad \urcorner\sim\alpha \in C_{\mathscr{L}}(\mathscr{A})$,

(iv) $\mathfrak{A}(\mathscr{S}(\mathscr{A}))$ *is isomorphic either to* $\mathfrak{B}_0$ *or to* $\mathfrak{C}_0$ (see V 3), i.e. $\mathfrak{A}(\mathscr{S}(\mathscr{A}))$ *is a non-degenerate quasi-pseudo-Boolean algebra which has at most three elements,*

(v) $\mathscr{S}(\mathscr{A})$ *has an adequate* $\mathbf{L}_{\mathscr{N}}$*-model in* $\mathfrak{C}_0$,

(vi) $\mathscr{S}(\mathscr{A})$ *is consistent and each* $\mathbf{L}_{\mathscr{N}}$*-model of* $\mathscr{S}(\mathscr{A})$ *in* $\mathfrak{C}_0$ *is adequate.*

By 7.4, V 4.22 and the definition of $\nabla_{\mathscr{A}}$, conditions (i), (ii), (iii) are equivalent. Since, by VIII 13.11, $\mathfrak{A}(\mathscr{S}(\mathscr{A}))$ is isomorphic to the quotient algebra $\mathfrak{A}(\mathscr{S})/\nabla_{\mathscr{A}}$, it follows from 7.4 and V 4.22 that (iv) is equivalent to (i). Observe that $\mathfrak{B}_0$ is a subalgebra of $\mathfrak{C}_0$. Hence, every $\mathbf{L}_{\mathscr{N}}$-model of $\mathscr{S}(\mathscr{A})$ in $\mathfrak{B}_0$ can be treated as an $\mathbf{L}_{\mathscr{N}}$-model of $\mathscr{S}(\mathscr{A})$ in $\mathfrak{C}_0$. It follows that (iv) implies (v) by 6.1 and VIII 13.11. If (v) holds, then by 6.2 $\mathscr{S}(\mathscr{A})$ is consistent. Let v_0 be an adequate $\mathbf{L}_{\mathscr{N}}$-model of $\mathscr{S}(\mathscr{A})$ in $\mathfrak{C}_0$. Then, by the definition of the operations in $\mathfrak{C}_0$ (see V 3), for each formula α of $\mathscr{L}$ exactly one of the following conditions is satisfied:

(d) $\alpha_{\mathfrak{C}_0}(v_0) = V, \quad \neg\alpha_{\mathfrak{C}_0}(v_0) = \Lambda,$

(e) $\alpha_{\mathfrak{C}_0}(v_0) = a, \quad \neg\alpha_{\mathfrak{C}_0}(v_0) = V,$

(f) $\alpha_{\mathfrak{C}_0}(v_0) = \Lambda, \quad \neg\alpha_{\mathfrak{C}_0}(v_0) = V.$

Since v_0 is an adequate $\mathbf{L}_{\mathcal{N}}$-model of $\mathscr{S}(\mathscr{A})$, it follows from (d), (e), (f) that for each formula α of $\mathscr{L}$ exactly one of the formulas α, $\neg\alpha$ is a theorem of $\mathscr{S}(\mathscr{A})$. Let v be an arbitrary $\mathbf{L}_{\mathcal{N}}$-model of $\mathscr{S}(\mathscr{A})$ in $\mathfrak{C}_0$. If α is a theorem of $\mathscr{S}(\mathscr{A})$, then by 6.3, $\alpha_{\mathfrak{C}_0}(v) = V$. If α is not a theorem of $\mathscr{S}(\mathscr{A})$, then $\neg\alpha \in C_{\mathscr{L}}(\mathscr{A})$. Hence, $\neg\alpha_{\mathfrak{C}_0}(v) = V$ and, consequently, either $\alpha_{\mathfrak{C}_0}(v) = a$ or $\alpha_{\mathfrak{C}_0}(v) = \Lambda$. Thus, for any formula α of $\mathscr{L}$, $\alpha_{\mathfrak{C}_0}(v) = V$ if and only if $\alpha \in C_{\mathscr{L}}(\mathscr{A})$, i.e. v is an adequate $\mathbf{L}_{\mathcal{N}}$-model of $\mathscr{S}(\mathscr{A})$. We have just proved that (v) implies (vi). Suppose that (vi) holds. Since $\mathscr{S}(\mathscr{A})$ is consistent, by 6.2 there exists an $\mathbf{L}_{\mathcal{N}}$-model of $\mathscr{S}(\mathscr{A})$ in $\mathfrak{C}_0$ which contains $\mathfrak{B}_0$ as a subalgebra. Let v_0 be this $\mathbf{L}_{\mathcal{N}}$-model. Thus for each formula α of $\mathscr{L}$ exactly one of the conditions (d), (e), (f) is satisfied. By (vi) v_0 is adequate. Hence, by (d), (e), (f), for each formula α of $\mathscr{L}$ exactly one of the formulas α, $\neg\alpha$ is a theorem of $\mathscr{S}(\mathscr{A})$. Thus (vi) implies (ii), which completes the proof.

A consistent $\mathbf{L}_{\mathcal{N}}$-theory $\mathscr{S}(\mathscr{A}) = (\mathscr{L}, C_{\mathscr{L}}, \mathscr{A})$ is said to be *prime* provided that for any formulas α, β of $\mathscr{L}$ the following condition is satisfied:

(p) if $(\alpha \cup \beta) \in C_{\mathscr{L}}(\mathscr{A})$, then either $\alpha \in C_{\mathscr{L}}(\mathscr{A})$ or $\beta \in C_{\mathscr{L}}(\mathscr{A})$.

Since the set $\nabla_{\mathscr{A}}$ (of all $||\gamma||$ in $\mathfrak{A}(\mathscr{S})$ such that $\gamma \in C_{\mathscr{L}}(\mathscr{A})$) is a s.f.f.k. in $\mathfrak{A}(\mathscr{S})$, it follows from the above definition of prime $\mathbf{L}_{\mathcal{N}}$-theories and from the definition of a prime s.f.f.k. in quasi-pseudo-Boolean algebras (see V 4) that

7.8. *An $\mathbf{L}_{\mathcal{N}}$-theory $\mathscr{S}(\mathscr{A}) = (\mathscr{L}, C_{\mathscr{L}}, \mathscr{A})$ is prime if and only if the* s.f.f.k. $\nabla_{\mathscr{A}}$ *in $\mathfrak{A}(\mathscr{S})$ is prime.*

By 7.8, 7.4 and V 4.12 we obtain the following result.

7.9. *An $\mathbf{L}_{\mathcal{N}}$-theory $\mathscr{S}(\mathscr{A}) = (\mathscr{L}, C_{\mathscr{L}}, \mathscr{A})$ is prime if and only if it is irreducible.*

8. Prime $\mathbf{L}_{\mathcal{N}}$-theories. We recall that a consistent $\mathbf{L}_{\mathcal{N}}$-theory $\mathcal{S}(\mathcal{A}) = (\mathcal{L}, C_{\mathcal{L}}, \mathcal{A})$ is *prime* provided condition (p) in Sec. 7 is satisfied. By 7.9 and 7.5, for each consistent $\mathbf{L}_{\mathcal{N}}$-theory $\mathcal{S}(\mathcal{A}) = (\mathcal{L}, C_{\mathcal{L}}, \mathcal{A})$ there exists a prime $\mathbf{L}_{\mathcal{N}}$-theory $\mathcal{S}(\mathcal{A}') = (\mathcal{L}, C_{\mathcal{L}}, \mathcal{A}')$ such that $C_{\mathcal{L}}(\mathcal{A}) \subset C_{\mathcal{L}}(\mathcal{A}')$. By 7.9 and VIII 13.9 each maximal $\mathbf{L}_{\mathcal{N}}$-theory is prime. Let us note the following property of prime $\mathbf{L}_{\mathcal{N}}$-theories.

8.1. *An $\mathbf{L}_{\mathcal{N}}$-theory $\mathcal{S}(\mathcal{A}) = (\mathcal{L}, C_{\mathcal{L}}, \mathcal{A})$ of zero order is prime if and only if it is consistent and for any formulas α, β of $\mathcal{L}$*

(p′) *if* $\sim(\alpha \cap \beta) \in C_{\mathcal{L}}(\mathcal{A})$, *then either* $\sim \alpha \in C_{\mathcal{L}}(\mathcal{A})$ *or* $\sim \beta \in C_{\mathcal{L}}(\mathcal{A})$.

Suppose that $\mathcal{S}(\mathcal{A})$ is prime and $\sim(\alpha \cap \beta) \in C_{\mathcal{L}}(\mathcal{A})$. Then, by 6.4, 3.3 and 1 (A13), $(\sim \alpha \cup \sim \beta) \in C_{\mathcal{L}}(\mathcal{A})$. Hence, by 7 (p), either $\sim \alpha \in C_{\mathcal{L}}(\mathcal{A})$ or $\sim \beta \in C_{\mathcal{L}}(\mathcal{A})$. Suppose that the condition (p′) holds for all α, β in $\mathcal{L}$. Observe that 3.3, (A13), (A16) and 3.4 imply that $(\sim(\sim \alpha \cap \sim \beta) \equiv (\alpha \cup \beta)) \in C_{\mathcal{L}}(O)$. If $(\alpha \cup \beta) \in C_{\mathcal{L}}(\mathcal{A})$, then by 6.4 $\sim(\sim \alpha \cap \sim \beta) \in C_{\mathcal{L}}(\mathcal{A})$. Hence, by (p′), either $\sim \sim \alpha \in C_{\mathcal{L}}(\mathcal{A})$ or $\sim \sim \beta \in C_{\mathcal{L}}(\mathcal{A})$. Consequently, either $\alpha \in C_{\mathcal{L}}(\mathcal{A})$ or $\beta \in C_{\mathcal{L}}(\mathcal{A})$.

The property (p′) will be called the *property of constructible falsity*. Thus 8.1 characterizes prime $\mathbf{L}_{\mathcal{N}}$-theories as consistent $\mathbf{L}_{\mathcal{N}}$-theories with the property of constructible falsity. Now we are going to give another characterization of prime $\mathbf{L}_{\mathcal{N}}$-theories, analogous to that of prime intuitionistic theories in XI 10. To begin with, we shall define the *characteristic valuation for consistent $\mathbf{L}_{\mathcal{N}}$-theories.*

Let $\mathcal{S}(\mathcal{A}) = (\mathcal{L}, C_{\mathcal{L}}, \mathcal{A})$ be a consistent $\mathbf{L}_{\mathcal{N}}$-theory. By 6.1 and V 5.5 there exists a monomorphism h_0 from $\mathfrak{A}(\mathcal{S}(\mathcal{A}))$ into a quasi-pseudo-Boolean algebra $\mathfrak{B}(X)$ connected with a pseudo-Boolean algebra $\mathfrak{G}(X_1) = (G(X_1), X_1, \Rightarrow_1, \cup, \cap, \neg_1)$ of open subsets of a topological space X_1 with the interior operation $\mathbf{I}$. We can assume that $B(X)$ consists of all subsets of X satisfying the conditions V 3 ((b$_1$), (b$_2$), (b$_3$)).

It follows from VIII 9 (15) and VIII 3.2 that for the canonical valuation v^0 of $\mathcal{L}$ in $\mathfrak{A}(\mathcal{S}(\mathcal{A}))$ and for any formula α of $\mathcal{L}$

(1) $$h_0 ||\alpha||_{\mathcal{A}} = h_0\, \alpha_{\mathfrak{A}(\mathcal{S}(\mathcal{A}))}(v^0) = \alpha_{\mathfrak{B}(X)}(h_0 v^0).$$

Let $X_1^0 = X_1 \cup \{x_1\}$, where $x_1 \notin X$, be the one-point strong compactification of the space X_1 (see VI 8). With these hypotheses and notation the following theorem holds.

8.2. *The class* $G(X_1^0)$ *composed of all subsets* $Y \subset X_1^0$ *which belong to* $G(X_1)$ *and of* X_1^0 *is closed under the set-theoretical operations* $\cup$, $\cap$ *and* $\Rightarrow_1$, $\neg_1$ *defined as follows*:

(2) $$Y \Rightarrow_1 Z = \mathbf{I}^0((X_1^0 - Y) \cup Z) \quad \textit{for all } Y, Z \in G(X_1^0),$$

(3) $$\neg_1 Y = \mathbf{I}^0(X_1^0 - Y) = Y \Rightarrow_1 O, \quad \textit{for all } Y \in G(X_1^0),$$

where $\mathbf{I}^0$ *is the interior operation in* X_1^0. *Hence* $\mathfrak{G}(X_1^0) = (G(X_1^0), X_1^0, \Rightarrow_1, \cup, \cap, \neg_1)$ *is a pseudo-field of open subsets of* X_1^0.

The easy proof is left to the reader.

Let us put

(4) $X_2^0 = X_2 \cup \{x_2\}$, where $x_2 \neq x_1$, $x_2 \notin X$, $X_2 = g(X_1)$ and g satisfies the conditions V 3 ((7), (8)),

(5) $$X^0 = X_1^0 \cup X_2^0,$$

(6) $g^0(x) = g(x)$ for any $x \in X$, $g^0(x_1) = x_2$, $g^0(x_2) = x_1$.

Observe that, by (6) and V 3 ((8), (7), (3)),

(7) g^0 *is an involution of* X^0 *and* $g^0(x) = x$ *for all* $x \in X_1^0 \cap X_2^0$.

Since $X_1^0 \cap X_2^0 = X_1 \cap X_2$, it easily follows by (6) and V 3 (4), V 3 (7) that

(8) $$X_1^0 \cap X_2^0 \subset \bigcap_{Y \in G(X_1^0)} \big(Y \cup g^0(\mathbf{I}^0(X_1^0 - Y))\big).$$

Let $B(X^0)$ be the class of all subsets Y of X^0 satisfying the three conditions:

(b_1') $Y \cap X_1^0 \in G(X_1^0)$,

(b_2') $Y \cap X_2^0 = X_2^0 - g^0(Z)$ for some $Z \in G(X_1^0)$,

(b_3') $Y \cap X_1^0 \subset g^0(Y) \cap X_1^0$.

By V 3.4, 8.2, (4), (5), (7), (8) the following theorem holds.

8.3. $\mathfrak{B}(X^0) = (B(X^0), X^0, \Rightarrow, \cup, \cap, \rightarrow, \neg, \sim)$, *where* $\cup$, $\cap$ *are set-theoretical union and intersection and* $\Rightarrow$, $\rightarrow$, $\neg$, $\sim$ *are defined by*

(9) $$Y \Rightarrow Z = (Y \rightarrow Z) \cap (\sim Z \rightarrow \sim Y),$$

(10) $$Y \rightarrow Z = (Y \cap X_1^0 \Rightarrow_1 Z \cap X_1^0) \cup ((X_2^0 - g^0(Y \cap X_1^0)) \cup Z \cap X_2^0),$$

(11) $$\neg Y = Y \to O = (\neg_1(Y \cap X_1^0)) \cup (X_2^0 - g^0(Y \cap X_1^0)),$$

(12) $$\sim Y = X^0 - g^0(Y),$$

for all $Y, Z \in B(X^0)$, *is a quasi-pseudo-Boolean algebra connected with* $\mathfrak{G}(X_1^0)$.

8.4. *The class* $B(X^0)$ *consists of the empty set, of* X^0 *and of all sets* $G \cup \{x_2\}$, *where* $G \in B(X)$.

It is easy to verify that O, X^0 and each set $G \cup \{x_2\}$, where $G \in B(X)$, satisfies the conditions (b_1'), (b_2'), (b_3') and therefore belongs to $B(X^0)$. Conversely, suppose that $Y \subset X^0$ satisfies the conditions (b_1'), (b_2'), (b_3') and $Y \neq O$, $Y \neq X^0$. Observe that $x_1 \notin Y$. Indeed, if $x_1 \in Y$, then by (b_1') and 8.2, $Y \cap X_1^0 = X_1^0$. Hence, by (b_3') and (7), $X_2^0 = g^0(X_1^0) \subset Y \cap X_2^0$, i.e. $X_2^0 \subset Y$. Thus $Y = X^0$, which contradicts our hypothesis. We shall prove that $x_2 \in Y$. Since $\sim Y \neq O$, $\sim Y \neq X^0$ and $\sim Y$ satisfies the conditions (b_1'), (b_2'), (b_3'), we infer that $x_1 \notin \sim Y$. Hence, by (12), $x_1 \in g^0(Y)$. Consequently, by (6), (7), $x_2 \in Y$. Hence, $Y = G \cup \cup \{x_2\}$, where $G \subset X$. We shall prove that $G \in B(X)$, i.e. Y satisfies the conditions V 3 ((b_1), (b_2), (b_3)). It follows from (b_1') that $G \cap X_1 = Y \cap X_1^0 \in G(X_1^0)$. Hence, by 8.2, $G \cap X_1 \in G(X_1)$. Observe that

(13) $$Y \cap X_2^0 = (G \cup \{x_2\}) \cap X_2^0 = (G \cap X_2) \cup \{x_2\} \neq O.$$

On the other hand, by (b_2'), $Y \cap X_2^0 = X_2^0 - g^0(Z)$, where $Z \in G(X_1^0)$. Note that $Z \neq X_1^0$. Indeed, if $Z = X_1^0$, then $Y \cap X_2^0 = O$, which contradicts (13). Hence, by 8.2, $G \cap X_2 = G \cap X_2^0 = X_2 - g(Z)$, where $Z \in G(X_1)$. It easily follows from (b_3') that $G \cap X_1 \subset g(G) \cap X_1$. Thus $G \in B(X)$.

Note that the following equations hold in $B(X^0)$:

(14) if $Y \subset Z$, then $Y \to Z = X^0$ (by V 3 ((28), (24))),

(15) $$X^0 \to Y = Y \text{ (by V 1.3 (22))},$$

(16) $$(G \cup \{x_2\}) \to O = \begin{cases} X^0 & \text{if } X_1 \cap G = O, \\ G_1 \cup \{x_2\} & \text{if } X_1 \cap G \neq O, \end{cases}$$

where $G_1 = G \to O$ in $\mathfrak{B}(X)$ (by (10), 8.4, V 3 (10)),

(17) $$(G_1 \cup \{x_2\}) \to (G_2 \cup \{x_2\}) = \begin{cases} X^0 & \text{if } X_1 \cap G_1 \subset X_1 \cap G_2, \\ G \cup \{x_2\} & \text{if } X_1 \cap G_1 \not\subset X_1 \cap G_2, \end{cases}$$

where $G = G_1 \to G_2$ in $\mathfrak{B}(X)$ (by (10), 8.4, V 3 (10)),

(18) $\sim O = X^0, \quad \sim X^0 = O, \quad \sim(G\cup\{x_2\}) = (X-g(G))\cup\{x_2\}$

(by (12), 8.4, V 3 (9)),

(19) $O\cup Y = Y, \quad X^0\cup Y = X^0, \quad (G_1\cup\{x_2\})\cup(G_2\cup\{x_2\}) = (G_1\cup G_2)\cup\{x_2\},$

(20) $O\cap Y = O, \quad X^0\cap Y = Y, \quad (G_1\cup\{x_2\})\cap(G_2\cup\{x_2\}) = (G_1\cap G_2)\cup\{x_2\},$

(21) $$Y\Rightarrow Z = (Y\rightarrow Z)\cap(\sim Z\rightarrow \sim Y) \quad \text{(cf. (9))}.$$

It is easy to see that equations (9)–(12), 8.4, equations (14)–(21) and V 3.4 imply that the following holds.

8.5. *The mapping h defined by*

(22) $$h(Y) = Y\cap X \quad \textit{for every } Y\in B(X^0)$$

is an epimorphism from $\mathfrak{B}(X^0)$ *onto* $\mathfrak{B}(X)$.

8.6. *If* $Y\cup Z = X^0$ *for* $Y, Z\in B(X^0)$, *then either* $Y = X^0$ *or* $Z = X^0$.

This follows by 8.4, since X^0 is the only set in $B(X^0)$ which contains $\{x_1\}$.

The valuation v of $\mathscr{L}$ in $\mathfrak{B}(X^0)$ defined by

(23) $$v(p) = \begin{cases} X^0 & \text{if } v^0(p) = ||p||_{\mathscr{A}} = \vee, \\ O & \text{if } v^0(p) = ||p||_{\mathscr{A}} = \wedge, \\ h_0(v^0(p))\cup\{x_2\} & \text{if } v^0(p)\neq \vee \text{ and } v^0(p)\neq\wedge \end{cases}$$

will be called the *characteristic valuation for* $\mathscr{S}(\mathscr{A})$.

By (22) and (23) we have

(24) $$hv = h_0 v^0.$$

8.7. *For the characteristic valuation v of a consistent* $\mathbf{L}_{\mathcal{N}}$*-theory* $\mathscr{S}(\mathscr{A})$ *and for any formula* α *of* $\mathscr{S}(\mathscr{A})$

(25) *either* $\alpha_{\mathfrak{B}(X^0)}(v) = X^0$ *or* $\alpha_{\mathfrak{B}(X^0)}(v) = h_0||\alpha||_{\mathscr{A}}\cup\{x_2\}$ *or* $\alpha_{\mathfrak{B}(X^0)}(v) = h_0||\alpha||_{\mathscr{A}} = O$.

Moreover,

(26) *if* $\alpha_{\mathfrak{B}(X^0)}(v) = X^0$, *then* $h_0||\alpha||_{\mathscr{A}} = X$, i.e. $||\alpha||_{\mathscr{A}} = \vee$.

By 8.5, VIII 3.2, (24), (1) and VIII 9 (15)

$$h(\alpha_{\mathfrak{B}(X^0)}(v)) = \alpha_{\mathfrak{B}(X)}(hv) = \alpha_{\mathfrak{B}(X)}(h_0 v^0) = h_0(\alpha_{\mathfrak{A}(\mathcal{S}(\mathcal{A}))}(v^0)) = h_0 ||\alpha||_{\mathcal{A}}.$$

Hence, by (22)

$$\alpha_{\mathfrak{B}(X^0)}(v) \cap X = h_0 ||\alpha||_{\mathcal{A}}. \tag{27}$$

Since, by 8.4, $B(X^0)$ consists of the empty set, of X^0 and of all sets $G \cup \{x_2\}$ where $G \in B(X)$ and $\alpha_{\mathfrak{B}(X^0)}(v) \in B(X^0)$, equation (27) implies (25) and (26).

8.8. *If the characteristic valuation v for a consistent $\mathbf{L}_{\mathcal{N}}$-theory $\mathcal{S}(\mathcal{A}) = (\mathcal{L}, C_{\mathcal{L}}, \mathcal{A})$ is an $\mathbf{L}_{\mathcal{N}}$-model of $\mathcal{S}(\mathcal{A})$, then the theory $\mathcal{S}(\mathcal{A})$ is prime.*

Indeed, if $(\alpha \cup \beta) \in C_{\mathcal{L}}(\mathcal{A})$, then by 6.3

$$(\alpha \cup \beta)_{\mathfrak{B}(X^0)}(v) = \alpha_{\mathfrak{B}(X^0)}(v) \cup \beta_{\mathfrak{B}(X^0)}(v) = X^0. \tag{28}$$

Hence, by 8.6, either $\alpha_{\mathfrak{B}(X^0)}(v) = X^0$ or $\beta_{\mathfrak{B}(X^0)}(v) = X^0$. Suppose that the first condition is satisfied. Then, by (26), $h_0 ||\alpha||_{\mathcal{A}} = X$, i.e. $||\alpha||_{\mathcal{A}} = V$, which implies by VIII 9 (9) that $\alpha \in C_{\mathcal{L}}(\mathcal{A})$.

For any consistent $\mathbf{L}_{\mathcal{N}}$-theory $\mathcal{S}(\mathcal{A}) = (\mathcal{L}, C_{\mathcal{L}}, \mathcal{A})$, let $Z_{\mathcal{S}(\mathcal{A})}$ be the set of all formulas α of $\mathcal{L}$ which have the following property

(p) if $\alpha \in C_{\mathcal{L}}(\mathcal{A})$, then $\alpha_{\mathfrak{B}(X^0)}(v) = X^0$

for the characteristic valuation v of $\mathcal{S}(\mathcal{A})$.

8.9. *The set $Z_{\mathcal{S}(\mathcal{A})}$ satisfies the following conditions:*

(i) *each propositional variable p of $\mathcal{L}$ is in $Z_{\mathcal{S}(\mathcal{A})}$;*

(ii) *for each propositional variable p of $\mathcal{L}$, $\sim p \in Z_{\mathcal{S}(\mathcal{A})}$;*

(iii) *if $\alpha \in Z_{\mathcal{S}(\mathcal{A})}$, then $\sim \sim \alpha \in Z_{\mathcal{S}(\mathcal{A})}$;*

(iv) *for any formula α, $\neg\alpha \in Z_{\mathcal{S}(\mathcal{A})}$;*

(v) *if $\alpha \in Z_{\mathcal{S}(\mathcal{A})}$, then $\sim \neg\alpha \in Z_{\mathcal{S}(\mathcal{A})}$;*

(vi) *if $\alpha, \beta \in Z_{\mathcal{S}(\mathcal{A})}$, then $(\alpha \cap \beta) \in Z_{\mathcal{S}(\mathcal{A})}$;*

(vii) *if $\sim \alpha, \sim \beta \in Z_{\mathcal{S}(\mathcal{A})}$, then $\sim(\alpha \cup \beta) \in Z_{\mathcal{S}(\mathcal{A})}$;*

(viii) *if $\beta \in Z_{\mathcal{S}(\mathcal{A})}$, then $(\alpha \to \beta) \in Z_{\mathcal{S}(\mathcal{A})}$;*

(ix) *if $\alpha, \sim \beta \in Z_{\mathcal{S}(\mathcal{A})}$, then $\sim(\alpha \to \beta) \in Z_{\mathcal{S}(\mathcal{A})}$;*

(x) *if $\sim \alpha, \beta \in Z_{\mathcal{S}(\mathcal{A})}$, then $(\alpha \Rightarrow \beta) \in Z_{\mathcal{S}(\mathcal{A})}$.*

(i) follows from 6.1 (2) and (23). (ii) follows easily by 6.1 (2), III 3 (q_1), III 3.1 (2), (23) and (18).

PROOF OF (iii): If $\sim\sim\alpha \in C_{\mathcal{L}}(\mathcal{A})$, then, by 3.3 (A16) and 6.4, $\alpha \in C_{\mathcal{L}}(\mathcal{A})$. Since $\alpha \in Z_{\mathcal{S}(\mathcal{A})}$, $\alpha_{\mathfrak{B}(X^0)}(v) = X^0$. Hence, by III 3 ($q_1$), $\sim\sim\alpha_{\mathfrak{B}(X^0)}(v) = X^0$.

Proof of (iv). If $\neg\alpha \in C_{\mathcal{L}}(\mathcal{A})$, then by 6.3, $h_0||\neg\alpha||_{\mathcal{A}} = X$. Hence, by V 3 (22), $h_0||\alpha||_{\mathcal{A}} \to O = X$. Consequently, by V 3 (14) and V 3 (1), $X_1 = X \cap X_1 = (h_0||\alpha||_{\mathcal{A}} \to O) \cap X_1 = h_0||\alpha||_{\mathcal{A}} \cap X_1 \Rightarrow_1 O = \mathbf{I}(X_1 - (h_0||\alpha||_{\mathcal{A}} \cap X_1))$. Thus

$$h_0||\alpha||_{\mathcal{A}} \cap X_1 = O. \tag{29}$$

By (29) and (25), (26), we infer that either $\alpha_{\mathfrak{B}(X^0)}(v) = O$ or $\alpha_{\mathfrak{B}(X^0)}(v) = h_0||\alpha||_{\mathcal{A}} \cup \{x_2\}$, where $h_0||\alpha||_{\mathcal{A}} \in B(X)$. If the first condition holds, then, by (11) and (14), $\neg\alpha_{\mathfrak{B}(X^0)}(v) = X^0$. The second case also implies by (11), (16) and (29) that $\neg\alpha_{\mathfrak{B}(X^0)}(v) = X^0$.

PROOF OF (v): If $\alpha \in Z_{\mathcal{S}(\mathcal{A})}$ and $\sim\neg\alpha \in C_{\mathcal{L}}(\mathcal{A})$, then, by 3.3 (A15), (A6) and 6.4, $\alpha \in C_{\mathcal{L}}(\mathcal{A})$ and consequently $\alpha_{\mathfrak{B}(X^0)}(v) = X^0$. Hence, by V 1 (1) and V 1 (qpB_{10}), $X^0 \to \sim\neg\alpha_{\mathfrak{B}(X^0)}(v) = X^0$. Applying V 1.3 (23), we get $\sim\neg\alpha_{\mathfrak{B}(X^0)}(v) = X^0$.

PROOF OF (vi): If $(\alpha \cap \beta) \in C_{\mathcal{L}}(\mathcal{A})$, then, by 3.3 (A6), (A7) and 6.4, $\alpha \in C_{\mathcal{L}}(\mathcal{A})$ and $\beta \in C_{\mathcal{L}}(\mathcal{A})$. Since $\alpha, \beta \in Z_{\mathcal{S}(\mathcal{A})}$, we get $\alpha_{\mathfrak{B}(X^0)}(v) = X^0$ and $\beta_{\mathfrak{B}(X^0)}(v) = X^0$. Hence by V 1 ($qpB_1$), $(\alpha \cap \beta)_{\mathfrak{B}(X^0)}(v) = X^0$.

PROOF OF (vii): If $\sim(\alpha \cup \beta) \in C_{\mathcal{L}}(\mathcal{A})$, then, by 3.3 (A14), 6.4, $(\sim\alpha \cap \sim\beta) \in C_{\mathcal{L}}(\mathcal{A})$. Since $\sim\alpha, \sim\beta \in Z_{\mathcal{S}(\mathcal{A})}$, applying (vi) we get $(\sim\alpha \cap \sim\beta)_{\mathfrak{B}(X^0)}(v) = X^0$. Hence, by V 1.3 (38), $\sim(\alpha \cup \beta)_{\mathfrak{B}(X^0)}(v) = X^0$.

PROOF OF (viii): Suppose that $(\alpha \to \beta) \in C_{\mathcal{L}}(\mathcal{A})$ and $\beta \in Z_{\mathcal{S}(\mathcal{A})}$. If $\alpha \in C_{\mathcal{L}}(\mathcal{A})$, then, by 6.4, $\beta \in C_{\mathcal{L}}(\mathcal{A})$ and consequently $\beta_{\mathfrak{B}(X^0)}(v) = X^0$. Hence, by (14), $(\alpha \to \beta)_{\mathfrak{B}(X^0)}(v) = X^0$. If $\alpha \notin C_{\mathcal{L}}(\mathcal{A})$ and $\beta \in C_{\mathcal{L}}(\mathcal{A})$, then we apply the first argument. If $\alpha \notin C_{\mathcal{L}}(\mathcal{A})$ and $\beta \notin C_{\mathcal{L}}(\mathcal{A})$, then, by 6.3 and (1),

$$h_0||\alpha||_{\mathcal{A}} \neq X, \quad h_0||\beta||_{\mathcal{A}} \neq X, \quad h_0||\alpha||_{\mathcal{A}} \to h_0||\beta||_{\mathcal{A}} = X. \tag{30}$$

By (30), V 3 (14) and V 3 (1), we get

$$h_0||\alpha||_{\mathcal{A}} \cap X_1 \subset h_0||\beta||_{\mathcal{A}} \cap X_1. \tag{31}$$

It follows from (30), 8.7, 8.4, that either $\alpha_{\mathfrak{B}(X^0)}(v) = O$ or $\alpha_{\mathfrak{B}(X^0)}(v) = h_0||\alpha||_{\mathcal{A}} \cup \{x_2\}$, where $h_0||\alpha||_{\mathcal{A}} \in B(X)$, and that either $\beta_{\mathfrak{B}(X^0)}(v) = O$

or $\beta_{\mathfrak{B}(X^0)}(v) = h_0\|\beta\|_{\mathscr{A}} \cup \{x_2\}$, where $h_0\|\beta\|_{\mathscr{A}} \in B(X)$. If $\alpha_{\mathfrak{B}(X^0)}(v) = O$, then $(\alpha \to \beta)_{\mathfrak{B}(X^0)}(v) = X^0$, by (14). If $\beta_{\mathfrak{B}(X^0)}(v) = O$, then $h_0\|\beta\|_{\mathscr{A}} = O$ by (25). Hence, by (31), $h_0\|\alpha\|_{\mathscr{A}} \cap X_1 = O$. Applying (16), we infer that $(\alpha \to \beta)_{\mathfrak{B}(X^0)}(v) = (h_0\|\alpha\|_{\mathscr{A}} \cup \{x_2\}) \to O = X^0$. If $\alpha_{\mathfrak{B}(X^0)}(v) \neq O$ and $\beta_{\mathfrak{B}(X^0)}(v) \neq O$, then, by (17) and (31), $(\alpha \to \beta)_{\mathfrak{B}(X^0)}(v) = (h_0\|\alpha\|_{\mathscr{A}} \cup \{x_2\}) \to (h_0\|\beta\|_{\mathscr{A}} \cup \{x_2\}) = X^0$.

PROOF OF (ix): If $\sim(\alpha \to \beta) \in C_{\mathscr{L}}(\mathscr{A})$, then, by 3.3 (A12), (A6) and 6.4, $(\alpha \cap \sim \beta) \in C_{\mathscr{L}}(\mathscr{A})$. Since $\alpha, \sim \beta \in Z_{\mathscr{S}(\mathscr{A})}$, applying (vi) we get $(\alpha \cap \sim \beta)_{\mathfrak{B}(X^0)}(v) = X^0$. Hence, by V 1 (1), V 1 (qpB$_8$), V 1.3 (23), we get $\sim(\alpha \to \beta)_{\mathfrak{B}(X^0)}(v) = X^0$.

PROOF OF (x): If $\sim \alpha, \beta \in Z_{\mathscr{S}(\mathscr{A})}$, then, by (viii), $(\alpha \to \beta) \in Z_{\mathscr{S}(\mathscr{A})}$ and $(\sim \beta \to \sim \alpha) \in Z_{\mathscr{S}(\mathscr{A})}$. Consequently, by (vi), $((\alpha \to \beta) \cap \cap(\sim \beta \to \sim \alpha)) \in Z_{\mathscr{S}(\mathscr{A})}$. If $(\alpha \Rightarrow \beta) \in C_{\mathscr{L}}(\mathscr{A})$, then, by 3.3 (A17) and 6.4, $((\alpha \to \beta) \cap (\sim \beta \to \sim \alpha)) \in C_{\mathscr{L}}(\mathscr{A})$. Thus $((\alpha \to \beta) \cap \cap(\sim \beta \to \sim \alpha))_{\mathfrak{B}(X^0)}(v) = X^0$. Hence, by V 1 (qpB$_4$), $(\alpha \Rightarrow \beta)_{\mathfrak{B}(X^0)}(v) = X^0$.

8.10. *If a consistent* $\mathbf{L}_{\mathscr{N}}$*-theory* $\mathscr{S}(\mathscr{A}) = (\mathscr{L}, C_{\mathscr{L}}, \mathscr{A})$ *is prime, then the set* $Z_{\mathscr{S}(\mathscr{A})}$ *satisfies the following conditions*:

(xi) *if* $\alpha, \beta \in Z_{\mathscr{S}(\mathscr{A})}$, *then* $(\alpha \cup \beta) \in Z_{\mathscr{S}(\mathscr{A})}$;

(xii) *if* $\sim \alpha, \sim \beta \in Z_{\mathscr{S}(\mathscr{A})}$, *then* $\sim(\alpha \cap \beta) \in Z_{\mathscr{S}(\mathscr{A})}$;

(xiii) *if* $\alpha, \sim \beta \in Z_{\mathscr{S}(\mathscr{A})}$, *then* $\sim(\alpha \Rightarrow \beta) \in Z_{\mathscr{S}(\mathscr{A})}$.

PROOF OF (xi): If $(\alpha \cup \beta) \in C_{\mathscr{L}}(\mathscr{A})$ and $\mathscr{S}(\mathscr{A})$ is prime, then either $\alpha \in C_{\mathscr{L}}(\mathscr{A})$ or $\beta \in C_{\mathscr{L}}(\mathscr{A})$. Suppose that $\alpha \in C_{\mathscr{L}}(\mathscr{A})$. Since $\alpha \in Z_{\mathscr{S}(\mathscr{A})}$, $\alpha_{\mathfrak{B}(X^0)}(v) = X^0$. Consequently, $(\alpha \cup \beta)_{\mathfrak{B}(X^0)}(v) = X^0$.

PROOF OF (xii): If $\sim(\alpha \cap \beta) \in C_{\mathscr{L}}(\mathscr{A})$ and $\mathscr{S}(\mathscr{A})$ is prime, then by 8.1 at least one of the formulas $\sim \alpha, \sim \beta$ is a theorem of $\mathscr{S}(\mathscr{A})$. Suppose that $\sim \alpha \in C_{\mathscr{L}}(\mathscr{A})$. Since $\sim \alpha \in Z_{\mathscr{S}(\mathscr{A})}$, $\sim \alpha_{\mathfrak{B}(X^0)}(v) = X^0$. Hence $\sim \alpha_{\mathfrak{B}(X^0)}(v) \cup \sim \beta_{\mathfrak{B}(X^0)}(v) = X^0$. Consequently, by V 1.3 (39), $\sim(\alpha \cap \beta)_{\mathfrak{B}(X^0)}(v) = X^0$.

PROOF OF (xiii): Suppose that $\sim(\alpha \Rightarrow \beta) \in C_{\mathscr{L}}(\mathscr{A})$. By 6.3, V1 (qpB$_4$), V 1.3 (9), $(\sim(\alpha \Rightarrow \beta) \to \sim((\alpha \to \beta) \cap (\sim \beta \to \sim \alpha))) \in C_{\mathscr{L}}(\mathscr{A})$. Hence, by 6.4, $\sim((\alpha \to \beta) \cap (\sim \beta \to \sim \alpha)) \in C_{\mathscr{L}}(\mathscr{A})$. Since $\alpha, \sim \beta \in Z_{\mathscr{S}(\mathscr{A})}$, by 8.9 (ix), $\sim(\alpha \to \beta) \in Z_{\mathscr{S}(\mathscr{A})}$. On the other hand, since $\alpha, \sim \beta \in Z_{\mathscr{S}(\mathscr{A})}$, by 8.9 (iii) and 8.9 (ix), $\sim(\sim \beta \to \sim \alpha) \in Z_{\mathscr{S}(\mathscr{A})}$.

Consequently, by (xii), $\sim((\alpha \to \beta) \cap (\sim \beta \to \sim \alpha)) \in Z_{\mathcal{S}(\mathcal{A})}$. Thus $\sim((\alpha \to \beta) \cap (\sim \beta \to \sim \alpha))_{\mathfrak{B}(X^0)}(v) = X^0$. Hence, by V 1.3 (9), $\sim(\alpha \Rightarrow \beta)_{\mathfrak{B}(X^0)}(v) = X^0$.

8.11. *If a consistent* $\mathbf{L}_{\mathcal{N}}$*-theory* $\mathcal{S}(\mathcal{A}) = (\mathcal{L}, C_{\mathcal{L}}, \mathcal{A})$ *is prime, then the set F of all formulas of* $\mathcal{L}$ *is contained in* $Z_{\mathcal{S}(\mathcal{A})}$.

It is convenient to show that for any formula α of $\mathcal{L}$ the formulas α and $\sim \alpha$ are both in $Z_{\mathcal{S}(\mathcal{A})}$. The easy proof by induction on the length of α, based on 8.9 and 8.10, is left to the reader.

The chief theorem of this section follows from 8.8 and 8.11:

8.12. *A consistent* $\mathbf{L}_{\mathcal{N}}$*-theory* $\mathcal{S}(\mathcal{A}) = (\mathcal{L}, C_{\mathcal{L}}, \mathcal{A})$ *of zero order is prime if and only if the characteristic valuation* v *for* $\mathcal{S}(\mathcal{A})$ *is its* $\mathbf{L}_{\mathcal{N}}$*-model.*

Observe that the intersection, F^*, of all sets Z satisfying conditions (i)-(x) of 8.9 (where $Z_{\mathcal{S}(\mathcal{A})}$ should be replaced by Z) also satisfies these conditions. F^* is the least set of formulas of $\mathcal{L}$ satisfying (i)-(x). The following theorem follows immediately from 8.8 and 8.9:

8.13. *If the set* $\mathcal{A}$ *of axioms of a consistent* $\mathbf{L}_{\mathcal{N}}$ *theory* $\mathcal{S}(\mathcal{A}) = (\mathcal{L}, C_{\mathcal{L}}, \mathcal{A})$ *is contained in* F^**, then the theory* $\mathcal{S}(\mathcal{A})$ *is prime.*

This theorem gives a simple method of constructing of prime $\mathbf{L}_{\mathcal{N}}$-*theories*. For instance, if no axiom of a consistent $\mathbf{L}_{\mathcal{N}}$-theory $\mathcal{S}(\mathcal{A})$ contains the disjunction sign $\cup$, the strong negation sign $\sim$ and the implication sign $\Rightarrow$, then $\mathcal{S}(\mathcal{A})$ is prime.

Exercises

1. Let α and β be arbitrary formulas of a propositional calculus $\mathcal{S} = (\mathcal{L}, C_{\mathcal{L}})$ in $\mathbf{L}_{\mathcal{N}}$ without the signs $\Rightarrow$, $\to$, $\neg$. Prove that the formula $((\alpha \Rightarrow \beta) \cap (\beta \Rightarrow \alpha))$ is derivable in $\mathcal{S}$ if and only if $\alpha_{\mathfrak{C}_0}(v) = \beta_{\mathfrak{C}_0}(v)$ for every $v: V \to C_0$, where V is the set of all propositional variables of $\mathcal{L}$, $\mathfrak{C}_0$ is the three-element quasi-Boolean algebra (see III 3) and C_0 is the set of elements of $\mathfrak{C}_0$ (16). Apply V 5.7, 3.2 and Ex. 11 in Chap. III.

2. Prove that the pseudo-Boolean algebra $\mathfrak{G}(E)$ of all open subsets of n-dimensional Euclidean space E (or of the Cantor discontinuum E) satisfies the condition *Adq* as given in Sec. 5 (17). (See the proof of 5.2).

(16) Białynicki-Birula and Rasiowa [2].
(17) Białynicki-Birula and Rasiowa [2].

3. Prove that each quasi-pseudo-Boolean algebra $\mathfrak{B}(X)$ connected with the pseudo-Boolean algebra $\mathfrak{G}(E)$ of all open subsets of n-dimensional Euclidean space E (or the Cantor discontinuum E) and such that $E \cap g(E) = O$, $X = E \cup g(E)$ and $\mathfrak{B}(X)$ consists of all subsets of X satisfying the conditions (b_1), (b_2), (b_3) in V 3 is functionally free for the class of all quasi-pseudo-Boolean algebras, i.e. any two quasi-pseudo-Boolean polynomials are equal in each quasi-pseudo-Boolean algebra if and only if they are equal in $\mathfrak{B}(X)$ [(18)]. Apply Ex. 2 and 5.1.

4. Let $\mathscr{S}_{\mathscr{N}s} = (\mathscr{L}, C_{\mathscr{L}})$ be the propositional calculus obtained from $\mathscr{S}_{\mathscr{N}}$ (see Sec. 1) by adjoining as logical axioms all formulas of the form $(\alpha \cup \neg\alpha)$. Prove that the class of all $\mathscr{S}_{\mathscr{N}s}$-algebras coincides with the class of all semi-simple quasi-pseudo-Boolean algebras (see V Ex. 7) [(19)].

5. Let $\mathbf{L}_{\mathscr{N}s}$ be the logic determined by $\mathscr{S}_{\mathscr{N}s}$ (i.e. the class of all consistent propositional calculi in $\mathbf{S}$ logically equivalent to $\mathscr{S}_{\mathscr{N}s}$). Prove an analogue of 3.2 for propositional calculi in $\mathbf{L}_{\mathscr{N}s}$. Apply Ex. 4, VIII 8.1, V Ex. 9 and VIII 7.2.

6. Let $\mathscr{S} = (\mathscr{L}, C_{\mathscr{L}})$ be a propositional calculus in $\mathbf{L}_{\mathscr{N}s}$ (see Ex. 5). Prove that, for every formula α of $\mathscr{L}$, α is derivable in $\mathscr{S}$ if and only if $\alpha_{\mathfrak{C}_0}(v) = V$ for every valuation v of $\mathscr{L}$ in the three-element quasi-pseudo-Boolean algebra $\mathfrak{C}_0$ (see V 3) [(20)]. Apply Ex. 5 and V Ex. 8.

7. Let $\mathfrak{C}_0 = (C_0, V, \Rightarrow, \cup, \cap, \sim, \rightarrow, \neg)$ be the three-element quasi-pseudo-Boolean algebra (see V 3). Then $\mathfrak{C} = (C_0, V, \Rightarrow, \cup, \cap, \sim)$ is said to be the *three-element algebra of Łukasiewicz* [(21)]. Prove that for every formula α of a propositional calculus $\mathscr{S} \in \mathbf{L}_{\mathscr{N}s}$ (see Ex. 5) such that neither $\rightarrow$ nor $\neg$ appears in α the condition $\alpha_{\mathfrak{C}}(v) = V$ for every valuation v in $\mathfrak{C}$ is equivalent to the derivability of α in $\mathscr{S}$ [(22)] (i.e. α is derivable in $\mathscr{S}$ if and only if α is derivable in the three-valued propositional calculus of Łukasiewicz; see Wajsberg [1]).

(18) Białynicki-Birula and Rasiowa [2].

(19) Monteiro [2].

(20) Monteiro [2].

(21) This algebra is identical with the matrix for the three-valued propositional calculus of Łukasiewicz (see Łukasiewicz [2], [3]).

(22) Monteiro [2].

CHAPTER XIII

CLASSICAL LOGIC AND MODAL LOGIC

Introduction. Classical logic is expounded here from the algebraic point of view in a cursory manner since a comprehensive exposition is to be found in [MM].

The modal logic (1) discussed in the present chapter corresponds to the system S4 of Lewis and Langford [1]. It was originated as a result of a philosophical criticism concerning classical (material) implication and is sometimes called the logic of strict implication.

The modal propositional calculi have been examined by many authors (2). The first results serving to establish a connection between modal logic and topology are due to Tang Tsao-Chen [1]. This relationship was pointed out and developed by McKinsey in [1], and jointly with Tarski in [1], [3].

The algebraic treatment of modal predicate calculi (3) and modal elementary theories was developed in papers by the present author (4) and written jointly with Sikorski (5). An exposition of researches concerning an algebraic approach to modal logic is given in [MM]. For this reason it is presented here rather briefly, and restricted to modal propositional calculi and theories of zero order.

In formalized languages of propositional calculi of modal logic there appears, besides connectives occurring in formalized languages

(1) For the philosophical aspects of modal logic cf. Lewis and Langford [1], Beth [4], Bergmann [1].

(2) For various interpretations cf. McKinsey [1], Kripke [1], Hintikka [1], Montague [1], Kanger [1], Guillaume [1].

(3) For modal predicate calculi different from those discussed by the present author and Sikorski see Barcan [1] and Carnap [2].

(4) [2], [3], [8].

(5) See [3], [4], [5], [6].

of classical propositional calculi, the necessity sign **I**. Then the strict implication mentioned above can be defined by means of the equation $(\alpha \to \beta) \overset{\mathrm{df}}{=} \mathbf{I}(\alpha \Rightarrow \beta)$. This explains our remark about another name for this logic.

Modal logic, to be denoted by $\mathbf{L}_\lambda$, is characterized algebraically by the class of all topological Boolean algebras, i.e. the class of all $\mathscr{S}$-algebras for $\mathscr{S} \in \mathbf{L}_\lambda$ coincides with that of these algebras. Besides the general metalogical theorems of Chapter VIII, which can be formulated for modal propositional calculi and $\mathbf{L}_\lambda$-theories, certain other theorems are obtained by applying the results of the theory of topological Boolean algebras.

The completeness theorem for propositional calculi of $\mathbf{L}_\lambda$ is formulated in a much stronger form (5.2, 5 (vi)). It contains the following topological characterization of formulas derivable in any $\mathscr{S} \in \mathbf{L}_\lambda$: exactly those formulas are derivable which are valid in the topological field $\mathfrak{B}(E)$ of all subsets of the real line E (McKinsey and Tarski [1], [3]). The decidability of propositional calculi in $\mathbf{L}_\lambda$ follows from the completeness theorem, which also states the equivalence between the derivability of any formula in $\mathscr{S} \in \mathbf{L}_\lambda$ and the validity in all finite topological Boolean algebras whose the number of elements is restricted and depends on the structure of that formula. Lemma VI 9.1 on imbeddings for topological Boolean algebras plays an essential part in the proof.

Classical logic $\mathbf{L}_\varkappa$ is the part of $\mathbf{L}_\lambda$ corresponding to formulas in which the necessity sign does not occur (5.4), i.e. a formula without the necessity sign is derivable in $\mathscr{S} \in \mathbf{L}_\lambda$ if and only if it is derivable in classical propositional calculi. An analogous connection holds between certain $\mathbf{L}_\lambda$-theories and $\mathbf{L}_\varkappa$-theories (6.4).

Intuitionistic propositional calculi find their adequate interpretation in propositional calculi of $\mathbf{L}_\lambda$ (5.5). Analogously, $\mathbf{L}_\chi$-theories can be interpreted in an adequate manner in corresponding $\mathbf{L}_\lambda$-theories (6.5). This interpretation is based on the Theorem VI 10.2, which states a relationship between pseudo-Boolean algebras and topological Boolean algebras.

Disjunctions in modal propositional calculi have the following property: $(\mathbf{I}\alpha \cup \mathbf{I}\beta)$ is derivable if and only if at least one of the formulas $\mathbf{I}\alpha$, $\mathbf{I}\beta$ is derivable (5.6).

Among all $\mathbf{L}_\lambda$-models of $\mathbf{L}_\lambda$-theories we distinguish $\mathbf{I}$-semantic models, i.e. models in topological Boolean algebras such that for each element a either $\mathbf{I}a = \bigvee$ or $\mathbf{I}a = \bigwedge$. The consistency of an $\mathbf{L}_\lambda$-theory is equivalent to the existence of an $\mathbf{L}_\lambda$-model, in particular an $\mathbf{I}$-semantic model of that theory (6.2). The representation theorem for topological Boolean algebras leads to a stronger form of the theorem characterizing formulas which are theorems of a consistent $\mathbf{L}_\lambda$-theory (6.3).

Deductive filters coincide with $\mathbf{I}$-filters in topological Boolean algebras. The deduction theorems (6.7, 6.8) for $\mathbf{L}_\lambda$-theories have a modified form which follows from the theorems on the structure of $\mathbf{I}$-filters generated by an $\mathbf{I}$-filter and an element, and generated by a non-empty set of elements, respectively. Every consistent $\mathbf{L}_\lambda$-theory can be extended to a maximal one (6.11). A characterization of maximal $\mathbf{L}_\lambda$-theories is given by 6.12. Irreducible $\mathbf{L}_\lambda$-theories coincide with $\mathbf{I}$-prime ones, i.e. theories which are consistent and such that $(\mathbf{I}\alpha \cup \mathbf{I}\beta)$ is a theorem if and only if at least one of the formulas $\mathbf{I}\alpha$, $\mathbf{I}\beta$ is a theorem. A sufficient condition for an $\mathbf{L}_\lambda$-theory to be $\mathbf{I}$-prime, formulated by means of a characteristic valuation, is given in Section 7.

1. Propositional calculus $\mathscr{S}_\varkappa$ of classical logic. Let $\mathscr{S}_\varkappa = (\mathscr{L}, C_\mathscr{L})$ be the propositional calculus briefly described as follows. The alphabet $A^0 = (V, L_0, L_1, L_2, U)$ of $\mathscr{S}_\varkappa$ differs from the alphabet of $\mathscr{S}_\pi$ (see X 1) only in admitting in L_1 one element denoted by $-$ and called the *negation sign*. Let F be the set of all formulas over $\mathscr{A}^0$. By definition $\mathscr{L} = (A^0, F)$. Let us assume that the set $\mathscr{A}_l$ of logical axioms consists of all formulas of the form (A_1)–(A_8) in X 1 (where α, β, γ are any formulas in F) and, moreover, of all formulas

$$(A_9) \qquad ((\alpha \Rightarrow -\beta) \Rightarrow (\beta \Rightarrow -\alpha)),$$

$$(A_{10}) \qquad (-(\alpha \Rightarrow \alpha) \Rightarrow \beta),$$

$$(A_{11}) \qquad (\alpha \cup -\alpha),$$

for any α, β in F.

The consequence operation $C_\mathscr{L}$ in $\mathscr{L}$ is determined by the set $\mathscr{A}_l$ of logical axioms and by modus ponens as the only rule of inference.

The system $\mathcal{S}_\varkappa = (\mathcal{L}, C_\mathcal{L})$ is said to be a *propositional calculus of classical logic.*

By the same argument as used in the proof of XI 1.1 we get

1.1. *For any formulas* α, β *in* F

$$((\alpha \Rightarrow \beta) \Rightarrow (-\beta \Rightarrow -\alpha)) \in C_\mathcal{L}(O). \tag{1}$$

1.2. *The system* $\mathcal{S}_\varkappa$ *belongs to the class* **S** *of standard systems of implicative extensional propositional calculi.*

Clearly, conditions VIII 5 $((s_1), (s_2))$ are satisfied. By X 1.1 and X 1.2 conditions VIII 5 $((s_3)$–$(s_6), (s_8))$ hold. Condition VIII 5 (s_7) is satisfied by 1.1. This completes the proof.

The next theorem characterizes the class of all $\mathcal{S}_\varkappa$-algebras.

1.3. *The class of all* $\mathcal{S}_\varkappa$*-algebras coincides with the class of all Boolean algebras* (cf. VI 5).

The easy proof, based on XI 7.3 and VI 1 $((b_1), (b_2))$, is left to the reader.

The following theorem follows from 1.3 and VIII 6.1.

1.4. *If a formula* α *is derivable in* $\mathcal{S}_\varkappa$, *then*

$$\alpha_\mathfrak{A}(v) = V \tag{2}$$

for every valuation v *of* $\mathcal{L}$ *in any Boolean algebra* $\mathfrak{A} = (A, V, \Rightarrow, \cup, \cap, -)$.

By 1.4

1.5. *The propositional calculus* $\mathcal{S}_\varkappa$ *is consistent.*

2. Classical logic $\mathbf{L}_\varkappa$. By *classical logic* $\mathbf{L}_\varkappa$ we shall mean the class of all consistent standard systems of implicative extensional propositional calculi logically equivalent (see VIII 8) to $\mathcal{S}_\varkappa$. Thus, by 1.3, the class $\mathbf{K}_{\mathbf{L}_\varkappa}$ of all $\mathcal{S}$-algebras for $\mathcal{S} \in \mathbf{L}_\varkappa$ is the class of all Boolean algebras (see VI 1). Hence, by VIII 6.5 and VIII 6.6, we get

2.1. *For every* $\mathcal{S}$ *in* $\mathbf{L}_\varkappa$ *the algebra* $\mathfrak{A}(\mathcal{S})$ *is a non-degenerate Boolean algebra. Moreover,* $\mathfrak{A}(\mathcal{S})$ *is a free algebra in the class of all Boolean algebras.*

A formula α of $\mathcal{S}$ in $\mathbf{L}_\varkappa$ is said to be $\mathbf{L}_\varkappa$-*valid* (*valid*) provided that for any valuation v of the formalized language of $\mathcal{S}$ in any Boolean algebra $\mathfrak{B} = (B, V, \Rightarrow, \cup, \cap, -)$, $\alpha_\mathfrak{B}(v) = V$. A formula α of $\mathcal{S}$

in $\mathbf{L}_{\varkappa}$ is said to be *semantically valid* if for any valuation v of the formalized language of $\mathscr{S}$ in the two-element Boolean algebra $\mathfrak{B}_0$, $\alpha_{\mathfrak{B}_0}(v) = V$.

The next theorem characterizes all formulas derivable in any system $\mathscr{S}$ in $\mathbf{L}_{\varkappa}$.

2.2. *For any formula* α *of any system* $\mathscr{S} \in \mathbf{L}_{\varkappa}$ *the following conditions are equivalent*:

(i) α *is derivable in* $\mathscr{S}$,

(ii) α *is valid*,

(iii) α *is valid in every field of sets*,

(iv) $\alpha_{\mathfrak{A}(\mathscr{S})}(v^0) = V$ *for the canonical valuation* v^0 *in the algebra* $\mathfrak{A}(\mathscr{S})$ *of the system* $\mathscr{S}$,

(v) α *is semantically valid*.

The equivalence of conditions (i) and (v) is called the *completeness theorem* (6) *for propositional calculi of classical logic*.

The proof, similar to that of IX 8.4, is omitted (cf. [MM] VII 2.2).

3. $\mathbf{L}_{\varkappa}$-theories of zero order and their connection with filters. For any system $\mathscr{S} = (\mathscr{L}, C_{\mathscr{L}})$ in $\mathbf{L}_{\varkappa}$ and for every set $\mathscr{A}$ of formulas of $\mathscr{L}$ the system $\mathscr{S}(\mathscr{A}) = (\mathscr{L}, C_{\mathscr{L}}, \mathscr{A})$ is said to be an $\mathbf{L}_{\varkappa}$-*theory* (a *classical theory*) of zero order based on $\mathscr{S}$. Any valuation v of $\mathscr{L}$ in any non-degenerate Boolean algebra $\mathfrak{B} = (B, V, \Rightarrow, \cup, \cap, -)$ such that for all α in $\mathscr{A}$

$$(1) \qquad \alpha_{\mathfrak{B}}(v) = V$$

is said to be an $\mathbf{L}_{\varkappa}$-*model* (a *model*) of $\mathscr{S}(\mathscr{A})$. An $\mathbf{L}_{\varkappa}$-model of $\mathscr{S}(\mathscr{A})$ in the two-element Boolean algebra will be said to be a *semantic model*.

The following theorem follows from VIII 9.2, VIII 9.3 and the fact that the class of all $\mathscr{S}$-algebras for $\mathscr{S} \in \mathbf{L}_{\varkappa}$ coincides with the class of all Boolean algebras.

3.1. *For every consistent classical theory* $\mathscr{S}(\mathscr{A}) = (\mathscr{L}, C_{\mathscr{L}}, \mathscr{A})$ *of zero order the algebra* $\mathfrak{A}(\mathscr{S}(\mathscr{A}))$ (see VIII 9) *is a non-degenerate Boolean*

(6) This theorem is due to Post [1]. For another proof cf. Łukasiewicz [3], Hilbert and Ackermann [1]. The first proof by the Boolean method was found by the present author and Sikorski (see Rasiowa and Sikorski [1], p. 200). See also Łoś [1].

algebra. The canonical valuation v^0 *of* $\mathscr{L}$ *in* $\mathfrak{A}(\mathscr{S}(\mathscr{A}))$ (cf. VIII 9 (12)) *is a model of* $\mathscr{S}(\mathscr{A})$. *This model is adequate*, i.e. *for any formula* α *of* $\mathscr{L}$

(2) $$\alpha \in C_{\mathscr{L}}(\mathscr{A}) \text{ if and only if } \alpha_{\mathfrak{A}(\mathscr{S}(\mathscr{A}))}(v^0) = V.$$

Observe that

(3) $$\mathbf{L}_{\varkappa} \subset \mathbf{S}_{\nu}$$

since by 2.2, for any $\mathscr{S} \in \mathbf{L}_{\varkappa}$, all formulas of the form $(-(\alpha \Rightarrow \alpha) \Rightarrow \beta)$, where α, β are any formulas of $\mathscr{S}$, are derivable in $\mathscr{S}$.

Let us note the following theorem.

3.2. *For any classical theory* $\mathscr{S}(\mathscr{A}) = (\mathscr{L}, C_{\mathscr{L}}, \mathscr{A})$ *of zero order the following conditions are equivalent*:

(i) $\mathscr{S}(\mathscr{A})$ *is consistent*,

(ii) *for any formula* α *of* $\mathscr{S}(\mathscr{A})$ *either* $\alpha \notin C_{\mathscr{L}}(\mathscr{A})$ *or* $-\alpha \notin C_{\mathscr{L}}(\mathscr{A})$,

(iii) *there exists a model of* $\mathscr{S}(\mathscr{A})$,

(iv) *there exists an adequate model of* $\mathscr{S}(\mathscr{A})$,

(v) *there exists a model of* $\mathscr{S}(\mathscr{A})$ *in a field of sets*,

(vi) *there exists a semantic model of* $\mathscr{S}(\mathscr{A})$ (7).

The proof, similar to that of XI 9.2, is omitted (cf. [MM] VII 7.7).

The next theorem characterizes all theorems of any consistent classical theory of zero order.

3.3. *For any formula* α *of any consistent classical theory* $\mathscr{S}(\mathscr{A}) = (\mathscr{L}, C_{\mathscr{L}}, \mathscr{A})$ *of zero order the following conditions are equivalent*:

(i) $\alpha \in C_{\mathscr{L}}(\mathscr{A})$,

(ii) *every* $\mathbf{L}_{\varkappa}$*-model of* $\mathscr{S}(\mathscr{A})$ *is an* $\mathbf{L}_{\varkappa}$*-model of* α,

(iii) *every* $\mathbf{L}_{\varkappa}$*-model of* $\mathscr{S}(\mathscr{A})$ *in any field of sets is an* $\mathbf{L}_{\varkappa}$*-model of* α,

(iv) $\alpha_{\mathfrak{A}(\mathscr{S}(\mathscr{A}))}(v^0) = V$ *for the canonical valuation* v^0 *in the algebra* $\mathfrak{A}(\mathscr{S}(\mathscr{A}))$ *of* $\mathscr{S}(\mathscr{A})$,

(v) *every semantic model of* $\mathscr{S}(\mathscr{A})$ *is a semantic model of* α (8).

The proof, analogous to that of IX 9.2, is omitted (cf. [MM] VII 7.8).

(7) Cf. [MM]. The equivalence of the conditions (i), (vi) for uncountable theories was first proved by Malcev [1].

(8) Cf. [MM].

Since the only rule of inference of the propositional calculus $\mathscr{S}$ is modus ponens, it follows from VIII 12.3, VI 3.1 that

3.4. *For any set ∇ of elements in a Boolean algebra the following conditions are equivalent:*

(i) ∇ *is a deductive filter,*

(ii) ∇ *is an implicative filter,*

(iii) ∇ *is a filter.*

Let us note that the theorem concerning the connection between the kernels of epimorphisms from Boolean algebras onto similar algebras and filters (cf. VI 3.2, VI 1.2, IV 2.3) follows from 3.4, VIII 12.2 and VIII 13.10 and from the equational definability of Boolean algebras.

By 3.4, VIII 13.2 and VIII 13.4 we get two theorems analogous to X 6.2 and X 6.3, which state that there exists a natural one-one correspondence between filters in the Boolean algebra $\mathfrak{A}(\mathscr{S})$ of a system $\mathscr{S}$ in $\mathbf{L}_\varkappa$ and formalized $\mathbf{L}_\varkappa$-theories of zero order based on $\mathscr{S}$, two theories $\mathscr{S}(\mathscr{A})$ and $\mathscr{S}(\mathscr{A}')$ being identified if and only if they have the same set of theorems [9].

Deduction theorems analogous to IX 5.5, IX 5.6, X 6.4 can be proved by using a similar argument. Clearly, theorems corresponding to X 6.5–X 6.9 can also be analogously deduced. The following theorem follows from (3) and VIII 13.8.

3.5. *For every consistent classical theory $\mathscr{S}(\mathscr{A}) = (\mathscr{L}, C_\mathscr{L}, \mathscr{A})$ of zero order there exists a maximal classical theory $\mathscr{S}(\mathscr{A}') = (\mathscr{L}, C_\mathscr{L}, \mathscr{A}')$ such that $C_\mathscr{L}(\mathscr{A}) \subset C_\mathscr{L}(\mathscr{A}')$* [10].

It follows from a theorem analogous to X 6.7, 3.4, VIII 13.5, VIII 13.6 and VI 3.3 that

3.6. *For every classical theory $\mathscr{S}(\mathscr{A}) = (\mathscr{L}, C_\mathscr{L}, \mathscr{A})$ of zero order the following conditions are equivalent:*

(i) $\mathscr{S}(\mathscr{A})$ *is irreducible,*

[9] These results as well as all those given later in this section except the deduction theorems and 3.5 are due to Tarski [5], [7]. For the deduction theorems see Herbrand [1], [2] and Tarski [1].

[10] This theorem is due to A. Lindenbaum (see Tarski [1]).

(ii) $\mathscr{S}(\mathscr{A})$ *is prime,*

(iii) $\mathscr{S}(\mathscr{A})$ *is maximal.*

The next theorem characterizes maximal classical theories of zero order.

3.7. *The following conditions are equivalent for any classical theory* $\mathscr{S}(\mathscr{A}) = (\mathscr{L}, C_{\mathscr{L}}, \mathscr{A})$ *of zero order:*

(i) $\mathscr{S}(\mathscr{A})$ *is maximal,*

(ii) *for each formula* α *of* $\mathscr{L}$ *exactly one of the formulas* α, $-\alpha$ *is a theorem of* $\mathscr{S}(\mathscr{A})$,

(iii) $\mathfrak{A}(\mathscr{S}(\mathscr{A}))$ *is the two-element Boolean algebra,*

(iv) $\mathscr{S}(\mathscr{A})$ *has an adequate semantic model,*

(v) $\mathscr{S}(\mathscr{A})$ *is consistent and every semantic model of* $\mathscr{S}(\mathscr{A})$ *is adequate.*

The proof, analogous to that of XI 9.7 and based on 3.4, VI 3.3, 3.1, 3.2, 3.3 and VIII 13.1, VIII 13.6, is left to the reader.

4. Propositional calculus $\mathscr{S}_\lambda$ of modal logic. In this section we shall deal with the propositional calculus $\mathscr{S}_\lambda = (\mathscr{L}, C_{\mathscr{L}})$ defined as follows. The alphabet $A^0 = (V, L_0, L_1, L_2, U)$ is obtained from the alphabet of $\mathscr{S}_\varkappa$ (see Sec. 1) by adjoining to L_1 an element denoted by **I** and called the *necessity sign* (11). Let F be the set of all formulas over A^0.

The set $\mathscr{A}_l$ of logical axioms consists of all formulas of the form (A_1)–(A_8) in X 1, (A_9)–(A_{11}) in Section 1 (where α, β, γ are any formulas in the set F of formulas) and, moreover, of the following ones:

(A_{12}) $((\mathbf{I}\alpha \cap \mathbf{I}\beta) \Rightarrow \mathbf{I}(\alpha \cap \beta))$,

(A_{13}) $(\mathbf{I}\alpha \Rightarrow \alpha)$,

(A_{14}) $(\mathbf{I}\alpha \Rightarrow \mathbf{II}\alpha)$,

(A_{15}) $\mathbf{I}(\alpha \cup -\alpha)$.

(11) The formula $\mathbf{I}\alpha$ should be read: *it is necessary that* α; the formula $-\mathbf{I}-\alpha$ is usually denoted by $\mathbf{C}\alpha$ and read: *it is possible that* α; the formula $\mathbf{I}(\alpha \Rightarrow \beta)$ is usually denoted by $(\alpha \rightarrow \beta)$ and called the *strict implication of* α *and* β.

The consequence operation $C_{\mathscr{L}}$ in $\mathscr{L}$ is determined by the set $\mathscr{A}_l$ of logical axioms and by the following two rules of inference: *modus ponens* and

$$(\mathrm{r}_\lambda) \qquad \frac{(\alpha \Rightarrow \beta)}{(\mathbf{I}\alpha \Rightarrow \mathbf{I}\beta)} \quad \text{for any } \alpha, \beta \in F.$$

The system $\mathscr{S}_\lambda = (\mathscr{L}, C_{\mathscr{L}})$ is said to be a *propositional calculus of modal logic*.

4.1. *The system $\mathscr{S}_\lambda$ belongs to the class* **S** *of standard systems of implicative extensional propositional calculi.*

Obviously, conditions VIII 5 $((\mathrm{s}_1), (\mathrm{s}_2))$ are satisfied. By X 1.1 and X 1.2 conditions VIII 5 $((\mathrm{s}_3)$–$(\mathrm{s}_6), (\mathrm{s}_8))$ hold. It is easy to see that Theorem 1.1 holds for $\mathscr{S}_\lambda$. By the above remark and (r_λ) condition VIII 5 (s_7) is also satisfied. This completes the proof.

4.2. *The class of all $\mathscr{S}_\lambda$-algebras coincides with the class of all topological Boolean algebras* (see VI 5).

The proof by an easy verification, making use of 1.3, is left to the reader.

It follows from 4.2 and VIII 6.1 that

4.3. *If a formula α is derivable in $\mathscr{S}_\lambda$, then*

$$(1) \qquad \alpha_{\mathfrak{A}}(v) = \vee$$

for every valuation v of $\mathscr{L}$ in any topological Boolean algebra $\mathfrak{A} = (A, \vee, \Rightarrow, \cup, \cap, -, \mathbf{I})$.

It follows from 4.3 that

4.4. *The propositional calculus $\mathscr{S}_\lambda$ is consistent.*

5. Modal logic $\mathbf{L}_\lambda$. By *modal logic* $\mathbf{L}_\lambda$ we shall mean the class of all consistent standard systems of implicative extensional propositional calculi logically equivalent to $\mathscr{S}_\lambda$. On account of 4.2 the class $\mathbf{K}_{\mathbf{L}_\lambda}$ of all $\mathscr{S}$-algebras for $\mathscr{S} \in \mathbf{L}_\lambda$ is the class of all topological Boolean algebras. Hence, by VIII 6.5 and VIII 6.6,

5.1. *For every $\mathscr{S} \in \mathbf{L}_\lambda$, the algebra $\mathfrak{A}(\mathscr{S})$ is a non-degenerate topological Boolean algebra. Moreover, $\mathfrak{A}(\mathscr{S})$ is a free algebra in the class of all topological Boolean algebras.*

A formula α in $\mathscr{S} \in \mathbf{L}_\lambda$ is said to be $\mathbf{L}_\lambda$-*valid* if for any valuation v of the formalized language of $\mathscr{S}$ in any topological Boolean algebra $\mathfrak{A} = (A, V, \Rightarrow, \cup, \cap, -, \mathbf{I})$, $\alpha_{\mathfrak{A}}(v) = V$.

The next theorem characterizes all formulas derivable in any system $\mathscr{S}$ of modal logic $\mathbf{L}_\lambda$.

5.2. *For any formula* α *of any system* $\mathscr{S} \in \mathbf{L}_\lambda$ *the following conditions are equivalent*:

(i) α *is derivable in* $\mathscr{S}$,

(ii) α *is* $\mathbf{L}_\lambda$-*valid*,

(iii) α *is valid in every topological field* $\mathfrak{B}(X)$,

(iv) $\alpha_{\mathfrak{A}(\mathscr{S})}(v^0) = V$ *for the canonical valuation* v^0 *in the algebra* $\mathfrak{A}(\mathscr{S})$ *of the system* $\mathscr{S}$,

(v) α *is valid in every topological Boolean algebra* $\mathfrak{A}$ *with at most* 2^{2^r} *elements, where r is the number of all subformulas of* α (12).

The equivalence of (i) and (ii) follows from VIII 8.1 and is called the *completeness theorem for propositional calculi of modal logic*. Obviously, (ii) implies (iii) and by the representation theorem VI 7.1, (iii) implies (iv). By VIII 7.2, (i) follows from (iv). Clearly, (ii) implies (v). The easy proof that (v) implies (i), based on VI 9.1 and similar to the proof of X 3.4, is left to the reader.

Let us note that each of conditions (i)-(v) is equivalent to the following one

(vi) $\alpha_{\mathfrak{B}(X)}(v) = X$ *for every valuation* v *in the topological field* $\mathfrak{B}(X)$ *of all subsets of a dense-in-itself metric space* $X \neq O$ (in particular, of an n-dimensional Euclidean space X) (13).

The following corollary follows from the equivalence of the conditions (i) and (v).

5.3. *Every propositional calculus* $\mathscr{S}$ *of modal logic* $\mathbf{L}_\lambda$ *is decidable* (14).

(12) McKinsey [1], McKinsey and Tarski [1], [3]. The first idea of a connection between modal propositional calculus and topology is due to Tang Tsao-Chen [1], see also Dugundji [1].

(13) McKinsey and Tarski [1], [3].

(14) McKinsey [1].

The next theorem states the connection between propositional calculi of modal logic and propositional calculi of classical logic.

5.4. *Let* $\mathscr{S} = (\mathscr{L}, C_{\mathscr{L}})$ *be a propositional calculus of modal logic and let* $\mathscr{S}_0 = (\mathscr{L}_0, C_{\mathscr{L}_0})$ *be a propositional calculus of classical logic such that* $\mathscr{L}$ *is an extension of* $\mathscr{L}_0$. *Then for every formula* α *of* $\mathscr{L}_0$ *the following conditions are equivalent*:

(i) α *is derivable in* $\mathscr{S}$,

(ii) α *is derivable in* $\mathscr{S}_0$ (15).

The proof, analogous to that of X 3.7 and based on 5.2, the definition of topological Boolean algebras (see VI 5), and 2.2, is left to the reader.

Let $\mathscr{S} = (\mathscr{L}, C_{\mathscr{L}})$ be a propositional calculus of modal logic and let $\mathscr{L}_0$ be the language obtained from $\mathscr{L}$ by the elimination of the sign $\mathbf{I}$ and by the replacement of the negation sign $-$ by the sign $\urcorner$ of intuitionistic negation. Let $\mathscr{S}_0 = (\mathscr{L}_0, C_{\mathscr{L}_0})$ be a propositional calculus of intuitionistic logic. In order to establish the connection between $\mathscr{S}$ and $\mathscr{S}_0$ let us define a mapping $\mathfrak{f}$ which to every formula α of $\mathscr{L}_0$ assigns a formula $\mathfrak{f}\alpha$ of $\mathscr{L}$. The mapping $\mathfrak{f}$ is defined by induction on the length of formulas as follows:

(1) $\mathfrak{f}p = \mathbf{I}p$ for every propositional variable p,

(2) $\mathfrak{f}(\alpha \Rightarrow \beta) = \mathbf{I}(\mathfrak{f}\alpha \Rightarrow \mathfrak{f}\beta)$,

(3) $\mathfrak{f}(\alpha \cup \beta) = (\mathfrak{f}\alpha \cup \mathfrak{f}\beta)$,

(4) $\mathfrak{f}(\alpha \cap \beta) = (\mathfrak{f}\alpha \cap \mathfrak{f}\beta)$,

(5) $\mathfrak{f}\urcorner\alpha = \mathbf{I}-\mathfrak{f}\alpha$,

where α, β denote formulas in $\mathscr{L}_0$.

With these hypotheses and notation the following theorem holds.

5.5. *For any formula* α *of* $\mathscr{L}_0$ *the following conditions are equivalent*:

(i) α *is derivable in the propositional calculus* $\mathscr{S}_0 = (\mathscr{L}_0, C_{\mathscr{L}_0})$ *of intuitionistic logic*,

(ii) $\mathfrak{f}\alpha$ *is derivable in the propositional calculus* $\mathscr{S} = (\mathscr{L}, C_{\mathscr{L}})$ *of modal logic* (16).

(15) Lewis and Langford [1].
(16) McKinsey and Tarski [3].

For any topological Boolean algebra $\mathfrak{A}$, let $\mathfrak{G}(\mathfrak{A})$ be the pseudo-Boolean algebra of all open elements in $\mathfrak{A}$ (see VI 10). It is easy to prove that for every formula α of $\mathscr{L}_0$ and for every valuation v of $\mathscr{L}$ in $\mathfrak{A}$

(6) $$\alpha_{\mathfrak{G}(\mathfrak{A})}(\mathbf{I}v) = \mathfrak{f}\alpha_{\mathfrak{A}}(v),$$

where $\mathbf{I}v(p) = \mathbf{I}(v(p))$ for every propositional variable p. The proof by induction on the length of α, based on VI 10.1 and VI 10 (1), is omitted. If $\mathfrak{f}\alpha \notin C_{\mathscr{L}}(O)$, then by 5.2 there exist a topological Boolean algebra $\mathfrak{A}$ and a valuation v of $\mathscr{L}$ in $\mathfrak{A}$ such that $\mathfrak{f}\alpha_{\mathfrak{A}}(v) \neq V$. Hence, by (6), $\alpha_{\mathfrak{G}(\mathfrak{A})}(\mathbf{I}v) \neq V$. Thus, by XI 8.2, $\alpha \notin C_{\mathscr{L}_0}(O)$. Conversely, if $\alpha \notin C_{\mathscr{L}_0}(O)$, then by XI 8.2 and VI 10.2 there exist a pseudo-Boolean algebra $\mathfrak{G}(\mathfrak{A})$ of all open elements in a topological Boolean algebra $\mathfrak{A}$ and a valuation v of $\mathscr{L}_0$ in $\mathfrak{G}(\mathfrak{A})$ such that $\alpha_{\mathfrak{G}(\mathfrak{A})}(v) \neq V$. Clearly, $\mathbf{I}v = v$ for v considered as a valuation of $\mathscr{L}$ in $\mathfrak{A}$. Hence, by (6), $\mathfrak{f}\alpha_{\mathfrak{A}}(v) \neq V$. Consequently, by 5.2, $\mathfrak{f}\alpha \notin C_{\mathscr{L}}(O)$.

The following theorem, analogous to X 4.1, holds for the propositional calculi of modal logic.

5.6. *Let $\mathscr{S} = (\mathscr{L}, C_{\mathscr{L}})$ be a propositional calculus of modal logic. Then, for any formulas α, β of $\mathscr{L}$,*

(7) $(\mathbf{I}\alpha \cup \mathbf{I}\beta) \in C_{\mathscr{L}}(O)$ *if and only either* $\alpha \in C_{\mathscr{L}}(O)$ *or* $\beta \in C_{\mathscr{L}}(O)$ [17].

To begin with, observe that, by 5.2 and VI 5 (i_4), VI 5 (3),

(8) $\alpha \in C_{\mathscr{L}}(O)$ *if and only if* $\mathbf{I}\alpha \in C_{\mathscr{L}}(O)$.

Moreover, on account of 5.2 and the axioms (A_3), (A_4) (see Sec .1), the formulas $(\mathbf{I}\alpha \Rightarrow (\mathbf{I}\alpha \cup \mathbf{I}\beta))$, $(\mathbf{I}\beta \Rightarrow (\mathbf{I}\alpha \cup \mathbf{I}\beta))$ are both derivable in $\mathscr{S}$. Hence, by (8) and modus ponens, if either $\alpha \in C_{\mathscr{L}}(O)$ or $\beta \in C_{\mathscr{L}}(O)$ then $(\mathbf{I}\alpha \cup \mathbf{I}\beta) \in C_{\mathscr{L}}(O)$.

By 5.1 and VI 7.1 there exists a monomorphism h from $\mathfrak{A}(\mathscr{S})$ into a topological field $\mathfrak{B}(X)$ of all subsets of a topological space X. By VIII 3.2, for the canonical valuation v^0 of $\mathscr{L}$ in $\mathfrak{A}(\mathscr{S})$ and for any formula γ of $\mathscr{L}$,

(9) $$h\gamma_{\mathfrak{A}(\mathscr{S})}(v^0) = \gamma_{\mathfrak{B}(X)}(hv^0).$$

[17] McKinsey and Tarski [1], [3]. For the proof, which is a modification of that by McKinsey and Tarski, see Rasiowa and Sikorski [5], [MM].

Let X_0 be the one-point strong compactification of X (see VI 8) and let $\mathfrak{B}(X_0)$ be the topological field of all subsets of X_0. It is easy to see that the mapping h_0, defined as

(10) $$h_0(Y) = Y \cap X \quad \text{for any } Y \subset X_0,$$

is an epimorphism from $\mathfrak{B}(X_0)$ onto $\mathfrak{B}(X)$. Let v be a valuation of $\mathscr{L}$ in $\mathfrak{B}(X_0)$ satisfying the following condition

(11) $$h_0\big(v(p)\big) = h\big(v^0(p)\big) \quad \text{for every propositional variable } p.$$

Thus $h_0 v = hv^0$. Suppose that $(\mathbf{I}\alpha \cup \mathbf{I}\beta) \in C_{\mathscr{L}}(O)$. Then by 5.2

(12) $$(\mathbf{I}\alpha \cup \mathbf{I}\beta)_{\mathfrak{B}(X_0)}(v) = \mathbf{I}\alpha_{\mathfrak{B}(X_0)}(v) \cup \mathbf{I}\beta_{\mathfrak{B}(X_0)}(v) = X_0.$$

Since X_0 is strongly compact, it follows from (12) that at least one of the summands is equal to X_0. Suppose that $\mathbf{I}\alpha_{\mathfrak{B}(X_0)}(v) = X_0$. Hence, $\alpha_{\mathfrak{B}(X_0)}(v) = X_0$. Thus by (10), (11) and VIII 3.2

(13)
$$X = h_0(X_0) = h_0\, \alpha_{\mathfrak{B}(X_0)}(v) = \alpha_{\mathfrak{B}(X)}(h_0 v) = \alpha_{\mathfrak{B}(X)}(hv^0) = h\alpha_{\mathfrak{A}(\mathscr{S})}(v^0).$$

Since h is a monomorphism, it follows from (13) that $\alpha_{\mathfrak{A}(\mathscr{S})}(v^0) = \bigvee$ and consequently, by 5.2, α is derivable in $\mathscr{S}$.

6. $\mathbf{L}_\lambda$-theories of zero order and their connection with I-filters. For any propositional calculus $\mathscr{S} = (\mathscr{L}, C_{\mathscr{L}})$ in $\mathbf{L}_\lambda$ and for any set $\mathscr{A}$ of formulas of $\mathscr{L}$, the system $\mathscr{S}(\mathscr{A}) = (\mathscr{L}, C_{\mathscr{L}}, \mathscr{A})$ is said to be an $\mathbf{L}_\lambda$-*theory based on* $\mathscr{S}$. Any valuation v of $\mathscr{L}$ in any non-degenerate topological Boolean algebra $\mathfrak{A} = (A, \bigvee, \Rightarrow, \cup, \cap, -, \mathbf{I})$ such that for all α in $\mathscr{A}$

(1) $$\alpha_{\mathfrak{A}}(v) = \bigvee$$

is said to be an $\mathbf{L}_\lambda$-*model* of $\mathscr{S}(\mathscr{A})$. An $\mathbf{L}_\lambda$-model of $\mathscr{S}(\mathscr{A})$ in a non-degenerate topological Boolean algebra $\mathfrak{A}$ such that, for all a in $\mathfrak{A}$, either $\mathbf{I}a = \bigvee$ or $\mathbf{I}a = \bigwedge$, is said to be an $\mathbf{I}$-*semantic model* of $\mathscr{S}(\mathscr{A})$.

The following theorem results from VIII 9.2, VIII 9.3 and the fact that the class of all $\mathscr{S}$-algebras for $\mathscr{S}$ in $\mathbf{L}_\lambda$ coincides with the class of all topological Boolean algebras.

6.1. *For every consistent* $\mathbf{L}_\lambda$-*theory* $\mathscr{S}(\mathscr{A}) = (\mathscr{L}, C_{\mathscr{L}}, \mathscr{A})$ *of zero order the algebra* $\mathfrak{A}(\mathscr{S}(\mathscr{A}))$ (see VIII 9) *is a non-degenerate topological Boolean algebra. The canonical valuation* v^0 *of* $\mathscr{L}$ *in* $\mathfrak{A}(\mathscr{S}(\mathscr{A}))$ (cf. VIII

9 (12)) *is an* $\mathbf{L}_\lambda$*-model of* $\mathcal{S}(\mathcal{A})$. *This* $\mathbf{L}_\lambda$*-model is adequate*, i.e. *for any formula* α *of* $\mathcal{L}$

(2) $\alpha \in C_{\mathcal{L}}(\mathcal{A})$ *if and only if* $\alpha_{\mathfrak{A}(\mathcal{S}(\mathcal{A}))}(v^0) = \vee$.

Clearly, by 5.2, for any formulas α, β of the formalized language $\mathcal{L}$ of $\mathcal{S} = (\mathcal{L}, C_{\mathcal{L}})$ in $\mathbf{L}_\lambda$, the formula $(-(\alpha \Rightarrow \alpha) \Rightarrow \beta)$ is derivable in $\mathcal{S}$. Hence, each system $\mathcal{S}$ in $\mathbf{L}_\lambda$ belongs to the class $\mathbf{S}_\nu$ of standard systems of implicative extensional propositional calculi with semi-negation (see VIII 10). Thus

(3) $$\mathbf{L}_\lambda \subset \mathbf{S}_\nu.$$

The following theorem characterizes consistent $\mathbf{L}_\lambda$-theories.

6.2. *For any* $\mathbf{L}_\lambda$*-theory* $\mathcal{S}(\mathcal{A}) = (\mathcal{L}, C_{\mathcal{L}}, \mathcal{A})$ *of zero order the following conditions are equivalent*:

(i) $\mathcal{S}(\mathcal{A})$ *is consistent*,

(ii) *for any formula* α *in* $\mathcal{L}$, *either* $\alpha \notin C_{\mathcal{L}}(\mathcal{A})$ *or* $-\alpha \notin C_{\mathcal{L}}(\mathcal{A})$,

(iii) *there exists an* $\mathbf{L}_\lambda$*-model of* $\mathcal{S}(\mathcal{A})$,

(iv) *there exists an* $\mathbf{L}_\lambda$*-model of* $\mathcal{S}(\mathcal{A})$ *in a topological field of sets*,

(v) *there exists an* $\mathbf{I}$*-semantic model of* $\mathcal{S}(\mathcal{A})$.

The equivalence of conditions (i), (ii), (iii) and (iv) follows from (3), VIII 10.2 and the representation theorem VI 7.1. Since (v) implies (iii), it is sufficient to prove that (iii) implies (v). Suppose that a valuation v of $\mathcal{L}$ in a non-degenerate topological Boolean algebra $\mathfrak{A}$ is an $\mathbf{L}_\lambda$-model of $\mathcal{S}(\mathcal{A})$. By VI 6.7 there exists a maximal $\mathbf{I}$-filter ∇ in $\mathfrak{A}$. By VI 6.11 the quotient algebra $\mathfrak{A}/\nabla$ is non-degenerate and has the property that, for each $||a||$ in $\mathfrak{A}/\nabla$, either $\mathbf{I}||a|| = \vee$ or $\mathbf{I}||a|| = \wedge$. Let h be the epimorphism from $\mathfrak{A}$ onto $\mathfrak{A}/\nabla$ (cf. VI 6.10). Then, by VIII 3.2, the valuation hv is an $\mathbf{I}$-semantic model of $\mathcal{S}(\mathcal{A})$.

The following theorem characterizes the theorems of any consistent $\mathbf{L}_\lambda$-theory of zero order.

6.3. *For any formula* α *of any consistent* $\mathbf{L}_\lambda$*-theory of zero order the following are equivalent*:

(i) $\alpha \in C_{\mathcal{L}}(\mathcal{A})$,

(ii) *every* $\mathbf{L}_\lambda$*-model of* $\mathcal{S}(\mathcal{A})$ *is an* $\mathbf{L}_\lambda$*-model of* α,

(iii) *every* $\mathbf{L}_\lambda$-*model of* $\mathscr{S}(\mathscr{A})$ *in any non-degenerate topological field of sets* $\mathfrak{B}(X)$ *is an* $\mathbf{L}_\lambda$-*model of* α,

(iv) $\alpha_{\mathfrak{A}(\mathscr{S}(\mathscr{A}))} v^0 = V$ *for the canonical valuation* v^0 *in the algebra* $\mathfrak{A}(\mathscr{S}(\mathscr{A}))$ *of* $\mathscr{S}(\mathscr{A})$.

Conditions (i), (ii) and (iv) are equivalent by VIII 9.5. Clearly, (ii) implies (iii) and by VI 7.1, (iii) implies (ii).

The next two theorems state connections between $\mathbf{L}_\lambda$-theories of zero order and classical or intuitionistic theories.

6.4. *Let* $\mathscr{S}(\mathscr{A}) = (\mathscr{L}, C_{\mathscr{L}}, \mathscr{A})$ *be an* $\mathbf{L}_\lambda$-*theory and let* $\mathscr{S}_0 = (\mathscr{L}_0, C_{\mathscr{L}_0})$ *be a propositional calculus of the classical logic* $\mathbf{L}_\varkappa$ *such that* $\mathscr{L}$ *is an extension of* $\mathscr{L}_0$. *If each formula* α *in* $\mathscr{A}$ *is a formula of* $\mathscr{L}_0$, *then for every formula* β *of* $\mathscr{L}_0$ *the following conditions are equivalent*:

(i) β *is a theorem of* $\mathscr{S}(\mathscr{A})$,

(ii) β *is a theorem of the classical theory* $\mathscr{S}_0(\mathscr{A}) = (\mathscr{L}_0, C_{\mathscr{L}_0}, \mathscr{A})$.

The easy proof, based on 3.3, 6.3 and the definition of topological Boolean algebras (see VI 5), is left to the reader.

Let $\mathscr{S}_0(\mathscr{A}) = (\mathscr{L}_0, C_{\mathscr{L}_0}, \mathscr{A})$ be an intuitionistic theory of zero order and let $\mathscr{L}$ be the formalized language obtained from $\mathscr{L}_0$ by the replacement of the sign $\urcorner$ by the negation sign $-$ and by adjoining the necessity sign **I**. Let $\mathfrak{f}$ be the mapping defined in Sec. 5 by (1)-(5) assigning to every formula α of $\mathscr{L}_0$ a formula $\mathfrak{f}\alpha$ of $\mathscr{L}$. Let $\mathfrak{f}\mathscr{A}$ be the set consisting of all $\mathfrak{f}\alpha$ for α in $\mathscr{A}$. With these hypotheses and notation the following theorem holds.

6.5. *For every formula* α *of* $\mathscr{L}_0$, α *is a theorem of the* $\mathbf{L}_\chi$-*theory* $\mathscr{S}_0(\mathscr{A}) = (\mathscr{L}_0, C_{\mathscr{L}_0}, \mathscr{A})$ *if and only if* $\mathfrak{f}\alpha$ *is a theorem of the* $\mathbf{L}_\lambda$-*theory* $\mathscr{S}(\mathfrak{f}\mathscr{A}) = (\mathscr{L}, C_{\mathscr{L}}, \mathfrak{f}\mathscr{A})$, *where* $\mathscr{S} = (\mathscr{L}, C_{\mathscr{L}})$ *is a propositional calculus in* $\mathbf{L}_\lambda$.

The proof, similar to that of 5.5, based on 6.3 and XI 9.3 instead of 5.2 and XI 8.2, respectively, is omitted.

Let us note the following theorem, which is the basis for the investigations concerning the connection between $\mathbf{L}_\lambda$-theories of zero order and **I**-filters.

6.6. *For every set* ∇ *of elements in a topological Boolean algebra* $\mathfrak{A} = (A, V, \Rightarrow, \cup, \cap, -, \mathbf{I})$, ∇ *is a deductive filter if and only if* ∇ *is an* **I**-*filter*.

Since in $\mathscr{S}_\lambda$ there are two rules of inference, modus ponens and (r_λ) (cf. Sec. 4), it follows from VIII 12.3 that ∇ is a deductive filter if and only if it is an implicative filter and the following condition is satisfied:

(4) $$\text{if} \quad a \Rightarrow b \in \nabla, \quad \text{then} \quad \mathbf{I}a \Rightarrow \mathbf{I}b \in \nabla.$$

Note that the notion of an implicative filter in a topological Boolean algebra coincides with the notion of a filter (see VI 6). Thus to prove 6.6 it is sufficient to show that for any implicative filter ∇ condition (4) is equivalent to the following

(5) $$\text{if} \quad a \in \nabla, \quad \text{then} \quad \mathbf{I}a \in \nabla.$$

Suppose that (4) holds and $a \in \nabla$. Then, by VI 1 (2), $a = V \Rightarrow a \in \nabla$. Consequently, by (4), $V \Rightarrow \mathbf{I}a = \mathbf{I}V \Rightarrow \mathbf{I}a \in \nabla$. Hence, by (f_1), (f_2) in II 1, $\mathbf{I}a \in \nabla$. Conversely, if (5) holds and $a \Rightarrow b \in \nabla$, then $\mathbf{I}(a \Rightarrow b) \in \nabla$. By VI 6 (4)

(6) $$\mathbf{I}(a \Rightarrow b) \leqslant \mathbf{I}a \Rightarrow \mathbf{I}b.$$

Since $\mathbf{I}(a \Rightarrow b) \in \nabla$ and ∇ is a filter, it follows from (6) and III 1 (f_2) that $\mathbf{I}a \Rightarrow \mathbf{I}b \in \nabla$.

By VI 6.9 and VI 6.10 the kernels of epimorphisms from topological Boolean algebras onto similar algebras are $\mathbf{I}$-filters and, conversely, every $\mathbf{I}$-filter is the kernel of an epimorphism determined by itself. The above statements follow also from 6.6, VIII 12.2, VIII 13.10 and the equational definability of the class of all topological Boolean algebras.

By 6.6, VIII 13.2 and VIII 13.4 we get two theorems, which state that there exists a natural one-one correspondence between $\mathbf{I}$-filters in the topological Boolean algebra $\mathfrak{A}(\mathscr{S})$ of a propositional calculus $\mathscr{S} = (\mathscr{L}, C_{\mathscr{L}})$ in $\mathbf{L}_\lambda$ and formalized $\mathbf{L}_\lambda$-theories of zero order based on $\mathscr{S}$, two theories $\mathscr{S}(\mathscr{A})$ and $\mathscr{S}(\mathscr{A}')$ being identified if and only if they have the same sets of theorems, i.e. $C_{\mathscr{L}}(\mathscr{A}) = C_{\mathscr{L}}(\mathscr{A}')$.

Let us note that for any system $\mathscr{S}$ in $\mathbf{L}_\lambda$ there holds a theorem analogous to IX 5.4 (where "implicative filter" is replaced by "I-filter"). The deduction theorem corresponding to IX 5.5 has the following formulation.

6.7. *For any formulas α, β in the formalized language $\mathscr{L}$ of an arbitrary system $\mathscr{S} = (\mathscr{L}, C_{\mathscr{L}})$ in $\mathbf{L}_\lambda$ and for every set $\mathscr{A}$ of formulas in $\mathscr{L}$:*

(7) $$\beta \in C_{\mathscr{L}}(\mathscr{A} \cup \{\alpha\}) \quad \textit{if and only if} \quad (\mathbf{I}\alpha \Rightarrow \beta) \in C_{\mathscr{L}}(\mathscr{A}).$$

To begin with, let us note that by 6.3, VI 5 (i_4), VI 5 (3), for any set $\mathscr{B}$ of formulas of $\mathscr{L}$ and for every formula γ of $\mathscr{L}$

$$(8) \qquad \gamma \in C_{\mathscr{L}}(\mathscr{B}) \quad \textit{if and only if} \quad \mathbf{I}\gamma \in C_{\mathscr{L}}(\mathscr{B}).$$

If $(\mathbf{I}\alpha \Rightarrow \beta) \in C_{\mathscr{L}}(\mathscr{A})$, then $(\mathbf{I}\alpha \Rightarrow \beta) \in C_{\mathscr{L}}(\mathscr{A} \cup \{\alpha\})$. By (8), $\mathbf{I}\alpha \in C_{\mathscr{L}}(\mathscr{A} \cup \{\alpha\})$. Hence, by modus ponens, $\beta \in C_{\mathscr{L}}(\mathscr{A} \cup \{\alpha\})$. Conversely, if $\beta \in C_{\mathscr{L}}(\mathscr{A} \cup \{\alpha\})$, then $||\beta||$ belongs to the $\mathbf{I}$-filter $\nabla_{\mathscr{A} \cup \{\alpha\}}$ in $\mathfrak{A}(\mathscr{S})$ composed of all $||\gamma||$ such that $\gamma \in C_{\mathscr{L}}(\mathscr{A} \cup \{\alpha\})$. By a theorem analogous to IX 5.4, $||\beta||$ belongs to the $\mathbf{I}$-filter generated by $\nabla_{\mathscr{A}}$ and $||\alpha||$. Consequently, by VI 6.2, there exists a $||\gamma|| \in \nabla_{\mathscr{A}}$ such that $\mathbf{I}||\alpha|| \cap \mathbf{I}||\gamma|| \leqslant ||\beta||$. Hence, by VI 1.2 and IV 1 (r), $\mathbf{I}||\gamma|| \leqslant \mathbf{I}||\alpha|| \Rightarrow ||\beta||$. Since $\nabla_{\mathscr{A}}$ is an $\mathbf{I}$-filter, $\mathbf{I}||\gamma|| \in \nabla_{\mathscr{A}}$. Thus, $\mathbf{I}||\alpha|| \Rightarrow ||\beta|| = ||(\mathbf{I}\alpha \Rightarrow \beta)|| \in \nabla_{\mathscr{A}}$, i.e. $(\mathbf{I}\alpha \Rightarrow \beta) \in C_{\mathscr{L}}(\mathscr{A})$.

The deduction theorem, analogous to X 6.4, has the following formulation.

6.8. *For every formula α in the formalized language $\mathscr{L}$ of any system $\mathscr{S} = (\mathscr{L}, C_{\mathscr{L}})$ in $\mathbf{L}_\lambda$ and for each non-empty set $\mathscr{A}$ of formulas of $\mathscr{L}$ the condition $\alpha \in C_{\mathscr{L}}(\mathscr{A})$ is equivalent to the condition that there exist formulas $\alpha_1, \ldots, \alpha_n$ in $\mathscr{A}$ such that*

$$(9) \qquad ((\mathbf{I}\alpha_1 \cap \ldots \cap (\mathbf{I}\alpha_{n-1} \cap \mathbf{I}\alpha_n)\ldots) \Rightarrow \alpha) \in C_{\mathscr{L}}(O).$$

The easy proof, analogous to that of X 6.4 and based on (8), the theorem corresponding to X 6.2 and VI 6.1, is omitted.

By 6.6 the notions of an irreducible deductive filter and of a maximal deductive filter in a topological Boolean algebra coincide with the notions of an irreducible $\mathbf{I}$-filter and a maximal $\mathbf{I}$-filter, respectively.

The following two theorems follow from 6.6, VIII 13.5, VIII 13.6 and VIII 13.7.

6.9. *An $\mathbf{L}_\lambda$-theory $\mathscr{S}(\mathscr{A}) = (\mathscr{L}, C_{\mathscr{L}}, \mathscr{A})$ is irreducible (maximal) if and only if the $\mathbf{I}$-filter $\nabla_{\mathscr{A}}$ in $\mathfrak{A}(\mathscr{S})$ consisting of all $||\alpha||$ in $\mathfrak{A}(\mathscr{S})$ such that $\alpha \in C_{\mathscr{L}}(\mathscr{A})$ is irreducible (maximal).*

6.10. *For every consistent $\mathbf{L}_\lambda$-theory $\mathscr{S}(\mathscr{A}) = (\mathscr{L}, C_{\mathscr{L}}, \mathscr{A})$ there exists an irreducible $\mathbf{L}_\lambda$-theory $\mathscr{S}(\mathscr{A}) = (\mathscr{L}, C_{\mathscr{L}}, \mathscr{A}')$ such that $C_{\mathscr{L}}(\mathscr{A}) \subset C_{\mathscr{L}}(\mathscr{A}')$.*

The following theorem follows from (3) and VIII 13.8.

6.11. *For every consistent* $\mathbf{L}_\lambda$*-theory* $\mathscr{S}(\mathscr{A}) = (\mathscr{L}, C_{\mathscr{L}}, \mathscr{A})$ *there exists a maximal* $\mathbf{L}_\lambda$*-theory* $\mathscr{S}(\mathscr{A}') = (\mathscr{L}, C_{\mathscr{L}}, \mathscr{A}')$ *such that* $C_{\mathscr{L}}(\mathscr{A}) \subset C_{\mathscr{L}}(\mathscr{A}')$.

The next theorem characterizes maximal $\mathbf{L}_\lambda$-theories.

6.12. *The following conditions are equivalent for every* $\mathbf{L}_\lambda$*-theory* $\mathscr{S}(\mathscr{A}) = (\mathscr{L}, C_{\mathscr{L}}, \mathscr{A})$ *of zero order*:

(i) $\mathscr{S}(\mathscr{A})$ *is maximal*,

(ii) *for each formula* α *exactly one of the formulas* $\mathbf{I}\alpha$, $-\mathbf{I}\alpha$ *is a theorem of* $\mathscr{S}(\mathscr{A})$,

(iii) $\mathfrak{A}(\mathscr{S}(\mathscr{A}))$ *is isomorphic to a non-degenerate topological Boolean algebra* $\mathfrak{A}$ *such that for each element* a *of* $\mathfrak{A}$ *either* $\mathbf{I}a = \bigvee$ *or* $\mathbf{I}a = \bigwedge$,

(iv) $\mathscr{S}(\mathscr{A})$ *has an adequate* $\mathbf{I}$*-semantic model*,

(v) $\mathscr{S}(\mathscr{A})$ *is consistent and each* $\mathbf{I}$*-semantic model of* $\mathscr{S}(\mathscr{A})$ *is adequate.*

By VIII 13.1, 6.6, 6.9, VI 6.11 and VIII 13.11 conditions (i), (ii) and (iii) are equivalent. By 6.1, (iii) implies (iv). Suppose that (iv) holds. Then by 6.2, $\mathscr{S}(\mathscr{A})$ is consistent. Observe that if v is an $\mathbf{I}$-semantic model of $\mathscr{S}(\mathscr{A})$ in $\mathfrak{A}$, then for each formula α of $\mathscr{L}$

(10) either $\mathbf{I}\alpha_{\mathfrak{A}}(v) = \bigvee$ or $-\mathbf{I}\alpha_{\mathfrak{A}}(v) = \bigvee$.

Since the same condition is satisfied for an adequate $\mathbf{I}$-semantic model of $\mathscr{S}(\mathscr{A})$, we infer that for each formula α of $\mathscr{L}$ either $\mathbf{I}\alpha \in C_{\mathscr{L}}(\mathscr{A})$ or $-\mathbf{I}\alpha \in C_{\mathscr{L}}(\mathscr{A})$. Suppose that v is an arbitrary $\mathbf{I}$-semantic model of $\mathscr{S}(\mathscr{A})$ in $\mathfrak{A}$. If $\alpha \in C_{\mathscr{L}}(\mathscr{A})$, then by 6.3, $\alpha_{\mathfrak{A}}(v) = \bigvee$. If $\alpha \notin C_{\mathscr{L}}(\mathscr{A})$, then by (8), $\mathbf{I}\alpha \notin C_{\mathscr{L}}(\mathscr{A})$. Consequently, $-\mathbf{I}\alpha \in C_{\mathscr{L}}(\mathscr{A})$. Hence, by 6.3, $-\mathbf{I}\alpha_{\mathfrak{A}}(v) = \bigvee$. Thus $\mathbf{I}\alpha_{\mathfrak{A}}(v) = \bigwedge$, which implies that $\alpha_{\mathfrak{A}}(v) \neq \bigvee$. Consequently, the $\mathbf{I}$-semantic model v is adequate. It suffices to show that (v) implies (ii). Suppose that (v) holds. Since $\mathscr{S}(\mathscr{A})$ is consistent, by 6.2 there exists an $\mathbf{I}$-semantic model v of $\mathscr{S}(\mathscr{A})$ in $\mathfrak{A}$. By (v) it is adequate. Hence, by (10), condition (ii) is satisfied. This completes the proof.

7. I-prime $\mathbf{L}_\lambda$-theories. A consistent $\mathbf{L}_\lambda$-theory $\mathscr{S}(\mathscr{A}) = (\mathscr{L}, C_{\mathscr{L}}, \mathscr{A})$ of zero order is said to be $\mathbf{I}$-*prime* if for any formulas α, β of $\mathscr{L}$ the following condition is satisfied:

(Ip) if $(\mathbf{I}\alpha \cup \mathbf{I}\beta) \in C_{\mathscr{L}}(\mathscr{A})$, then either $\mathbf{I}\alpha \in C_{\mathscr{L}}(\mathscr{A})$ or $\mathbf{I}\beta \in C_{\mathscr{L}}(\mathscr{A})$.

It follows from the above definition, the definition of **I**-prime filters in topological Boolean algebras, 6.6 and VIII 13.1 that

7.1. *An* $\mathbf{L}_\lambda$-*theory* $\mathscr{S}(\mathscr{A}) = (\mathscr{L}, C_{\mathscr{L}}, \mathscr{A})$ *is* **I**-*prime if and only if the* **I**-*filter* $\nabla_{\mathscr{A}}$ *in* $\mathfrak{A}(\mathscr{S})$, *which consists of all* $||\alpha||$ *in* $\mathfrak{A}(\mathscr{S})$ *such that* $\alpha \in C_{\mathscr{L}}(\mathscr{A})$, *is* **I**-*prime.*

The following corollary follows from 7.1, 6.9 and VI 6.4.

7.2. *An* $\mathbf{L}_\lambda$-*theory of zero order is* **I**-*prime if and only if it is irreducible.*

It follows from 7.2 and VIII 13.9 that

7.3. *Every maximal* $\mathbf{L}_\lambda$-*theory of zero order is* **I**-*prime.*

We now give a sufficient condition for consistent $\mathbf{L}_\lambda$-theories to be **I**-prime. The notion of the characteristic valuation for consistent $\mathbf{L}_\lambda$-theories will provide the starting point.

Let $\mathscr{S}(\mathscr{A}) = (\mathscr{L}, C_{\mathscr{L}}, \mathscr{A})$ be a consistent $\mathbf{L}_\lambda$-theory. By 6.1 and VI 7.1 there exists a monomorphism h from the algebra $\mathfrak{A}(\mathscr{S}(\mathscr{A}))$ into the topological field $\mathfrak{B}(X)$ of all subsets of a topological space X. It follows from VIII 9 (15) and VIII 3.2 that for the canonical valuation v^0 of $\mathscr{L}$ in $\mathfrak{A}(\mathscr{S}(\mathscr{A}))$ and for each formula α of $\mathscr{L}$

$$(1) \qquad h||\alpha||_{\mathscr{A}} = h\alpha_{\mathfrak{A}(\mathscr{S}(\mathscr{A}))}(v^0) = \alpha_{\mathfrak{B}(X)}(hv^0).$$

Let $X_0 = X \cup \{x_0\}$, where $x_0 \notin X$, be the one-point strong compactification of the space X (see VI 8). Setting

$$(2) \qquad g(Y) = X \cap Y \quad \text{for each subset } Y \text{ of } X_0,$$

we define an epimorphism from the topological field $\mathfrak{B}(X_0)$ of all subsets of X_0 onto $\mathfrak{B}(X)$. The valuation v of $\mathscr{L}$ in $\mathfrak{B}(X_0)$, defined by:

$$(3) \qquad v(p) = \begin{cases} X_0 & \text{if} \quad v^0(p) = ||p||_{\mathscr{A}} = V, \\ h(v^0(p)) & \text{if} \quad v^0(p) = ||p||_{\mathscr{A}} \neq V \end{cases}$$

for each propositional variable p, will be called the *characteristic valuation for* $\mathscr{S}(\mathscr{A})$.

By (2) and (3) we get

$$(4) \qquad gv = hv^0.$$

7.4. *For the characteristic valuation* v *of a consistent* $\mathbf{L}_\lambda$-*theory* $\mathscr{S}(\mathscr{A}) = (\mathscr{L}, C_{\mathscr{L}}, \mathscr{A})$ *and for every formula* α *of* $\mathscr{L}$

$$(5) \qquad \textit{either} \quad \alpha_{\mathfrak{B}(X_0)}(v) = h||\alpha||_{\mathscr{A}} \cup \{x_0\} \quad \textit{or} \quad \alpha_{\mathfrak{B}(X_0)}(v) = h||\alpha||_{\mathscr{A}}.$$

Moreover,

(6) *if* $\alpha_{\mathfrak{B}(X_0)}(v) = X_0$, *then* $h||\alpha||_{\mathscr{A}} = X$, i.e. $||\alpha||_{\mathscr{A}} = V$.

By VIII 3.2, (4), (1) and VIII 9 (13)

$$g\alpha_{\mathfrak{B}(X_0)}(v) = \alpha_{\mathfrak{B}(X)}(gv) = \alpha_{\mathfrak{B}(X)}(hv^0) = h\alpha_{\mathfrak{A}(\mathscr{S}(\mathscr{A}))}(v^0) = h||\alpha||_{\mathscr{A}}.$$

Hence, by (2),

$$\alpha_{\mathfrak{B}(X_0)}(v) \cap X = h||\alpha||_{\mathscr{A}}. \tag{7}$$

Since, for every subset Y of X_0, either $Y \subset X$ or $Y = Z \cup \{x_0\}$ where $Z \subset X$, $\alpha_{\mathfrak{B}(X_0)}(v)$ is a subset of X_0, $h||\alpha||_{\mathscr{A}} \subset X$, equation (7) implies (5) and (6).

7.5. *If the characteristic valuation* v *for a consistent* $\mathbf{L}_\lambda$-*theory is an* $\mathbf{L}_\lambda$-*model of* $\mathscr{S}(\mathscr{A})$, *then* $\mathscr{S}(\mathscr{A})$ *is* $\mathbf{I}$-*prime.*

Indeed, if $(\mathbf{I}\alpha \cup \mathbf{I}\beta) \in C_{\mathscr{L}}(\mathscr{A})$, then by 6.3

$$\mathbf{I}\alpha_{\mathfrak{B}(X_0)}(v) \cup \mathbf{I}\beta_{\mathfrak{B}(X_0)}(v) = (\mathbf{I}\alpha \cup \mathbf{I}\beta)_{\mathfrak{B}(X_0)}(v) = X_0. \tag{8}$$

Since both sumands on the left side of equation (8) are open subsets of a strongly compact topological space X_0, it follows from (8) that either $\mathbf{I}\alpha_{\mathfrak{B}(X_0)}(v) = X_0$ or $\mathbf{I}\beta_{\mathfrak{B}(X_0)}(v) = X_0$. Hence, by (6), either $||\mathbf{I}\alpha||_{\mathscr{A}} = V$ or $||\mathbf{I}\beta||_{\mathscr{A}} = V$, which implies by VIII 9 (9) that either $\mathbf{I}\alpha \in C_{\mathscr{L}}(\mathscr{A})$ or $\mathbf{I}\beta \in C_{\mathscr{L}}(\mathscr{A})$.

Given a consistent $\mathbf{L}_\lambda$-theory $\mathscr{S}(\mathscr{A}) = (\mathscr{L}, C_{\mathscr{L}}, \mathscr{A})$, let $\mathscr{Z}$ be the set of all formulas α of $\mathscr{L}$ which have the property

$$\text{if} \quad \alpha \in C_{\mathscr{L}}(\mathscr{A}), \quad \text{then} \quad \alpha_{\mathfrak{B}(X_0)}(v) = X_0 \tag{9}$$

for the characteristic valuation v for $\mathscr{S}(\mathscr{A})$.

7.6. *The set* $\mathscr{Z}$ *satisfies the following conditions*:

(i) *each propositional variable* p *is in* $\mathscr{Z}$,

(ii) *for each formula* β *of* $\mathscr{L}$, $-\mathbf{I}\beta$ *is in* $\mathscr{Z}$,

(iii) *if* α *is in* $\mathscr{Z}$, *then* $\mathbf{I}\alpha$ *is also in* $\mathscr{Z}$,

(iv) *if* β, γ *are in* $\mathscr{Z}$, *then* $(\beta \cap \gamma)$ *is also in* $\mathscr{Z}$,

(v) *if* γ *is in* $\mathscr{Z}$, *then for each formula* β *of* $\mathscr{L}$, $(\mathbf{I}\beta \Rightarrow \gamma)$ *is in* $\mathscr{Z}$.

The easy proof is left to the reader.

Since the intersection F_0 of all sets $\mathscr{X}$ of formulas in $\mathscr{L}_0$ satisfying conditions (i)-(v) also satisfies these conditions, we get the following theorem.

7.7. *If* $\mathscr{S}(\mathscr{A}) = (\mathscr{L}, C_{\mathscr{L}}, \mathscr{A})$ *is a consistent* $\mathbf{L}_\lambda$*-theory and* $\mathscr{A} \subset F_0$, *then* $\mathscr{S}(\mathscr{A})$ *is* $\mathbf{I}$*-prime.*

Indeed, by (9), the characteristic valuation for $\mathscr{S}(\mathscr{A})$ is an $\mathbf{L}_\lambda$-*model of* $\mathscr{S}(\mathscr{A})$, which implies by 7.5 that $\mathscr{S}(\mathscr{A})$ is $\mathbf{I}$-prime.

Now let F_1 be the intersection of all sets $\mathscr{X}$ of formulas of $\mathscr{L}$ satisfying conditions (i)-(v) and the following one:

(vi) *if* α, β *are in* $\mathscr{X}$, *then* $(\mathbf{I}\alpha \cup \mathbf{I}\beta) \in \mathscr{X}$.

7.8. *If* $\mathscr{S}(\mathscr{A}) = (\mathscr{L}, C_{\mathscr{L}}, \mathscr{A})$ *is a consistent* $\mathbf{L}_\lambda$*-theory of zero order and* $\mathscr{A} \subset F_1$, *then* $\mathscr{S}(\mathscr{A})$ *is* $\mathbf{I}$*-prime if and only if the characteristic valuation* v *for* $\mathscr{S}(\mathscr{A})$ *is an* $\mathbf{L}_\lambda$*-model of* $\mathscr{S}(\mathscr{A})$.

The sufficiency follows from 7.5. If $\mathscr{S}(\mathscr{A})$ is $\mathbf{I}$-prime, then the set $\mathscr{X}$ of all formulas of $\mathscr{L}$ which satisfy condition (9) also has the property (vi). Hence, every formula in F_1 satisfies (9). Since $\mathscr{A} \subset F_1$, the characteristic valuation v for $\mathscr{S}(\mathscr{A})$ is an $\mathbf{L}_\lambda$-model of $\mathscr{S}(\mathscr{A})$, which completes the proof.

Exercises

1. Let $\mathfrak{A} = (A, V, \cup, \cap, \sim)$ be a non-degenerate quasi-Boolean algebra and let ∇ be the filter generated by the set $\{a \cup \sim a\}_{a \in A}$. Prove that if $\nabla \neq A$, then for every formula α of a classical propositional calculus $\mathscr{S} = (\mathscr{L}, C_{\mathscr{L}})$, such that $\Rightarrow$ does not appear in α the following conditions are equivalent: (i) α is derivable in $\mathscr{S}$, (ii) for every valuation v: $V \to A$ (where V is the set of all propositional variables of $\mathscr{L}$), $\alpha_{\mathfrak{A}}(v) \in \nabla$ [18].

2. By an A-lattice we shall mean an algebra $\mathfrak{B} = (B, \cup, \cap, -)$ such that $(B, \cup, \cap)$ is a distributive lattice and $-$ is an anti-automorphism of $(B, \cup, \cap)$, i.e. a one-one mapping of B onto itself such that $-(a \cup b) = -a \cap -b$ for all a, b in B. Let ∇ be the filter in $(B, \cup, \cap)$ generated by all elements $a \cup -^k a$, where $k = 1, 3, 5, \ldots$ and $a \in B$. Prove that if $\nabla \neq B$, then for each formula α of a classical propositional calculus $\mathscr{S} = (\mathscr{L}, C_{\mathscr{L}})$ such that $\Rightarrow$ does not appear in α the following conditions are equivalent: (i) α is derivable in $\mathscr{S}$, (ii) for every valuation v: $V \to B$, where V is the set of all propositional variables of $\mathscr{L}$, $\alpha_{\mathfrak{B}}(v) \in \nabla$ [19].

(18) Monteiro [1].
(19) Henkin [3].

3. Let $\mathscr{S}_{\lambda 5}$ be the propositional calculus obtained from $\mathscr{S}_{\lambda}$ by adding to the logical axioms of $\mathscr{S}_{\lambda}$ all formulas of the form $(-\mathbf{I}-\alpha \Rightarrow \mathbf{I}-\mathbf{I}-\alpha)$ for any formula α of $\mathscr{L}$ [20]. Prove that the class of all $\mathscr{S}_{\lambda 5}$-algebras coincides with the class of all topological Boolean algebras in which every closed element is open [21].

4. Prove that for every formula α of $\mathscr{S}_{\lambda}$, α is derivable in $\mathscr{S}_{\lambda 5}$ if and only if $\mathbf{I}-\mathbf{I}-\mathbf{I}\alpha$ is derivable in $\mathscr{S}_{\lambda}$; moreover, if α is derivable in $\mathscr{S}_{\lambda 5}$, then $-\mathbf{I}-\alpha$ is derivable in $\mathscr{S}_{\lambda}$ [22].

5. Prove that the formula $(\mathbf{I}-p \cup \mathbf{I}-\mathbf{I}-p)$, where p is a propositional variable, is derivable in $\mathscr{S}_{\lambda 5}$, but neither $-p$ nor $-\mathbf{I}-p$ is derivable in $\mathscr{S}_{\lambda 5}$, i.e. Theorem 5.6 does not hold for $\mathscr{S}_{\lambda 5}$ [23].

6. Let $\mathscr{A}$ be the set of all formulas of $\mathscr{S}_{\lambda}$ of the following form: $\mathbf{I}-\mathbf{I}-\mathbf{I}(-\mathbf{I}-\alpha \Rightarrow \alpha)$ for any α in the formalized language of $\mathscr{S}_{\lambda}$. Prove that the $\mathbf{L}_{\lambda}$-theory $\mathscr{S}_{\lambda}(\mathscr{A})$ is $\mathbf{I}$-prime [24].

7. Let f be the mapping which with every formula α of $\mathscr{S}_{\varkappa}$ associates a formula $f\alpha$ of $\mathscr{S}_{\lambda 5}$ (see Ex. 3) defined by induction on the length of α as follows: $fp = \mathbf{I}p$,

$$f(\alpha \Rightarrow \beta) = \mathbf{I}(f\alpha \Rightarrow f\beta),\ f(\alpha \cup \beta) = (f\alpha \cup f\beta),\ f(\alpha \cap \beta) = (f\alpha \cap f\beta),\ f-\alpha = \mathbf{I}-f\alpha.$$

Prove that for every formula α of $\mathscr{S}_{\varkappa}$, α is derivable in $\mathscr{S}_{\varkappa}$ if and only if $f\alpha$ is derivable in $\mathscr{S}_{\lambda 5}$ [25].

8. Let $\mathscr{A}$ be the set of all formulas of $\mathscr{S}_{\lambda}$ of the following form: $(\mathbf{I}(\mathbf{I}\alpha \Rightarrow \mathbf{I}\beta) \cup \mathbf{I}(\mathbf{I}\beta \Rightarrow \mathbf{I}\alpha))$, for any formulas α, β of $\mathscr{S}_{\lambda}$. Let $\mathscr{S}$ be the propositional calculus as defined in XI Ex. 10 and let $\mathfrak{f}$ be the mapping from the set of all formulas of $\mathscr{S}$ into the set of all formulas of $\mathscr{S}_{\lambda}$ defined in Sec. 5. Prove that for each formula α, of $\mathscr{S}$, α is derivable in $\mathscr{S}$ if and only if $\mathfrak{f}\alpha$ is a theorem of $\mathscr{S}_{\lambda}(\mathscr{A})$ [26].

[20] Lewis and Langford [1].
[21] McKinsey [1], McKinsey and Tarski [3].
[22] Matsumoto [1].
[23] McKinsey and Tarski [3].
[24] McKinsey and Tarski [3].
[25] Dummet and Lemmon [1].
[26] Dummet and Lemmon [1]. $\mathscr{S}_{\lambda}(\mathscr{A})$ is the propositional calculus S 4.3.

CHAPTER XIV

MANY-VALUED LOGICS

Introduction. Philosophical problems arising from the idea that there exist statements which are neither true nor false, led to the formulation of many-valued logics by Łukasiewicz [1], [2], [3] (see also Łukasiewicz and Tarski [1]). Independently Post [1] constructed his m-valued logics. Infinite-valued logics will not be dealt with in this chapter, which is concerned with those m-valued logics, $m = 2, 3, \ldots$, which correspond, from the algebraic point of view, to Post algebras.

The m-valued propositional calculi of Łukasiewicz and Post were not constructed as formalized axiomatic deductive systems (1) but were built up by means of the truth-table method, which can briefly be described as follows. Let $\wedge = e_0 \leqslant e_1 \leqslant \ldots \leqslant e_{m-1} = \vee$ be a chain and let $E = \{e_0, \ldots, e_{m-1}\}$. Consider E as the set of all admissible logical values, where e_0 and e_{m-1} correspond falsehood and truth, respectively. Let $\mathscr{L}$ be a formalized language of zero order. Let us assign to each i-argument connective o $(i = 0, 1, 2)$ of $\mathscr{L}$ an i-argument operation o_E: $E^i \to E$ on E, which has been defined in advance. Thus we start by defining an abstract algebra $\mathfrak{A} = (E, \vee, o_{1E}, \ldots, o_{kE})$, where $o_{1E}, \ldots, o_{kE}$ are operations assigned to all the connectives $o_1, \ldots$ $\ldots, o_k$, respectively, which appear in $\mathscr{L}$. The algebra $\mathfrak{A}$ determines a consequence operation $C_{\mathscr{L}}$ (2) in $\mathscr{L}$ as follows: 1) the set $C_{\mathscr{L}}(O)$ con-

(1) The first axiomatic system of the three-valued propositional calculus of Łukasiewicz is due to Wajsberg [1]. For the axiomatic systems of m-valued propositional calculi of Łukasiewicz see Rosser and Turquette [1]. For the axiomatization of functionally complete m-valued propositional calculi of Łukasiewicz see Słupecki [1], [2]. For an algebraic approach to m-valued logics of Łukasiewicz see Moisil [2].

(2) Sometimes the consequence operation is determined analogously by assuming in the condition 1) that $\alpha_{\mathfrak{A}}(v) \in \{e_k, \ldots, e_{m-1}\}$, where k is a fixed number $0 < k \leqslant m-1$, and in the condition 2) that $\beta_{\mathfrak{A}}(v) \in \{e_k, \ldots, e_{m-1}\}$ for every valuation v such that $\alpha_{\mathfrak{A}}(v) \in \{e_k, \ldots, e_{m-1}\}$ whenever $\alpha \in \mathscr{A}$.

sists of all formulas α of $\mathscr{L}$ such that $\alpha_{\mathfrak{A}}(v) = e_{m-1} = V$ for each valuation v of $\mathscr{L}$ in $\mathfrak{A}$, 2) for every set $\mathscr{A}$ of formulas of $\mathscr{L}$, $C_{\mathscr{L}}(\mathscr{A})$ consists of all formulas in $C_{\mathscr{L}}(O)$ and all formulas β satisfying the condition $\beta_{\mathfrak{A}}(v) = V$ for each valuation v of $\mathscr{L}$ in $\mathfrak{A}$ such that $\alpha_{\mathfrak{A}}(v) = V$ whenever $\alpha \in \mathscr{A}$.

An m-valued propositional calculus $\mathscr{S} = (\mathscr{L}, C_{\mathscr{L}})$, defined by this method, will be said to be *functionally complete* if the abstract algebra $\mathfrak{A}$ determining $\mathscr{S}$ is functionally complete, i.e., if each i-argument mapping $o\colon E_{\rm j}^i \to E$, $i = 0, 1, \ldots$, can be defined by the operations in this algebra. The m-valued propositional calculi to be discussed in this chapter are functionally complete.

Using a modification of the Gentzen method, Kirin [1], [2] constructed deductive formalized systems of m-valued propositional calculi and also of m-valued predicate calculi, and proved that Post algebras of order m play the same role for his propositional calculi as do $\mathscr{S}$-algebras for the propositional calculi discussed in this book. Independently, Rousseau [1], [3] applied Gentzen's method to construct formalized systems of m-valued propositional calculi and predicate calculi, but the connection between his systems and Post algebras was not established in those papers.

The first standard systems of m-valued propositional calculi corresponding to Post algebras were constructed by Rousseau [2], [4] (3). Predicate calculi constructed on the basis of these propositional calculi were studied by Rasiowa [9], [10], [11], [12].

The aim of the present chapter is an exposition of the algebraic treatment of Rousseau's m-valued propositional calculi, m-valued logics determined by them, and formalized theories of zero order based on these logics.

The m-valued logics considered here are denoted by $\mathbf{L}_m$ ($m = 2, 3, \ldots$). Propositional calculi of $\mathbf{L}_m$, $m > 2$, coincide with functionally complete m-valued propositional calculi of Łukasiewicz (as examined by Rosser and Turquette) with one distinguished value in the following sense: all connectives occurring in $\mathscr{S} \in \mathbf{L}_m$ are definable in Łukasiewicz's

(3) These results were presented by Rousseau at the Seminar on Foundations of Mathematics at the Institute of Mathematics of the Polish Academy of Sciences in November 1966.

propositional calculi and conversely. Thus this relationship is the same as, for example, that between classical propositional calculi with negation and implication as the only connectives on one hand, and classical propositional calculi with negation and disjunction as the only connectives, on the other hand.

Besides general metalogical theorems of Chapter VIII, which can in particular be formulated for propositional calculi in $\mathbf{L}_m$ and for $\mathbf{L}_m$-theories, certain other theorems are obtained by applying the theory of Post algebras, in particular the representation Theorem VII 6.1 and theorems on D-filters.

By Theorem 3.2, exactly those formulas are derivable in $\mathscr{S} \in \mathbf{L}_m$ which are valid in the m-element Post algebra $\mathfrak{P}_m$. Consequently, those formulas, in which no other connectives and propositional constants occur except $\Rightarrow$, $\cup$, $\cap$, $\urcorner$, are derivable in $\mathscr{S} \in \mathbf{L}_m$ which are valid in the m-element pseudo-Boolean algebra which is a chain. The formulas derivable in propositional calculi of $\mathbf{L}_m$ coincide also with those which are valid in every Post field of sets (of order m).

Among all $\mathbf{L}_m$-models of $\mathbf{L}_m$-theories we distinguish m-valued semantic models, i.e. models in the m-element Post algebra $\mathfrak{P}_m$. They play an important role in metalogical investigations. For instance, the consistency of any $\mathbf{L}_m$-theory is equivalent to the existence of an m-valued semantic model for that theory (4.2). A formula α is a theorem of a consistent $\mathbf{L}_m$-theory if and only if each m-valued semantic model of that theory is also an $\mathbf{L}_m$-model of α (4.3).

Deductive filters in Post algebras coincide with D-filters. Two deduction theorems (4.5, 4.6) have a modified form, connected with the structure of D-filters generated by a D-filter and an element, and generated by a non-empty set of elements, respectively. Irreducible $\mathbf{L}_m$-theories coincide with prime $\mathbf{L}_m$-theories and also with maximal ones. Thus each consistent $\mathbf{L}_m$-theory can be extended to a maximal one. A characterization of maximal $\mathbf{L}_m$-theories analogous to that of classical theories is formulated in 4.11.

1. Propositional calculus $\mathscr{S}_m$ [4] of m-valued logic. For each pos-

[4] See Rousseau [2], [4]. For the sake of uniformity the case $m = 2$ is here neither omitted nor simplified.

itive integer $m \geqslant 2$, let $\mathcal{S}_m = (\mathcal{L}, C_{\mathcal{L}})$ be a propositional calculus described briefly as follows. The alphabet $A^0 = (V, L_0, L_1, L_2, U)$ of $\mathcal{S}_m$ is obtained from the alphabet of $\mathcal{S}_\chi$ (see XI 7) by adjoining to L_1 $m-1$ unary propositional connectives denoted by $D_1, \ldots, D_{m-1}$ and by admitting in L_0 m propositional constants denoted by $e_0, \ldots, e_{m-1}$.

Let F be the set of all formulas over A^0. By definition $\mathcal{L} = (A^0, F)$. For any formulas α, β of $\mathcal{L}$ we shall write for brevity $(\alpha \equiv \beta)$ instead of $((\alpha \Rightarrow \beta) \cap (\beta \Rightarrow \alpha))$.

Let us assume that the set $\mathcal{A}_l$ of logical axioms consists of all formulas of the form (A_1)-(A_8) in X 1, (A_9), (A_{10}) in XI 7, where α, β, γ are any formulas of the language $\mathcal{L}$ defined above, and, moreover, of all formulas of the following form:

(A_{11}) $\quad (D_i(\alpha \cup \beta) \equiv (D_i\alpha \cup D_i\beta)), \; i = 1, \ldots, m-1,$

(A_{12}) $\quad (D_i(\alpha \cap \beta) \equiv (D_i\alpha \cap D_i\beta)), \; i = 1, \ldots, m-1,$

(A_{13}) $\quad (D_i(\alpha \Rightarrow \beta) \equiv ((\ldots (D_1\alpha \Rightarrow D_1\beta) \cap \ldots) \cap (D_i\alpha \Rightarrow D_i\beta))),$
$\quad i = 1, \ldots, m-1,$

(A_{14}) $\quad (D_i \neg \alpha \equiv \neg D_1 \alpha), \; i = 1, \ldots, m-1,$

(A_{15}) $\quad (D_i D_j \alpha \equiv D_j \alpha), \; i, j = 1, \ldots, m-1,$

(A_{16}) $\quad D_i e_j$ for $i \leqslant j$ and $\neg D_i e_j$ for $i > j$, $i = 1, \ldots, m-1,$
$\quad j = 0, \ldots, m-1,$

(A_{17}) $\quad (\alpha \equiv (\ldots (D_1 \alpha \cap e_1) \cup \ldots \cup (D_{m-1}\alpha \cap e_{m-1}))),$

(A_{18}) $\quad (D_1 \alpha \cup \neg D_1 \alpha).$

Let (r_m) be the following rule of inference

$$(r_m) \qquad \frac{\alpha}{D_{m-1}\alpha} \quad \text{for every } \alpha \text{ in } F.$$

The consequence operation $C_{\mathcal{L}}$ is determined by the set $\mathcal{A}_l$ of logical axioms and by the set of rules of inference made up of modus ponens and the rule (r_m). The system $\mathcal{S}_m = (\mathcal{L}, C_{\mathcal{L}})$ is said to be a *propositional calculus of m-valued logic*.

1.1. *For all formulas* $\alpha, \beta \in F$ *and for every set* $\mathcal{A} \subset F$,

(1) *if* $(\alpha \Rightarrow \beta) \in C_{\mathcal{L}}(\mathcal{A})$, *then* $(D_i\alpha \Rightarrow D_i\beta) \in C_{\mathcal{L}}(\mathcal{A})$,
$i = 1, \ldots, m-1.$

Suppose that $(\alpha \Rightarrow \beta) \in C_{\mathscr{L}}(\mathscr{A})$. Then, by (r_m), $D_{m-1}(\alpha \Rightarrow \beta) \in C_{\mathscr{L}}(\mathscr{A})$. Hence, by (A_{13}), (A_6), (A_7) and modus ponens, we get $(D_i\alpha \Rightarrow D_i\beta) \in C_{\mathscr{L}}(\mathscr{A})$, $i = 1, \ldots, m-1$.

1.2. *For any formulas α, β in F*

(2) $$((\alpha \Rightarrow \beta) \Rightarrow (\neg\beta \Rightarrow \neg\alpha)) \in C_{\mathscr{L}}(O).$$

The proof analogous to that of XI 1.1 (iii) is omitted.

1.3. *The system $\mathscr{S}_m = (\mathscr{L}, C_{\mathscr{L}})$ belongs to the class* **S** *of standard systems of implicative extensional propositional calculi.*

Conditions VIII 5 $((s_1), (s_2))$ are obviously satisfied. Conditions VIII 5 $((s_3)$-$(s_6), (s_8))$ hold by X 1.1 and X 1.2. By 1.1, 1.2 and modus ponens condition VIII 5 (s_7) is also satisfied. This completes the proof.

2. $\mathscr{S}_m$-algebras. The main purpose of this section is to prove the following theorem:

2.1. *The class of all $\mathscr{S}_m$-algebras coincides with the class of all Post algebras of order m* [5].

Suppose that an abstract algebra $\mathfrak{A} = (A, V, \Rightarrow, \cup, \cap, \neg, D_1, \ldots, D_{m-1}, e_0, \ldots, e_{m-1})$ is an $\mathscr{S}_m$-algebra. It is easy to show, by a similar argument to that used in the proof of X 2.1, that $(A, V, \Rightarrow, \cup, \cap)$ is a relatively pseudo-complemented lattice. Moreover, by VIII 6 (a_1) and the axioms (A_9), (A_{10}), we infer that, for all a, b in A,

(1) $$(a \Rightarrow \neg b) \Rightarrow (b \Rightarrow \neg a) = V,$$

(2) $$\neg(a \Rightarrow a) \Rightarrow b = V.$$

It follows from (1) and IV 1.3 ((2), (3)) that

(3) $$a \Rightarrow \neg b = b \Rightarrow \neg a.$$

Thus, $(A, V, \Rightarrow, \cup, \cap, \neg)$ is a pseudo-Boolean algebra (see IV 6), i.e. condition VII 1 (p_0) holds. Note that in every relatively pseudo-complemented lattice for all elements a, b we have

(4) $$a \cap b = V \quad \text{if and only if} \quad a = V \text{ and } b = V.$$

It follows from VII 6 (a_1), the axioms (A_{11})-(A_{18}), (4), IV 1.3 ((2), (3)), IV 6 (6) and, IV 6 (1) that conditions (p_1)-(p_8) in VII 1 are also satisfied. Thus $\mathfrak{A}$ is a Post algebra of order m.

[5] Rousseau [2], [4].

Now suppose that $\mathfrak{A}$ is a Post algebra of order m. By II 2 $((\mathrm{p}_1), (\mathrm{p}_2))$, IV 1.4, IV 6 ((c), (s)), IV 1 (1) the condition VIII 6 (a_1) holds for the axioms (A_1)-(A_{10}) in $\mathscr{A}_I$ (see Sec. 1), since $(A, \vee, \Rightarrow, \cup, \cap, \neg)$ is a pseudo-Boolean algebra. This condition is also satisfied for the axioms (A_{11})–(A_{18}) by VII 1 $((\mathrm{p}_0)$–$(\mathrm{p}_8))$, IV 1 (1), (4), IV 6 (1), IV 6 (6).

Observe that, by IV 6 and IV 1.4,

(5) $(A, \vee, \Rightarrow)$ is a positive implication algebra.

Hence, by II 2 (1), condition VIII 6 (a_2) holds for modus ponens. It follows from VII 1 (p_6) and VII 1 (6) that $D_{m-1}(\vee) = \vee$. Hence condition VIII 6 (a_2) is also satisfied for (r_m). The conditions VIII 6 $((\mathrm{a}_3), (\mathrm{a}_4))$ hold by (5), II 2.2, II 2 (p_3). Thus $\mathfrak{A}$ is an $\mathscr{S}_m$-algebra.

The next statement follows from 2.1 and VIII 6.1.

2.2. *If α is a formula derivable in $\mathscr{S}_m$, then for every valuation v of $\mathscr{L}$ in any Post algebra $\mathfrak{P} = (P, \vee, \Rightarrow, \cup, \cap, \neg, D_1, \ldots, D_{m-1}, e_0, \ldots, e_{m-1})$*

(6) $$\alpha_{\mathfrak{P}}(v) = \vee.$$

By 2.2

2.3. *The propositional calculus $\mathscr{S}_m$ is consistent.*

Indeed, no propositional variable is derivable in $\mathscr{S}_m$.

3. m-valued logic $\mathbf{L}_m$. By 1.3, for each positive integer $m \geqslant 2$, the system $\mathscr{S}_m$ belongs to the class $\mathbf{S}$ of standard systems of implicative extensional propositional calculi and by 2.3 it is consistent. Let $\mathbf{L}_m$ $(m \geqslant 2)$ be the class of all consistent systems in $\mathbf{S}$ logically equivalent to $\mathscr{S}_m$. The class $\mathbf{L}_m$ will be said to be *m-valued logic*.

By the definition of logically equivalent systems in $\mathbf{S}$ (see VIII 8) and by 2.1, the class $\mathbf{K}_{\mathbf{L}_m}$ of all $\mathscr{S}$-algebras for $\mathscr{S}$ in $\mathbf{L}_m$ coincides with the class of all Post algebras of order m. Hence, by VIII 6.5 and VIII 6.6, we obtain the following corollary.

3.1. *For every $\mathscr{S}$ in $\mathbf{L}_m$, the algebra $\mathfrak{A}(\mathscr{S})$ is a non-degenerate Post algebra of order m. Moreover, it is a free algebra in the class $\mathbf{K}_{\mathbf{L}_m}$ of all Post algebras of order m.*

A formula α of any system $\mathscr{S} = (\mathscr{L}, C_{\mathscr{L}})$ in $\mathbf{L}_m$ is said to be $\mathbf{L}_m$-*valid* provided for every valuation v of $\mathscr{L}$ in any Post algebra $\mathfrak{P}$ of order m

(1) $$\alpha_{\mathfrak{P}}(v) = V.$$

The following theorem characterizes all formulas derivable in any system $\mathscr{S} = (\mathscr{L}, C_{\mathscr{L}})$ in $\mathbf{L}_m$.

3.2. *For any formula α of any system $\mathscr{S} = (\mathscr{L}, C_{\mathscr{L}})$ of the m-valued logic $\mathbf{L}_m$ the following conditions are equivalent:*

(i) *α is derivable in $\mathscr{S}$,*

(ii) *α is $\mathbf{L}_m$-valid,*

(iii) *α is valid in every Post field $\mathfrak{P}(X)$ (of order m) of sets,*

(iv) *$\alpha_{\mathfrak{A}(\mathscr{S})}(v^0) = V$ for the canonical valuation v^0 of $\mathscr{L}$ in the Post algebra $\mathfrak{A}(\mathscr{S})$ of $\mathscr{S}$,*

(v) *α is valid in the m-element Post algebra $\mathfrak{P}_m$ of order m* [(6)].

Conditions (i) and (ii) are equivalent by VIII 8.1. Clearly, (ii) implies (iii). By 3.1 and the representation theorem VII 6.1, (iii) implies (iv). By VIII 7.2, (iv) implies (i). Since (ii) implies (v), it suffices to show that (v) implies (i).

Suppose that $\alpha \notin C_{\mathscr{L}}(O)$. Thus (see VIII 6 (8)) the element $||\alpha||$ of the Post algebra $\mathfrak{A}(\mathscr{S})$ satisfies the condition

(2) $$||\alpha|| \neq V.$$

Since by VII 1.2 (9), $D_{m-1}(||\alpha||) \leqslant ||\alpha||$, it follows from (2) that

(3) $$D_{m-1}(||\alpha||) \neq V.$$

Let ∇_0 be the D-filter consisting of the one element V. By (3), $D_{m-1}(||\alpha||) \notin \nabla_0$. Hence, by VII 3.8 and VII 3.4, there exists a prime D-filter ∇ such that $||\alpha|| \notin \nabla$. By VII 4.5, the quotient algebra $\mathfrak{A}(\mathscr{S})/\nabla$ is the m-element Post algebra of order m. Let h be the epimorphism from $\mathfrak{A}(\mathscr{S})$ onto $\mathfrak{A}(\mathscr{S})/\nabla$ determined by ∇. Since $||\alpha|| \notin \nabla$, we infer by VII 4.2 that

(4) $$h||\alpha|| \neq V \quad \text{in } \mathfrak{A}(\mathscr{S})/\nabla.$$

(6) Rousseau [2], [4].

Let v^0 be the canonical valuation of $\mathscr{L}$ in $\mathfrak{A}(\mathscr{S})$. Then, by VIII 3.2, VIII 9 (3) and (4), for the valuation hv^0 of $\mathscr{L}$ in $\mathfrak{A}(\mathscr{S})/\nabla$ we get

$$\alpha_{\mathfrak{A}(\mathscr{S})/\nabla}(hv^0) = h\alpha_{\mathfrak{A}(\mathscr{S})}(v^0) = h\|\alpha\| \neq V.$$

This completes the proof that (v) implies (i).

The equivalence of conditions (i) and (v) is called the *completeness theorem for the propositional calculi of m-valued logic* $\mathbf{L}_m$.

The following corollary follows from this completeness theorem.

3.3. *Every propositional calculus* $\mathscr{S}$ *of m-valued logic* $\mathbf{L}_m$ *is decidable.*

Many examples of formulas derivable in any system $\mathscr{S}$ of m-valued logic can be given by using the completeness theorem.

4. $\mathbf{L}_m$-theories of zero order and their connection with D-filters. Let $\mathscr{S} = (\mathscr{L}, C_{\mathscr{L}})$ be a propositional calculus of the m-valued logic $\mathbf{L}_m$. For every set $\mathscr{A}$ of formulas of $\mathscr{L}$ the system $\mathscr{S}(\mathscr{A}) = (\mathscr{L}, C_{\mathscr{L}}, \mathscr{A})$ is said to be an $\mathbf{L}_m$*-theory of zero order based on* $\mathscr{S}$.

By an $\mathbf{L}_m$*-model* of an $\mathbf{L}_m$-theory $\mathscr{S}(\mathscr{A}) = (\mathscr{L}, C_{\mathscr{L}}, \mathscr{A})$ we mean any valuation v of $\mathscr{L}$ in any non-degenerate Post algebra $\mathfrak{P} = (P, V, \Rightarrow, \cup, \cap, \neg, D_1, \ldots, D_{m-1}, e_0, \ldots, e_{m-1})$ of order m such that for every α in $\mathscr{A}$

(1) $$\alpha_{\mathfrak{P}}(v) = V.$$

Every $\mathbf{L}_m$-model of an $\mathbf{L}_m$-theory $\mathscr{S}(\mathscr{A})$ in the m-element Post algebra $\mathfrak{P}_m$ of order m (see VII 2) will be called an *m-valued semantic model.*

The following theorem follows directly from VIII 9.2, VIII 9.3 and 2.1.

4.1. *For every consistent* $\mathbf{L}_m$*-theory* $\mathscr{S}(\mathscr{A}) = (\mathscr{L}, C_{\mathscr{L}}, \mathscr{A})$ *of zero order the algebra* $\mathfrak{A}(\mathscr{S}(\mathscr{A}))$ (see VIII 9) *is a non-degenerate Post algebra of order m. The canonical valuation* v^0 *of* $\mathscr{L}$ *in* $\mathfrak{A}(\mathscr{S}(\mathscr{A}))$ *is an* $\mathbf{L}_m$*-model of* $\mathscr{S}(\mathscr{A})$*. Moreover, this* $\mathbf{L}_m$*-model is adequate for* $\mathscr{S}(\mathscr{A})$, i.e. *for every formula* α *of* $\mathscr{L}$

(2) $$\alpha \in C_{\mathscr{L}}(\mathscr{A}) \quad \textit{if and only if} \quad \alpha_{\mathfrak{A}(\mathscr{S}(\mathscr{A}))}(v^0) = V.$$

Observe that

(3) $$\mathbf{L}_m \subset \mathbf{S}_\nu,$$

since it is easy to verify by using 3.2 (the equivalence of (i) and (ii)) that for any $\mathscr{S}$ in $\mathbf{L}_m$ all formulas of the form $(\neg(\alpha \Rightarrow \alpha) \Rightarrow \beta)$ (where, α, β are any formulas of $\mathscr{S}$) are derivable in $\mathscr{S}$.

The next theorem characterizes consistent $\mathbf{L}_m$-theories of zero order.

4.2. *For any* $\mathbf{L}_m$*-theory* $\mathscr{S}(\mathscr{A}) = (\mathscr{L}, C_{\mathscr{L}}, \mathscr{A})$ *of zero order the following conditions are equivalent*:

(i) $\mathscr{S}(\mathscr{A})$ *is consistent,*

(ii) *for any formula* α *of* $\mathscr{S}(\mathscr{A})$ *either* $\alpha \notin C_{\mathscr{L}}(\mathscr{A})$ *or* $\neg\alpha \notin C_{\mathscr{L}}(\mathscr{A})$,

(iii) *there exists an* $\mathbf{L}_m$*-model of* $\mathscr{S}(\mathscr{A})$,

(iv) *there exists a model of* $\mathscr{S}(\mathscr{A})$ *in a Post field (of order* m*) of sets,*

(v) *there exists an* m*-valued semantic model of* $\mathscr{S}(\mathscr{A})$.

It follows from (3), VIII 10.2, 4.1 and the representation theorem VII 6.1 that conditions (i)–(iv) are equivalent. Since (v) implies (iii), it is sufficient to prove that (iii) implies (v). Suppose that a valuation v of $\mathscr{L}$ in a non-degenerate Post algebra $\mathfrak{P}$ of order m is an $\mathbf{L}_m$-model of $\mathscr{S}(\mathscr{A})$. It follows from VII 3.6 and VII 3.4 that there exists a prime D-filter ∇ in $\mathfrak{P}$. By VII 4.5, the quotient algebra $\mathfrak{P}/\nabla$ is the m-element Post algebra of order m. Let h be the epimorphism from $\mathfrak{P}$ onto $\mathfrak{P}/\nabla$ determined by ∇. It follows from VIII 3.2 and (1) that the valuation hv of $\mathscr{L}$ in $\mathfrak{A}/\nabla$ is an m-valued semantic model of $\mathscr{S}(\mathscr{A})$.

The next theorem gives a characterization of the theorems of any consistent $\mathbf{L}_m$-theory of zero order.

4.3. *For each formula* α *of a consistent* $\mathbf{L}_m$*-theory* $\mathscr{S}(\mathscr{A}) = (\mathscr{L}, C_{\mathscr{L}}, \mathscr{A})$ *of zero order the following conditions are equivalent*:

(i) $\alpha \in C_{\mathscr{L}}(\mathscr{A})$,

(ii) *every* $\mathbf{L}_m$*-model of* $\mathscr{S}(\mathscr{A})$ *is an* $\mathbf{L}_m$*-model of* α,

(iii) *every* $\mathbf{L}_m$*-model of* $\mathscr{S}(\mathscr{A})$ *in any non-degenerate Post field* $\mathfrak{P}(X)$ *(of order* m*) of sets is an* $\mathbf{L}_m$*-model of* α,

(iv) $\alpha_{\mathfrak{A}(\mathscr{S}(\mathscr{A}))}(v^0) = \bigvee$ *for the canonical calculation* v^0 *of* $\mathscr{L}$ *in the Post algebra* $\mathfrak{A}(\mathscr{S}(\mathscr{A}))$ *of the theory* $\mathscr{S}(\mathscr{A})$,

(v) $\alpha_{\mathfrak{P}_m}(v) = \bigvee$ *for every* m*-valued semantic model* v *of* $\mathscr{S}(\mathscr{A})$.

Conditions (i), (ii), (iii) and (iv) are equivalent by VIII 9.5 and the representation theorem VII 6.1. Since (ii) implies (v), it is sufficient to prove that (v) implies (i). If $\alpha \notin C_{\mathscr{L}}(\mathscr{A})$, then the element $||\alpha||_{\mathscr{A}}$ of $\mathfrak{A}(\mathscr{S}(\mathscr{A}))$ is different from the unit element V (see VIII 9 (9)). Hence, by VII 1.2 (9), $D_{m-1}(||\alpha||_{\mathscr{A}}) \neq V$. Let ∇_0 be the D-filter in $\mathfrak{A}(\mathscr{S}(\mathscr{A}))$ consisting of the one element V. Then $D_{m-1}(||\alpha||_{\mathscr{A}}) \notin \nabla_0$. Hence, by VII 3.8 and VII 3.4, there exists a prime D-filter ∇ such that $||\alpha||_{\mathscr{A}} \notin \nabla$. By VII 4.5 the quotient algebra $\mathfrak{A}(\mathscr{S}(\mathscr{A}))/\nabla$ is the m-element Post algebra of order m. Let h be the epimorphism from $\mathfrak{A}(\mathscr{S}(\mathscr{A}))$ onto $\mathfrak{A}(\mathscr{S}(\mathscr{A}))/\nabla$ determined by ∇. Since $||\alpha||_{\mathscr{A}} \notin \nabla$, we infer by VII 4.2 that

$$(4) \qquad h(||\alpha||_{\mathscr{A}}) \neq V \text{ in } \mathfrak{A}(\mathscr{S}(\mathscr{A}))/\nabla.$$

By VIII 3.2 and (1) the valuation hv^0 of $\mathscr{L}$ in $\mathfrak{A}(\mathscr{S}(\mathscr{A}))/\nabla$ (where v^0 is the canonical valuation of $\mathscr{L}$ in $\mathfrak{A}(\mathscr{S}(\mathscr{A}))$) is an m-valued semantic model of $\mathscr{S}(\mathscr{A})$. On the other hand, by VIII 3.2, VIII 9 (13) and (4), $\alpha_{\mathfrak{A}(\mathscr{S}(\mathscr{A}))/\nabla}(hv^0) = h\alpha_{\mathfrak{A}(\mathscr{S}(\mathscr{A}))}(v^0) = h||\alpha||_{\mathscr{A}} \neq V$.

The following theorem is the basis for the study of the connection between $\mathbf{L}_m$-theories of zero order and D-filters.

4.4. *For every set* ∇ *of elements in a Post algebra* $\mathfrak{P} = (P, V, \Rightarrow, \cup, \cap, \neg, D_1, \ldots, D_{m-1}, e_0, \ldots, e_{m-1})$ *of order* m *the following conditions are equivalent*:

(i) ∇ *is a* D*-filter*,

(ii) ∇ *is a deductive filter*.

Recall that the following rules of inference are adopted in $\mathscr{S}_m$: modus ponens and (r_m) (see Sec. 1). Consequently, it follows from VIII 12.3 that ∇ is a deductive filter if and only if it is an implicative filter and for each element a

$$(5) \qquad \text{if} \quad a \in \nabla, \quad \text{then} \quad D_{m-1}(a) \in \nabla.$$

Since the notion of an implicative filter in Post algebras coincides with the notion of a filter, conditions (i) and (ii) are equivalent by the definition of D-filters given in VII 3 and VII 1.2 (7).

By VII 4.1 and 4.2 the kernels of epimorphisms from Post algebras of order m onto similar algebras are D-filters and, conversely, every D-filter is the kernel of the epimorphism determined by itself. This follows also from 4.4, VIII 12.2, VIII 13.10 and the equational definability of the class of Post algebras of order m.

By 4.4, VIII 13.2 and VIII 13.4 we get two theorems which state that there exists a natural one-one correspondence between D-filters in the Post algebra $\mathfrak{A}(\mathscr{S})$ (of order m) of a propositional calculus $\mathscr{S} = (\mathscr{L}, C_{\mathscr{L}})$ in $\mathbf{L}_m$ and the formalized $\mathbf{L}_m$-theories of zero order based on $\mathscr{S}$, two theories $\mathscr{S}(\mathscr{A})$ and $\mathscr{S}(\mathscr{A}')$ being identified if and only if $C_{\mathscr{L}}(\mathscr{A}) = C_{\mathscr{L}}(\mathscr{A}')$.

Let us note that for any system $\mathscr{S}$ in $\mathbf{L}_m$ there holds a theorem analogous to IX 5.4. Clearly, in its formulation the notion of a filter is replaced by the notion of a D-filter. The deduction theorem corresponding to IX 5.5 has the following form.

4.5. *For any formulas α, β in the formalized language $\mathscr{L}$ of a system $\mathscr{S} = (\mathscr{L}, C_{\mathscr{L}})$ in $\mathbf{L}_m$ and for every set $\mathscr{A}$ of formulas of $\mathscr{L}$*

(6) $\quad \beta \in C_{\mathscr{L}}(\mathscr{A} \cup \{\alpha\})$ *if and only if* $(D_{m-1}\alpha \Rightarrow \beta) \in C_{\mathscr{L}}(\mathscr{A})$.

The following remark will be useful for the proof of 4.5 and also later:

(7) *for each formula γ of $\mathscr{L}$, $\gamma \in C_{\mathscr{L}}(\mathscr{A})$ if and only if $D_{m-1}\gamma \in C_{\mathscr{L}}(\mathscr{A})$.*

Indeed, by 3.2, VII 1.2 (9), VII 1 (p_0), IV 6 and IV 1.3 (2), for any formula γ of $\mathscr{L}$, $(D_{m-1}\gamma \Rightarrow \gamma) \in C_{\mathscr{L}}(O)$. Thus, if $D_{m-1}\gamma \in C_{\mathscr{L}}(\mathscr{A})$, then $\gamma \in C_{\mathscr{L}}(\mathscr{A})$ (cf. VIII 5 (s_4)). On the other hand, since in each Post algebra of order m we have $D_{m-1}(V) = V$ (see VII 1 (p_6), VII 1.2 (6)), it follows from 4.3, that the condition $\gamma \in C_{\mathscr{L}}(\mathscr{A})$ implies that $D_{m-1}\gamma \in C_{\mathscr{L}}(\mathscr{A})$.

To prove 4.5 suppose that $(D_{m-1}\alpha \Rightarrow \beta) \in C_{\mathscr{L}}(\mathscr{A}) \subset C_{\mathscr{L}}(\mathscr{A} \cup \{\alpha\})$. By (7), $D_{m-1}\alpha \in C_{\mathscr{L}}(\mathscr{A} \cup \{\alpha\})$. Hence, by VIII 5 ($s_4$), $\beta \in C_{\mathscr{L}}(\mathscr{A} \cup \{\alpha\})$. Conversely, if $\beta \in C_{\mathscr{L}}(\mathscr{A} \cup \{\alpha\})$, then the element $||\beta||$ of $\mathfrak{A}(\mathscr{S})$ belongs to the D-filter $\nabla_{\mathscr{A} \cup \{\alpha\}}$ composed of all $||\gamma||$ such that $\gamma \in C_{\mathscr{L}}(\mathscr{A} \cup \{\alpha\})$. By a theorem analogous to IX 5.4, $||\beta||$ belongs to the D-filter generated by the D-filter $\nabla_{\mathscr{A}}$ and $||\alpha||$. Hence, by VII 3.2, there exists a $||\gamma|| \in \nabla_{\mathscr{A}}$ such that $D_{m-1}(||\alpha||) \cap D_{m-1}(||\gamma||) \leqslant ||\beta||$. This condition is equivalent to $D_{m-1}(||\gamma||) \leqslant D_{m-1}(||\alpha||) \Rightarrow ||\beta||$ (see VII 1 (p_0), IV 6, IV 1 (r)). Since $\nabla_{\mathscr{A}}$ is a D-filter and $||\gamma|| \in \nabla_{\mathscr{A}}$, we infer that $D_{m-1}(||\gamma||) \in \nabla_{\mathscr{A}}$. Consequently, $||(D_{m-1}\alpha \Rightarrow \beta)|| = D_{m-1}(||\alpha||) \Rightarrow ||\beta|| \in \nabla_{\mathscr{A}}$. Thus $(D_{m-1}\alpha \Rightarrow \beta) \in C_{\mathscr{L}}(\mathscr{A})$.

The deduction theorem, analogous to X 6.4, has the following form.

4.6. *For each formula* α *of the formalized language* $\mathscr{L}$ *of a system* $\mathscr{S} = (\mathscr{L}, C_{\mathscr{L}})$ *in* $\mathbf{L}_m$ *and for each non-empty set* $\mathscr{A}$ *of formulas of* $\mathscr{L}$ *the condition* $\alpha \in C_{\mathscr{L}}(\mathscr{A})$ *is equivalent to the condition that there exist formulas* $\alpha_1, \ldots, \alpha_n$ *in* $\mathscr{A}$ *such that*

$$(8) \qquad ((D_{m-1}\alpha_1 \cap \ldots \cap (D_{m-1}\alpha_{n-1} \cap D_{m-1}\alpha_n) \ldots) \Rightarrow \alpha) \in C_{\mathscr{L}}(O).$$

The easy proof, analogous to that of X 6.4 and based on (7), the theorem corresponding to X 6.2 and VII 3.1, is omitted.

By 4.4 the notions of an irreducible deductive filter and a maximal deductive filter in Post algebras coincide with the notions of an irreducible D-filter and a maximal D-filter, respectively.

The following theorem follows from 4.4, VIII 13.5 and VIII 13.6.

4.7. *An* $\mathbf{L}_m$*-theory* $\mathscr{S}(\mathscr{A}) = (\mathscr{L}, C_{\mathscr{L}}, \mathscr{A})$ *is irreducible (maximal) if and only if the* D*-filter* $\nabla_{\mathscr{A}}$ *(consisting of all elements* $||\alpha||$ *in* $\mathfrak{A}(\mathscr{S})$ *such that* $\alpha \in C_{\mathscr{L}}(\mathscr{A}_{\mathscr{L}})$*) is irreducible (maximal).*

The next theorem follows from (3) and VIII 13.8.

4.8. *For every consistent* $\mathbf{L}_m$*-theory* $\mathscr{S}(\mathscr{A}) = (\mathscr{L}, C_{\mathscr{L}}, \mathscr{A})$ *of zero order there exists a maximal* $\mathbf{L}_m$*-theory* $\mathscr{S}(\mathscr{A}') = (\mathscr{L}, C_{\mathscr{L}}, \mathscr{A}')$ *such that* $C_{\mathscr{L}}(\mathscr{A}) \subset C_{\mathscr{L}}(\mathscr{A}')$.

A consistent $\mathbf{L}_m$-theory $\mathscr{S}(\mathscr{A}) = (\mathscr{L}, C_{\mathscr{L}}, \mathscr{A})$ is said to be *prime* if for any formulas α, β of $\mathscr{L}$:

(9) if $(\alpha \cup \beta) \in C_{\mathscr{L}}(\mathscr{A})$, then either $\alpha \in C_{\mathscr{L}}(\mathscr{A})$ or $\beta \in C_{\mathscr{L}}(\mathscr{A})$.

It follows from the above definition and the definition of prime D-filters in Post algebras (see VII 3), 4.4, VIII 13.1 that

4.9. *An* $\mathbf{L}_m$*-theory* $\mathscr{S}(\mathscr{A}) = (\mathscr{L}, C_{\mathscr{L}}, \mathscr{A})$ *is prime if and only if the* D*-filter* $\nabla_{\mathscr{A}}$ *in* $\mathfrak{A}(\mathscr{S})$ *is prime.*

The following corollary follows from 4.7, 4.9 and VII 3.4.

4.10. *For every* $\mathbf{L}_m$*-theory* $\mathscr{S}(\mathscr{A}) = (\mathscr{L}, C_{\mathscr{L}}, \mathscr{A})$ *of zero order the following conditions are equivalent:*

(i) $\mathscr{S}(\mathscr{A})$ *is maximal,*

(ii) $\mathscr{S}(\mathscr{A})$ *is irreducible,*

(iii) $\mathscr{S}(\mathscr{A})$ *is prime.*

A characterization of maximal $\mathbf{L}_m$-theories is given by the following theorem.

4.11. *The following conditions are equivalent for every* $\mathbf{L}_m$*-theory* $\mathcal{S}(\mathcal{A}) = (\mathcal{L}, C_{\mathcal{L}}, \mathcal{A})$ *of zero order*:

(i) $\mathcal{S}(\mathcal{A})$ *is maximal,*

(ii) *for each formula* α *of* $\mathcal{L}$ *exactly one of the formulas* $D_{m-1}\alpha$, $\neg D_{m-1}\alpha$ *is a theorem of* $\mathcal{S}(\mathcal{A})$,

(iii) $\mathfrak{A}(\mathcal{S}(\mathcal{A}))$ *is isomorphic to the* m*-element Post algebra of order* m,

(iv) $\mathcal{S}(\mathcal{A})$ *has an adequate* m*-valued semantic model,*

(v) $\mathcal{S}(\mathcal{A})$ *is consistent and each* m*-valued semantic model of* $\mathcal{S}(\mathcal{A})$ *is adequate.*

By VIII 13.1, 4.4, 4.7, VII 3.4 conditions (i) and (ii) are equivalent. By 4.7, VII 3.4, VII 4.5 and VIII 13.11 conditions (i), (iii) are equivalent. By 4.1, (iii) implies (iv). Suppose that (iv) holds. Then, by 4.2, $\mathcal{S}(\mathcal{A})$ is consistent and has an m-valued semantic model. Observe that if v is an m-valued semantic model of $\mathcal{S}(\mathcal{A})$, then (see VII 2) for each formula α of $\mathcal{L}$

(10) either $D_{m-1}\alpha_{\mathfrak{P}_m}(v) = V$ or $\neg D_{m-1}\alpha_{\mathfrak{P}_m}(v) = V$.

Since the same condition is satisfied for an adequate m-valued semantic model of $\mathcal{S}(\mathcal{A})$, we infer that for each formula α of $\mathcal{L}$ either $D_{m-1}\alpha \in C_{\mathcal{L}}(\mathcal{A})$ or $\neg D_{m-1}\alpha \in C_{\mathcal{L}}(\mathcal{A})$. Let v be an arbitrary m-valued semantic model of $\mathcal{S}(\mathcal{A})$. If $\alpha \in C_{\mathcal{L}}(\mathcal{A})$, then by 4.3, $\alpha_{\mathfrak{P}_m}(v) = V$. If $\alpha \notin C_{\mathcal{L}}(\mathcal{A})$, then by (7), $D_{m-1}\alpha \notin C_{\mathcal{L}}(\mathcal{A})$. Hence, $\neg D_{m-1}\alpha \in C_{\mathcal{L}}(\mathcal{A})$. By 4.3, $\neg D_{m-1}\alpha_{\mathfrak{P}_m}(v) = V$. Hence (see VII 2), $D_{m-1}\alpha_{\mathfrak{P}_m}(v) = e_0$ and consequently, $\alpha_{\mathfrak{P}_m}(v) \neq e_{m-1} = V$. Thus the m-valued semantic model v is adequate. We have just proved that (iv) implies (v). It suffices to show that (v) implies (ii). Suppose that (v) holds. Since $\mathcal{S}(\mathcal{A})$ is consistent, by 4.2 there exists an m-valued semantic model v of $\mathcal{S}(\mathcal{A})$. By (v) it is adequate. Hence, by (10) and 4.2 (ii), for each formula α of $\mathcal{L}$ exactly one of the formulas $D_{m-1}\alpha$, $\neg D_{m-1}\alpha$ is a theorem of $\mathcal{S}(\mathcal{A})$. Thus (v) implies (ii), which completes the proof of 4.11.

Exercises

1. Let $\mathfrak{A} = (A, V, o_A)$ be an abstract algebra such that $A = \{V, a_1, \ldots, a_n\}$ and $o_A\colon A \times A \to A$. Consider a formalized language $\mathscr{L}$, whose alphabet contains exactly one binary connective o and no others. A formula α of $\mathscr{L}$ is said to be *valid* in $\mathfrak{A}$ provided for every valuation v of $\mathscr{L}$ in $\mathfrak{A}$, $\alpha_{\mathfrak{A}}(v) = V$. Prove that the set of all formulas of $\mathscr{L}$, valid in $\mathfrak{A}$, is empty if and only if there exists a non-empty set $X \subset A - \{V\}$ such that for all $x, y \in X$, $o_A(x, y) \in X$ (7).

2. The following algebra $\mathfrak{A}_m = (E, V, \rightarrow_E, \sim_E)$, where $m \geqslant 3$, $E = \{e_0, \ldots, e_{m-1}\}$, $V = e_{m-1}$, $e_i \rightarrow_E e_j = e_{m-1}$ if $i \leqslant j$, $e_i \rightarrow_E e_j = e_{(m-1)-i+j}$ if $i > j$, and $\sim_E e_i = e_{(m-1)-i}$ for $i, j \in \{0, 1, \ldots, m-1\}$, defines implication and negation in the m-valued logic of Łukasiewicz. Define these connectives by means of the connectives occurring in $\mathscr{S}_m$.

3. Prove that for each formula α of the formalized language of $\mathscr{S}_\chi$ (see XI 1) the following conditions are equivalent: (i) α is derivable in each propositional calculus $\mathscr{S}_m$, $m = 2, 3, \ldots$; (ii) α is derivable in the propositional calculus obtained from $\mathscr{S}_\chi$ by adjoining to the logical axioms all formulas of the form $((\beta \Rightarrow \gamma) \cup (\gamma \Rightarrow \beta))$. Apply: Ex. 10 in Chap. XI, 3.2.

4. Consider the propositional calculus $\mathscr{S}_{m\chi}$ obtained from $\mathscr{S}_m$, $m = 2, 3, \ldots$, by the elimination of the axiom scheme (A_{18}) in Sec. 1. Prove that the class of all $\mathscr{S}_{m\chi}$-algebras coincides with the class of all pseudo-Post algebras of order m (see Ex. 3 in Chap. VII). Define the logic $\mathbf{L}_{m\chi}$ as the class of all consistent propositional calculi in $\mathscr{S}$ logically equivalent to $\mathscr{S}_{m\chi}$, $m = 2, 3, \ldots$, and examine the properties of the propositional calculi in $\mathbf{L}_{m\chi}$ and of the $\mathbf{L}_{m\chi}$-theories of zero order (8).

5. Prove that for each formula α of $\mathscr{S}_\chi$ (see XI 1), α is derivable in $\mathscr{S}_\chi$ if and only if α is derivable in every $\mathscr{S}_{m\chi}$, $m = 2, 3, \ldots$, as defined in Ex. 4 (9).

(7) Rasiowa [1].

(8) Those propositional calculi were constructed and examined by Rousseau [2], [4].

(9) Rousseau [4].

SUPPLEMENT

FIRST ORDER PREDICATE CALCULI OF NON-CLASSICAL LOGICS

Introduction. Every logic **L**, as defined in Chapter VIII 8, determines a class of first order predicate calculi of **L**, by a method which is analogous to that used in the case of classical logic. This method can be described briefly as follows. For any logic **L** we define a class of first order formalized languages asociated with **L**. The alphabets of these languages contain the same connectives as the alphabets of the propositional calculi of **L** and, moreover, free individual variables, bound individual variables, functors, predicates, quantifiers and parentheses. The terms and formulas are constructed of elements of the alphabet, as usual. Given a first order formalized language $\mathscr{L}^*$ associated with **L**, we introduce a consequence operation $C_{\mathscr{L}^*}$ in $\mathscr{L}^*$ by choosing a set $\mathscr{A}_l^*$ of logical axioms and a set of rules of inference. The set $\mathscr{A}_l^*$ consists of all formulas in $\mathscr{L}^*$ which are substitutions of axioms of a propositional calculus $\mathscr{S}$ in **L**. The set of rules of inference is composed of rules of inference corresponding to those adopted in $\mathscr{S}$, the rule of substitution for free individual variables and four well-known rules for quantifiers: the rule of introduction of an existential quantifier, the rule of introduction of a universal quantifier, the rule of elimination of an existential quantifier, the rule of elimination of a universal quantifier. For any set $\mathscr{A}$ of formulas of $\mathscr{L}^*$, $C_{\mathscr{L}^*}(\mathscr{A})$ is the least set of formulas of $\mathscr{L}^*$ containing the union of sets $\mathscr{A}_l^*$ and $\mathscr{A}$, and closed under the rules of inference. The system $\mathscr{S}^* = (\mathscr{L}^*, C_{\mathscr{L}^*})$ is then called a predicate calculus of **L**. For any set $\mathscr{A}$ of formulas of $\mathscr{L}^*$, the system $\mathscr{S}^*(\mathscr{A}) = (\mathscr{L}^*, C_{\mathscr{L}^*}, \mathscr{A})$ is called an elementary **L**-theory based on $\mathscr{S}^*$. The Supplement is intended as an outline of a general algebraic approach [1] to predicate calculi determined by a given logic **L** and of elementary **L**-theories. This approach is analogous to that expounded in Chapter VIII concerning the algebraization of propositional calculi of a logic **L** and of **L**-theories of zero order. The exposition is a slight modification of that presented in [MM].

The idea of a generalization of the algebraic treatment of formulas of propositional calculi to formulas of predicate calculi is due to Mostowski [2], who introduced an algebraic interpretation of formulas of intuitionistic predicate calculi. Mostowski proposed to interpret: (i) free individual variables as variables ranging over a non-empty set J; (ii) m-argument predicates ($m = 1, 2, \ldots$) as mappings from J^m into a complete pseudo-Boolean algebra $\mathfrak{A}$; (iii) propositional connectives as corresponding algebraic operations in $\mathfrak{A}$; (iv) quantifiers $\bigcup\xi$, $\bigcap\xi$ as least upper bounds and greatest lower bounds in $\mathfrak{A}$, respectively.

Mostowski [2] proved that for any formula α of an intuitionistic predicate calculus, if α is derivable, then every interpretation of α in every set $J \neq O$ and in every complete pseudo-Boolean algebra $\mathfrak{A}$ is identically equal to the unit element V of $\mathfrak{A}$. The converse statement was proved by the present author [2]. An analogous algebraic characterization of formulas derivable in modal predicate calculi of $\mathbf{L}_\lambda$ was also presented in the above-mentioned paper (Rasiowa [2]). In this characterization complete topological Boolean algebras were used instead of complete pseudo-Boolean algebras. The assumption that the lattices under consideration are complete was made only in order to ensure that all the least upper bounds and greatest lower bounds appearing in an interpretation of formulas exist. However, it may happen that, by a certain interpretation of m-argument predicates as mappings from J^m into a suitable lattice $\mathfrak{A}$, all the least upper bounds and all the greatest lower bounds appearing in the interpretation of formulas exist in spite of the fact that $\mathfrak{A}$ is incomplete. This holds with a certain interpretation of formulas in a set $J \neq O$ and in the algebra $\mathfrak{A}$ of the given predicate calculus. The construction of the algebra of a predicate calculus is the same as that of the algebra of any propositional cal-

[1] This approach is a generalization of that indicated by Mostowski [2], Henkin [2], and developed by Rasiowa [2], [3], [5], [8], [9], [10], Rasiowa and Sikorski [3], [4], [5], [MM].

culus. The quantifiers appearing in formulas of a predicate calculus determine in its algebra the least upper bounds and the greatest lower bounds. This enables us to interpret these quantifiers as the corresponding least upper bounds and greatest lower bounds in the algebra of the predicate calculus, in spite of the fact that this algebra is not a complete lattice.

This idea was used independently by Henkin [2] and the present author [2] in their proofs of theorems concerning an algebraic characterization of formulas derivable in predicate calculi of positive implicative logic and of intuitionistic and modal logic, respectively. This idea enables us also to formulate and prove the completeness theorem for predicate calculi of any logic **L**. The completeness theorem states that, for any formula α of a predicate calculus $\mathscr{S}^*$ of **L**, α is derivable in $\mathscr{S}^*$ if and only if for every interpretation of the formalized language of $\mathscr{S}^*$ in any set $J \neq O$ and in any $\mathscr{S}$-algebra $\mathfrak{A}$ (for propositional calculi $\mathscr{S}$ in **L**) the interpretation of α is identically equal to the unit element of $\mathfrak{A}$.

It may happen that the completeness theorem for predicate calculi of a given logic **L** can be formulated in a stronger way by restricting the class of all $\mathscr{S}$-algebras for $\mathscr{S}$ in **L** to some of its subclasses or even to a unique $\mathscr{S}$-algebra. For instance in the case of intuitionistic predicate calculi the class of all pseudo-Boolean algebras can be restricted to the class of all complete pseudo-Boolean algebras and also to the class of all pseudo-Boolean algebras of all open subsets of any topological space (Rasiowa-Sikorski [3]). It is also possible to replace the class of all pseudo-Boolean algebras by one pseudo-Boolean algebra of all open subsets of a set of irrational numbers (Sikorski [5]). The same is true for predicate calculi of positive logic (Rasiowa-Sikorski [3], Sikorski [5]). The case of modal predicate calculi of $\mathbf{L}_\lambda$ is analogous to that of intuitionistic predicate calculi. The class of all topological Boolean algebras can be restricted to the class of all complete topological Boolean algebras and also to the class of all topological Boolean algebras of all subsets of any topological space (Rasiowa [2], Rasiowa-Sikorski [3]). Moreover, it is possible to replace the class of all topological Boolean algebras by a unique topological Boolean algebra of all subsets of a set of irrational numbers (Sikorski [5]). Formulating the completeness theorem for predicate calculi of minimal logic, we can restrict the class of all contrapositionally complemented lattices to the class of all contrapositionally complemented lattices of all open subsets of any topological space (Rasiowa-Sikorski [3]). The situation is analogous in the case of predicate calculi of positive logic with semi-negation: the class of all semi-complemented lattices can be replaced by the class of all semi-complemented lattices of all open subsets of any topological space (Rasiowa-Sikorski [3]). For predicate calculi of constructive logic with strong negation the class of all quasi-pseudo-Boolean algebras can be restricted to the class of all quasi-pseudo-Boolean algebras of sets satisfying an additional condition (Białynicki-Birula and Rasiowa [2]). In the case of m-valued logics the class of all Post algebras of order m can be replaced by the m-element Post algebra $\mathfrak{P}_m$ (Rasiowa [9]).

The notion of an algebraic model, as introduced by the present author [3], is a generalization of that of a semantic model for elementary theories of classical logic and can be applied to any elementary **L**-theory. Algebraic models of any elementary **L**-theory will be called **L**-*models* in this book. A theorem common to all elementary **L**-theories states that every consistent **L**-theory has an adequate **L**-model. Moreover, if **L** is a logic with semi-negation, then the existence of an **L**-model for an **L**-theory is equivalent to its consistency. Much more specialized theorems concerning **L**-models can be formulated in a similar way, as has been indicated above by considering various formulations of the completeness theorem.

The application of the algebraic method in the methodology of predicate calculi and of elementary **L**-theories enables us to reduce metalogical problems to problems concerning the theory of $\mathscr{S}$-algebras (for $\mathscr{S}$ in **L**) with least upper bounds and greatest lower bounds. Several results in this field pertaining to logics $\mathbf{L}_\pi$, $\mathbf{L}_\mu$, $\mathbf{L}_\nu$, $\mathbf{L}_\chi$, $\mathbf{L}_N$, $\mathbf{L}_\varkappa$, $\mathbf{L}_\lambda$ and $\mathbf{L}_m$ are to be found e.g. in [MM] and in the following papers: Rasiowa-Sikorski [3], [4], [5], Białynicki-Birula and Rasiowa [2], Rasiowa [3]-[12]. The exposition of results concerned with particular cases of these logics requires a broad development of the theories of corresponding classes of algebras, which exceeds the framework of this book.

The Supplement contains only some theorems common to all predicate calculi of any logic **L** and for all elementary **L**-theories. For the sake of simplicity, certain algebraic aspects are omitted. First-order formalized languages connected with a logic **L** are described in Section 1. First order predicate calculi of a logic **L** and elementary **L**-theories are introduced in Sections 2, 3. An algebraic treatment of terms and of formulas is discussed in Sections 4, 5. The construction of the algebra of an elementary **L**-theory, the concepts of **L**-model and **L**-validity and some related problems are presented in Sections 6-10. Section 11 contains the completeness theorem for the first order predicate calculi of any logic **L**. The theorem on the existence of **L**-model for consistent elementary **L**-theories and also a theorem which characterizes formulas derivable in these theories by means of **L**-models are proved in Section 12.

The general method of the algebraization of logic, as indicated in Chapter VIII and also in the Supplement, explained by means of examples of propositional calculi belonging to particular logics, which have been discussed in Chapters IX–XIV, and extended by exercises attached to each Chapter, provides the reader with an apparatus sufficient for his own study of the subject.

1. Formalized languages of first order. Let **L** be a fixed logic in the sense of the definition given in VIII 8, i.e. a class of all consistent logically equivalent standard systems of implicative extensional propositional calculi. The logic **L** determines uniquely the sets L_0, L_1, L_2 of propositional constants, unary propositional connectives and binary propositional connectives in the alphabet $A^0 = (V, L_0, L_1, L_2, U)$ of the formalized language $\mathscr{L} = (A^0, F)$ of any propositional calculus $\mathscr{S} = (\mathscr{L}, C_{\mathscr{L}})$ in **L**.

By an *alphabet of a first order formalized language associated with the logic* **L** we shall mean an ordered system

(1) $$A^* = (V_f, V_b, \{\Phi_i\}_{i\in I_0}, \{P_i\}_{i\in I}, L_0, L_1, L_2, Q, U)$$

where

1° I_0 denotes the set of all non-negative integers and I denotes the set of all positive integers,

2° V_f, V_b denote enumerable sets,

3° Φ_i, $i \in I_0$, and P_i, $i \in I$, denote finite (possibly empty) or enumerable sets,

4° the union of the sets $P_1, P_2, \ldots$ is not empty,

5° Q denotes a set consisting of two elements,

6° U denotes a set consisting of two elements,

7° all the sets $V_f, V_b, \Phi_0, \Phi_1, \Phi_2, \ldots, P_1, P_2, \ldots, L_0, L_1, L_2, Q, U$ are disjoint.

Elements in V_f are called *free individual variables* and denoted by x, y, z (with indices if necessary).

Elements in V_b are called *bound individual variables* and denoted by ξ, η, ζ (with indices if necessary).

Elements in the union Φ of the sets Φ_i, $i \in I_0$, are called *functors* and denoted by φ, ψ (with indices if necessary). Elements in Φ_0 are also called *individual constants* and elements in Φ_i, $i \in I_0$, are also called *i-argument functors*.

Elements in the union P of the sets $P_1, P_2, \ldots$ are called *predicates* and denoted by ϱ, σ (with indices if necessary). Elements in P_i, $i \in I$, are also called *i-argument predicates*.

Elements in Q are denoted by $\bigcup$, $\bigcap$ and called the *existential quantifier sign* and the *universal quantifier sign*, respectively.

Elements in U are denoted by $($, $)$ and called *parentheses in* A^*.

Elements in the union of the sets $V_f, V_b, \Phi_0, \Phi_1, \ldots, P_1, P_2, \ldots$ $\ldots, L_0, L_1, L_2, Q, U$ are called *signs of the alphabet* A^*.

The set of all *terms over* A^* is the least set T of finite sequences of signs in A^* such that:

(t_1) T contains the union of the sets V_f and Φ_0,

(t_2) if $\varphi \in \Phi_i$, $i > 0$ and $\tau_1, \ldots, \tau_i \in T$, then $\varphi(\tau_1, \ldots, \tau_i) \in T$.

Terms will be denoted by τ with indices if necessary.

The set of all *formulas over* A^* is the least set F^* of finite sequences of signs in A^* such that:

(f_1) if $\varrho \in P_i$ and $\tau_1, \ldots, \tau_i \in T$, then $\varrho(\tau_1 \ldots \tau_i) \in F^*$,

(f_2) if $e \in L_0$, then $e \in F^*$,

(f_3) if $o \in L_1$ and $\alpha \in F^*$, then $o\alpha \in F^*$,

(f_4) if $o \in L_2$ and $\alpha, \beta \in F^*$, then $(\alpha o \beta) \in F^*$,

(f_5) if $\alpha(x)$ (where x is a free individual variable) is in F^*, then for each bound individual variable ξ that does not appear in $\alpha(x)$ the sequences $\bigcup \xi \alpha(\xi)$, $\bigcap \xi \alpha(\xi)$, where $\alpha(\xi)$ denotes the sequence obtained from $\alpha(x)$ by the simultaneous replacement of every occurrence of x by ξ, are in F^*.

The formulas mentioned in (f_1) and (f_2) are called *atomic*.

Formulas will be denoted by α, β, γ, δ (with indices if necessary).

Note that the set F^* of all formulas over A^* is enumerable.

The system $\mathscr{L}^* = (A^*, T, F^*)$ will be called a *first order formalized language associated with the logic* **L**.

Clearly, the following simple statement, analogous to VIII 2.1, holds.

1.1. *Let* $\mathscr{L}^* = (A^*, T, F^*)$ *be a first order formalized language associated with the logic* **L** *and let* $L_0 = \{e_0, \ldots, e_{m-1}\}$, $L_1 = \{o^1, \ldots, o^s\}$, $L_2 = \{\Rightarrow, o_1, \ldots, o_t\}$ $(m, s, t = 0, 1, 2, \ldots)$. *Then the system*

$$\mathfrak{F}^* = (F^*, \Rightarrow, o_1, \ldots, o_t, o^1, \ldots, o^s, e_0, \ldots, e_{m-1}) \tag{2}$$

is an abstract algebra with $t+1$ *binary operations* $\Rightarrow, o_1, \ldots, o_t$, s *unary operations* $o^1, \ldots, o^s$ *and* m *zero-argument operations* $e_0, \ldots, e_{m-1}$ *defined by* (1), (2), (3) in VIII 2. *Moreover, the algebra* $\mathfrak{F}^*$ *is similar to the algebra of formulas of the formalized language* $\mathscr{L} = (V, F)$ *of each propositional calculus* $\mathscr{S} = (\mathscr{L}, C_{\mathscr{L}})$ *in* **L**.

The algebra $\mathfrak{F}^*$ is called the *algebra of formulas of* $\mathscr{L}^*$.

Let $\mathscr{L} = (A^0, F)$ be the formalized language of a propositional calculus $\mathscr{S}$ in **L** and let $\mathscr{L}^* = (A^*, T, F^*)$ be a first order formalized language associated with **L**. By a *substitution from* $\mathscr{L}$ *into* $\mathscr{L}^*$ we shall mean any mapping $\mathfrak{s}\colon V \to F^*$, where V denotes the set of all propositional variables in A^0. Thus any substitution $\mathfrak{s}$ from $\mathscr{L}$ into $\mathscr{L}^*$ assigns to each propositional variable p of $\mathscr{L}$ a formula $\mathfrak{s}p$ in $\mathscr{L}^*$.

It follows from VIII 2.3, I 4.4 and 1.1 that

1.2. *Any substitution* $\mathfrak{s}$ *from* $\mathscr{L}$ *into* $\mathscr{L}^*$ *can be extended in a unique way to a homomorphism from the algebra* $\mathfrak{F}$ *of formulas of* $\mathscr{L}$ *into the algebra* $\mathfrak{F}^*$ *of formulas of* $\mathscr{L}^*$, *which therefore will be denoted by the same letter* $\mathfrak{s}$.

2. First order predicate calculi of a logic L. Let **L** be a fixed logic and let $\mathscr{S} = (\mathscr{L}, C_{\mathscr{L}})$, where $\mathscr{L} = (A^0, F)$, be a propositional calculus in **L**. Consider a first order formalized language $\mathscr{L}^* = (A^*, T, F^*)$ associated with **L**. The propositional calculus $\mathscr{S}$ determines a consequence operation $C_{\mathscr{L}^*}$ in $\mathscr{L}^*$ in the following way.

Suppose that $\mathscr{A}_l$ is the set of logical axioms of $\mathscr{S}$. Then the set $\mathscr{A}_l^*$ of logical axioms in the language $\mathscr{L}^*$ is defined as follows: for each formula α^* in $\mathscr{L}^*$

(1) $\alpha^* \in \mathscr{A}_l^*$ if and only if $\alpha^* = \mathfrak{s}\alpha$ for some α in $\mathscr{A}_l$ and for a certain substitution $\mathfrak{s}$ from $\mathscr{L}$ into $\mathscr{L}^*$.

Each rule of inference $(r): P \to F$, where $P \subset F^n$, of the propositional calculus $\mathscr{S}$ determines the corresponding rule of inference (r^*): $P^* \to F^*$, where $P^* \subset F^{*n}$ in the language $\mathscr{L}^*$ defined as follows:

$(r^*)(\alpha_1^*, \ldots, \alpha_n^*) = \beta^*$ if and only if the following conditions are satisfied:

(2) 1° $\alpha_i^* = \mathfrak{s}\alpha_i$, $i = 1, \ldots, n$, and $\beta^* = \mathfrak{s}\beta$ for some formulas α_i, β in F and a certain substitution $\mathfrak{s}$ from $\mathscr{L}$ into $\mathscr{L}^*$;

2° $(r)(\alpha_1, \ldots, \alpha_n) = \beta$.

If $\{(r_1), \ldots, (r_k)\}$ is the set of rules of inference of $\mathscr{S}$, then the set of the rules of inference adopted in $\mathscr{L}^*$ consists of the corresponding rules of inference $(r_1^*), \ldots, (r_k^*)$ defined above and, moreover, of the five rules of inference (r_s^*), (r_{ie}^*), (r_{iu}^*), (r_{ee}^*), (r_{eu}^*) which will be defined below.

Rule of substitution for free individual variables

Let $\alpha(x_1, \ldots, x_m)$ be a formula of $\mathscr{L}^*$ such that $x_1, \ldots, x_m$ are some of the free individual variables in that formula, and let $\alpha(\tau_1, \ldots, \tau_m)$ be the formula obtained from $\alpha(x_1, \ldots, x_m)$ by the simultaneous replacement of each occurrence of x_i ($i = 1, \ldots, m$) by a term τ_i. Then

$$(r_s^*) \quad \frac{\alpha(x_1, \ldots, x_m)}{\alpha(\tau_1, \ldots, \tau_m)}.$$

Rule of introduction of the existential quantifier

Let $\alpha(x)$, β be formulas of $\mathscr{L}^*$ and suppose that β contains no occurrence of x and $\alpha(x)$ contains no quantifier binding ξ. Then

$$(r_{ie}^*) \quad \frac{(\alpha(x) \Rightarrow \beta)}{(\bigcup \xi \alpha(\xi) \Rightarrow \beta)}.$$

Rule of introduction of the universal quantifier

Let α be a formula of $\mathscr{L}^*$ which contains no occurrence of a free individual variable x and let $\beta(x)$ be a formula of $\mathscr{L}^*$ which contains no quantifier binding the bound individual variable ξ. Then

$$(r_{iu}^*) \quad \frac{(\alpha \Rightarrow \beta(x))}{(\alpha \Rightarrow \bigcap \xi \beta(\xi))}.$$

Rule of elimination of the existential quantifier

Let $\alpha(x)$, β be formulas of $\mathscr{L}^*$ such that the bound individual variable ξ does not appear in $\alpha(x)$. Then

$$(r_{ee}^*) \quad \frac{(\bigcup \xi \alpha(\xi) \Rightarrow \beta)}{(\alpha(x) \Rightarrow \beta)}.$$

Rule of elimination of the universal quantifier

Let α, $\beta(x)$ be formulas of $\mathscr{L}^*$ and suppose that the bound individual variable ξ does not appear in $\beta(x)$. Then

$$(r_{eu}^*) \quad \frac{(\alpha \Rightarrow \bigcap \xi \beta(\xi))}{(\alpha \Rightarrow \beta(x))}.$$

Now we define the consequence operation $C_{\mathscr{L}^*}$ in $\mathscr{L}^*$ as follows: for each set $\mathscr{A}^* \subset F^*$, the set $C_{\mathscr{L}^*}(\mathscr{A}^*)$ is the least set of formulas in F^* containing $\mathscr{A}^* \cup \mathscr{A}_l^*$ and closed under the rules of inference (r_s^*), (r_{ie}^*), (r_{iu}^*), (r_{ee}^*) (r_{eu}^*), (r_i^*), $i = 1, \ldots, k$.

The system $\mathscr{S}^* = (\mathscr{L}^*, C_{\mathscr{L}^*})$ will be said to be a *first order predicate calculus of the logic* **L** (more exactly: a *calculus connected with* $\mathscr{S} = (\mathscr{L}, C_{\mathscr{L}})$.

By a *formal proof in* $\mathscr{S}^*$ *of a formula* α *from a set* $\mathscr{A}^* \subset F^*$ we mean any finite sequence $\alpha_1, \ldots, \alpha_n$ of formulas of $\mathscr{L}^*$ such that

(p_1) $\alpha_1 \in \mathscr{A}^* \cup \mathscr{A}_l^*$,

(p_2) for each $1 < i \leqslant n$, either $\alpha_i \in \mathscr{A}^* \cup \mathscr{A}_l^*$ or α_i is the conclusion of one of the rules of inference adopted in $\mathscr{S}^*$, of which some of the formulas $\alpha_1, \ldots, \alpha_{i-1}$ are premises,

(p_3) $\alpha_n = \alpha$.

If there exists a formal proof in $\mathscr{S}^*$ of a formula α from a set $\mathscr{A}^* \subset F^*$, then we write

(3) $$\mathscr{A}^* \vdash \alpha.$$

In particular, if $\mathscr{A}^* = O$, we write instead of $O \vdash \alpha$,

(4) $$\vdash \alpha$$

and say that α *is derivable in* $\mathscr{S}^*$.

It is easy to show (cf. VIII 5 (8)) that for every set $\mathscr{A}^* \subset F^*$

(5) $$C_{\mathscr{L}^*}(\mathscr{A}^*) = \{\alpha \in F^*: \mathscr{A}^* \vdash \alpha\}.$$

In particular,

(6) $$C_{\mathscr{L}^*}(O) = \{\alpha \in F^*: \vdash \alpha\}.$$

2.1. *Let* $\mathscr{S}^* = (\mathscr{L}^*, C_{\mathscr{L}^*})$ *be a first order predicate calculus associated with a propositional calculus* $\mathscr{S} = (\mathscr{L}, C_{\mathscr{L}})$ *of a logic* **L**. *Then for every formula* α *of* $\mathscr{L}$ *and every substitution* $\mathfrak{s}$ *from* $\mathscr{L}$ *into* $\mathscr{L}^*$, *the condition* $\alpha \in C_{\mathscr{L}}(O)$ *implies that* $\mathfrak{s}\alpha \in C_{\mathscr{L}^*}(O)$.

If $\alpha \in C_{\mathscr{L}}(O)$, then there exists a formal proof $\alpha_1, \ldots, \alpha_n$ of α in $\mathscr{S}$. By (1) and (2), $\mathfrak{s}\alpha_1, \ldots, \mathfrak{s}\alpha_n$ is a formal proof of $\mathfrak{s}\alpha$ in $\mathscr{S}^*$. Thus, by (5), $\mathfrak{s}\alpha \in C_{\mathscr{L}^*}(O)$.

2.2. *Let* $\mathscr{S}^* = (\mathscr{L}^*, C_{\mathscr{L}^*})$ *be a first order predicate calculus associated with a propositional calculus* $\mathscr{S} = (\mathscr{L}, C_{\mathscr{L}})$ *of a logic* **L**. *Then for any formulas* $\alpha, \beta, \gamma, \delta$ *of* $\mathscr{L}^*$ *and each set* $\mathscr{A}$ *of formulas of* $\mathscr{L}^*$ *the following conditions are satisfied:*

(i) $(\alpha \Rightarrow \alpha) \in C_{\mathscr{L}^*}(O)$,

(ii) *if* $\alpha, (\alpha \Rightarrow \beta) \in C_{\mathscr{L}^*}(\mathscr{A})$, *then* $\beta \in C_{\mathscr{L}^*}(\mathscr{A})$,

(iii) *if* $(\alpha \Rightarrow \beta), (\beta \Rightarrow \gamma) \in C_{\mathscr{L}^*}(\mathscr{A})$, *then* $(\alpha \Rightarrow \gamma) \in C_{\mathscr{L}^*}(\mathscr{A})$,

(iv) *if* $\alpha \in C_{\mathscr{L}^*}(\mathscr{A})$, *then* $(\beta \Rightarrow \alpha) \in C_{\mathscr{L}^*}(\mathscr{A})$,

(v) *if* $(\alpha \Rightarrow \beta), (\beta \Rightarrow \alpha) \in C_{\mathscr{L}^*}(\mathscr{A})$, *then for each unary connective* o *of* $\mathscr{L}^*$, $(o\alpha \Rightarrow o\beta) \in C_{\mathscr{L}^*}(\mathscr{A})$,

(vi) *if* $(\alpha \Rightarrow \beta), (\beta \Rightarrow \alpha), (\gamma \Rightarrow \delta), (\delta \Rightarrow \gamma) \in C_{\mathscr{L}^*}(\mathscr{A})$, *then for each binary connective* o *of* $\mathscr{L}^*$, $((\alpha o \gamma) \Rightarrow (\beta o \delta)) \in C_{\mathscr{L}^*}(\mathscr{A})$.

Suppose that p is a propositional variable of $\mathscr{L}$. By VIII 5 (s_3), the formula $(p \Rightarrow p) \in C_{\mathscr{L}}(O)$. Let $\mathfrak{s}$ be a substitution from $\mathscr{L}$ into $\mathscr{L}^*$ such that $\mathfrak{s}p = \alpha$. Then $(\alpha \Rightarrow \alpha) = \mathfrak{s}(p \Rightarrow p)$. Consequently, by 2.1, $(\alpha \Rightarrow \alpha) \in C_{\mathscr{L}^*}(O)$, i.e. (i) holds.

Now, let p, q be any two different propositional variables of $\mathscr{L}$. Clearly,

(7) $$p, (p \Rightarrow q) \in C_{\mathscr{L}}(\{p, (p \Rightarrow q)\}).$$

Hence, by VIII 5 (s_4), $q \in C_{\mathscr{L}}(\{p, (p \Rightarrow q)\})$. Consequently, there exists a formal proof $\alpha_1, \ldots, \alpha_n$ of q in $\mathscr{S}$ from $\{p, (p \Rightarrow q)\}$. Thus

1° $\alpha_1 \in \mathscr{A}_l \cup \{p, (p \Rightarrow q)\}$, where $\mathscr{A}_l$ is the set of logical axioms of $\mathscr{S}$,

2° for each $1 < i \leqslant n$, either $\alpha_i \in \mathscr{A}_l \cup \{p, (p \Rightarrow q)\}$ or α_i is the conclusion of one of the rules of inference of $\mathscr{S}$ of which some of the formulas $\alpha_1, \ldots, \alpha_{i-1}$ are premises,

3° $\alpha_n = q$.

Suppose that for some formulas α, β of $\mathscr{L}^*$ and a set $\mathscr{A}$ of formulas of $\mathscr{L}^*$, we have α, $(\alpha \Rightarrow \beta) \in C_{\mathscr{L}^*}(\mathscr{A})$. Let $\beta_1, \ldots, \beta_k$ and $\gamma_1, \ldots, \gamma_m$ be formal proofs of α and $(\alpha \Rightarrow \beta)$ in $\mathscr{S}^*$ from $\mathscr{A}$, respectively. Let $\mathfrak{s}$ be a substitution from $\mathscr{L}$ into $\mathscr{L}^*$ such that $\mathfrak{s}p = \alpha$ and $\mathfrak{s}q = \beta$. By (1), (2), 1°, 2°, 3°, the sequence obtained from

$$\beta_1, \ldots, \beta_k, \gamma_1, \ldots, \gamma_m, \mathfrak{s}\alpha_1, \ldots, \mathfrak{s}\alpha_n$$

by the elimination of the formulas α and $(\alpha \Rightarrow \beta)$ from $\mathfrak{s}\alpha_1, \ldots, \mathfrak{s}\alpha_n$ is a formal proof of β in $\mathscr{S}^*$ from the set $\mathscr{A}$. Thus $\beta \in C_{\mathscr{L}^*}(\mathscr{A})$, which completes the proof of (ii). The proofs of conditions (iii)–(vi) are similar. Instead of VIII 5 (s_4) we apply VIII 5 ((s_5)–(s_8)), respectively.

2.3. *Let* $\mathscr{S}^* = (\mathscr{L}^*, C_{\mathscr{L}^*})$ *be a first order predicate calculus of a logic* **L**. *Then, for each formula* $\alpha(x)$ *of* $\mathscr{L}^*$ *such that the bound individual variable* ξ *does not appear in* $\alpha(x)$,

(i) $$\left(\bigcap \xi\, \alpha(\xi) \Rightarrow \alpha(\tau)\right) \in C_{\mathscr{L}^*}(O),$$

and

(ii) $$\left(\alpha(\tau) \Rightarrow \bigcup \xi\, \alpha(\xi)\right) \in C_{\mathscr{L}^*}(O),$$

for each term τ *of* $\mathscr{L}^*$.

The easy proof, based on 2.2 (i) and the rules of inference (r^*_{eu}), (r^*_{ee}), (r^*_s) in Section 1, is omitted.

3. Elementary L-theories. Let $\mathscr{S}^* = (\mathscr{L}^*, C_{\mathscr{L}^*})$ be a first order predicate calculus of a logic **L** and let $\mathscr{A}$ be a set of formulas of $\mathscr{L}^*$. The system

(1) $$\mathscr{S}^*(\mathscr{A}) = (\mathscr{L}^*, C_{\mathscr{L}^*}, \mathscr{A})$$

will be said to be an *elementary* **L**-*theory based on the system* $\mathscr{S}^*$. The set $\mathscr{A}$ is then called the *set of specific axioms of* $\mathscr{S}^*(\mathscr{A})$ and the formulas in $C_{\mathscr{L}^*}(\mathscr{A})$ are called *theorems of* $\mathscr{S}^*(\mathscr{A})$. If $\mathscr{A}$ is empty, the elementary theory $(\mathscr{L}^*, C_{\mathscr{L}^*}, O)$ will be identified with the system $\mathscr{S}^* = (\mathscr{L}^*, C_{\mathscr{L}^*})$.

An elementary **L**-theory $\mathscr{S}^*(\mathscr{A})$ is said to be *consistent* if there exists a formula α of $\mathscr{L}^*$ such that $\alpha \notin C_{\mathscr{L}^*}(\mathscr{A})$.

An elementary **L**-theory $\mathscr{S}^*(\mathscr{A})$ is said to be *maximal* provided it is consistent and for each consistent elementary **L**-theory $\mathscr{S}^*(\mathscr{A}')$ based on the same system $\mathscr{S}^*$ the condition $C_{\mathscr{L}^*}(\mathscr{A}) \subset C_{\mathscr{L}^*}(\mathscr{A}')$ implies that $C_{\mathscr{L}^*}(\mathscr{A}) = C_{\mathscr{L}^*}(\mathscr{A}')$.

An elementary **L**-theory $\mathscr{S}^*(\mathscr{A})$ is said to be *irreducible* provided it is consistent and for any two consistent elementary **L**-theories $\mathscr{S}^*(\mathscr{A}')$ and $\mathscr{S}^*(\mathscr{A}'')$ based on the same system $\mathscr{S}^*$ the condition $C_{\mathscr{L}^*}(\mathscr{A}) = C_{\mathscr{L}^*}(\mathscr{A}') \cap C_{\mathscr{L}^*}(\mathscr{A}'')$ implies that either $C_{\mathscr{L}^*}(\mathscr{A}) = C_{\mathscr{L}^*}(\mathscr{A}')$ or $C_{\mathscr{L}^*}(\mathscr{A}) = C_{\mathscr{L}^*}(\mathscr{A}'')$.

4. The algebra of terms [2]. Let $\mathscr{L}^* = (A^*, T, F^*)$ be a first order formalized language associated with a logic **L**.

It follows immediately from the definition of the set T of all terms over A^* (see Sec. 1) that

4.1. *The system* $\mathfrak{T} = (T, \{\varphi\}_{\varphi\in\Phi})$, *where* Φ *is the set of all functors in* A^*, *is an abstract algebra if for each* i-*argument functor* φ *in* Φ $(i = 0, 1, \ldots)$ *the term* $\varphi(\tau_1 \ldots \tau_i)$ *is considered as the result of the operation* φ *performed on the terms* $\tau_1, \ldots, \tau_i$.

The algebra $\mathfrak{T}$ will be called the *algebra of terms* of the language $\mathscr{L}^*$.

Let $\mathbf{K}_\Phi$ be the class of all abstract algebras $(J, \{o_\varphi\}_{\varphi\in\Phi})$ such that for each i-argument functor φ in Φ, o_φ is an i-argument operation on J $(i = 0, 1, \ldots)$. Obviously $\mathfrak{T} \in \mathbf{K}_\Phi$.

4.2. *The algebra* $\mathfrak{T} = (T, \{\varphi\}_{\varphi\in\Phi})$ *of terms of* $\mathscr{L}^*$ *is a free algebra in the class* $\mathbf{K}_\Phi$, *the set* V_f *of all free individual variables of* $\mathscr{L}^*$ *being the set of free generators.*

In fact, each mapping $v\colon V_f \to J$ can be extended to a homomorphism h from $\mathfrak{T}$ into $(J, \{o_\varphi\}_{\varphi\in\Phi})$ defined by induction on the length of the terms as follows:

(1) $$hx = vx \quad \text{for each } x \text{ in } V_f,$$

[2] The exposition in this section is the same as in [MM].

(2) if $h\tau_1, \ldots, h\tau_i$ are defined and φ is an i-argument functor in Φ, then $h\varphi(\tau_1, \ldots, \tau_i) = o_\varphi(h\tau_1, \ldots, h\tau_i)$.

By a *substitution in* $\mathcal{L}^*$ we mean any mapping $\mathfrak{s}$: $V_f \to T$ from the set V_f of all free individual variables of $\mathcal{L}^*$ into the set T of all terms of $\mathcal{L}^*$.

Since any substitution $\mathfrak{s}$ in $\mathcal{L}^*$ can be treated as a mapping from the set of generators of the algebra $\mathfrak{T}$ of terms of $\mathcal{L}^*$ into the set of all elements of $\mathfrak{T}$, it follows from 4.2 and I 4.4 that the following holds:

4.3. *Any substitution* $\mathfrak{s}$ *in* $\mathcal{L}^*$ *can be extended in a unique way to an endomorphism from the algebra* $\mathfrak{T}$ *of terms of* $\mathcal{L}^*$ *into* $\mathfrak{T}$, *which therefore will be denoted by the same letter* $\mathfrak{s}$. *Thus for every* i*-argument functor* φ *of* $\mathcal{L}^*$ *and any terms* $\tau_1, \ldots, \tau_i$ *we have*

(3) $$\mathfrak{s}\varphi(\tau_1 \ldots \tau_i) = \varphi(\mathfrak{s}\tau_1 \ldots \mathfrak{s}\tau_i).$$

5. Realizations of terms [3]. Let $\mathcal{L}^* = (A^*, T, F^*)$ be a first order formalized language connected with a logic **L**. By a *realization of the terms* of $\mathcal{L}^*$ in a non-void set J we mean any mapping R which to every i-argument functor φ of $\mathcal{L}^*$ assigns an i-argument operation φ_R on J, i.e. a mapping

(1) $$\varphi_R: J^i \to J, \quad i = 0, 1, \ldots$$

Observe that each realization R of the terms in a non-void set J assigns to the algebra $\mathfrak{T} = (T, \{\varphi\}_{\varphi\in\Phi})$ of terms of $\mathcal{L}^*$ an algebra $\mathfrak{T}_R = (J, \{\varphi_R\}_{\varphi\in\Phi})$ which belongs to the class $\mathbf{K}_\Phi$ defined in Section 4.

By a *valuation* of $\mathcal{L}^*$ in $J \neq O$ we understand any mapping v: $V_f \to J$ from the set V_f of free individual variables of $\mathcal{L}^*$ into J.

Let R be a realization of the terms of $\mathcal{L}^*$ in a set $J \neq O$. Then every valuation v of $\mathcal{L}^*$ in J can be treated as a mapping from the set of generators of the algebra $\mathfrak{T}$ of the terms of $\mathcal{L}^*$ into the algebra $\mathfrak{T}_R$ defined above. Thus by 4.2 and I 4.4 we get the following:

5.1. *Every valuation* v *of* $\mathcal{L}^*$ *in* $J \neq O$ *can be uniquely extended to a homomorphism* v_R *from the algebra* $\mathfrak{T}$ *of the terms of* $\mathcal{L}^*$ *into the*

[3] See Rasiowa [3]. The exposition in this section is the same as in [MM].

algebra $\mathfrak{T}_R$ *determined by a realization* R *of the terms of* $\mathscr{L}^*$ *in the same set* J. *Thus we have*

(2) $v_R(x) = vx$ *for each free individual variable* x *of* $\mathscr{L}^*$,

(3) $v_R(\varphi(\tau_1 \dots \tau_i)) = \varphi_R(v_R(\tau_1), \dots, v_R(\tau_i))$ *for each* i*-argument functor* φ *of* $\mathscr{L}^*$ *and any terms* $\tau_1, \dots, \tau_i$ *of* $\mathscr{L}^*$.

It follows from 5.1 that for each term τ, $v_R(\tau)$ is a well-defined element of J. If τ is fixed and v is variable, then $v_R(\tau)$ is a function of $v \in J^{V_f}$. Let us denote this mapping by τ_R. Clearly,

(4) $$\tau_R\colon J^{V_f} \to J$$

is defined by the following equation

(5) $$\tau_R(v) = v_R(\tau).$$

It follows from (2), (3) and (5) that

(6) $$x_R(v) = vx \quad \text{for each } x \text{ in } V_f,$$

(7) $\varphi(\tau_1 \dots \tau_i)_R(v) = \varphi_R(\tau_{1R}(v), \dots, \tau_{iR}(v))$ for each i-argument functor φ of $\mathscr{L}^*$ and any terms $\tau_1, \dots, \tau_i$ of $\mathscr{L}^*$.

Equations (6), (7) can be conceived to be another inductive definition of the mapping τ_R.

Observe that any given realization R of the terms of $\mathscr{L}^*$ in $J \neq O$ assigns to each term τ of $\mathscr{L}^*$ its realization τ_R, which is a mapping from J^{V_f} into J.

Note the following simple result.

5.2. *For every fixed valuation* $v\colon V_f \to J$ *and every realization* R *of* $\mathscr{L}^*$ *in* J, $\tau_R(v)$ *considered as a function of* τ *is a homomorphism from the algebra* $\mathfrak{T}$ *of the terms of* $\mathscr{L}^*$ *into the algebra* $\mathfrak{T}_R$ *determined by* R *such that* $x_R(v) = vx$ *for each free individual variable* x *of* $\mathscr{L}^*$.

Let R be a realization of the terms of $\mathscr{L}^*$ in a set $J \neq O$. Then each substitution $\mathfrak{s}$ in $\mathscr{L}^*$ (see Sec. 4) determines a mapping $\mathfrak{s}_R\colon J^{V_f} \to J^{V_f}$, which to every valuation v *assigns* the valuation $\mathfrak{s}_R v$ defined as follows:

(8) $$\mathfrak{s}_R v(x) = \mathfrak{s} x_R(v) \quad \text{for each } x \text{ in } V_f.$$

Note the following simple fact.

5.3. *For every substitution* $\mathfrak{s}$ *in* $\mathscr{L}^*$, *every term* τ *of* $\mathscr{L}^*$ *and every valuation* v *of* $\mathscr{L}^*$

(9) $$\mathfrak{s}\tau_R(v) = \tau_R(\mathfrak{s}_R v).$$

The easy proof, similar to that of VIII 3.2 and based on 4.3, 5.2, 4.2 and I 4.4, is omitted.

By the *canonical realization of the terms* of $\mathscr{L}^*$ we mean the realization R in the set T of terms of $\mathscr{L}^*$ defined as follows:

(10) $\varphi_R(\tau_1, \ldots, \tau_i) = \varphi(\tau_1 \ldots \tau_i)$ for each i-argument functor φ of $\mathscr{L}^*$ and any terms $\tau_1, \ldots, \tau_i$ in T.

Observe that each valuation $v\colon V_f \to T$ of $\mathscr{L}^*$ in the set T of terms of $\mathscr{L}^*$ is a substitution in $\mathscr{L}^*$ (see Sec. 4). By the *canonical valuation* of $\mathscr{L}^*$ in T we shall mean the valuation v^0 defined as follows:

(11) $v^0(x) = x$ for every free individual variable x in V_f.

Let us note the following result.

5.4. *If* R *is the canonical realization of the terms of* $\mathscr{L}^*$, *then for each term* τ *in* T *and every valuation* $v\colon V_f \to T$ *in the set* T *of terms of* $\mathscr{L}^*$ *we have*

(12) $$\tau_R(v) = v\tau.$$

In particular, for the canonical valuation v^0,

(13) $$\tau_R(v^0) = \tau.$$

Both sides of (12), considered as functions of τ, are endomorphisms from the algebra $\mathfrak{T}$ of terms of $\mathscr{L}^*$ into itself (see 5.2 and 4.3). Since these endomorphisms are equal on the set V_f of generators of $\mathfrak{T}$, they are equal, by I 4.4. The equation (13) follows directly from (12).

6. Implicative algebras with generalized joins and meets. Let $\mathfrak{A} = (A, V, \Rightarrow)$ be an implicative algebra (see II 1). By II 1.1, the binary relation $\leqslant$ on A defined by the equivalence

(1) $$a \leqslant b \quad \text{if and only if} \quad a \Rightarrow b = V$$

is an ordering on A and V is the greatest element in the ordered set $(A, \leqslant)$.

Let $\{a_t\}_{t\in T}$ be an indexed set of elements in A. If the least upper bound of $\{a_t\}_{t\in T}$ (the greatest lower bound of $\{a_t\}_{t\in T}$) exists in $(A, \leqslant)$, it will be denoted by

$$(2) \qquad (\mathfrak{A}) \bigcup_{t\in T} a_t \qquad ((\mathfrak{A}) \bigcap_{t\in T} a_t)$$

and called the *generalized join* (the *generalized meet*). Sometimes we shall omit the sign $(\mathfrak{A})$ in (2). It follows from the above definition that $a = \bigcup_{t\in T} a_t$ $\left(a = \bigcap_{t\in T} a_t\right)$ in $\mathfrak{A}$ if and only if the following conditions are satisfied:

$$(3) \qquad a_t \leqslant a \quad (a \leqslant a_t) \qquad \text{for each } t \in T,$$

(4) if $c \in A$ and $a_t \leqslant c$ $(c \leqslant a_t)$ for each $t \in T$, then $a \leqslant c$ $(c \leqslant a)$.

If for every set $\{a_t\}_{t\in T}$ of elements in A there exists a generalized join and meet in $\mathfrak{A}$, then $\mathfrak{A}$ will be said to be a *complete implicative algebra*. It follows from this definition and III 1.2 that every complete implicative algebra is a lattice.

If an implicative algebra $\mathfrak{A}' = (A', V, \Rightarrow)$ is a subalgebra of an implicative algebra $\mathfrak{A} = (A, V, \Rightarrow)$, then it is necessary to distinguish generalized joins and meets in $\mathfrak{A}'$ from the generalized joins and meets in $\mathfrak{A}$. Indeed, we have

$$(5) \qquad (\mathfrak{A}) \bigcup_{t\in T} a_t \leqslant (\mathfrak{A}') \bigcup_{t\in T} a_t,$$

and

$$(6) \qquad (\mathfrak{A}') \bigcap_{t\in T} a_t \leqslant (\mathfrak{A}) \bigcap_{t\in T} a_t,$$

whenever the generalized joins and meets concerned exist. In general, the sign $\leqslant$ in (5) and (6) cannot be replaced by $=$.

Let $\mathfrak{A} = (A, V, \Rightarrow)$ and $\mathfrak{B} = (B, V, \Rightarrow)$ be implicative algebras and let h be a homomorphism from $\mathfrak{A}$ into $\mathfrak{B}$. Observe that

$$(7) \qquad \text{if} \quad a \leqslant b, \quad \text{then} \quad h(a) \leqslant h(b) \qquad \text{for any } a, b \text{ in } A.$$

Indeed, if $a \leqslant b$, then $a \Rightarrow b = V$. Hence, $h(a) \Rightarrow h(b) = h(a \Rightarrow b) = h(V) = V$. Thus $h(a) \leqslant h(b)$.

Suppose that there exists a generalized join $(\mathfrak{A}) \bigcup_{t\in T} a_t = a$ (a generalized meet $(\mathfrak{A}) \bigcap_{t\in T} a_t = a$). Then, by (3) and (7),

$$h(a_t) \leqslant h(a) \quad (h(a) \leqslant h(a_t)) \qquad \text{for each } t \in T.$$

Hence, by (4),

$$(8) \qquad (\mathfrak{B}) \bigcup_{t\in T} h(a_t) \leqslant h(a) \quad \big(h(a) \leqslant (\mathfrak{B}) \bigcap_{t\in T} h(a_t)\big)$$

if the generalized join (meet) under consideration in $\mathfrak{B}$ exists. The symbol $\leqslant$ in (8) cannot, in general, be replaced by $=$. If

$$(\mathfrak{B}) \bigcup_{t\in T} h(a_t) = h(a) \quad \big((\mathfrak{B}) \bigcap_{t\in T} h(a_t) = h(a)\big),$$

then we say that the *homomorphism* h *preserves the generalized join* $(\mathfrak{A}) \bigcup_{t\in T} a_t$ (the *generalized meet* $(\mathfrak{A}) \bigcap_{t\in T} a_t$).

Let $\mathbf{K_L}$ be the class of all $\mathscr{S}$-algebras for each propositional calculus $\mathscr{S}$ of a logic $\mathbf{L}$. It follows from VIII 6.2 that, for every algebra $\mathfrak{A} = (A, V, \Rightarrow, o_1, \ldots, o_t, o^1, \ldots, o^s, e_0, \ldots, e_{m-1})$ in $\mathbf{K_L}$, the algebra $(A, V, \Rightarrow)$ is an implicative algebra. Therefore, we can consider in $\mathfrak{A}$ generalized joins and meets. An algebra $\mathfrak{A}$ in $\mathbf{K_L}$ will be said to be *complete* if $(A, V, \Rightarrow)$ is a complete implicative algebra. Similarly, all the notions introduced in this section which concern generalized joins and meets can be extended in a natural way to algebras in $\mathbf{K_L}$ and will be applied later to those algebras.

7. The algebra of an elementary L-theory. Consider an elementary $\mathbf{L}$-theory $\mathscr{S}^*(\mathscr{A}) = (\mathscr{L}^*, C_{\mathscr{L}^*}, \mathscr{A})$ based on a first order predicate calculus $\mathscr{S}^* = (\mathscr{L}^*, C_{\mathscr{L}^*})$ of a logic $\mathbf{L}$. Let

$$\mathfrak{F}^* = (F^*, \Rightarrow, o_1, \ldots, o_t, o^1, \ldots, o^s, e_0, \ldots, e_{m-1})$$

be the algebra of formulas of $\mathscr{L}^*$ (see 1.1).

By 2.2 ((i), (iii)), the binary relation $\leqslant_{\mathscr{A}}$ on F^* defined by the equivalence

$$\alpha \leqslant_{\mathscr{A}} \beta \quad \text{if and only if} \quad (\alpha \Rightarrow \beta) \in C_{\mathscr{L}^*}(\mathscr{A})$$

is a quasi-ordering on F^*. Hence, by I 3.2 and 2.2 ((v), (vi)), the binary relation $\approx_{\mathscr{A}}$ on F^* defined as

$$(2) \qquad \alpha \approx_{\mathscr{A}} \beta \quad \text{if and only if} \quad (\alpha \Rightarrow \beta), (\beta \Rightarrow \alpha) \in C_{\mathscr{L}^*}(\mathscr{A})$$

is a congruence relation in the algebra $\mathfrak{F}^*$. Moreover, the relation $\leqslant_{\mathscr{A}}$ on $F^*/_{\approx\mathscr{A}}$ defined by the equivalence

$$(3) \qquad ||\alpha||_{\mathscr{A}} \leqslant_{\mathscr{A}} ||\beta||_{\mathscr{A}} \quad \text{if and only if} \quad (\alpha \Rightarrow \beta) \in C_{\mathscr{L}^*}(\mathscr{A})$$

is an ordering on $F^*/_{\approx\mathscr{A}}$.

By (2), 2.2 (iv) and 2.2 (ii), any formulas α, β of $\mathscr{L}^*$ such that $\alpha \in C_{\mathscr{L}^*}(\mathscr{A})$ and $\beta \in C_{\mathscr{L}^*}(\mathscr{A})$ determine the same equivalence class in $F^*/\approx_{\mathscr{A}}$ (see the proof of VIII 9 (8)), which will be denoted by V. By definition we have

(4) $$V = ||\alpha||_{\mathscr{A}} \quad \text{if and only if} \quad \alpha \in C_{\mathscr{L}^*}(\mathscr{A}).$$

The algebra

$$\mathfrak{A}(\mathscr{S}^*(\mathscr{A})) = (F^*/_{\approx_{\mathscr{A}}}, V, \Rightarrow, o_1, \ldots, o_t, o^1, \ldots, o^s, ||e_0||_{\mathscr{A}}, \ldots, ||e_{m-1}||_{\mathscr{A}}),$$

where the operations $\Rightarrow, o_1, \ldots, o_t, o^1, \ldots, o^s$ are defined by the equations:

(5) $$||\alpha||_{\mathscr{A}} \Rightarrow ||\beta||_{\mathscr{A}} = ||(\alpha \Rightarrow \beta)||_{\mathscr{A}},$$

(6) $$||\alpha|| \; o_i \; ||\beta||_{\mathscr{A}} = ||(\alpha \; o_i \; \beta)||_{\mathscr{A}}, \quad i = 1, \ldots, t,$$

(7) $$o^i||\alpha||_{\mathscr{A}} = ||o^i\alpha||_{\mathscr{A}}, \quad i = 1, \ldots, s,$$

will be said to be *the algebra of the elementary* **L**-theory $\mathscr{S}^*(\mathscr{A})$ (4).

If $\mathscr{A} = O$, we shall omit the subscript $\mathscr{A}$, i.e. we shall write $\leqslant$, $\approx$, $||\alpha||$, $\mathfrak{A}(\mathscr{S}^*)$ instead of $\leqslant_{\mathscr{A}}$, $\approx_{\mathscr{A}}$, $||\alpha||_{\mathscr{A}}$, $\mathfrak{A}(\mathscr{S}^*(\mathscr{A}))$, respectively and $\mathfrak{A}(\mathscr{S}^*)$ will be called *the algebra of the predicate calculus* $\mathscr{S}^*$.

7.1. *The algebra* $\mathfrak{A}(\mathscr{S}^*(\mathscr{A}))$ *of an elementary* **L**-*theory* $\mathscr{S}^*(\mathscr{A})$ *belongs to the class* $\mathbf{K}_{\mathbf{L}}$ *of all* $\mathscr{S}$-*algebras for* $\mathscr{S}$ *in* **L**. *Moreover,* $\mathfrak{A}(\mathscr{S}^*(\mathscr{A}))$ *is non-degenerate if and only if* $\mathscr{S}^*(\mathscr{A})$ *is consistent.*

Suppose that $\mathscr{S}^* = (\mathscr{L}^*, C_{\mathscr{L}^*})$ is associated with a propositional calculus $\mathscr{S} = (\mathscr{L}, C_{\mathscr{L}})$ in **L**. We shall prove that $\mathfrak{A}(\mathscr{S}^*(\mathscr{A}))$ is an $\mathscr{S}$-algebra (see VIII 6).

Let v be a valuation of $\mathscr{L}$ in $\mathfrak{A}(\mathscr{S}^*(\mathscr{A}))$ and let $\mathfrak{s}$ be a substitution from $\mathscr{L}$ into $\mathscr{L}^*$ satisfying the following condition

(8) $||\mathfrak{s}p||_{\mathscr{A}} = v(p)$ for each propositional variable p of $\mathscr{L}$.

With these hypotheses, for each formula α of $\mathscr{L}$

(9) $$\alpha_{\mathfrak{A}(\mathscr{S}^*(\mathscr{A}))}(v) = ||\mathfrak{s}\alpha||_{\mathscr{A}}.$$

The easy proof by induction on the length of formulas, based on (8), VIII 3 (3), I 4 (6), VIII 2 ((8), (9), (10)) is omitted.

If a formula α of $\mathscr{L}$ belongs to the set $\mathscr{A}_l$ of logical axioms of $\mathscr{S}$, then by 2 (1) for each substitution $\mathfrak{s}$ from $\mathscr{L}$ into $\mathscr{L}^*$ $\mathfrak{s}\alpha$ belongs to

(4) This notion was first introduced by A. Lindenbaum and (in a slightly different form) by A. Tarski.

the set $\mathscr{A}_l^*$ of logical axioms of $\mathscr{S}^*$. Thus $\mathfrak{s}\alpha \in C_{\mathscr{L}^*}(\mathscr{A})$. Hence, by (4) and (9), $\alpha_{\mathfrak{A}(\mathscr{S}^*(\mathscr{A}))}(v) = V$, i.e. condition VIII 6 (a_1) is satisfied.

Suppose that a rule of inference (r) of $\mathscr{S}$ assigns to the premises $\alpha_1, \ldots, \alpha_k$ the conclusion β. Let v be a valuation of $\mathscr{L}$ in $\mathfrak{A}(\mathscr{S}^*(\mathscr{A}))$ such that

(10) $$\alpha_{i\mathfrak{A}(\mathscr{S}^*(\mathscr{A}))}(v) = V, \qquad i = 1, \ldots, k.$$

Let $\mathfrak{s}$ be a substitution from $\mathscr{L}$ into $\mathscr{L}^*$ satisfying (8). Then, by (9) and (10), we get, $||\mathfrak{s}\alpha_i||_{\mathscr{A}} = V$, $i = 1, \ldots, k$. Consequently, by (4), $\mathfrak{s}\alpha_i \in C_{\mathscr{L}^*}(\mathscr{A})$ for each $i = 1, \ldots, k$. By 2 (2), the rule of inference (r^*) of $\mathscr{S}^*$ which corresponds to (r) in $\mathscr{S}$ assigns to $\mathfrak{s}\alpha_1, \ldots, \mathfrak{s}\alpha_k$ the formula $\mathfrak{s}\beta$. Thus $\mathfrak{s}\beta \in C_{\mathscr{L}^*}(\mathscr{A})$. Hence, by (4), $||\mathfrak{s}\beta||_{\mathscr{A}} = V$. Applying (9), we get $\beta_{\mathfrak{A}(\mathscr{S}^*(\mathscr{A}))}(v) = V$. We have just proved that the condition VIII 6 (a_2) is also satisfied. By I 4 (6), (4), 2.2 (iii), and (2), the conditions VIII 6 ((a_3), (a_4)) also hold. Thus $\mathfrak{A}(\mathscr{S}^*(\mathscr{A}))$ is an $\mathscr{S}$-algebra.

The second part of 7.1 follows from (4).

The following result follows from 7.1 and VIII 6.2.

7.2. *If* $\mathfrak{A}(\mathscr{S}^*(\mathscr{A})) = (F^*/\approx_{\mathscr{A}}, V, \Rightarrow, o_1, \ldots, o_t, o^1, \ldots, o^s, e_0, \ldots, e_{m-1})$ *is the algebra of an elementary* **L**-*theory* $\mathscr{S}^*(\mathscr{A})$, *then* $(F^*/\approx_{\mathscr{A}}, V, \Rightarrow)$ *is an implicative algebra.*

Let $\mathfrak{s}\colon V_f \to T$ be a substitution in $\mathscr{L}^*$ (see Sec. 4). Then for each formula α of $\mathscr{L}^*$, $\mathfrak{s}\alpha$ will denote the formula obtained from α by the simultaneous replacement of each occurrence of every free individual variable x in α by $\mathfrak{s}x$ respectively. If $\alpha(x_1, \ldots, x_m)$ is a formula of $\mathscr{L}^*$ whose free individual variables are $x_1, \ldots, x_m$, then $\mathfrak{s}\alpha(x_1, \ldots, x_m)$ will also be denoted suggestively by

$$\alpha(\mathfrak{s}x_1, \ldots, \mathfrak{s}x_m).$$

The next theorem states the existence of some generalized joins and meets in the algebra of each elementary **L**-theory.

7.3. *In the algebra* $\mathfrak{A}(\mathscr{S}^*(\mathscr{A}))$ *of an elementary* **L**-*theory* $\mathscr{S}^*(\mathscr{A}) = (\mathscr{L}^*, C_{\mathscr{L}^*}, \mathscr{A})$, *where* $\mathscr{L}^* = (A^*, T, F^*)$, *there exist the generalized joins and meets*

(16) $$\bigcup_{\tau \in T} ||\mathfrak{s}_\tau \alpha(x_0)||_{\mathscr{A}} = ||\mathfrak{s} \textstyle\bigcup \xi \alpha(\xi)||_{\mathscr{A}},$$

(17) $$\bigcap_{\tau \in T} ||\mathfrak{s}_\tau \alpha(x_0)||_{\mathscr{A}} = ||\mathfrak{s} \textstyle\bigcap \xi \alpha(\xi)||_{\mathscr{A}},$$

for each formula $\alpha(x_0)$ *in which the bound individual variable* ξ *does not appear, and each substitution* $\mathfrak{s}\colon V_f \to T$ *in* $\mathscr{L}^*$, *where* $\mathfrak{s}_\tau\colon V_f \to T$ *is the substitution in* $\mathscr{L}^*$ *defined as follows*:

(18) $\mathfrak{s}_\tau x_0 = \tau$ *and* $\mathfrak{s}_\tau x = \mathfrak{s}x$ *for each* $x \neq x_0$, $x \in V_f$, $\tau \in T$.

In particular, if $\mathfrak{s}x = x$ for each free individual variable $x \in V_f$, we obtain from (16), (17) and (18) the following equations

(Q)
$$\bigcup_{\tau\in T} ||\alpha(\tau)||_{\mathscr{A}} = ||\bigcup \xi\, \alpha(\xi)||_{\mathscr{A}},$$
$$\bigcap_{\tau\in T} ||\alpha(\tau)||_{\mathscr{A}} = ||\bigcap \xi\, \alpha(\xi)||_{\mathscr{A}}\ ^{(5)}.$$

The *generalized joins and meets* (Q) will be said to *correspond to the quantifiers*.

Let $\alpha(x_0, x_1, \ldots, x_m)$ be a formula of $\mathscr{L}^*$ whose free individual variables are $x_0, x_1, \ldots, x_m$ and in which ξ does not occur. Let $\mathfrak{s}\colon V_f \to T$ be an arbitrary substitution in $\mathscr{L}^*$. By 2.3 (ii), and the rule of substitution (r_s^*) (see Sec. 2),

$(\alpha(\tau, \mathfrak{s}x_1, \ldots, \mathfrak{s}x_m) \Rightarrow \bigcup \xi\, \alpha(\xi, \mathfrak{s}x_1, \ldots, \mathfrak{s}x_m)) \in C_{\mathscr{L}^*}(\mathscr{A})$ for each $\tau \in T$. Thus $(\mathfrak{s}_\tau \alpha(x_0, x_1 \ldots, x_m) \Rightarrow \mathfrak{s}\bigcup \xi \alpha(\xi, x_1, \ldots, x_m)) \in C_{\mathscr{L}^*}(\mathscr{A})$. Applying (3), we get

(19) $$||\mathfrak{s}_\tau \alpha(x_0, x_1 \ldots, x_m)||_{\mathscr{A}} \leqslant_{\mathscr{A}} ||\mathfrak{s}\bigcup \xi\, \alpha(\xi, x_1, \ldots, x_m)||_{\mathscr{A}}.$$

Now let us suppose that

(20) $$||\mathfrak{s}_\tau \alpha(x_0, x_1 \ldots, x_m)||_{\mathscr{A}} \leqslant_{\mathscr{A}} ||\gamma||_{\mathscr{A}}, \quad \text{for each } \tau \in T.$$

Hence, by (3) and (18), $(\alpha(\tau, \mathfrak{s}x_1, \ldots, \mathfrak{s}x_m) \Rightarrow \gamma) \in C_{\mathscr{L}^*}(\mathscr{A})$ for each term τ in T. Thus there exists a free individual variable x such that x appear neither in γ nor in $\mathfrak{s}x_i$, $i = 1, \ldots, m$, and $(\alpha(x, \mathfrak{s}x_1, \ldots, \mathfrak{s}x_m) \Rightarrow \gamma) \in C_{\mathscr{L}^*}(\mathscr{A})$. Applying the rule of inference (r_{ie}^*) (see Sec. 2), we get $(\bigcup \xi\, \alpha(\xi, \mathfrak{s}x_1, \ldots, \mathfrak{s}x_m) \Rightarrow \gamma) \in C_{\mathscr{L}^*}(\mathscr{A})$. Hence, by (3),

(21) $$||\mathfrak{s}\bigcup \xi\, \alpha(\xi, x_1, \ldots, x_m)||_{\mathscr{A}} \leqslant_{\mathscr{A}} ||\gamma||_{\mathscr{A}}.$$

Thus (20) implies (21). Since (19) holds, we get (16). The proof of (17) is analogous and based on 2.3 (i) and (r_{iu}^*) instead of 2.3 (ii) and (r_{ie}^*), respectively.

(5) Henkin [2], Rasiowa [2], [3], see also Chandrasekharan [1].

8. Realizations of first order formalized languages associated with a logic L. Let $\mathscr{L}^* = (A^*, T, F^*)$ be a first order formalized language associated with a fixed logic **L** and let $\mathbf{K_L}$ be the class of all $\mathscr{S}$-algebras for $\mathscr{S} \in \mathbf{L}$.

By a *partial realization of* $\mathscr{L}^*$ in a set $J \neq O$ and a non-degenerate abstract algebra

$$\mathfrak{A} = (A, V, \Rightarrow, o_1, \ldots, o_t, o^1, \ldots, o^s, e_0, \ldots, e_{m-1}) \text{ in } \mathbf{K_L}$$

we shall mean any mapping R such that

1° to every i-argument functor φ of $\mathscr{L}^*$ R assigns an i-argument function φ_R on J, i.e. a mapping

(1) $$\varphi_R\colon J^i \to J, \quad i = 0, 1, \ldots;$$

2° to every i-argument predicate ϱ of $\mathscr{L}^*$ R assigns an i-argument function defined on J with values in A, i.e. a mapping

(2) $$\varrho_R\colon J^i \to A, \quad i = 1, 2, \ldots$$

Observe that the partial realization R restricted to the set of all functors of $\mathscr{L}^*$ is the realization of the terms of $\mathscr{L}^*$ in the sense of the definition given in Section 5. For every term $\tau \in T$, τ_R will denote the mapping defined in 5 (5).

We recall that by a *valuation of* $\mathscr{L}^*$ in a set $J \neq O$ we mean any mapping $v\colon V_f \to J$ which assigns to each free individual variable x in V_f an element vx in J.

Given a partial realization R of $\mathscr{L}^*$ in $\mathfrak{A} \in \mathbf{K_L}$ and $J \neq O$, we shall consider each formula α of $\mathscr{L}^*$ as a mapping

(3) $$\alpha_R\colon J^{V_f} \to A$$

(i.e. a mapping which assigns to each valuation $v\colon V_f \to J$ an element of A) defined by induction on the length of α as follows:

1° if α is an atomic formula of the form $\varrho(\tau_1 \ldots \tau_i)$, then

(4) $$\alpha_R(v) = \varrho_R\big(\tau_{1R}(v), \ldots, \tau_{iR}(v)\big) \quad \text{for each } v\colon V_f \to J;$$

2° if α is a propositional constant e_i $(i = 0, \ldots, m-1)$, then

(5) $$\alpha_R(v) = e_i \in A, \quad \text{for each } v\colon V_f \to J;$$

3° if α is the formula $(\beta \Rightarrow \gamma)$ and $\beta_R(v)$, $\gamma_R(v)$ are defined, then

(6) $$\alpha_R(v) = \beta_R(v) \Rightarrow \gamma_R(v);$$

4° if α is the formula $(\beta o_i \gamma)$, and $\beta_R(v)$, $\gamma_R(v)$ are defined, then

$$\alpha_R(v) = \beta_R(v) o_i \gamma_R(v), \quad i = 1, \ldots, t; \tag{7}$$

5° if α is the formula $o^i\beta$ and $\beta_R(v)$ is defined, then

$$\alpha_R(v) = o^i \beta_R(v), \quad i = 1, \ldots, s; \tag{8}$$

6° if for a formula $\beta(x_0)$ in which the bound individual variable ξ does not appear and for a valuation $v: V_f \to J$ the mapping $\beta(x_0)_R$ is defined for every valuation $v_j: V_f \to J$, $j \in J$, such that

$$v_j x_0 = j \quad \text{and} \quad v_j x = vx \quad \text{for} \quad x \neq x_0,\ x \in V_f,$$

and α is the formula $\bigcup \xi \beta(\xi)$, then

$$\begin{aligned} &\alpha_R(v) = \textstyle\bigcup_{j \in J} \beta(x_0)_R(v_j) \text{ if this generalized join exists in } \mathfrak{A}, \\ &\alpha_R(v) \text{ is not defined otherwise.} \end{aligned} \tag{9}$$

7° if for a formula $\beta(x_0)$ in which the bound individual variable ξ does not appear and for a valuation $v: V_f \to J$ the mapping $\beta(x_0)_R$ is defined for every valuation $v_j: V_f \to J$, $j \in J$ such that

$$v_j x_0 = j \quad \text{and} \quad v_j x = vx \quad \text{for} \quad x \neq x_0,\ x \in V_f,$$

and α is the formula $\bigcap \xi \beta(\xi)$, then

$$\begin{aligned} &\alpha_R(v) = \textstyle\bigcap_{j \in J} \beta(x_0)_R(v_j) \text{ if this generalized meet exists in } \mathfrak{A}, \\ &\alpha_R(v) \text{ is not defined otherwise.} \end{aligned} \tag{10}$$

If for each formula α of $\mathscr{L}^*$ the mapping α_R [6] is defined, then the partial realization R is said to be a *realization of* $\mathscr{L}^*$.

Observe that every partial realization R of $\mathscr{L}^*$ in any complete algebra $\mathfrak{A}$ in $\mathbf{K}_{\mathrm{L}}$ and a set $J \neq O$ is a realization of $\mathscr{L}^*$. In particular, if $\mathfrak{A}$ in $\mathbf{K}_{\mathrm{L}}$ is a finite lattice with respect to the ordering relation $\leqslant$ determined by $\Rightarrow$ and $\vee$ (cf. 6 (1)), then $\mathfrak{A}$ is a complete algebra in $\mathbf{K}_{\mathrm{L}}$.

Let R be a realization of $\mathscr{L}^*$ in a non-degenerate algebra $\mathfrak{A}$ in $\mathbf{K}_{\mathrm{L}}$ and $J \neq O$. For every substitution $\mathfrak{s}: V_f \to T$ in $\mathscr{L}^*$ and every formula α of $\mathscr{L}^*$ let $\mathfrak{s}\alpha$ have the meaning explained in Section 7 (see p. 365). Moreover, let $\mathfrak{s}_R: J^{V_f} \to J^{V_f}$ be the mapping defined in 5 (8). With these hypotheses and notation the following theorem holds:

[6] The interpretation of formulas of the first order formalized languages as mappings with values ranging over complete pseudo-Boolean algebras was first used by Mostowski [2]. Henkin [2] generalized this notion to the case of arbitrary positive implication algebras.

8.1. *For every substitution* $\mathfrak{s}$ *in* $\mathscr{L}^*$ *every valuation* v *of* $\mathscr{L}^*$ *in* J *and every formula* α *of* $\mathscr{L}^*$

$$(11) \qquad \alpha_R(\mathfrak{s}_R v) = \mathfrak{s}\alpha_R(v).$$

The proof by induction on the length of α, based on 5.3 and (4)–(10), is left to the reader.

Now let $\mathscr{L} = (A^0, F)$ be the formalized language of a propositional calculus $\mathscr{S} = (\mathscr{L}, C_{\mathscr{L}})$ in $\mathbf{L}$ and let $\mathfrak{s}$ be a substitution from $\mathscr{L}$ into $\mathscr{L}^*$ (see Sec. 1). Consider a realization R of $\mathscr{L}^*$ in a non-degenerate algebra $\mathfrak{A}$ in $\mathbf{K}_{\mathbf{L}}$ and in a set $J \neq O$. For every valuation v of $\mathscr{L}^*$ in J let $\mathfrak{s}_R v$ be the valuation of $\mathscr{L}$ in $\mathfrak{A}$ defined as follows:

(12) $\quad \mathfrak{s}_R v(p) = \mathfrak{s}p_R(v) \quad$ for each propositional variable p of $\mathscr{L}$.

In this situation the following holds.

8.2. *For each formula* α *of* $\mathscr{L}$

$$(13) \qquad \alpha_{\mathfrak{A}}(\mathfrak{s}_R v) = \mathfrak{s}\alpha_R(v).$$

The easy proof by induction on the length of α, based on (12), (5)–(8), VIII 3 (3), 1.2, is left to the reader.

9. Canonical realizations for elementary L-theories. Let $\mathscr{S}^*(\mathscr{A}) = (\mathscr{L}^*, C_{\mathscr{L}^*}, \mathscr{A})$, where $\mathscr{L}^* = (\mathscr{A}^*, T, F^*)$, be a consistent elementary $\mathbf{L}$-theory. Let

$$\mathfrak{A}(\mathscr{S}^*(\mathscr{A})) = (F^*/_{\approx_{\mathscr{A}}}, V, \Rightarrow, o_1, \ldots, o_t, o^1, \ldots, o^s, e_0, \ldots, e_{m-1})$$

be the algebra of $\mathscr{S}^*(\mathscr{A})$. By 7.1, $\mathfrak{A}(\mathscr{S}^*(\mathscr{A}))$ is a non-degenerate algebra in the class $\mathbf{K}_{\mathbf{L}}$ of all $\mathscr{S}$-algebras for $\mathscr{S} \in \mathbf{L}$.

We define a partial realization R of $\mathscr{L}^*$ in $\mathfrak{A}(\mathscr{S}^*(\mathscr{A}))$ and in the set T of all terms of $\mathscr{L}^*$ as follows:

(1) $\quad \varphi_R(\tau_1, \ldots, \tau_i) = \varphi(\tau_1 \ldots \tau_i)$ for each i-argument functor φ of $\mathscr{L}^*$ and any terms $\tau_1, \ldots, \tau_i$ in T.

(2) $\quad \varrho_R(\tau_1, \ldots, \tau_i) = ||\varrho(\tau_1 \ldots \tau_i)||_{\mathscr{A}}$ for each i-argument predicate ϱ of $\mathscr{L}^*$ and any terms $\tau_1, \ldots, \tau_i$ in T.

Observe that R restricted to the set of all functors of $\mathscr{L}^*$ is the canonical realization of terms (see Sec. 5).

Every valuation $v\colon V_f \to T$ of $\mathscr{L}^*$ in the set T is a substitution in $\mathscr{L}^*$. Thus, for every term τ, $v\tau$ is the term obtained from τ by the simultaneous replacement of each free individual variable x in τ by vx. Similarly, for each formula α of $\mathscr{L}^*$, $v\alpha$ is the formula obtained from α by the simultaneous replacement of each free individual variable x in α by vx.

9.1. *For each formula α of $\mathscr{L}^*$ and every valuation $v\colon V_f \to T$, $\alpha_R(v)$ is defined and the following equation holds*:

$$\alpha_R(v) = ||v\alpha||_{\mathscr{A}}\text{[7]}. \tag{3}$$

The proof is by induction on the length of α.

If α is an atomic formula $\varrho(\tau_1 \dots \tau_i)$, then by 8 (4), 5.4 and (2), $\alpha_R(v) = \varrho_R(\tau_{1R}(v), \dots, \tau_{iR}(v)) = \varrho_R(v\tau_1, \dots, v\tau_i) = ||\varrho(v\tau_1 \dots v\tau_i)||_{\mathscr{A}} = ||v\alpha||_{\mathscr{A}}$, i.e. (3) holds for each valuation v.

If α is a propositional constant e_i, then by 8 (5),

$$\alpha_R(v) = ||e_i||_{\mathscr{A}} = ||ve_i||_{\mathscr{A}} = ||v\alpha||_{\mathscr{A}} \quad \text{for each valuation } v.$$

If α is the formula $(\beta \Rightarrow \gamma)$, $\beta_R(v)$ and $\gamma_R(v)$ are defined for each valuation v and $\beta_R(v) = ||v\beta||_{\mathscr{A}}$, $\gamma_R(v) = ||v\gamma||_{\mathscr{A}}$, then by 8 (6) and 7 (5), $\alpha_R(v) = \beta_R(v) \Rightarrow \gamma_R(v) = ||v\beta||_{\mathscr{A}} \Rightarrow ||v\gamma||_{\mathscr{A}} = ||(v\beta \Rightarrow v\gamma)||_{\mathscr{A}} = ||v(\beta \Rightarrow \gamma)||_{\mathscr{A}} = ||v\alpha||_{\mathscr{A}}$, for each valuation v. A similar proof for α equal to $(\beta o_i \gamma)$ and $o^i\beta$ is left to the reader.

If for a formula $\beta(x_0)$, in which the bound individual variable ξ does not appear, $\beta(x_0)_R(v)$ is defined for each valuation v and $\beta(x_0)_R(v) = ||v\beta(x_0)||_{\mathscr{A}}$, then for α equal to $\bigcup\xi\beta(\xi)$ $\left(\bigcap\xi\beta(\xi)\right)$ we get by 8 (9) and 7.3 (16) (by 8 (10) and 7.3 (17)) the equations

$$\begin{aligned} \alpha_R(v) &= \bigcup\xi\beta(\xi)_R(v) = \bigcup_{\tau\in T}\beta(x_0)_R(v_\tau) = \bigcup_{\tau\in T}||v_\tau\beta(x_0)||_{\mathscr{A}} = \\ &= ||v\bigcup\xi\beta(\xi)||_{\mathscr{A}} = ||v\alpha||_{\mathscr{A}} \\ \Big(\alpha_R(v) &= \bigcap\xi\beta(\xi)_R(v) = \bigcap_{\tau\in T}\beta(x_0)_R(v_\tau) = \bigcap_{\tau\in T}||v_\tau\beta(x_0)||_{\mathscr{A}} = \\ &= ||v\bigcap\xi\beta(\xi)||_{\mathscr{A}} = ||v\alpha||_{\mathscr{A}}\Big) \end{aligned}$$

for each valuation v.

It follows from 9.1 that the partial realization R defined by (1) and (2) is a realization. It will be called the *canonical realization for the elementary* **L**-*theory* $\mathscr{S}(\mathscr{A})$.

[7] Canonical realizations and Theorem 9.1 were first used by Henkin [2] and independently (in a slightly different form) by Rasiowa [2].

The following property of the canonical realization for $\mathcal{S}(\mathcal{A})$ follows directly from 9.1.

9.2. *For each formula* α *of* $\mathcal{L}^*$ *and the canonical valuation* v^0 (see 5 (11)),

$$\alpha_R(v^0) = ||\alpha||_{\mathcal{A}}.$$

10. L-models. In this section let **L** be a fixed logic, $\mathcal{L}^* = (A^*, T, F^*)$ a fixed first order formalized language associated with **L**, and $\mathbf{K_L}$ the class of all $\mathcal{S}$-algebras for $\mathcal{S}$ in **L**.

Consider a non-degenerate algebra $\mathfrak{A}$ in $\mathbf{K_L}$ and let R be a realization of $\mathcal{L}^*$ in $\mathfrak{A}$ and in a set $J \neq O$. A valuation $v\colon V_f \to J$ of $\mathcal{L}^*$ in J is said to *satisfy a formula* α of $\mathcal{L}^*$ if

$$(1) \qquad \alpha_R(v) = \bigvee.$$

A formula α is said to be *satisfiable in R* if there exists a valuation v of $\mathcal{L}^*$ in J which satisfies α, i.e. which meets condition (1). A formula α is said to be *valid in R* provided that every valuation v of $\mathcal{L}^*$ in J satisfies α. In that case the realization R is called an **L**-*model for the formula* α (more exactly: an **L**-model for α in the algebra $\mathfrak{A}$ and in the set J). The realization R is said to be an **L**-*model*[8] *for a set* $\mathcal{A} \subset F^*$ if R is an **L**-model for every formula α in $\mathcal{A}$.

A formula α of $\mathcal{L}^*$ is said to be **L**-*valid* provided that every realization of $\mathcal{L}^*$ in every non-degenerate algebra $\mathfrak{A} \in \mathbf{K_L}$ and in every set $J \neq O$ is an **L**-model for α.

10.1. *If a formula* α^* *of* $\mathcal{L}^*$ *is a logical axiom of a first order predicate calculus* $\mathcal{S}^* = (\mathcal{L}^*, C_{\mathcal{L}^*})$ *of the logic* **L**, *then* α^* *is* **L**-*valid.*

Suppose that $\mathcal{S}^*$ is associated with a propositional calculus $\mathcal{S} = (\mathcal{L}, C_{\mathcal{L}})$ in **L**. Let $\mathcal{A}_l$ be the set of logical axioms of $\mathcal{S}$ and let $\mathcal{A}_l^*$ be the set of logical axioms of $\mathcal{S}^*$. If $\alpha^* \in \mathcal{A}_l^*$, then by 2 (1), $\alpha^* = \mathfrak{s}\alpha$ for some $\alpha \in \mathcal{A}_l$ and a substitution $\mathfrak{s}$ from $\mathcal{L}$ into $\mathcal{L}^*$. Let R be a realization of $\mathcal{L}^*$ in an algebra $\mathfrak{A} \in \mathbf{K_L}$ and a set $J \neq O$. For every valuation v of $\mathcal{L}^*$ in J, let $\mathfrak{s}_R v$ be the valuation of $\mathcal{L}$ in $\mathfrak{A}$ defined as follows:

$$(2) \qquad \mathfrak{s}_R v(p) = \mathfrak{s}p_R(v) \quad \text{for each propositional variable } p \text{ of } \mathcal{L}.$$

[8] This general notion was first introduced by Rasiowa [3] and called an *algebraic model*. For the notions of **L**-satisfiability and **L**-validity see Rasiowa and Sikorski [3].

It follows from 8.2 and VIII 8.1 that

$$\mathfrak{s}\alpha_R(v) = \alpha_{\mathfrak{A}}(\mathfrak{s}_R v) = V.$$

Thus α^* is valid in R. Since R is an arbitrary realization of $\mathscr{L}^*$, α^* is **L**-valid.

10.2. *The set $F^*(R)$ of all formulas of $\mathscr{L}^*$ which are valid in a realization R of $\mathscr{L}^*$ (in a non-degenerate algebra $\mathfrak{A} \in \mathbf{K}_\mathbf{L}$ and a set $J \neq O$) is closed under the rules of inference of every first order predicate calculus $\mathscr{S}^* = (\mathscr{L}^*, C_{\mathscr{L}^*})$ of the logic* **L**.

Suppose that $\mathscr{S}^* = (\mathscr{L}^*, C_{\mathscr{L}^*})$ is associated with a propositional calculus $\mathscr{S} = (\mathscr{L}, C_{\mathscr{L}})$ in **L**.

If (r^*) is a rule of inference of $\mathscr{S}^*$ which corresponds to a rule of inference (r) of $\mathscr{S}$ and (r^*) $(\alpha_1^*, \ldots, \alpha_k^*) = \beta^*$, then by 2 (2),

(3) $\alpha_i^* = \mathfrak{s}\alpha_i$, $i = 1, \ldots, k$, and $\beta^* = \mathfrak{s}\beta$ for some formulas α_i, β of $\mathscr{L}$ and a substitution $\mathfrak{s}$ from $\mathscr{L}$ into $\mathscr{L}^*$,

and

(4) $$(r)\ (\alpha_1, \ldots, \alpha_k) = \beta.$$

Assume that for a valuation v of $\mathscr{L}^*$ in J

(5) $$\alpha_{iR}^*(v) = V, \quad i = 1, \ldots, k.$$

Let $\mathfrak{s}_R v$ be the valuation of $\mathscr{L}$ in $\mathfrak{A}$ defined by (2). Thus, by 8.2, (3) and (5), $\alpha_{i\mathfrak{A}}(\mathfrak{s}_R v) = \mathfrak{s}\alpha_{iR}(v) = \alpha_{iR}^*(v) = V$, $i = 1, \ldots, k$. Since $\mathfrak{A}$ is an $\mathscr{S}$-algebra, the above condition implies (see VIII 6 (a_2)) by (4) that $\beta_{\mathfrak{A}}(\mathfrak{s}_R v) = V$. Hence, by (3) and 8.2,

(6) $$\beta_R^*(v) = \mathfrak{s}\beta_R(v) = \beta_{\mathfrak{A}}(\mathfrak{s}_R v) = V.$$

Thus (5) implies (6), which completes the proof that $F^*(R)$ is closed under the rule (r^*).

Proof for (r_s^*): Suppose that $\alpha(x_1, \ldots, x_m)$ is in $F^*(R)$. We shall prove that for any terms $\tau_1, \ldots, \tau_m$ the formula $\alpha(\tau_1, \ldots, \tau_m)$ is in $F^*(R)$. Let $\mathfrak{s}$ be the substitution in $\mathscr{L}^*$ defined as follows: $\mathfrak{s}x_i = \tau_i$, $i = 1, \ldots, m$, $\mathfrak{s}x = x$ for each free individual variable $x \neq x_i$. Thus $\alpha(\tau_1, \ldots, \tau_m)$ is the formula $\mathfrak{s}\alpha(x_1, \ldots, x_m)$. Given an arbitrary valuation v of $\mathscr{L}^*$ in J, let $\mathfrak{s}_R v$ be the valuation of $\mathscr{L}^*$ in J defined in 5 (8). Then by 8.1, $\alpha(\tau_1, \ldots, \tau_m)_R(v) = \mathfrak{s}\alpha(x_1, \ldots, x_m)_R(v) = \alpha(x_1, \ldots, x_m)_R(\mathfrak{s}_R v) = V$, for every valuation v of $\mathscr{L}^*$ in J. Thus $\alpha(\tau_1, \ldots, \tau_m)$ is in $F^*(R)$.

PROOF FOR (r^*_{ie}): Suppose that a formula $(\alpha(x_0) \Rightarrow \beta)$ is in $F^*(R)$, x_0 does not appear in β and the bound individual variable ξ does not appear in $\alpha(x_0)$. Let v be an arbitrary valuation of $\mathscr{L}^*$ in J and for each $j \in J$, let v_j be the valuation defined as follows:

(7) $\quad v_j x_0 = j, \quad v_j x = vx \quad$ for every free individual variable $x \neq x_0$.

Since x_0 does not occur in β, we get $\beta_R(v) = \beta_R(v_j)$, $j \in J$. Consequently, since $(\alpha(x_0) \Rightarrow \beta)$ is in $F^*(R)$, we obtain

$$\alpha(x_0)_R(v_j) \Rightarrow \beta_R(v) = \alpha(x_0)_R(v_j) \Rightarrow \beta_R(v_j) = (\alpha(x_0) \Rightarrow \beta)_R(v_j) = V$$

for each $j \in J$. Thus $\alpha(x_0)_R(v_j) \leqslant \beta_R(v)$, for each $j \in J$. Hence, by 6 (4), $\bigcup_{j \in J} \alpha(x_0)_R(v_j) \leqslant \beta_R(v)$. Applying 8 (9), we get $\bigcup \xi \alpha(\xi)_R(v) \leqslant \beta_R(v)$, i.e. $(\bigcup \xi \alpha(\xi) \Rightarrow \beta)_R(v) = V$. Thus the formula $(\bigcup \xi \alpha(\xi) \Rightarrow \beta)$ is in $F^*(R)$.

PROOF FOR (r^*_{iu}): This is analogous to the proof above.

PROOF FOR (r^*_{ee}): Let $\alpha(x_0)$, β be formulas of $\mathscr{L}^*$ such that the bound individual variable ξ does not appear in $\alpha(x_0)$. Suppose that the formula $(\bigcup \xi \alpha(\xi) \Rightarrow \beta)$ is in $F^*(R)$. Thus, by 8 (9),

$$\bigcup_{j \in J} \alpha(x_0)_R(v_j) \Rightarrow \beta_R(v) = (\bigcup \xi \alpha(\xi) \Rightarrow \beta)_R(v) = V$$

for each valuation v of $\mathscr{L}^*$ in J and the corresponding valuations v_j, $j \in J$, defined by (7). Hence, $\bigcup_{j \in J} \alpha(x_0)_R(v_j) \leqslant \beta_R(v)$. Consequently, by 6 (3), $\alpha(x_0)_R(v_j) \leqslant \beta_R(v)$, $j \in J$. In particular, for $j_0 = vx_0$ we have $v_{j_0} = v$. Thus $\alpha(x_0)_R(v) \leqslant \beta_R(v)$. Hence $(\alpha(x_0) \Rightarrow \beta)_R(v) = \alpha(x_0)_R(v) \Rightarrow \beta_R(v) = V$ for every valuation v, i.e. $(\alpha(x_0) \Rightarrow \beta)$ is in $F^*(R)$.

PROOF FOR (r^*_{eu}): This is analogous to that for (r^*_{ee}).

The next theorem easily follows from 10.1 and 10.2.

10.3. *If a formula α of $\mathscr{L}^*$ is derivable in a first order predicate calculus $\mathscr{S}^* = (\mathscr{L}^*, C_{\mathscr{L}^*})$ of the logic* **L**, *then α is* **L**-*valid.*

This will be applied to the proof of the next theorem, which states the consistency of the first order predicate calculi of the logic **L**.

10.4. *Each first order predicate calculus $\mathscr{S}^* = (\mathscr{L}^*, C_{\mathscr{L}^*})$ of the logic* **L** *is consistent.*

Since the set of all predicates in the alphabet A^* of $\mathscr{L}^*$ is not empty (see sec. 1), there exists an m-argument predicate σ in A^*. The atomic formula $\sigma(x_1 \ldots x_m)$, where $x_1, \ldots, x_m$ are any free individual variables in A^*, is not derivable in $\mathscr{S}^*$. To prove this it is sufficient, on account of 10.3, to give a realization R of $\mathscr{L}^*$ in an algebra $\mathfrak{A} \in \mathbf{K}_\mathbf{L}$ and in a set $J \neq O$, and a valuation v of $\mathscr{L}^*$ in J such that $\sigma(x_1 \ldots x_m)_R(v) \neq V$. Let $\mathfrak{A}$ be an arbitrary non-degenerate algebra in $\mathbf{K}_\mathbf{L}$ and let A be the set of all elements in $\mathfrak{A}$. Let $J = \{j\}$, i.e. J is an one-element set. Consider the following partial realization R of $\mathscr{L}^*$ in $\mathfrak{A}$ and J:

(8) $\varphi_R(j_1, \ldots, j_i) = j$ for every i-argument functor φ in A^* $(i = 0, 1, \ldots)$ and $j_1 = \ldots = j_i = j$,

(9) $\varrho_R(j_1, \ldots, j_i) = a \in A$, $a \neq V$, for every i-argument predicate ϱ in A^* $(i = 1, 2, \ldots)$ and $j_1 = \ldots = j_i = j$.

It is easy to verify (making use of 8 ((4)–(10))) that R is a realization of $\mathscr{L}^*$ and that for the unique valuation v of L^* in J ($v(x) = j$, for each free individual variable x in A^*), $\sigma(x_1 \ldots x_m)_R(v) = a \neq V$.

A realization R of $\mathscr{L}^*$ in a non-degenerate algebra $\mathfrak{A}$ in $\mathbf{K}_\mathbf{L}$ and a set $J \neq O$ will be said to be an **L**-*model for an elementary* **L**-*theory* $\mathscr{S}^*(\mathscr{A}) = (\mathscr{L}^*, C_{\mathscr{L}^*}, \mathscr{A})$ if R is an **L**-model for the set $\mathscr{A}$ of specific axioms of $\mathscr{S}^*(\mathscr{A})$.

The following theorem is a generalization of 10.3.

10.5. *Let R be an* **L**-*model for an elementary* **L**-*theory* $\mathscr{S}^*(\mathscr{A}) = (\mathscr{L}^*, C_{\mathscr{L}^*}, \mathscr{A})$. *Then R is an* **L**-*model for each formula* $\alpha \in C_{\mathscr{L}^*}(\mathscr{A})$ (i.e. *for each theorem of the theory* $\mathscr{S}^*(\mathscr{A})$).

Let $F^*(R)$ be the set of all formulas of $\mathscr{L}^*$ which are valid in R. By the definition of an **L**-model for $\mathscr{S}^*(\mathscr{A})$, $\mathscr{A} \subset F^*(R)$. By 10.1, the set $\mathscr{A}_l^*$ of logical axioms of $\mathscr{S}^* = (\mathscr{L}^*, C_{\mathscr{L}^*})$ is also contained in $F^*(R)$. Thus $\mathscr{A}_l^* \cup \mathscr{A} \subset F^*(R)$. By 10.2 the set $F^*(R)$ is closed under the rules of inference of $\mathscr{S}^* = (\mathscr{L}^*, C_{\mathscr{L}^*})$. Hence (see Sec. 2), $C_{\mathscr{L}^*}(\mathscr{A}) \subset F^*(R)$.

11. The completeness theorem for the first order predicate calculi of a logic L. Let **L** be a fixed logic and let $\mathscr{S}^* = (\mathscr{L}^*, C_{\mathscr{L}^*})$ be an arbitrary first order predicate calculus of the logic **L**. It follows from 10.3 that every formula α of $\mathscr{L}^*$ derivable in $\mathscr{S}^*$ is **L**-valid. The converse statement is also true. The following theorem is called the *completeness theorem.*

11.1. *A formula* α *of a first order predicate calculus* $\mathcal{S}^* = (\mathcal{L}^*, C_{\mathcal{L}^*})$ *of the logic* **L** *is derivable in* $\mathcal{S}^*$ *if and only if it is* **L**-*valid.*

Theorem 11.1 is a part of the following theorem:

11.2. *For every formula* α *of a first order predicate calculus* $\mathcal{S}^* = (\mathcal{L}^*, C_{\mathcal{L}^*})$ *of the logic* **L** *the following conditions are equivalent*:

(i) α *is derivable in* $\mathcal{S}^*$,

(ii) α *is* **L**-*valid,*

(iii) $\alpha_R(v^0) = V$ *for the canonical realization* R *for* $\mathcal{S}^*$ *and the canonical valuation* v^0 *of* $\mathcal{L}^*$ *in the set* T *of all terms of* $\mathcal{L}^*$ (9).

Condition (i) implies (ii) by 10.3. Clearly, (ii) implies (iii). Suppose that $\alpha \notin C_{\mathcal{L}^*}(O)$. Hence, by 7 (4), $||\alpha|| \neq V$ in the algebra $\mathfrak{A}(\mathcal{S}^*)$ of the predicate calculus $\mathcal{S}^*$. On the other hand, by 9.2, $\alpha_R(v^0) = ||\alpha||$ for the canonical realization R for $\mathcal{S}^*$ and the canonical valuation v^0 of $\mathcal{L}^*$ in T. Consequently, $\alpha_R(v^0) \neq V$, which completes the proof that (iii) implies (i).

This theorem completely characterizes derivable formulas in any first order predicate calculus of the logic **L**.

12. The existence of L-models for consistent elementary L-theories. In this section let **L** be a fixed logic. Consider an arbitrary consistent elementary **L**-theory $\mathcal{S}^*(\mathcal{A}) = (\mathcal{L}^*, C_{\mathcal{L}^*}, \mathcal{A})$. An **L**-model R for $\mathcal{S}^*(\mathcal{A})$ will be said to be *adequate* provided that for each formula α of $\mathcal{L}^*$, α is a theorem of $\mathcal{S}^*(\mathcal{A})$ if and only if α is valid in R.

12.1. *For every consistent elementary* **L**-*theory* $\mathcal{S}^*(\mathcal{A}) = (\mathcal{L}^*, C_{\mathcal{L}^*}, \mathcal{A})$ *the canonical realization* R *for* $\mathcal{S}^*(\mathcal{A})$ *is an adequate* **L**-*model for* $\mathcal{S}^*(\mathcal{A})$. *Moreover, for every formula* α *of* $\mathcal{L}^*$

(1) $$\alpha \in C_{\mathcal{L}^*}(\mathcal{A}) \quad \textit{if and only if} \quad \alpha_R(v^0) = V,$$

where v^0 *is the canonical valuation of* $\mathcal{L}^*$ *in the set* T *of all terms.*(10)

(9) Theorem 11.2 is a generalization of Henkin's result (see Henkin [2]) and Rasiowa's result [2].

(10) Theorem 12.1 is a generalization of a theorem by Rasiowa [2].

If $\mathscr{S}^*(\mathscr{A})$ is consistent, then, by 7.1, the algebra $\mathfrak{A}(\mathscr{S}^*(\mathscr{A}))$ is a non-degenerate algebra in the class $\mathbf{K_L}$ of all $\mathscr{S}$-algebras for $\mathscr{S} \in \mathbf{L}$. Let R be the canonical realization for $\mathscr{S}^*(\mathscr{A})$ in $\mathfrak{A}(\mathscr{S}^*(\mathscr{A}))$ and the set T of all terms of $\mathscr{L}^*$. By 9.1, for each formula α of $\mathscr{L}^*$ and every valuation v of $\mathscr{L}^*$ in T (i.e. for every substitution v in $\mathscr{L}^*$)

$$(2) \qquad \alpha_R(v) = ||v\alpha||_{\mathscr{A}}.$$

In particular, for the canonical valuation v^0 (see 9.2),

$$(3) \qquad \alpha_R(v^0) = ||\alpha||_{\mathscr{A}}.$$

Suppose that $\alpha \in \mathscr{A}$ and v is an arbitrary valuation of $\mathscr{L}^*$ in T, i.e. a substitution in $\mathscr{L}^*$. Applying the rule of substitution (r_s^*) (see Sec. 2), we get $v\alpha \in C_{\mathscr{L}^*}(\mathscr{A})$. By 7 (4)

$$(4) \qquad ||\beta||_{\mathscr{A}} = V \text{ in } \mathfrak{A}(\mathscr{S}^*(\mathscr{A})) \quad \text{if and only if} \quad \beta \in C_{\mathscr{L}^*}(\mathscr{A}),$$

for each formula β of $\mathscr{L}^*$. Consequently, $||v\alpha||_{\mathscr{A}} = V$. Hence, by (2), $\alpha_R(v) = V$. Thus R is an $\mathbf{L}$-model for $\mathscr{S}^*(\mathscr{A})$. The equivalence (1) follows from (3) and (4). If $\alpha \in C_{\mathscr{L}^*}(\mathscr{A})$, then, by 10.5, α is valid in R. If $\alpha \notin C_{\mathscr{L}^*}(\mathscr{A})$, then, by (1), $\alpha_R(v^0) \neq V$. Thus R is an adequate $\mathbf{L}$-model for $\mathscr{S}^*(\mathscr{A})$.

We recall (see VIII 10) that $\mathbf{S}_\nu$ is the class of all standard systems of implicative extensional propositional calculi with semi-negation.

12.2. *If* $\mathbf{L} \subset \mathbf{S}_\nu$, *then the existence of an* $\mathbf{L}$-*model for an elementary* $\mathbf{L}$-*theory* $\mathscr{S}^*(\mathscr{A}) = (\mathscr{L}^*, C_{\mathscr{L}^*}, \mathscr{A})$ *implies that* $\mathscr{S}^*(\mathscr{A})$ *is consistent.*

The above theorem is a part of the following theorem:

12.3. *If* $\mathbf{L} \subset \mathbf{S}_\nu$, *then for any elementary* $\mathbf{L}$-*theory* $\mathscr{S}^*(\mathscr{A}) = (\mathscr{L}^*, C_{\mathscr{L}^*}, \mathscr{A})$ *the following conditions are equivalent*:

(i) $\mathscr{S}^*(\mathscr{A})$ *is consistent,*

(ii) *there exists an* $\mathbf{L}$-*model for* $\mathscr{S}^*(\mathscr{A})$,

(iii) *for any formula* α *of* $\mathscr{L}^*$ *either* $\alpha \notin C_{\mathscr{L}^*}(\mathscr{A})$ *or* $o\alpha \notin C_{\mathscr{L}^*}(\mathscr{A})$, *where* o *is the connective of semi-negation in* $\mathscr{L}^*$ [11].

(i) implies (ii) by 12.1. Let R be an $\mathbf{L}$-model for $\mathscr{S}^*(\mathscr{A})$ and let v be a valuation of $\mathscr{L}^*$. If $\alpha \in C_{\mathscr{L}^*}(\mathscr{A})$ and $o\alpha \in C_{\mathscr{L}^*}(\mathscr{A})$, then, by 10.5, $\alpha_R(v) = V$ and $o\alpha_R(v) = V$. On the other hand, by VIII 10.1, $oV \neq V$

[11] See Rasiowa and Sikorski [3].

in every non-degenerate $\mathcal{S}$-algebra for $\mathcal{S} \in \mathbf{S}_\nu$, which contradicts the previous conclusion. Thus (ii) implies (iii). Clearly, (iii) implies (i), which completes the proof.

The following theorem characterizes all theorems of any consistent elementary $\mathbf{L}$-theory.

12.4. *For every formula α of a consistent elementary $\mathbf{L}$-theory $\mathcal{S}^*(\mathcal{A}) = (\mathcal{L}^*, C_{\mathcal{L}^*}, \mathcal{A})$ the following conditions are equivalent*:

(i) *α is a theorem of $\mathcal{S}^*(\mathcal{A})$,*

(ii) *α is valid in every $\mathbf{L}$-model for $\mathcal{S}^*(\mathcal{A})$,*

(iii) *$\alpha_R(v^0) = V$ in the canonical realization R for $\mathcal{S}^*(\mathcal{A})$ and the canonical valuation v^0 of $\mathcal{L}^*$ in the set T of terms of $\mathcal{L}^*$* (12).

(i) implies (ii) by 10.5. (ii) implies (iii), since the canonical realization is an $\mathbf{L}$-model (see 12.1). (iii) implies (i) by (1).

Exercises

1. Let $\mathcal{S}^*(\mathcal{A}) = (\mathcal{L}^*, C_{\mathcal{L}^*}, \mathcal{A})$ be a consistent elementary $\mathbf{L}$-theory and let $\mathbf{K}_\mathbf{L}$ be the class of all $\mathcal{S}$-algebras for $\mathcal{S}$ in $\mathbf{L}$. Let h be a homomorphism from $\mathfrak{A}(\mathcal{S}^*(\mathcal{A}))$ into a complete algebra $\mathfrak{A}$ in $\mathbf{K}_\mathbf{L}$ preserving the generalized joins and meets (Q) corresponding to the quantifiers (see Sec. 7). If R is the canonical realization for $\mathcal{S}^*(\mathcal{A})$, then the following equations define another realization, denoted by hR of $\mathcal{L}^*$ in the set T of all terms and in $\mathfrak{A}$: (i) $\varphi_{hR} = \varphi_R$ for each functor φ of $\mathcal{L}^*$; (ii) $\varrho_{hR}(\tau_1, \ldots, \tau_k) = h\varrho_R(\tau_1, \ldots, \tau_k) = h||\varrho(\tau_1, \ldots \tau_k)||_{\mathcal{A}}$ for each k-argument predicate ϱ of $\mathcal{L}^*$ ($k = 1, 2, \ldots$) and any terms $\tau_1, \ldots, \tau_k$. Prove that for every formula α of $\mathcal{L}^*$ and every valuation (substitution) $v\colon V_f \to T$, where V_f is the set of all free individual variables in $\mathcal{L}^*$, $\alpha_{hR}(v) = h\alpha_R(v) = h||v\alpha||_{\mathcal{A}}$ (see Sec. 8, 9.1).

2. Prove that for each formula α of a consistent elementary $\mathbf{L}_\varkappa$-theory $\mathcal{S}^*(\mathcal{A}) = (\mathcal{L}^*, C_{\mathcal{L}^*}, \mathcal{A})$ the following conditions are equivalent (13): (i) $\alpha \in C_{\mathcal{L}^*}(\mathcal{A})$; (ii) α is valid in every $\mathbf{L}_\varkappa$-model of $\mathcal{S}^*(\mathcal{A})$; (iii) α is valid in every $\mathbf{L}_\varkappa$-model of $\mathcal{S}^*(\mathcal{A})$ in the field $\mathfrak{B}(X)$ of all subsets of any set $X \neq O$; (iv) $\alpha_R(v^0) = V$ for the canonical realization R for $\mathcal{S}^*(\mathcal{A})$ and the canonical valuation v^0; (v) α is valid in every $\mathbf{L}_\varkappa$-model of $\mathcal{S}^*(\mathcal{A})$ in the two-element Boolean algebra. Formulate this theorem in the case of $\mathcal{A} = O$ (i.e. the completeness theorem for the first order predicate calculi of the logic $\mathbf{L}_\varkappa$). Apply 11.2, Ex. 1, VI Ex. 4, VI Ex. 5 to prove the theorem (14).

(12) This theorem is a generalization of a theorem by Rasiowa [2].

(13) The equivalence of (i) and (v) is the Lőwenheim-Skolem-Gödel theorem. For that formulation see Rasiowa and Sikorski [3], [MM].

(14) For the proof see Rasiowa and Sikorski [1], [2], [3], [MM].

3. Does an analogue of the theorem formulated in Ex. 2 hold for consistent elementary $L_{\varkappa\iota}$-theories?

4. Prove that for each formula α of a consistent elementary L_λ-theory the following conditions are equivalent [15]: (i) $\alpha \in C_{\mathscr{L}^*}(\mathscr{A})$; (ii) α is valid in every L_λ-model of $\mathscr{S}^*(\mathscr{A})$; (iii) α is valid in every L_λ-model of $\mathscr{S}^*(\mathscr{A})$ in the topological Boolean algebra $\mathfrak{B}(X)$ of all subsets of any topological space $X \neq O$; (iv) $\alpha_R(v^0) = V$ for the canonical realization R for $\mathscr{S}^*(\mathscr{A})$ and the canonical valuation v^0. Formulate this theorem in the case of $\mathscr{A} = O$ (i.e. the completeness theorem for the first order predicate calculi of the logic L_λ). Apply 11.2, Ex. 1, VI Ex. 6 to prove the theorem.

5. Prove that for each formula α of a consistent elementary L_χ-theory $\mathscr{S}^*(\mathscr{A}) = (\mathscr{L}^*, C_{\mathscr{L}^*}, \mathscr{A})$ the following conditions are equivalent [16]: (i) $\alpha \in C_{\mathscr{L}^*}(\mathscr{A})$; (ii) α is valid in every L_χ-model of $\mathscr{S}^*(\mathscr{A})$; (iii) α is valid in every L_χ-model of $\mathscr{S}^*(\mathscr{A})$ in the pseudo-Boolean algebra $\mathfrak{G}(X)$ of all open subsets of any topological space $X \neq O$; (iv) $\alpha_R(v^0) = V$ for the canonical realization R for $\mathscr{S}^*(\mathscr{A})$ and the canonical valuation v^0. Formulate this theorem in the case of $\mathscr{A} = O$ (i.e. the completeness theorem for the first order predicate calculi of the logic L_χ). Apply 11.2, Ex. 1, VI Ex. 8 in the proof.

6. Does an analogue of the theorem formulated in Ex. 5 hold for consistent elementary $L_{\pi\iota}$-theories?

7. Prove analogues of the theorem formulated in Ex. 5 for consistent elementary L_π-theories, L_μ-theories, L_ν-theories [17]. Apply 11.2, Ex. 1, VI Ex. 9 in the proof.

8. Prove that for each formula α of a consistent elementary $L_{\mathscr{N}}$-theory $\mathscr{S}^*(\mathscr{A}) = (\mathscr{L}^*, C_{\mathscr{L}^*}, \mathscr{A})$ the following conditions are equivalent: (i) $\alpha \in C_{\mathscr{L}^*}(\mathscr{A})$; (ii) α is valid in every $L_{\mathscr{N}}$-model of $\mathscr{S}^*(\mathscr{A})$; (iii) α is valid in every $L_{\mathscr{N}}$-model of $\mathscr{S}^*(\mathscr{A})$ in any non-degenerate quasi-pseudo-Boolean algebra of sets; (iv) $\alpha_R(v^0) = V$ for the canonical realization R for $\mathscr{S}^*(\mathscr{A})$ and the canonical valuation v. Formulate this theorem in the case of $\mathscr{A} = O$ (i.e. the completeness theorem for the first order predicate calculi of the logic $L_{\mathscr{N}}$) [18]. Apply 11.2, V 5.6 to prove it.

9. Prove that for each formula α of a consistent elementary L_m-theory $\mathscr{S}^*(\mathscr{A}) = (\mathscr{L}^*, C_{\mathscr{L}^*}, \mathscr{A})$ the following conditions are equivalent: (i) $\alpha \in C_{\mathscr{L}^*}(\mathscr{A})$; (ii) α is valid in every L_m-model of $\mathscr{S}^*(\mathscr{A})$; (iii) α is valid in every L_m-model of $\mathscr{S}^*(\mathscr{A})$ in any non-degenerate Post field; (iv) $\alpha_R(v^0) = V$ for the canonical realization R for $\mathscr{S}^*(\mathscr{A})$ and the canonical valuation v^0; (v) α is valid in every L_m-model of $\mathscr{S}^*(\mathscr{A})$ in the m-element Post algebra $\mathfrak{P}_m$ [19]. Formulate this theorem in the case

(15) Rasiowa and Sikorski [3], Rasiowa [2], [3], [8].
(16) Rasiowa and Sikorski [3], Rasiowa [2], [3].
(17) Rasiowa and Sikorski [3], Rasiowa [3].
(18) Białynicki-Birula and Rasiowa [2].
(19) The equivalence of conditions (i) and (v) was proved by Rasiowa [9].

of $\mathscr{A} = O$ (i.e. the completeness theorem for the first order predicate calculi of the logic $\mathbf{L}_m$). Apply 11.2, Ex. 1, VII Ex. 5, VII Ex. 6 to prove it.

10. Let $\mathscr{S}^* = (\mathscr{L}^*, C_{\mathscr{L}^*})$ be a predicate calculus of the logic $\mathbf{L}_\pi$ ($\mathbf{L}_\mu$, $\mathbf{L}_\nu$, $\mathbf{L}_\chi$). Prove that for any formulas α, β of $\mathscr{L}^*$ the following conditions are equivalent: (i) $(\alpha \cup \beta) \in C_{\mathscr{L}^*}(O)$; (ii) either $\alpha \in C_{\mathscr{L}^*}(O)$ or $\beta \in C_{\mathscr{L}^*}(O)$ (20). Apply a method analogous to that used in the proofs of X 4.1, XI 2.6, XI 5.6, XI 8.6. Similarly prove that $\bigcup \xi \alpha(\xi) \in C_{\mathscr{L}^*}(O)$ if and only if there exists a term τ in $\mathscr{L}^*$ such that $\alpha(\tau) \in C_{\mathscr{L}^*}(O)$ (21).

11. Let $\mathscr{S}^* = (\mathscr{L}^*, C_{\mathscr{L}^*})$ be a predicate calculus of the logic $\mathbf{L}_{\mathscr{N}}$. Prove that: (i) $\sim(\alpha \cap \beta) \in C_{\mathscr{L}^*}(O)$ if and only if either $\sim \alpha \in C_{\mathscr{L}^*}(O)$ or $\sim \beta \in C_{\mathscr{L}^*}(O)$; (ii) $\sim \bigcap \xi \alpha(\tau) \in C_{\mathscr{L}^*}(O)$ if and only if there exists a term τ in $\mathscr{L}^*$ such that $\sim \alpha(\tau) \in C_{\mathscr{L}^*}(O)$ (22). Apply a method similar to that used in the proof of X 4.1.

12. Let $\mathscr{S}^* = (\mathscr{L}^*, C_{\mathscr{L}^*})$ be a predicate calculus of the logic $\mathbf{L}_\lambda$. Prove that: (i) $(\mathbf{I}\alpha \cup \mathbf{I}\beta) \in C_{\mathscr{L}^*}(O)$ if and only if either $\alpha \in C_{\mathscr{L}^*}(O)$ or $\beta \in C_{\mathscr{L}^*}(O)$; (ii) $\bigcup \xi \mathbf{I} \alpha(\xi) \in C_{\mathscr{L}^*}(O)$ if and only if there exists a term τ in $\mathscr{L}^*$ such that $\alpha(\tau) \in C_{\mathscr{L}^*}(O)$ (23). Apply a method analogous to that used in the proof of XIII 5.6.

(20) Rasiowa and Sikorski [5].
(21) Rasiowa and Sikorski [5].
(22) Białynicki-Birula and Rasiowa [2].
(23) Rasiowa and Sikorski [5].

BIBLIOGRAPHY

Abbott, J. C.

[1] *Semi-Boolean algebra,* Matematički Vesnik, 4 (1967), pp. 177–198.

Barcan, R. C.

[1] *A functional calculus of the first order based on strict implication,* The Journal of Symbolic Logic 11 (1946), pp. 1–16.

Bergmann, G.

[1] *The philosophical significance of modal logic,* Mind 69 (1960), pp. 466–485.

Beth, E. W.

[1] *A topological proof of the theorem of Löwenheim-Skolem-Gödel,* Konikl. Nederl. Akademie van Wetenschappen, Amsterdam, Proceedings, Series A, 54 No. 5, and Indagationes Math. 13 No. 5 (1951), pp. 436–444.
[2] *Semantic entailment and formal derivability,* Mededelingen der Kon. Ned. Akad. v. Wet., new series, 18 No. 13 (1955), pp. 309–342.
[3] *Semantic construction of intuitionistic logic,* ibid., 19 No. 11 (1956), pp. 357–388.
[4] *The foundations of mathematics,* Amsterdam 1959.

Białynicki-Birula, A.

[1] *Remarks on quasi-Boolean algebras,* Bull. Ac. Pol. Sc., Cl. III, 5 (1957), pp. 615–619.

Białynicki-Birula, A., and Rasiowa, H.

[1] *On the representation of quasi-Boolean algebras,* ibid., 5 (1957), pp. 259–261.
[2] *On constructible falsity in the constructive logic with strong negation,* Colloquium Mathematicum 6 (1958), pp. 287–310.

Birkhoff, G.

[1] *On the combination of subalgebras,* Proc. Camb. Phil. Soc. 29 (1933), pp. 441–464.
[2] *Lattice theory,* revised edition, New York 1948 (repr. 1960).

Brignole, D., and Monteiro, A.

[1] *Caractérisation des algèbres de Nelson par des égalités,* I, II, Proc. Japan. Ac. 43 (1967), pp. 279–283, 284–285.

Brouwer, L. E. J.

[1] *De onbetrouwbaarheid der logische principes*, Tijdschrift voor wijsbegeerte 2 (1908), pp. 152–158.

Carnap, R.

[1] *Logische Syntax der Sprache*, Wien 1934. English translation (enlarged) *The logical syntax of languages*, New York–London 1937.

[2] *Modalities and quantifications*, The Journal of Symbolic Logic 11 (1946), pp. 33–64.

Čech, E.

[1] *On bicompact spaces*, Ann. of Math. 38 (1937), pp. 823–844.

Chandrasekharan, K.

[1] *The logic of the intuitionistic mathematics*, Mat. Student 9 (1941), pp. 143–154.

Chang, C. C., and Horn, A.

[1] *Prime ideals characterization of generalized Post algebras, Lattice theory*, Proceedings of Symp. in pure mathematics 2 (1961), pp. 43–48.

Chateaubriand, O., and Monteiro, A.

[1] *Les algèbres De Morgan libres*, Notas de Logica Matematica 26 (1969), Instituto de Matematica, Universidad Nacional del Sur, Bahia Blanca.

Church, A.

[1] *Introduction to mathematical logic* I, Princeton 1956.

Diego, A.

[1] *Sobre algebras de Hilbert*, Notas de Logica Matematica, Instituto de Matematica, Universidad Nacional del Sur, Bahia Blanca 12 (1965).

[2] *Les algèbres de Hilbert*, Paris 1966.

Dugundji, J.

[1] *Note on a property of matrices for Lewis and Langford's calculi of propositions*, The Journal of Symbolic Logic 5 (1940), pp. 150–151.

Dummet, M.

[1] *A propositional calculus with denumerable matrix*, The Journal of Symbolic Logic 24 (1959), pp. 97–106.

Dummett, M., and Lemmon, E. J.

[1] *Modal logics between* S4 *and* S5, Zeitschrift für mathematische Logik und Grundlagen der Mathematik 5 (1959), pp. 250–264.

Dwinger, P.

[1] *Notes on Post algebras* I, II, Indagationes Math. 28 (1966), pp. 462–478.

[2] *Generalized Post algebras*, Bull. Ac. Pol. Sc., Ser. Sci. Math. Astr. Phys. 16 (1968), pp. 559–563.

[3] *Ideals in generalized Post algebras*, ibid. 17 (1969), pp. 483–486.

Dyson, V. H., and Kreisel, G.

[1] *Analysis of Beth's semantic construction of intuitionistic logic*, Applied Mathematics and Statistics Laboratories, Stanford University, Technical report No. 3, Jan. 27, 1961 (mimeographed), pp. 39–65.

Epstein, G.

[1] *The lattice theory of Post algebras*, Trans. Amer. Math. Soc. 95 (1960), pp. 300–317.

Gastaminza, M. L., and Gastaminza, S.

[1] *Characterization of De Morgan lattices by the operations of implication and negation*, Notas de logica matematica, Instituto de Matematica, Universidad Nacional del Sur, Bahia Blanca; Proc. Japan Ac. 44 (1968), p. 659.

Gentzen, G.

[1] *Untersuchungen über das logische Schliessen*, Mathematische Zeitschrift 39 (1934–5), pp. 176–210, 405–431.

Glivenko, V.

[1] *Sur quelques points de la logique de M. Brouwer*, Académie Royale de Belgique, Bulletins de la classe des sciences, ser. 5, 15 (1929), pp. 183–188.

Gödel, K.

[1] *Die Vollständigkeit der Axiome des logischen Funktionenkalküls*, Monatshefte für Mathematik und Physik 37 (1930), pp. 349–360.

[2] *Zum intuitionistischen Aussagenkalkül, Ergebnisse eines mathematischen Kolloquiums*, 4 (for 1931–2, pub. 1933), p. 40.

[3] *Zur intuitionistischen Arithmetik und Zahlentheorie*, ibid., pp. 34–38.

[4] *Eine Interpretation des intuitionistischen Aussagenkalküls*, ibid., pp. 39–40.

Guillaume, M.

[1] *Rapports entre calculs propositionnels modaux et topologie impliqués par certaines extensions de la méthode des tableaux sémantiques. Système* S4 *de Lewis*, C. R. Acad. des Sc., 246 (1958).

Halmos, P. R.

[1] *Algebraic Logic* (I, Compos. Math. 12 (1955), pp. 217–249; II, Fund. Math. 43 (1956), pp. 255–325; III, Trans. Amer. Math. Soc. 83 (1956), pp. 430–470; IV, Trans. Amer. Math. Soc. 85 (1957), pp. 1–27).

Hanf, W.

[1] *Models of languages with infinitely long expressions*, Intern. Congr. for Logic,

Methodology and Philosophy of Sciences, Abstract of contributed papers. Stanford University 1960 (mimeographed), p. 24.

Halldén, S.

[1] *A pragmatic approach to modal logic*, Acta Phil. Fennica, Fasc. 16 (1963), pp. 53–63.

Henkin, L.

[1] *The completeness of the first order functional calculus*, The Journal of Symbolic Logic 14 (1949), pp. 159–166.
[2] *An algebraic characterization of quantifiers*, Fund. Math. 37 (1950), pp. 63–74.
[3] *A class of non-normal models for classical sentential logic*, The Journal of Symbolic Logic 28 (1963), p. 300.

Henkin, L., and Tarski, A.

[1] *Cylindrical algebras*, Summaries of talks presented at the Summer Institute of Symbolic Logic in 1957, at Cornell Univ. mimeographed, Vol. III, pp. 332–340.

Herbrand, J.

[1] *Sur la théorie de la démonstration*, Comptes rendus hebdomadaires des séances de l'Académie des Sciences (Paris) 186 (1928), pp. 1274–1276.
[2] *Recherches sur la théorie de la démonstration*, Travaux de la Société des Sciences et des Lettres de Varsovie, Cl. III, 33 (1930).

Heyting, A.

[1] *Die formalen Regeln der intuitionistischen Logik*, Sitzungsberichte der Preussischen Akademie der Wissenschaften, Phys. mathem. Klasse, 1930, pp. 42–56.
[2] *Die formalen Regeln der intuitionistischen Mathematik*, ibid., pp. 57–71, 158–169.
[3] *Intuitionism, an Introduction*, Studies in Logic and Foundations of Mathematics, Amsterdam 1956.

Hilbert, D., and Ackerman, W.

[1] *Grundzüge der theoretischen Logik*, Berlin 1928 (3rd ed. Berlin, Göttingen, Heidelberg 1949).

Hilbert, D., and Bernays, P.

[1] *Grundlagen der Mathematik*, vol. I, Berlin 1934 (repr. Ann Arbor, Mich., 1944), vol. II, Berlin, 1939 (repr. Ann Arbor, Mich. 1944).

Hintikka, J.

[1] *The modes of modality*, Acta Phil. Fennica, Fasc. 16 (1963), pp. 65–81.

Horn, A.

[1] *The separation theorem of intuitionist propositional calculus*, The Journal of Symbolic Logic 27 (1962), pp. 391–399.

[2] *Logic with truth values in a linearly ordered Heyting algebra*, The Journal of Symbolic Logic 34 (1969), pp. 395–408.

Hosoi, T., and Ono, H.

[1] *Intermediate propositional logics* (A survey), Journ. of Tsuda College, to appear.

Jaśkowski, S.

[1] *Recherches sur le système de la logique intuitioniste*, Actes du Congrès Intern. de Phil. Scientifique, VI Philosophie des mathématiques. Actualités scientifiques et industrielles 393, Paris 1936, pp. 58–61.

Johansson, I.

[1] *Der Minimalkalkül, ein reduzierter intuitionistischer Formalismus*, Compositio Math. 4 (1936).

Kalman, J. A.

[1] *Lattices with involution*, Trans. Amer. Math. Soc. 87 (1958), pp. 485–491.

Kanger, S.

[1] *Provability in Logic*, Acta univ. Stockholmiensis, Stockholm, Studies in Philosophy 1 (1957), pp. 1–47.

Karp, C. R.

[1] *Languages with expressions of infinite length*, Studies in Logic and Foundations of Mathematics, Amsterdam 1964.

Kelley, J. L.

[1] *General topology*, 1957.

Kirin, V. G.

[1] *Gentzen's method for the many-valued propositional calculi*, Zeitschrift für Math. Logik und Grundlagen der Math. 12 (1966), pp. 317–332.

[2] *Post algebras as semantic bases of some many-valued logics*, Fund. Math. 63 (1968), pp. 279–294.

Kleene, S. C.

[1] *On the interpretation of intuitionistic number theory*, Trans. of the Amer. Math. Soc. 61 (1947), pp. 307–368.

[2] *Introduction to metamathematics*, New York–Toronto 1952.

Kleene, S. C., and Vesley, R. E.

[1] *The foundations of intuitionistic mathematics, especially in relation to recursive functions*, Studies in Logic and Foundations of Mathematics, Amsterdam 1965.

Kreisel, G., and Putnam, H.

[1] *Eine Unableitbarkeitsbeweismethode für den intuitionistischen Aussagenkalkül*, Arch. f Math. Log. 3 (1957), pp.74–78.

Kripke, S. A.

[1] *Semantical considerations on modal logic*, Acta Phil. Fennica, Fasc. 16 (1963), pp. 83–94.

Kuratowski, C.

[1] *Une méthode d'élimination des nombres transfinis des raisonnements mathématiques*, Fund. Math. 3 (1922), pp. 76–108.
[2] *Topologie* I, Warszawa 1958, and II, Warszawa 1961.

Lewis, C. I., and Langford, C. H.

[1] *Symbolic logic*, New York 1932.

Łoś, J.

[1] *An algebraic proof of completeness for the two-valued propositional calculus*, Colloquium Mathematicum 2 (1951), pp. 236–240.
[2] *Algebraic treatment of the methodology of elementary deductive systems*, Studia Logica 2 (1955), pp. 151–212.

Łukasiewicz, J.

[1] *Treść wykładu pożegnalnego wygłoszonego w auli Uniwersytetu Warszawskiego dnia* 7 *marca* 1918, Warszawa 1918.
[2] *O logice trójwartościowej*, Ruch Filozoficzny 5 (1920), pp. 169–170.
[3] *Elementy logiki matematycznej*, Warszawa 1929 (repr. Warszawa 1958).
[4] *W obronie logistyki*, Studia Gnesnensia 15 (1937).

Łukasiewicz, J., and Tarski, A.

[1] *Untersuchungen über den Aussagenkalkül*, Comptes-rendus des séances de la Société de Sciences et de Lettres de Varsovie, Cl. III, 23 (1930), pp. 30–50.

Malcev, A. I.

[1] *Untersuchungen aus dem Gebiete der mathematischen Logik*, Matematičeskij Sbornik 1 (1936), pp. 323–336.
[2] *Iterativnye algebry i mnogoobrazija Posta*, Algebra i Logika, Novosibirsk, 1966, vol. 5, pp. 8–24.
[3] *Ob odnom usilenii teorem Slupeckogo i Jablonskogo*, Algebra i Logika, Novosibirsk, 6 (1967), pp. 61–75.

Markov, A. A.

[1] *Konstruktivnaja logika*, Usp. Mat. Nauk 5 (1950), pp. 187–188.

Matsumoto, K.

[1] *Reduction theorem in Lewis' sentential calculi,* Mathematica Japonica 3 (1955), pp. 133–135.

McKinsey, J. C. C.

[1] *A solution of the decision problem for the Lewis systems* S. 2 *and* S.4 *with an application to topology*, The Journal of Symbolic Logic 6 (1941), pp. 117–134.

McKinsey, J. C. C., and Tarski, A.

[1] *The algebra of topology*, Annals of Mathematics 45 (1944), pp. 141–191.
[2] *On closed elements in closure algebras,* ibid. 47 (1946), pp. 122–162.
[3] *Some theorems about the sentential calculi of Lewis and Heyting*, The Journal of Symbolic Logic 13 (1948), pp. 1–15.

Moisil, G. C.

[1] *Recherches sur l'algèbre de la logique,* Annales Sc. de l'Univ. de Jassy 22 (1935), pp. 1–117.
[2] *Essais sur les logiques non chrysippiennes,* Buçarest 1972.

Montague, R.

[1] *Syntactical treatments of modality, with corollaries on reflexion principles and finite axiomatizability*, Acta Phil. Fennica 16 (1963), pp. 151–167.

Monteiro, A.

[1] *Matrices de Morgan caractéristiques pour le calcul propositionnel classique*, Anais da Academia Brasileira de Ciências, 1960, pp. 1–7.
[2] *Les algèbres de Nelson semi-simple*, Notas de Logica Matematica, Inst. de Mat. Universidad Nacional del Sur, Bahia Blanca.
[3] *Construction des algèbres de Nelson finies*, Bull. Ac. Pol. Sc. Cl. III, 11 (1963), pp. 359–362.

Monteiro, L., and Picco, D.

[1] *Les réticulés DeMorgan et l'opération de Sheffer*, Bull. Ac. Pol. Sc. Cl. III, 11 (1963), pp. 355–358.

Mostowski, A.

[1] *Logika matematyczna*, Warszawa–Wrocław 1948.
[2] *Proofs of non-deducibility in intuitionistic functional calculus,* The Journal of Symbolic Logic 13 (1948), pp. 204–207.

Nelson, D.

[1] *Recursive functions and intuitionistic number theory*, Trans. of the Amer. Math. Soc. 61 (1947), pp. 307–368.
[2] *Constructible falsity*, The Journal od Symbolic Logic 14 (1949), pp. 16–26.

Nöbeling, G.

[1] *Grundlagen der analytischen Topologie*, Berlin–Götingen–Heidelberg 1954.

Post, E. L.

[1] *Introduction to a general theory of elementary propositions*, Amer. Journ. Mathem. 43 (1921), pp. 165–185.

Rasiowa, H.

[1] *Sur certaines matrices logiques*, Comptes-rendues de la Soc. Pol. Math. 20 (1947), pp. 402–403.

[2] *Algebraic treatment of the functional calculi of Heyting and Lewis*, Fund. Math. 38 (1951), pp. 99–126.

[3] *Algebraic models of axiomatic theories*, ibid. 41 II (1954), pp. 291–310.

[4] *Constructive theories*, Bull. Ac. Pol. Sc. Cl. III, 2 (1954), pp. 121–124.

[5] *Sur la méthode algébraique dans la méthodologie des systèmes déductifs élémentaires*, Bull. Math. Soc. Sc. Math. Phys. de la R.P.R. 1 (49), No. 2 (1957), pp. 223–231.

[6] *Algebraische Charakterisierung der intuitionistischen Logik mit starker Negation*, Constructivity in Mathematics, Proc. of the Coll. held at Amsterdam 1957, Studies in Logic and the Foundations of Mathematics, 1959, pp. 234–240.

[7] *N-lattices and constructive logic with strong negation*, Fund. Math. 46 (1958), pp. 61–80.

[8] *On modal theories*, Acta Phil. Fennica 16 (1963), pp. 201–214.

[9] *A theorem on the existence of prime filters in Post algebras and the completeness theorem for some many-valued predicate calculi*, Bull. Ac. Pol. Sc. Sér. Sci. Math. Astr. Phys. 17 (1969), pp. 347–354.

[10] *Ultraproducts of m-valued models and a generalization of the Löwenheim-Skolem-Gödel-Malcev theorem for theories based on m-valued logics*, ibid 18 (1970), pp. 415–420.

[11] *The Craig interpolation theorem for m-valued predicate calculi*, ibid. 20 (1972), pp. 341–346.

[12] *Post algebras as a semantic foundation of many-valued logics*, MAA Studies in Mathematics, to appear.

Rasiowa, H., and Sikorski, R.

[1] *A proof of the completeness theorem of Gödel*, Fund. Math. 37 (1950), pp. 193–200.

[2] *A proof of the Skolem-Löwenheim theorem*, ibid. 38 (1951), pp. 230–232.

[3] *Algebraic treatment of the notion of satisfiability*, ibid. 40 (1953), pp. 62–95.

[4] *On satisfiability and deducibility in non-classical functional calculi*, Bull. Ac. Pol. Sc. Cl. III 1 (1953), pp. 229–231.

[5] *On existential theorems in non-classical functional calculi*, Fund. Math. 41 (1954), pp. 21–28.
[6] *The Mathematics of Metamathematics*, Warszawa 1963 , 3rd ed. 1970.

Reichbach, J.

[1] *On the completeness of the functional calculus of first order*, Studia Logica 2 (1955), pp. 245–250.

Rieger, L.

[1] *On the lattice theory of Brouwerian propositional logic*, Acta facultatis rerum naturalium Univ. Carolinae 189 (1949), pp. 1–40.
[2] *On conutable generalized σ-algebras, with a new proof of Gödel's completeness theorem*, Czech. Mathem. Journ. 1 (76), (1951), pp. 29–40.

Rosenbloom, P. C.

[1] *Post algebras* I. *Postulates and general theory*, Amer. Journ. of Mathem. 64 (1942), pp. 167–188.

Rosser, J. B., and Turquette, A. R.

[1] *Many-valued logics*, Studies in Logic and Foundations of Mathematics, Amsterdam 1952.

Rousseau, G.

[1] *Sequents in many-valued logic*, I, Fund. Math. 60 (1967), pp. 23–33.
[2] *Logical systems with finitely many truth-values*, Bull. Ac. Pol. Sc., Ser. Sci. Math. Astr. Phys. 17 (1969), pp. 189–194.
[3] *Sequents in many-valued logic*, II, Fund. Math. 67 (1970), pp. 125–131.
[4] *Post algebras and pseudo-Post algebras*, ibid. 67 (1970), pp. 133–145.

Savicka, H.

[1] *On some properties of generalized Post algebras*, Bull. Ac. Pol Sci., Ser. Sci. Math. Astr. Phys. 19 (1971), pp. 267–2€9.

Sholander, M.

[1] *Postulates for distributive lattices*, Canad. Journ. Math. 3 (1951), pp. 28–30.

Sikorski, R.

[1] *Closure algebras*, Fund. Math. 36 (1949), pp. 165–206.
[2] *Products of abstract algebras*, ibid. 39 (1952), pp. 211–228.
[3] *Boolean algebras*, Berlin–Göttingen–Heidelberg 1960.
[4] *Products of generalized algebras and products of realizations*, Colloquium Math. 10 (1963), pp. 1–13.

[5] *Some applications of interior mappings*, Fund. Math. 45 (1958), pp. 200–212.

Słupecki, J.

[1] *Dowód aksjomatyzowalności pełnych systemów wielowartościowych rachunków zdań*, Comptes-rendus de la Soc. des Sciences et des Lettres de Varsovie 32 (1939) pp. 102.

[2] *Pełny trójwartościowy rachunek zdań*, Ann. Univ. Mariae Curie Skłodowska, sectio F I, 3 (1946), p. 193.

Stone, M. H.

[1] *Boolean algebras and their relation to topology*, Proc. Nat. Ac. Sci. 20 (1934), pp. 197–202.

[2] *The theory of representation for Boolean algebras*, Trans. of the Amer. Math. Soc. 40 (1936), pp. 37–111.

[3] *Applications of the theory of Boolean rings to general topology*, ibid. 41 (1937), pp. 321–364.

[4] *Topological representation of distributive lattices and Brouwerian logics*, Čas. Mat. Fys. 67 (1937), pp. 1–25.

Tarski, A.

[1] *Über einige fundamentale Begriffe der Metamathematik*, Comptes-rendus des Séances de la Soc. des Sciences et des Lettres de Varsovie, Cl. III, 23 (1930), pp. 22–29.

[2] *Fundamentale Begriffe der Methodologie der deduktiven Wissenschaften*, I, Monatshefte für Mathematik und Physik 37 (1930), pp. 361–404.

[3] *Untersuchungen über den Aussagenkalkül*, Ergebnisse eines Math. Koll. 2 (1932), pp. 13–14.

[4] *Pojęcie prawdy w językach nauk dedukcyjnych*, Travaux de la Soc. des Sci. et des Lettres de Varsovie, Cl. III, 34 (1933).

[5] *Grundzüge des Systemenkalküls*, I, Fund. Math. 25 (1935), pp. 503–526.

[6] *Über die Erweiterungen der unvollständigen Systeme des Aussagenkalküls*, Ergebnisse eines Math. Koll. 7 (1936), pp. 283–401.

[7] *Grundzüge des Systemenkalküls*, II, Fund. Math. 26 (1936), pp. 283–301.

[8] *Der Aussagenkalkül und die Topologie*, Fund. Math. 31 (1938), pp. 103–134.

Thomason, R. H.

[1] *A semantical study of contructible falsity*, Zeitschr. für math. Logik und Grundl. der Math. 15 (1969), pp. 247-257.

Traczyk, T.

[1] *On axioms and some properties of Post algebras*, Bull. Ac. Pol. Sc. Cl. III, 10 (1962), pp. 509–512.

[2] *Some theorems on independence in Post algebras*, ibid. 11 (1963), pp. 3–8.

[3] *Axioms and some properties of Post algebras*, Colloquium Math. 10 (1963), pp. 198–209.

[4] *A generalization of Loomis-Sikorski's theorem*, Colloquium Math. 12 (1964), pp. 155–161.

[5] *An equational definition of a class of Post algebras*, Bull. Ac. Pol. Sc. Cl. III, 12 (1964), pp. 147–149.

[6] *Weak isomorphism of Boolean and Post algebras*, Colloquium Math. 13 (1965), pp. 159–194.

[7] *On Post algebras with uncountable chain of constants. Algebras of homomorphisms*, Bull. Ac. Pol. Sc. Cl. III, 15 (1967), pp. 673–680.

[8] *Prime ideals in generalized Post algebras*, ibid. 16 (1968), pp. 369–373.

Tsao-Chen, Tang

[1] *Algebraic postulates and a geometric interpretation for the Lewis calculus of strict implication*, Bull. of the Amer. Math. Soc. 44 (1938), pp. 737–744.

von Bummert, J.

[1] *Quasideduktive Systeme und S-algebren*, I, II, Archiv für Mathematische Logik und Grundlagenforschung 11 (1967), pp. 56–67, 101–112.

Vorobiev, N. N.

[1] *Konstruktivnoje isčislenie vyskasivanij s silnym otrizaniem*, Dokl. Akad. Nauk SSSR 85 (1952), pp. 456–468.

[2] *Problema vyvodimosti w konstruktivnom isčislenii vyskasivanij s silnym otrizaniem*, ibid. pp. 689–692.

Wajsberg, M.

[1] *Aksjomatyzacja trójwartościowego rachunku zdań*, Comptes-rendus de la Soc. des Sciences et des Lettres de Varvosie, Cl. III, 24 (1931), pp. 126–148.

[2] *Untersuchungen über den Aussagenkalkül von A. Heyting*, Wiadomości Matem. 46 (1938), pp. 45–101.

Włodarska, E.

[1] *On representation of Post algebras preserving some infinite joins and meets*, Bull. Ac. Pol. Sc., Ser. Sci. Math. Astr. Phys. 18 (1970), pp. 49–54.

Zorn, M.

[1] *A remark on method in transfinite algebra*, Bull. of the Amer. Math. Soc. 41 (1935), pp. 667–670.

LIST OF SYMBOLS

AUTHOR INDEX

SUBJECT INDEX